ADVANCES IN CYCLIC NUCLEOTIDE RESEARCH

VOLUME 6

Advances in Cyclic Nucleotide Research

Series Editors:

Paul Greengard, New Haven, Connecticut
G. Alan Robison, Houston, Texas

International Advisory Board:

Bruce Breckenridge, *Piscataway, New Jersey*
R. W. Butcher, *Worcester, Massachusetts*
E. Costa, *Washington, D.C.*
George I. Drummond, *Calgary, Alberta, Canada*
Nelson Goldberg, *Minneapolis, Minnesota*
Joel G. Hardman, *Nashville, Tennessee*
Oscar Hechter, *Chicago, Illinois*
David M. Kipnis, *St. Louis, Missouri*
Edwin G. Krebs, *Davis, California*
Thomas A. Langan, *Denver, Colorado*
Joseph Larner, *Charlottesville, Virginia*
Grant W. Liddle, *Nashville, Tennessee*
Yasutomi Nishizuka, *Kobe, Japan*
Ira H. Pastan, *Bethesda, Maryland*
Th. Posternak, *Geneva, Switzerland*
Theodore W. Rall, *Cleveland, Ohio*
Martin Rodbell, *Bethesda, Maryland*
Charles G. Smith, *Tuckahoe, New York*

Advances in Cyclic Nucleotide Research

Volume 6

EDITORS:

Paul Greengard, Ph.D.
Professor of Pharmacology
Yale University School of Medicine
New Haven, Connecticut

G. Alan Robison, Ph.D.
Professor of Pharmacology
University of Texas Medical School
Houston, Texas

Raven Press, Publishers ▪ New York

© 1975 by Raven Press Books, Ltd. All rights reserved. This book is protected by copyright. No part of it may be reproduced, stored in a retrieval system, or transmitted, in any form or by any means, electronic, mechanical, photocopying, recording, or otherwise, without the prior written permission of the publisher.

Made in the United States of America

International Standard Book Number 0-89004-042-7
Library of Congress Catalog Card Number 71-181305

ISBN outside North and South America only: 0-7204-4406-3

Preface

Two volumes of *Advances in Cyclic Nucleotide Research* are being published this year. The first of these, Volume 5, contains the proceedings of the Second International Conference on Cyclic AMP, which was held last summer in Vancouver, Canada, on July 8–11, 1974. The present volume contains five comprehensive reviews on subjects not previously covered. The literature on cyclic nucleotides continues to explode, sometimes in ways that could have been predicted from earlier reviews, but often in new and surprising directions. Some of the present reviews raise serious questions about the role of cyclic nucleotides in certain biological processes. We hope that these papers will serve as a counterbalance to the uncritical manner with which cyclic nucleotides are sometimes invoked to explain biological phenomena. We trust that the reviews in this volume will also serve to help readers keep abreast of current developments, and to stimulate further progress in the field.

Paul Greengard
G. Alan Robison
(January 1975)

Contents

1   The Interaction of Cyclic Nucleotides and Calcium in the Control of Cellular Activity
Michael J. Berridge

99   Cyclic AMP and Adrenocortical Function
Ian D. K. Halkerston

137   The Role of Cyclic AMP in Gonadal Function
John M. Marsh

201   Cyclic AMP and the Physiology of the Islets of Langerhans
W. Montague and S. L. Howell

245   Cyclic Nucleotides in Cultured Cells
F. J. Chlapowski, L. A. Kelly, and R. W. Butcher

339   Author Index

351   Subject Index

Contributors

Michael J. Berridge
Agricultural Research Council Unit of Invertebrate Chemistry and Physiology, Department of Zoology, University of Cambridge, Downing Street, Cambridge CB2 3EJ, England

R. W. Butcher
Department of Biochemistry, University of Massachusetts Medical School, 55 Lake Avenue North, Worcester, Massachusetts 01605

F. J. Chlapowski
Department of Biochemistry, University of Massachusetts Medical School, 55 Lake Avenue North, Worcester, Massachusetts 01605

Ian D. K. Halkerston
Department of Biochemistry, University of Massachusetts Medical School, 55 Lake Avenue North, Worcester, Massachusetts 01605

S. L. Howell
School of Biological Sciences, University of Sussex, Brighton BN1 9QG, Sussex, England

L. A. Kelly
Department of Biochemistry, University of Massachusetts Medical School, 55 Lake Avenue North, Worcester, Massachusetts 01605

JOHN M. Marsh
The Endocrine Laboratory, University of Miami School of Medicine, Miami, Florida 33152

W. Montague
Department of Chemical Pathology and Diabetic Department, King's College Hospital Medical School, Denmark Hill, London SE5 8RX, England

The Interaction of Cyclic Nucleotides and Calcium in the Control of Cellular Activity

Michael J. Berridge

Agricultural Research Council Unit of Invertebrate Chemistry and Physiology, Department of Zoology, University of Cambridge, Downing Street, Cambridge CB2 3EJ, England

CONTENTS

I. Introduction 2

II. General Features of Calcium and Cyclic Nucleotide Metabolism . 3
    A. Regulation of Intracellular Calcium Concentration 4
    B. Regulation of Intracellular Cyclic Nucleotide Levels 9
    C. The Second Messenger Action of Calcium and the Cyclic Nucleotides 11
    D. Feedback Mechanisms Operating Between Calcium and the Cyclic Nucleotides 12
    E. Mono- and Bidirectional Control Systems 15

III. Monodirectional Systems 16
    A. Nervous System 16
    B. Adrenal Medulla 18
    C. Anterior Pituitary 21
    D. Endocrine Pancreas and Insulin Secretion 23
    E. Exocrine Pancreas 26
    F. Insect Salivary Gland 30
    G. Mammalian Salivary Gland 34
    H. Toad Bladder 38
    I. Adrenal Cortex 40
    J. Photoreceptors 43
    K. Summary of Monodirectional Systems 45

IV. Bidirectional Systems 46
    A. Smooth Muscle 46
    B. Heart 50
    C. Melanophores 52
    D. Blood Platelets 54
    E. Mast Cells 58

F. Liver 60
G. Summary of Bidirectional Systems 62

V. Control of Cell Division 64
   A. The Interactions of Cyclic Nucleotides and Calcium in
      Stimulus-Division Coupling 64
   B. Cultured Cells 68
   C. Lymphocytes 73
   D. Salivary Glands 75
   E. Liver Regeneration 76

VI. Summary and Conclusions 78

VII. Acknowledgments 79

VIII. References 80

I. INTRODUCTION

Calcium and the cyclic nucleotides (cyclic AMP and cyclic GMP) are the main components of an internal signaling system which regulates the activity of most cells. The primary function of these intracellular signals is to mediate the cell's response to a wide range of external stimuli. The plasma membrane usually acts as a transducer where the incoming signals are received and transformed into these internal signals (second messengers) which are ultimately responsible for adjusting cellular activity. The first indication that the level of an internal signal might regulate cellular activity came from studies on the control of muscle contraction, where calcium has a second messenger function during stimulus-contraction coupling (Sandow, 1970). Calcium also has a central role in stimulus-secretion coupling during the release of granules from a number of different cells (Douglas and Rubin, 1961; Rubin, 1970). Despite evidence that it functioned as an intermediary in both contraction and secretion, the importance of calcium as a second messenger was not fully appreciated, perhaps in part because attention was diverted to the cyclic nucleotides.

The discovery that cyclic AMP mediated the metabolic effects of glucagon captured the imagination of a number of investigators and initiated an explosive research effort which soon revealed that cyclic AMP was present in almost every cell. The ubiquitous nature of cyclic AMP contributed to the view that it had a universal role as a second messenger even in some of the systems where calcium had previously been implicated. The recent interest in cyclic GMP has further strengthened the notion that cyclic nucleotides play a central role in regulating cellular activity. Since so much effort has been focused on the cyclic nucleotides, it seems important to redress the balance by returning calcium to the second messengers' arena. This chapter assesses the relative importance of both calcium and the cyclic nucleotides as second messengers in the control of a wide range of cellular events.

Rasmussen (1970) first stressed the importance of the interactions between cyclic AMP and calcium during hormone action. Much of the evidence was based on information derived from studying the mode of action of calcitonin and parathyroid hormone. In these examples an interaction between the cyclic nucleotides and calcium is not too surprising, because these hormones are concerned with regulating calcium homeostasis. This work is discussed in several recent reviews (Rasmussen, 1970, 1971; Rasmussen and Nagata, 1970; Rasmussen, Kurokawa, Mason, and Goodman, 1971; Borle, 1973) and therefore will not be dealt with in detail. Attention will be concentrated on cellular processes not directly related to whole-body calcium homeostasis.

The available information on the interactions between cyclic nucleotides and calcium is very fragmentary. Research has concentrated on either the cyclic nucleotides or on calcium, and there have been few combined studies designed to unravel how these different second messengers may interact with each other during cell activation. For this reason, it is premature to make too many generalizations, and I have concentrated on trying to establish the exact role of these second messengers in some selected examples. Some of the models which are presented are clearly speculative in nature, but it is hoped that they will draw attention to the cooperative interactions which exist between calcium and the cyclic nucleotides.

This chapter consists of three main parts. The first deals with some general properties of calcium and the cyclic nucleotides as second messengers (Section II). In the second part, the role of these second messengers in cell activation is analyzed in detail by considering a number of separate examples (Sections III and IV). Some general conclusions which emerge from studying these various control systems are used in the last part (Section V) to analyze how these second messengers may function in the control of cell division. It is concluded that calcium may be the primary intracellular mitogenic signal during stimulus-division coupling. This is not to deny the importance of the cyclic nucleotides, which may play an essential role as modulators of the calcium signal.

II. GENERAL FEATURES OF CALCIUM AND CYCLIC NUCLEOTIDE METABOLISM

A basic feature of second messenger homeostasis, which is applicable to both calcium and the cyclic nucleotides, is the very rapid turnover of these intracellular messengers. Their concentrations are the result of a dynamic balance between processes that are continuously supplying and those that are removing the signal. An important consequence of having a rapid turnover of the second messengers is that it permits the signaling system to respond rapidly to small changes in the signal input arriving at the cell. The fact that oscillations in cyclic AMP level can be detected during the course of a single heartbeat (Brooker, 1973) emphasizes the responsiveness of the cyclic nucleotide system. The processes responsible for generating the intracellular signals, e.g., adenylyl cyclase or calcium influx, are usually stimulated during cell activation, but there are examples where the rate of removal of these signals can also be regulated. Some of the important feedback mechanisms which exist between calcium and cyclic AMP may operate by modulating the processes responsible for removing these second messengers. These feedback mechanisms will be described in detail later, but first it is essential to outline the major processes responsible for maintaining the intracellular level of these various second messengers.

A. Regulation of Intracellular Calcium Concentration

Free calcium in the cytoplasmic compartment originates from two important reservoirs—one extracellular, the other intracellular (a and b in Fig. 1). The extracellular calcium concentration bathing most cells is much higher (10^{-3} M) than the free intracellular concentration (10^{-5} to 10^{-7} M). This intracellular level is based on measurements on muscle and nerve cells (Hodgkin and Keynes, 1957; Portzehl, Caldwell, and Ruegg, 1964; Luxoro and Yañez, 1968), but is thought to be applicable to other cell types. There is thus a large concentration gradient promoting calcium entry into the cell, which is augmented by an electrical gradient since the majority of cells are electronegative. Despite this very large electrochemical gradient favoring calcium entry, the influx of calcium into most cells is relatively low, particularly at rest. The surface membrane of most cells is relatively impermeable to calcium, which thus limits its entry. As a consequence of this low membrane permeability, many cells at rest are relatively insensitive to variations in the level of external calcium. During stimulation, however, they become much more sensitive to external calcium, which suggests that an increase in calcium permeability may be a basic feature of cell activation in many cells.

It is important to remember that the external calcium which enters a cell, especially during a phasic response such as an action potential, may come from a restricted area near the plasma membrane (Langer, 1973). The calcium level immediately adjacent to the membrane may be different from that in the surrounding medium because most cells are covered with complex surface coats which usually carry a net negative charge. Muller and Finkelstein (1974) speculated that the ability of magnesium to block certain calcium-mediated events, such as transmitter release, may depend upon its ability to displace calcium from this area near the surface of the membrane. Some of the other calcium inhibitors (e.g., lanthanum, manganese, cobalt, ruthenium red, D-600, tetracaine) may act by blocking the subsequent influx of calcium across the plasma membrane (a in Fig. 1).

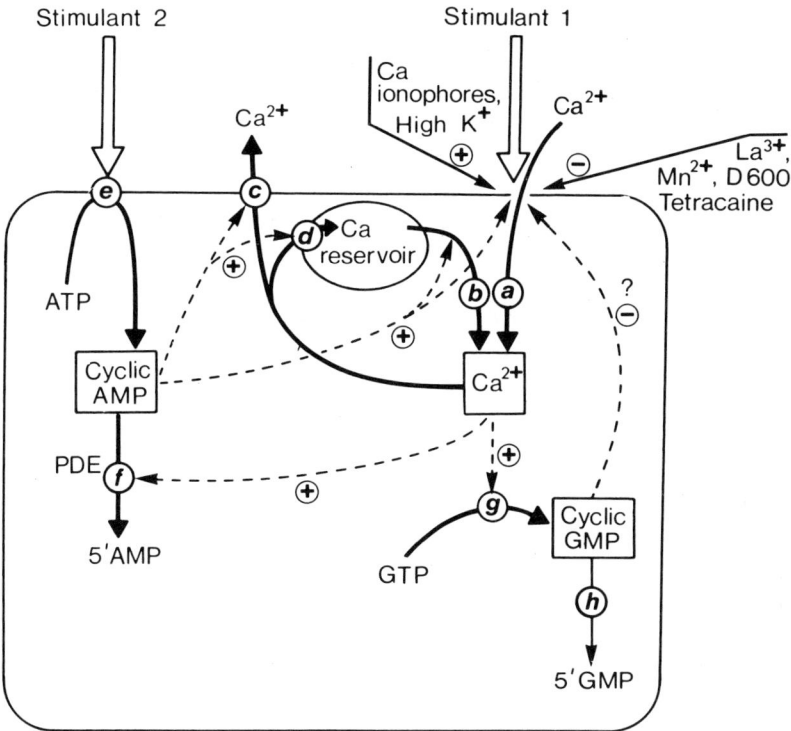

FIG. 1. Summary of the major processes responsible for regulating the intracellular levels of calcium and the cyclic nucleotides. Some stimulants (1) increase calcium influx (a), and may also increase the release of calcium from intracellular reservoirs (b) by a mechanism which is still unknown. Some of the agents which can inhibit or augment calcium entry are included. Calcium is removed from the cytoplasm by being pumped either out of the cell (c) or into the calcium reservoir (d). Other stimulants (2) raise the intracellular level of cyclic AMP by stimulating adenylyl cyclase (e). Methylxanthines may influence the level of cyclic AMP by inhibiting the hydrolysis of cyclic AMP by a phosphodiesterase (f). Cyclic GMP is synthesized by guanylyl cyclase (g), which is located primarily in the cytoplasm, and is hydrolyzed by a phosphodiesterase (h). The activity of guanylyl cyclase may depend upon the intracellular level of calcium.

The broken lines inside the cell illustrate some of the feedback relationships which exist between these different second messengers.

For simplicity, most of the following diagrams have retained the same format as that laid out in this general diagram.

Control over the calcium permeability of the plasma membrane is one of the most important factors responsible for regulating the intracellular level of calcium. There appear to be at least two main ways of altering calcium permeability. In the first case, the stimulant (hormone, neurotransmitter, etc.) interacts with a membrane receptor to induce a change in membrane permeability with little or no change in membrane potential. Examples include the action of ACTH on the adrenal cortex (Section III,I), 5-HT on insect salivary gland (Section III,F), and the action of either acetylcholine or pancreozymin on the exocrine pancreas

(Section III,E). Changes in the resting membrane potential seem to have little effect on the action of the stimulant, which is still capable, in some cases, of activating cells which have been fully depolarized by high potassium solutions. In the second case, the action of the stimulant is tightly coupled to changes in the membrane potential, which is usually depolarized to induce an increase in calcium permeability. This effect of depolarization may account for the ability of high potassium media to activate a wide range of cellular processes. By depolarizing the membrane, the high potassium medium induces an influx of calcium, thus effectively bypassing the normal stimulant. This ability to stimulate cellular activity experimentally using high potassium media has been used extensively and warrants further discussion.

The action of high potassium on membrane permeability has been studied in detail in squid axon by Baker, Meves, and Ridgway (1973). During treatment with high potassium, there was a very rapid phasic increase in the intracellular level of calcium which then settled down to an equilibrium level slightly above that of the control. This equilibrium level was relatively independent of the initial potassium concentration, whereas the extent of the initial phasic response was directly related to the external potassium concentration. During the course of a normal action potential, a small quantity of calcium enters with the sodium through a channel which is open for 500 μsec and is sensitive to tetrodotoxin (TTX), but most enters through a "late calcium channel" which is open for much longer (100 msec) and is blocked by manganese, cobalt, and D-600 (Baker, Hodgkin, and Ridgway, 1971). The phasic increase in calcium concentration (lasting for approximately 2 min) produced by prolonged depolarization with high potassium is thought to represent calcium entering through this late calcium channel (Baker et al., 1973). The high rate of calcium entry is not maintained because there is an inactivation of the calcium channel which curtails the rapid influx of calcium. This inactivation probably prevents the cell from being swamped with calcium flooding into the cell down the very large electrochemical gradient described earlier. The problem of regulating calcium influx will be discussed further in a later section on the interactions between cyclic nucleotides and calcium. High potassium media can thus prolong the normal but temporary phase of calcium entry which occurs during the depolarizing phase of the action potential. Likewise, high potassium can activate a wide range of cellular processes including the release of neurohormones (Douglas and Poisner, 1964; Maddrell and Gee, 1974), release of hormones from the anterior pituitary (Samli and Geschwind, 1968; MacLeod and Lehmeyer, 1970) and β cells (Grodsky and Bennett, 1966), muscle contraction (Durbin and Jenkinson, 1961), and fluid secretion by insect salivary glands (Berridge, Lindley, and Prince, 1975c). In all these examples, external calcium is essential for the high potassium effect and provides strong evidence that extracellular calcium plays an important role in cell activation.

Another artificial technique for introducing calcium into cells is to use divalent cation ionophores such as A 23187 or X-537 A (Caswell and Pressman, 1972;

Reed and Lardy, 1972; Scarpa and Inesi, 1972). Although these ionophores do not discriminate among different divalent cations, they are proving of great interest in that they can mimic a wide range of natural stimulants. As with high potassium stimulation, the action of these ionophores depends upon external calcium. The cell processes stimulated by these ionophores are described in later sections and include the following:

1. Fluid secretion by insect salivary glands (Prince, Rasmussen, and Berridge, 1973);
2. Potassium and, to a lesser extent, enzyme release by the parotid gland (Selinger, Eimerl, and Schramm, 1974);
3. Release of histamine from mast cells (Foreman, Mongar, and Gomperts, 1973; Cochrane and Douglas, 1974);
4. Release of granules from blood platelets (Feinman and Detwiler, 1974);
5. Aggregation of blood platelets (Yuen and Macey, 1974);
6. Release of vasopressin from the neurohypophysis (Nakazato and Douglas, 1974; Russell, Hansen, and Thorn, 1974);
7. Contraction of skeletal, cardiac, and smooth muscle (Levy, Cohen, and Inesi, 1973);
8. Fertilization in the sea urchin egg (Steinhardt and Epel, 1974);
9. DNA synthesis and transformation of lymphocytes (Maino, Green, and Crumpton, 1974);
10. Release of catecholamines from the adrenal medulla (Garcia, Kirpekar, and Prat, 1975); and
11. Secretion of amylase by the pancreas (Williams and Lee, 1974).

When using these calcium ionophores on intact cells, it is important to remember that they may raise the intracellular level of calcium by releasing it from the mitochondria or sarcoplasmic reticulum. Ionophores are certainly capable of releasing calcium from isolated mitochondria (Reed and Lardy, 1972) or from fragments of the sarcoplasmic reticulum (Caswell and Pressman, 1972; Scarpa and Inesi, 1972). Studies on various muscles (Levy et al., 1973) or intact insect salivary glands (see Section III,F for details) indicate that A 23187 may also release calcium from intracellular reservoirs. Feinman and Detwiler (1974) also found that A 23187 could stimulate granule release from blood platelets in the absence of external calcium. These ionophores apparently can cross the plasma membrane to influence the distribution of calcium across the various intracellular reservoirs.

Release of calcium from intracellular reservoirs such as the sarcoplasmic reticulum or mitochondria may be an important source of calcium during normal cell activation (*b* in Fig. 1) (Borle, 1973). The classic example is the release of calcium from the sarcoplasmic reticulum of skeletal muscle during stimulus-contraction coupling (Sandow, 1970). However, evidence is growing that intracellular calcium may also be important in stimulus-secretion coupling (Section III; Thorn, 1974) and in stimulus-division coupling (Section V). The way in which a stimulant, which acts on the surface of a cell, can influence the release of calcium from the internal reservoirs remains a mystery, even in the case of skeletal muscle which has been studied extensively. One possibility is that during stimulus-contraction coupling there is a direct movement of charge from the surface membrane across to the sarcoplasmic reticulum (Schneider and Chandler, 1973). The periodic densities which are found between these two membranes in both smooth and skeletal muscle (Franzini-Armstrong, 1970; Devine, Somlyo, and Somlyo, 1972) may provide an insulated space in which this charge transfer can occur. Another possibility is that during the action potential there is a small influx or release of "trigger calcium" which then induces a further regenerative release of calcium from the sarcoplasmic reticulum (Endo, Tanaka, and Ogawa, 1970; Ford and Podolsky, 1970). It is not clear whether these mechanisms, which have been proposed for skeletal muscle, are applicable to calcium release from the reservoirs of nonexcitable cells. The possible role of cyclic AMP as a regulator of calcium movement across intracellular membranes is described in Section II,D. Release of stored calcium brought about by an influx of sodium may account for stimulus-secretion coupling in the exocrine pancreas (Section III,E). There is clearly a need to learn more about how intracellular calcium movement is regulated.

The presence of intracellular calcium reservoirs can make cells temporarily independent of external calcium. If cells continue to function in the absence of external calcium, it may be erroneous to assume that calcium is not important, because the cells may be running on their internal reservoirs. This phenomenon has been well illustrated in studies on insect salivary glands (Prince and Berridge, 1973; Berridge, Lindley, and Prince, 1974). Utilization of intracellular calcium seems to be an important component of cellular control mechanisms and will be dealt with in detail during the discussion of the role of calcium in specific control mechanisms (Sections III and IV) including stimulus-division coupling (Section V).

The influx of calcium from the outside or from the internal reservoirs is counteracted by active uptake or extrusion mechanisms (c and d in Fig. 1). Calcium can be extruded from the cell by two separate mechanisms (Brinley, 1973). In various excitable cells, such as in nerve or heart muscle, calcium may leave the cell by a forced exchange with sodium (Reuter and Seitz, 1968; Blaustein and Hodgkin, 1969; Baker, 1970). In nonexcitable cells, calcium is extruded by means of a calcium pump which is probably equivalent to the Ca-activated ATPase found in most cells (Lee and Shin, 1969; Schatzmann and Vincenzi, 1969; Lamb and Lindsay, 1971; Perdue, 1971; Hurwitz, Fitzpatrick, Debbas, and Landon, 1973). The sarcoplasmic reticulum also has a calcium pump with properties resembling that of the plasma membrane. Warren, Toon, Birdsall, Lee, and Metcalfe (1974) have reconstructed the calcium pump from the sarcoplasmic reticulum and have found strong evidence that a single protein acts as both ATPase and ionophore. It is interesting to note that microsomes from nonexcitable cells, such as the liver and kidney, are not able to accumulate calcium, and such cells may lack the equivalent of the sarcoplasmic reticulum found in muscle (Drahota, Carafoli, Rossi, Gamble, and Lehninger, 1965). However, using pyroantimonate as a trapping agent, Hales, Luzio, Chandler, and Herman (1974) have localized calcium in the smooth endoplasmic reticulum; they suggest that it may be functionally analogous to the sarcoplasmic reticulum in skeletal muscle. Mitochondria are also capable of sequestering calcium against considerable gradients (Drahota et al., 1965) and may play an important role in regulating the intracellular level of calcium in nonexcitable cells (Borle, 1973). The calcium pump on mitochondria is different from that found on the cell surface. Calcium uptake by mitochondria is inhibited by oligomycin (Brierley, Murer, and Bachmann, 1964) or by an increase in sodium (Dransfeld, Greeff, Hess, and Schorn, 1967), both of which have no effect on the ATPase-dependent surface pump (Lee and Shin, 1969; Schatzmann and Vincenzi, 1969). Another interesting feature concerning calcium uptake by mitochondria is the report by Spencer and Bygrave (1972) that calcium may function as a regulator of itself because it can stimulate its own uptake into the mitochondria. These different calcium removal mechanisms are of crucial importance not only in determining the equilibrium level of calcium but also because they determine the rate at which a cell recovers after being stimulated.

The kinetics of calcium movement during cell activation have been studied by measuring calcium fluxes with ^{45}Ca (Thorn, 1974). In some cases, various stimulants appear to increase the influx (step a in Fig. 1) of ^{45}Ca because there is an increase in the amount of label which can be extracted from the cell. However, an increased uptake of ^{45}Ca may not necessarily be caused by an increase in influx; this increased uptake could also arise if the stimulant had no effect on step a but markedly increased steps b and d. Cittadini, Scarpa, and Chance (1973) have shown that most of the calcium entering Ehrlich ascites tumor cells is rapidly taken up by the mitochondria. By increasing the turnover of calcium between such intracellular reservoirs and the cytoplasm, a stimulant could cause an apparent increased influx of calcium without altering the permeability of the plasma membrane. Thorn (1974) has also stressed that the amount of calcium entering a cell during a stimulus is probably very small and thus difficult to detect. The efflux of ^{45}Ca from prelabeled cells has provided more reliable information on changes in calcium homeostasis (Borle, 1973). There are many examples where cell activation is associated with an increased efflux of calcium which is thought to reflect a mobilization of calcium from intracellular reservoirs (b in Fig. 1). Some of the calcium which is released into the cytoplasm will be rapidly removed from the cell by the pumps on the surface membrane (c in Fig. 1), and may account for the increase in calcium efflux.

B. Regulation of Intracellular Cyclic Nucleotide Levels

The intracellular level of cyclic nucleotides is determined by a dynamic balance between synthetic and degradative processes. In most cases, the change in cyclic nucleotide levels which occurs during cell activation is produced by stimulating adenylyl or guanylyl cyclase. However, there are examples, as in the eye (Section III,J), where cyclic nucleotide levels may be altered by a change in phosphodiesterase activity. The main point to establish, however, is the dynamic nature of the nucleotide level and how minor modifications of either the synthetic or degradative processes can cause marked changes in concentration.

Cyclic AMP is generated by an adenylyl cyclase which is usually associated with the plasma membrane (e in Fig. 1). The properties of this enzyme will not be considered in great detail, since it was the subject of a recent review by Perkins (1973). The original idea put forward by Robison, Butcher, and Sutherland (1967), that there is a catalytic unit (converts ATP to cyclic AMP) separate from a receptor unit (detects the incoming stimulant molecule), seems to be gaining ground. Evidence is growing for a transducer unit which holds together the receptor and catalytic subunits. In keeping with current concepts on membrane fluidity, it is conceivable that these various subunits float around separately within the membrane and that hormones act by bringing them together into a stable and functional unit (Perkins, 1973). The significance of membrane fluidity as a determinant of membrane properties is particularly important in our understanding of the etiology of cancer. The low levels of cyclic AMP found in many transformed (cancerous) cells may be caused by the greater fluidity of their membranes (Section V,B).

The role of guanylyl cyclase in the synthesis of cyclic GMP is described by Goldberg, O'Dea, and Haddox (1973) and by Hardman, Schultz, and Sutherland (1974). Most of the enzyme appears to be soluble, and there is no evidence that it is activated directly by extracellular stimulants. Those stimulants which are known to increase the cyclic GMP level in intact cells (e.g., acetylcholine) have no effect on the isolated guanylyl cyclase. Calcium is an essential requirement for activation of guanylyl cyclase in intact cells (Schultz, Hardman, Schultz, Baird, and Sutherland, 1973; Ferrendelli, Kinscherf, and Chang, 1973; Goldberg et al., 1973; Hardman et al., 1974). It is conceivable, therefore, that guanylyl cyclase activity is somehow regulated by the intracellular level of calcium (Fig. 1; Goldberg et al., 1973). In all the cases which have been examined so far, conditions which are known to raise the level of calcium also cause an increase in the level of cyclic GMP.

Specific enzymes are present for the degradation of both cyclic AMP and cyclic GMP (Appleman, Thompson, and Russell, 1973; Goldberg et al., 1973). In most cells there appears to be little direct regulation of phosphodiesterase activity by hormones or related factors. One notable exception is the light activation of cyclic GMP phosphodiesterase activity in photoreceptors (Section III,J). Hormones may regulate phosphodiesterase activity indirectly by changing the intracellular level of calcium which, in turn, may activate the enzyme (Fig. 1). Phosphodiesterase is inhibited by methylxanthines, which have been widely used as a method of artificially raising the intracellular level of the cyclic nucleotides. Theophylline and caffeine have been used extensively as inhibitors of phosphodiesterase (f in

Fig. 1), but it must be recognized that these drugs may have other effects. In particular, they are known to release calcium from the sarcoplasmic reticulum of muscles and may thus activate certain cells by raising the intracellular level of calcium rather than by elevating cyclic AMP.

C. Second Messenger Action of Calcium and The Cyclic Nucleotides

Both calcium and the cyclic nucleotides are present in most cells and, before they can be considered to function as second messengers, they should fulfill certain basic criteria. Four such criteria have been established for cyclic AMP (Robison, Butcher, and Sutherland, 1971). Such guidelines are not applicable to calcium, and it is difficult to construct a corresponding set of criteria because of the complicated nature of the calcium homeostatic mechanisms. However, there is one important criterion which should be satisfied when trying to prove the second messenger action of either calcium or the cyclic nucleotides. *It is essential to show that the proposed second messenger does directly stimulate a rate-limiting step of the effector system under investigation.* Before dealing with specific examples (Sections III and IV), it is important to have some idea of how these different second messengers mediate their effects within the cell.

Calcium has several characteristic actions. During stimulus-contraction coupling, it interacts with specific components of the contractile proteins. At rest, the myosin heads are prevented from interacting with actin by the strands of tropomyosin which are displaced out of the groove of the actin helix (Huxley, 1972; Parry and Squire, 1973). Calcium binds to troponin (Hitchcock, Huxley, and Szent-Györgyi, 1973), which somehow moves the tropomyosin into the groove and thus allows actin and myosin to interact. Calcium also stimulates the myosin ATPase which is essential for the repeated actin-myosin interactions. This stimulatory effect of calcium has been analyzed primarily in skeletal muscle, but is probably also applicable to cardiac and smooth muscle and may occur in a wide range of other cells as well. The presence of contractile elements, and in particular actin, has been described in a very wide range of cells, many of which are not obviously contractile (Adelstein, Conti, Johnson, Pastan, and Pollard, 1972; Bray, 1972; Yang and Perdue, 1972; Hanson, Repke, Katz, and Aledort, 1973; Huxley, 1973; Schroeder, 1973). In many of these systems, the functional organization of these contractile elements is still unclear, but they have been implicated in cellular motility (Huxley, 1973) and in the movement of various intracellular components including granules and chromosomes (Rebhun, 1972). Numerous authors (Douglas and Rubin, 1961; Rubin, 1970; Malaisse, Malaisse-Lagae, Walker, and Lacy, 1971; Rahwan and Borowitz, 1973; Rahwan, Borowitz, and Miya, 1973) have stressed the similarities between stimulus-contraction and stimulus-secretion coupling, but most of the similarities are centered around the processes of generating the calcium signal. Unfortunately, we still do not know how an increase in the intracellular level of calcium mediates the release of granules by exocytosis. Contractile microfilaments have been implicated in granule release (Stormorken, 1969; Rasmussen, 1970; Rubin, 1970; Orr, Hall, and Allison, 1972; Berl, Puszkin, and Nicklas, 1973; Rahwan and Borowitz, 1973), but conclusive evidence is lacking. Alternative theories concerning the mode of action of calcium on granule release do not invoke microfilaments (Matthews, 1970).

In many cases, the microfilaments (which can usually be equated with actin since they can be decorated with heavy meromyosin) are situated immediately below the plasma membrane and their ends might be attached to certain proteins in the membrane, especially in those cases where they are thought to mediate cell motility (Huxley, 1973). It is conceivable that the lateral migration of certain membrane components may be influenced through their attachment, or interaction, with the underlying web of filaments. If calcium is responsible for the contractile activity of these microfilaments, an increase in calcium concentration should be associated with greater membrane fluidity. Such a correlation is beginning to emerge from studies on transformed cells (Section V,B). Although these ideas are somewhat speculative, they do stress how a single action of calcium on various contractile elements might influence a wide range of cellular processes.

Another important action of calcium is to alter membrane permeability to ions. For example, an increase in the intracellular level of calcium can increase potassium permeability in red blood cells (Gárdos, 1958; Romero and Whittam, 1971) and in the ganglia of *Aplysia* (Meech, 1974). Calcium also has very marked effects on the chloride permeability of insect salivary glands (Section III,F). Some of the fluid secretory processes stimulated by calcium may thus be mediated through such increases in ionic permeability.

Finally, calcium can also act by altering enzyme activity. A classic example is its role in regulating phosphorylase activity in smooth muscle (Section IV,A).

The primary action of cyclic nucleotides is to stimulate protein kinases which phosphorylate various key proteins (Langan, 1973; Walsh and Ashby, 1973). Kuo (1974) demonstrated the existence of cyclic GMP-dependent protein kinase activity in a range of tissues from the guinea pig and rat. In those metabolic tissues where cyclic AMP has a direct second messenger action, it phosphorylates proteins which are rate limiting for various processes such as the breakdown of glycogen and lipid or the synthesis of a protein required for steroidogenesis in the adrenal cortex. In many other tissues, however, cyclic AMP plays an indirect role as a modulator of calcium homeostasis. This function may also be mediated by stimulating protein kinases. This action of cyclic AMP as a feedback regulator of calcium is extremely important for our understanding of cell activation and will be dealt with in greater detail in the following section on the interactions between the various second messengers.

D. Feedback Mechanisms Operating Between Calcium and the Cyclic Nucleotides

A very important component of intracellular control mechanisms is the feedback relationships operating between the different intracellular second messengers. These second messengers cannot be considered as independent entities because they are connected to each other by an intricate web of feedback relationships. It is essential to take these interrelationships into account when considering how the intracellular signaling system functions during cell activation.

Calcium may play an important role in regulating the level of cyclic nucleotides by activating phosphodiesterase (Fig. 1). Kakiuchi and associates (Kakiuchi and Yamazaki, 1970; Kakiuchi, Yamazaki, and Teshima, 1971; Kakiuchi, Yamazaki, Teshima, and Uenishi, 1973) have shown that, in the presence of an activator, very low levels of calcium (5×10^{-6} M) can markedly enhance brain phosphodiesterase activity. Calcium can also stimulate the phosphodiesterase of heart muscle (Teo and Wang, 1973), suggesting that this feedback mechanism may be of general significance. There are several instances in the literature where cell activation is associated with a fall in the level of cyclic AMP, which may be caused indirectly by calcium activating phosphodiesterase. Calcium may also contribute to a lowering of the cyclic AMP level by inhibiting adenylyl cyclase. Removal of such an inhibition could account for the increased cyclic AMP levels seen in various tissues when stimulated in calcium-free media (Nagata and Rasmussen, 1970; Prince, Berridge, and Rasmussen, 1972). Calcium can inhibit adenylyl cyclase in broken cell preparations (Streeto, 1969; Drummond and Duncan, 1970).

Perhaps the most significant feedback relationship is the ability of cyclic AMP to modulate calcium homeostasis (Fig. 1). Cyclic AMP can act by stimulating either the mechanisms responsible for removing calcium (c and d in Fig. 1), or it can increase the influx of extracellular (a in Fig. 1) or the release of intracellular calcium (b in Fig. 1). The stimulatory effect on calcium pumps has been particularly well characterized in cardiac muscle, where cyclic AMP stimulates the uptake of calcium by the sarcoplasmic reticulum (Entman, Levey, and Epstein, 1969). Cyclic AMP stimulates a protein kinase which phosphorylates a component of the sarcoplasmic reticulum, leading to a stimulation of calcium uptake (Kirchberger, Tada, Repke, and Katz, 1972; Tada, Kirchberger, Repke, and Katz, 1974b). Subsequent work has shown that the kinase does not phosphorylate the Ca-ATPase itself, but rather a smaller (22,000 dalton) subunit which is associated with the larger (100,000 dalton) carrier molecule (Tada, Kirchberger, Iorio, and Katz, 1974a). The smaller protein, called phospholamban, is thought to modulate the transport protein. A similar mechanism may exist in smooth muscle, where cyclic AMP stimulates calcium uptake into microsomal preparations (Section IV,A), but the mechanism has not been studied in the same detail as in heart muscle. In the case of smooth muscle, and perhaps in other cells as well, cyclic AMP may also stimulate the pump on the plasma membrane (c in Fig. 1). In both heart and smooth muscle, cyclic AMP stimulates relaxation through its ability to enhance the removal of calcium by stimulating these calcium pumps.

In many tissues, cyclic AMP can augment the calcium signal by stimulating the entry of calcium either from the outside (a in Fig. 1) or from intracellular reservoirs (b in Fig. 1). An effect of cyclic AMP on calcium entry has been postulated in nerves (Section III,A) and in the heart (Section IV,B). The ability of cyclic AMP to increase the mobilization of calcium from intracellular reservoirs has been well characterized in insect salivary glands (Section III,F), but is also thought to occur in a number of other cell types (see Section III). Stimulation of calcium release from intracellular reservoirs by cyclic AMP may account for the increased efflux of ^{45}Ca which is seen during activation of certain cells. As discussed earlier, the nature of this reservoir has not been firmly established, but likely candidates are either the endoplasmic reticulum or the mitochondria (Borle, 1973). Borle (1974) has evidence that the mitochondria may be important sources of calcium; he found that low concentrations of cyclic AMP (10^{-6} M) can stimulate a rapid release of calcium from the mitochondria of liver, kidney, or heart. The exact mechanism for this enhanced release has not been determined and, although cyclic AMP is generally considered to stimulate the efflux of calcium from the internal reservoir (*b* in Fig. 1), it could work equally as well by inhibiting the uptake process (*d* in Fig. 1). On the basis of his work with isolated mitochondria, Borle (1974) favors the former hypothesis.

The possibility that the intracellular calcium concentration may regulate the level of cyclic GMP through an action on guanylyl cyclase was discussed earlier. The exact function of cyclic GMP as a second messenger in cell activation is the subject of considerable speculation (Goldberg et al., 1973; Hardman et al., 1974). In many of the cells where there are increases in cyclic GMP associated with cell activation, there are parallel increases in the level of calcium which is directly responsible for stimulating the cell. For example, agents which stimulate contraction in smooth muscle increase the intracellular level of both cyclic GMP and calcium, but only the latter is directly concerned with activating the contractile machinery.

What is the function of the parallel increase in cyclic GMP? It has been suggested that cyclic GMP may have something to do with modulating calcium homeostasis (Schultz et al., 1973; Hardman et al., 1974). There is no direct evidence for such a notion but, considering the close relationship which exists between cyclic GMP and calcium (Goldberg et al., 1973), I think it is important to consider the possibility that cyclic GMP may exert feedback effects on the processes regulating the level of calcium. One possibility is that it operates in a feedback loop to control calcium influx (Fig. 1). During the action of acetylcholine on the heart, there is an increase in the level of cyclic GMP which is associated with a decrease in contractile force (George, Wilkerson, and Kodowitz, 1973). These authors suggest that one possible action of cyclic GMP is to decrease calcium entry. Schultz et al. (1973) have also speculated that cyclic GMP may act as a negative feedback signal to speed up the removal of calcium. If cells use the entry of extracellular calcium as a signaling device, they run the risk of being swamped with calcium, especially during prolonged stimulation, and it may be essential to have a feedback loop which will inactivate calcium entry. For example, the inactivation of the late calcium channel in the squid axon during prolonged depolarization (Baker et al., 1973) could conceivably be mediated by cyclic GMP. It is of interest that the addition of exogenous cyclic GMP to various cells often produces effects which are opposite to those seen with agents which, during the course of their action, normally induce an elevation in the intracellular level of cyclic GMP (Goldberg et al., 1973).

Much more work needs to be done on these feedback relationships, but the evidence accumulated so far clearly indicates that they play a vital role in regulating cellular activity. After analyzing the available information from a wide range of different control systems, it is evident that a major function of the cyclic nucleotides in many cells is to modulate the level of calcium. An exception to this general conclusion is found in the control of various cells specialized to carry out certain major metabolic events where cyclic AMP is found to have a direct second messenger function. Apart from these metabolic tissues, the primary intracellular second messenger in cells appears to be calcium, and the cyclic nucleotides seem to play a secondary role in either enhancing or dampening the calcium signal. These two different actions of cyclic AMP on calcium homeostasis have provided a basis for classifying control mechanisms into two main groups.

E. Mono- and Bidirectional Control Systems

Before considering some specific examples, it is important to realize that there are at least two basic types of control mechanisms. Goldberg, Haddox, Dunham, Lopez, and Hadden (1974) introduced the terms mono- and bidirectional to describe the main features of these two mechanisms. In monodirectional systems, control is exercised by a simple on-off principle. When the stimulant is present the cells are switched on, and when the stimulant is removed they become quiescent. In bidirectional systems, the control is more dynamic in that stimulation and recovery are mediated by separate stimulants. Cellular activity is often determined by a balance between the two opposing stimulants. Goldberg and his colleagues (1974) have set out to describe the intracellular basis of these control mechanisms in terms of interactions between cyclic AMP and cyclic GMP.

In monodirectional systems both nucleotides are thought to function in concert, whereas in bidirectional systems they antagonize each other (Goldberg et al., 1974). This antagonism between the two nucleotides has been incorporated into the Yin-Yang, or dualism hypothesis whereby cyclic AMP is postulated to stimulate the cell in one direction and cyclic GMP has the opposite effect (Goldberg et al., 1974). In many respects, this Yin-Yang hypothesis presupposes that these cyclic nucleotides have a direct action on the effector systems within the cell, but in many cases, such as in smooth muscle, this is manifestly not so. It seems essential to include calcium in our attempts to define the intracellular basis of these two control systems. The following definition of mono- and bidirectional control systems will provide a framework for a detailed discussion of specific examples of each system.

Monodirectional control systems are usually found in cells which are regulated by a single stimulant. If there are additional stimulants, these act in parallel with the primary stimulant. The appearance of the stimulant induces an increase in cell activity, and when the stimulant is removed the cell becomes quiescent of its own accord. The second messenger employed during cell activation is usually calcium, but in some cases, as in various metabolic tissues, it may be cyclic AMP. If cyclic AMP is formed during cell activation (either in response to the primary stimulant or due to the action of additional stimulants), it usually enhances the calcium signal.

Bidirectional control systems are usually found in cells which are regulated by two opposing stimulants. The appearance of one stimulant causes an increase in cell activity, whereas the second stimulant either mediates the recovery from or opposes the action of the first stimulant. In most cases, calcium is the second messenger for the first stimulant, whereas the antagonistic action of the second stimulant is apparently mediated by cyclic AMP, which acts by speeding up the removal of the calcium signal. In cases where cyclic AMP is the primary second messenger, as in liver and adipose tissue, the nature or even the existence of a second messenger mediating the action of the opposing stimulant (insulin) has not been established.

The basic difference between these two systems lies in the nature of the interaction between cyclic AMP and calcium. In monodirectional systems cyclic AMP enhances the calcium signal, whereas in bidirectional systems it opposes the calcium signal.

III. MONODIRECTIONAL SYSTEMS

A. Nervous System

Release of neurotransmitters during synaptic transmission, or the release of neurohormones from the terminals of neurosecretory cells, is regulated by calcium (Simpson, 1968; Rubin, 1970). The current hypothesis is that the release of vesicles from nerve terminals is sensitive to an increase in the intracellular level of calcium. Direct evidence for such a view was provided by Miledi (1973), who stimulated transmitter release by injecting calcium into the presynaptic nerve terminals of the giant synapse of the squid. There is much evidence to show that the calcium which triggers the release of vesicles comes from the outside and enters the cell as the action potential invades the presynaptic terminal. The important event during the action potential is the depolarization of the membrane. Artificial membrane depolarization produced by elevating the external potassium concentration will also open the calcium gate (see Section II,A) leading to a massive release of neurotransmitters or neurohormones (Douglas and Poisner, 1964; Maddrell and Gee, 1974). An example of the latter, which has been studied in some detail, is the release of vasopressin from the neurosecretory cells in the neurohypophysis (Thorn, 1974). Electrical stimulation or treatment with high potassium will release vasopressin as long as the outside medium contains calcium (Douglas and Poisner, 1964). The influx of calcium, as well as the release of vasopressin, is inhibited by agents such as manganese, lanthanum, and D600 (Fig. 1) which reduce the membrane permeability to calcium (Dreifuss, Grau, and Nordmann, 1973; Thorn, 1974). Tetrodotoxin, which blocks sodium movement, has no effect on either calcium movement or the release of vasopressin. Calcium entry into the presynaptic terminal during membrane depolarization may thus take place through the late calcium channel which has been described in squid axon (Baker et al., 1973). The calcium ionophores A 23187 and X-537 A can also stimulate release of vasopressin by a calcium-dependent process (Nakazato and Douglas, 1974; Russell et al., 1974). An influx of calcium is thus an essential component of the mechanisms regulating the secretory activity of a wide range of nerve cells.

Although cyclic AMP appears to play no direct role in the release mechanisms in the neurohypophysis (Thorn, 1974), there are indications that it may be important in modulating the sensitivity of certain nerve terminals, especially at neuromuscular junctions. It has been known for some time that catecholamines can facilitate neuromuscular transmission (Krnjević and Miledi, 1958). Norepinephrine can increase the quantal content of the end-plate potential (Jenkinson, Stamenović, and Whitaker, 1968). This action of norepinephrine is apparently mediated by cyclic AMP because it can be mimicked by either dibutyryl cyclic AMP or theophylline (Breckenridge, Burn, and Matschinsky, 1967; Goldberg and Singer, 1969). An earlier claim (Goldberg and Singer, 1969; Singer and Goldberg, 1970) that cyclic AMP may act by increasing the frequency of miniature end-plate potentials *(mepp)*, perhaps by altering the intracellular level of calcium, has not been confirmed (Miyamoto and Breckenridge, 1974; Wilson, 1974). However, Miyamoto and Breckenridge (1974) did find that both catecholamines and dibutyryl cyclic AMP could reverse neuromuscular fatigue after prolonged potassium stimulation by increasing *mepp* frequency. Cyclic AMP clearly does not play a direct role in the release of neurotransmitters, but there are indications that it can modulate the calcium signal by increasing calcium permeability (Miyamoto and Breckenridge, 1974). A similar effect of cyclic AMP on calcium permeability has been noted in heart muscle (Pappano, 1970). Wilson (1974) has also suggested that cyclic AMP may facilitate the release of transmitter by increasing its synthesis, mobilization, or storage rather than by adjusting the resting level of calcium.

Cyclic nucleotides may also play a role in the peripheral autonomic and central nervous systems. The levels of both cyclic AMP and cyclic GMP vary markedly during nervous activity, but the problem has been to establish whether these nucleotides are concerned with excitability or with the control of nonexcitable cells such as the glial elements. Detailed studies on Purkinje cells (Siggins, Hoffer, and Bloom, 1971*a;* Siggins, Oliver, Hoffer, and Bloom, 1971*b;* Siggins, Battenberg, Hoffer, Bloom, and Steiner, 1973) and on peripheral ganglia (Greengard and Kebabian, 1974) have shown that cyclic nucleotides do play a role in both pre- and postsynaptic events. The spontaneous activity of Purkinje cells is inhibited by norepinephrine released from adrenergic nerves originating from the locus coeruleus. Norepinephrine acts on the Purkinje cells to raise the intracellular level of cyclic AMP, which then leads to membrane hyperpolarization (Siggins et al., 1971b). The reason for this hyperpolarization is not clear. It may arise through an increase in electrogenic pump activity perhaps mediated by calcium, but it is not clear where this calcium comes from. It is conceivable that cyclic AMP may mobilize internal calcium.

A similar hyperpolarization is seen in the postganglionic neurons of the rabbit superior cervical ganglion after stimulation of inhibitory nerves (McAfee and Greengard, 1972). In this case, dopamine is the transmitter which induces the increase in cyclic AMP in the postganglionic neuron. Cyclic AMP may induce the hyperpolarization by phosphorylating specific components of the membrane. The release of dopamine from the interneuron may also be mediated, in part, by cyclic nucleotides. The inhibitory fibers release acetylcholine, which interacts with muscarinic receptors on the interneuron and leads to a release of dopamine. During this release of dopamine, there is an increase in the cyclic GMP level within the interneuron (Greengard and Kebabian, 1974). Whether cyclic GMP is directly responsible for releasing dopamine, or whether it develops secondarily due to an increase in calcium, has not been established. This latter possibility is much more likely because of the almost universal role of calcium in stimulus-secretion coupling. Ferrendelli et al. (1973) have also found that the marked increase in cyclic GMP seen in cerebellar slices during treatment with high potassium is abolished in calcium-free solution.

B. Adrenal Medulla

Release of catecholamines from chromaffin cells in the adrenal medulla has many features in common with the release of transmitters and hormones from nerve endings (Rahwan and Borowitz, 1973; Rubin, 1970). In particular, we see that calcium once again plays a central role in the link between stimulation and secretion (Douglas, 1968). The medulla has considerable historical significance, because it was while studying the release of catecholamines that Douglas and Rubin (1961) formulated their hypothesis concerning the role of calcium in "stimulus-secretion coupling." The chromaffin cells, which are clustered around the highly branched blood vessels in the adrenal medulla, are innervated by branches of the splanchnic nerve. The stimulus-secretion hypothesis attempts to explain how acetylcholine released from the splanchnic nerve leads to a rapid release of catecholamines from the chromaffin cells. The term "stimulus-secretion coupling" was coined to draw attention to the many similarities which exist between release of secretion granules and the phenomenon of "stimulus-contraction coupling" which occurs in muscle (Rubin, 1970). The main similarity between the two systems is the central role of calcium as a mediator between the external stimulus and the subsequent change in cellular activity.

The interaction of acetylcholine with the membrane receptors of chromaffin cells causes a change in ionic permeability which permits calcium to flood into the cell to activate the granule release mechanism (Fig. 2). As a consequence of

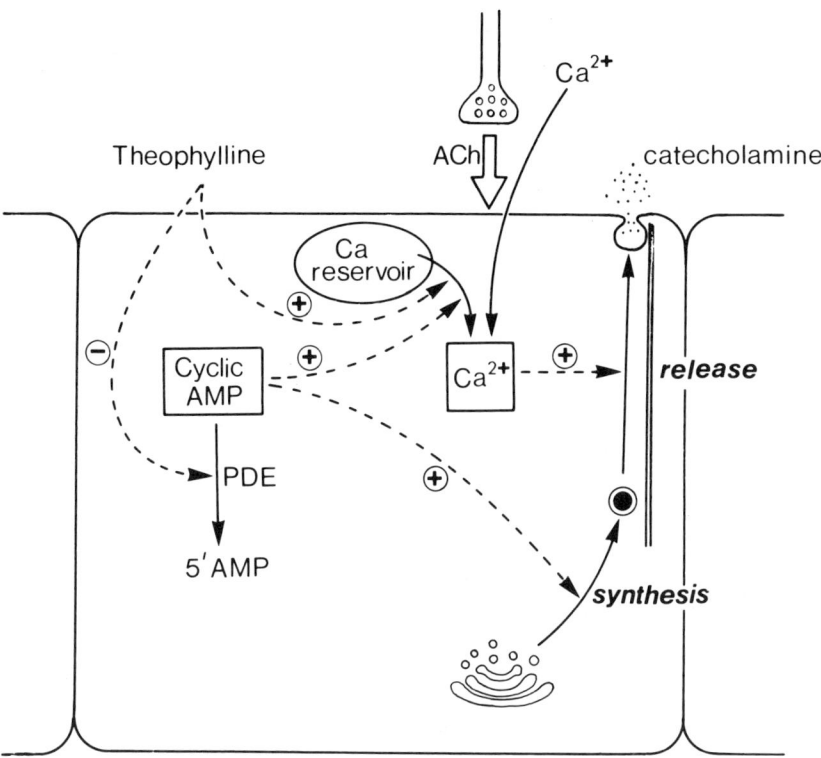

FIG. 2. The role of calcium in stimulus-secretion coupling in the adrenal medulla. Acetylcholine (ACh) released from the splanchnic nerve stimulates the entry of calcium, which then mediates the release of catecholamine granules by exocytosis. Cyclic AMP and theophylline may release calcium from intracellular reservoirs. Cyclic AMP may also speed up the synthesis of catecholamines.

these changes in permeability, there is a depolarization of the membrane (Douglas, Kanno, and Sampson, 1966). Artificial depolarization of the membrane with elevated potassium levels also leads to catecholamine release (Douglas and Rubin, 1963; Lastowecka and Trifaró, 1974). There are several ways of dissociating depolarization from secretion. For example, both acetylcholine and high potassium depolarize chromaffin cells in the absence of calcium, but there is no secretion (Douglas, Kanno, and Sampson, 1967). In sodium-free media, acetylcholine was capable of causing release even though the membrane potential remains above the resting level. The crucial event in stimulation, therefore, is not the depolarization, but the preceding changes in ionic permeability. Acetylcholine seems to act through a direct effect on the permeability of the membrane to both sodium and calcium (Douglas et al., 1967).

It is the influx of calcium which is important in stimulating secretion. Removing sodium or potassium has no immediate inhibitory effect, whereas removal of calcium blocks secretion stimulated with either acetylcholine or high potassium (Douglas and Rubin, 1961; Rubin, 1970). Indeed, secretion of catecholamines can be observed in a sucrose medium completely devoid of monovalent cations providing that calcium is present (Banks, Biggins, Bishop, Christian, and Currie, 1969). The magnitude of the secretory response is related to the external calcium concentration (Rubin, 1970). Direct evidence for calcium entry into chromaffin cells has been obtained from studies with ^{45}Ca. During the action of acetylcholine there is an increased influx of calcium (Douglas and Poisner, 1962) which can be blocked by tetracaine (Rubin, Feinstein, Jaanus, and Paimre, 1967). However, chemical determinations have shown that there is no net increase in the calcium content of the chromaffin cells (Rubin, 1970), suggesting that the increased influx must be matched by an increased efflux. Rubin et al. (1967) have found large exchanges of calcium during acetylcholine stimulation, and it may be necessary to consider the possible role of intracellular calcium reservoirs in the control of chromaffin cell activity. Both the chromaffin granules and the mitochondria contain large quantities of calcium (Borowitz, Fuwa, and Weiner, 1965) which may contribute to cell activation.

The existence of a functional intracellular pool of calcium has been revealed during studies with various agents which can stimulate secretion in the absence of external calcium. Theophylline, cyclic AMP, and dibutyryl cyclic AMP can stimulate secretion (Peach, 1972). This stimulatory effect of cyclic nucleotides persists in calcium-free conditions which completely block the stimulatory effects of both high potassium and nicotine. On the basis of these observations, Peach has suggested that cyclic AMP may stimulate secretion by releasing intracellular calcium (Fig. 2). Poisner (1973a,b) arrived at a similar conclusion from his observations that the phosphodiesterase inhibitors aminophylline or caffeine can stimulate catecholamine release independently of external calcium. Poisner took the precaution of including hexamethonium and atropine in his incubation media to guard against the possibility that these methylxanthines were acting indirectly by stimulating a release of acetylcholine from the nerves (Section III,A).

Bovine adrenal medulla has adenylyl cyclase activity, but there is no indication as yet about where this enzyme is located or how it is regulated *in vivo* (Hurko, Elster, and Wurtman, 1974). Jaanus and Rubin (1974) have demonstrated little correlation between catecholamine release (peak at 2 min) and the increase in cyclic AMP (peak at 8 min) which is obtained during stimulation with acetylcholine. They found no significant catecholamine secretion during application of either cyclic AMP or dibutyryl cyclic AMP and conclude, probably correctly, that this nucleotide is not directly concerned with the release of catecholamines. However, they raise the possibility that cyclic AMP may function earlier in the sequence, perhaps in the synthesis and packaging of the chromaffin granules. Guidotti and Costa (1973) had already observed that cyclic AMP may regulate catecholamine synthesis by stimulating tyrosine hydroxylase.

When considering the action of methylxanthines, we must not ignore the possibility that these agents may have a direct effect on calcium distribution across cell membranes (Fig. 2), as is thought to occur in muscle. Indeed, Rahwan et al. (1973) have shown that caffeine and chlorpromazine can stimulate a release of calcium from adrenal medullary mitochondria. Although there is considerable evidence for the existence of an intracellular reservoir of calcium which can be utilized under experimental conditions, it is not clear what role this internal calcium plays during the normal course of cell activation with acetylcholine. Cyclic AMP, by releasing intracellular calcium, could facilitate the release of catecholamine. Apart from the flux data discussed earlier (Rubin et al., 1967), studies on the kinetics of cell activation further suggest that there may be subtle adjustments of calcium homeostasis during the action of acetylcholine on chromaffin cells.

A characteristic feature of catecholamine release from the adrenal medulla is its phasic nature. During continuous treatment with acetylcholine or high potassium, there is a sudden and large output of catecholamines which lasts for a few minutes before declining exponentially to a lower plateau level which is slightly above the unstimulated level (Douglas and Rubin, 1963). The nature and time course of this phasic response is remarkably similar to the changes in intracellular calcium concentration observed in squid axons during potassium depolarization (Baker et al., 1973). As discussed earlier (Section II,A), Baker et al. (1973) suggest that the decline is caused by inactivation of the calcium channel, thus cutting down on calcium entry. The decline in catecholamine release which occurs during continuous stimulation of chromaffin cells may also be explained by a reduction in internal calcium concentration, because there are increases in secretion if the external calcium concentration is elevated during the plateau period (Douglas and Rubin, 1963). Whether calcium becomes limiting due to inactivation of calcium entry or to an increase in the rate of calcium removal, either by transfer into the intracellular reservoirs or back across the plasma membrane, remains to be seen. It is conceivable that cyclic AMP or cyclic GMP may play some role in the changes in calcium homeostasis which occur during prolonged stimulation. It is interesting to find that after treatment with acetylcholine there is an initial increase in the intracellular level of cyclic AMP which then returns rapidly to the resting level (Jaanus and Rubin, 1974). Clearly, we need more information about changes in cyclic AMP and cyclic GMP during cell activation before being able to speculate further on their possible roles in regulating either calcium homeostasis or catecholamine biosynthesis.

C. Anterior Pituitary

The anterior pituitary is a non-nervous tissue which secretes at least six different hormones. Each hormone comes from a specific cell type and is released in response to specific factors originating from the hypothalamic area. On the basis of available information we can consider the anterior pituitary as a whole because the underlying cellular control mechanisms appear to be very similar for all the different cell types and releasing factors. Both cyclic AMP and calcium have been implicated as second messengers regulating the release of hormones from the anterior pituitary.

During the action of the releasing factors on the anterior pituitary there is a marked increase in the intracellular level of cyclic AMP (Zor, Kaneko, Schneider, McCann, Lowe, Bloom, Borland, and Field, 1969; Zor, Kaneko, Schneider, McCann, and Field, 1970; Borgeat, Chavancy, Dupont, Labrie, Arimura, and Schally, 1972; Deery and Howell, 1973). Borgeat et al. (1972) used purified synthetic luteinizing hormone-releasing factor and observed a close correspondence between the release of luteinizing hormone and the increase in the intracellular level of cyclic AMP. Prostaglandins (PGE_1 and PGE_2), which are capable of stimulating the release of pituitary hormones (MacLeod and Lehmeyer, 1970), can also stimulate adenylyl cyclase to raise the intracellular level of cyclic AMP (Zor et al., 1969; Deery and Howell, 1973). Direct application of the di- or monobutyryl derivatives of cyclic AMP brings about the release of most of the hormones stored in the anterior pituitary (Fleischer, Donald, and Butcher, 1969; Wilber, Peake, and Utiger, 1969; MacLeod and Leymeyer, 1970; Lemay and Labrie, 1972). Theophylline, which can raise the cyclic AMP level in the anterior pituitary (Zor et al., 1970), can also stimulate the release mechanisms (Schofield, 1967; Wilber et al., 1969; MacLeod and Lehmeyer, 1970). In addition to cyclic AMP, Peake, Steiner, and Daughaday (1972) have implicated cyclic GMP in the control of growth hormone release. Cyclic GMP by itself can stimulate secretion. Hypothalamic extracts, containing growth hormone-releasing factor (GRF), increased the cyclic GMP level in the anterior pituitary. It remains to be seen whether GRF has a direct effect on guanylyl cyclase or whether it acts indirectly by raising the intracellular level of calcium which is thought to stimulate the formation of cyclic GMP in other systems (Section II).

The possibility that calcium may regulate hormone release was arrived at indirectly by stimulating the anterior pituitary with high potassium media. The release of hormone by high potassium solutions was totally dependent on external calcium (Vale and Guillemin, 1967; Samli and Geschwind, 1968; Katsumi, Kamberi, and McCann, 1969; Jutisz and de la Llosa, 1970). The high potassium media depolarized the plasma membrane of the pituitary cells (Martin, York, and Kraicer, 1973), and this change was thought to mimic the action of the normal releasing factors. Membrane potential was not tightly coupled to secretion (Martin et al., 1973) because depolarization was still apparent in calcium-free solutions which prevented hormone secretion. The high potassium seems to act by increasing the permeability of the membrane to calcium.

Although the effect of high potassium has an absolute requirement for calcium, there are certain contradictions concerning the calcium requirement for the action of the releasing factors. In some cases, calcium is necessary (Vale, Burgus, and Guillemin, 1967; Samli and Geschwind, 1968; Jutisz and de la Llosa, 1970), whereas in others, external calcium is not essential for secretion to occur (Katsumi et al., 1969; Milligan and Kraicer, 1974). Measurements of increased calcium influx during stimulation of the anterior pituitary with either high potassium (Milligan and Kraiser, 1969) or the natural releasing factors (Milligan, Kraicer, Fawcett, and Illner, 1972) would tend to support the idea that external calcium plays a crucial role in stimulus-secretion coupling. However, as outlined earlier (Section II,A), we must be careful about equating an increased uptake of ^{45}Ca with an increase in the permeability of the basal plasma membrane to calcium. On the basis of available evidence, therefore, we cannot be certain about the exact source of calcium during stimulus-secretion coupling in the anterior pituitary. Milligan and Kraicer (1974) have suggested that there may be different pools of calcium and that a redistribution of intracellular calcium is probably the primary event during the release process. Such a hypothesis would neatly explain the apparent role of both cyclic AMP and calcium in the control of hormone release. Under normal conditions, the releasing factors may interact with their specific receptors to increase the intracellular level of cyclic AMP which, in turn, would stimulate the release of calcium from intracellular reservoirs. It is of interest that the mitochondria, which are possible candidates for this reservoir, have high protein kinase activity (Lemaire, Pelletier, and Labrie, 1971). The possibility that the releasing factors increase the influx of external calcium either directly, or indirectly through the action of cyclic AMP, is still an open question. An increase in the intracellular level of calcium would then trigger the release of secretion, as is thought to occur in other cells (Rubin, 1970). High potassium had no effect on the level of cyclic AMP in the anterior pituitary (Zor et al., 1970), but was capable of stimulating hormone release. During this stimulation with high potassium, therefore, the cyclic AMP step is bypassed because depolarization can lead directly to an increased influx of calcium.

This attempt to unify the studies on cyclic AMP and calcium into a simple model is obviously very speculative, but it does indicate that these two second messengers may interact with each other to mediate hormone release from the anterior pituitary.

D. Endocrine Pancreas and Insulin Secretion

Insulin release from β cells in the islets of Langerhans is stimulated by an increase in the plasma level of glucose. There is a simple feedback system operating between the metabolite and the hormone which regulates its metabolism. When considered in this way, glucose may be thought of as a hormone. Indeed, the β cells have a glucoreceptor which responds specifically to glucose and some closely related sugars. Apart from glucose, other natural stimuli (e.g., glucagon) can also stimulate insulin release, and their mode of action must be incorporated into any model which sets out to explain the control of insulin secretion. We will begin with the mode of action of glucose which is the primary stimulus regulating insulin secretion (see also review by Montague and Howell, *This Volume*).

The ability of glucose to stimulate the release of insulin is apparently mediated

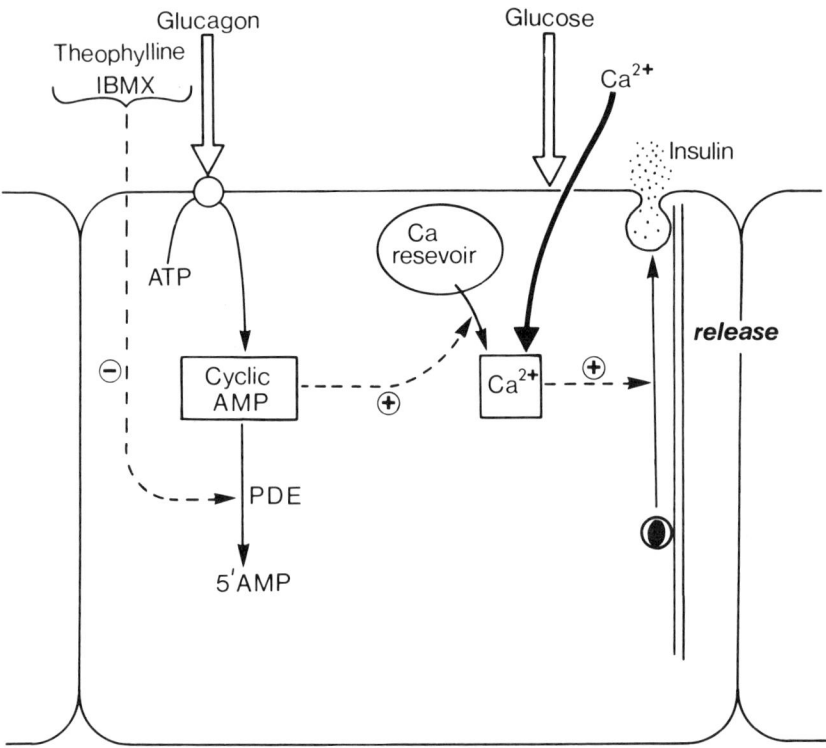

FIG. 3. Stimulus-secretion coupling in β cells. Insulin release by exocytosis is stimulated by calcium, which enters during the action of glucose. Glucagon raises the intracellular level of cyclic AMP, which may release calcium from intracellular reservoirs and thus potentiate the action of glucose.

by calcium (Fig. 3). External calcium is a prerequisite for insulin release (Rubin, 1970). Malaisse-Lagae and Malaisse (1971) found that glucose increased the influx of ^{45}Ca into β cells. There was also a linear relationship between calcium uptake and the rate of insulin release. Furthermore, the local anesthetic tetracaine can inhibit both the influx of calcium and the release of insulin (Brisson, Camu, Malaisse-Lagae, and Malaisse, 1971). The fact that high potassium can also stimulate secretion (Grodsky and Bennett, 1966; Hales and Milner, 1968) by a calcium-dependent process further suggests that calcium entry across the plasma membrane may be an essential step during insulin secretion.

There is considerable evidence from electrophysiological studies to show that glucose acts by increasing the permeability of the cell surface to calcium (Dean and Matthews, 1970a,b). An increase in glucose concentration depolarizes the membrane of β cells (the potential decreased 8.6 mV for each 10-fold increase in glucose concentration). Of even greater interest was the observation that above a critical external glucose concentration, which corresponded with the normal plasma glucose level where insulin begins to be released, there were frequent bursts of action potentials (Dean and Matthews, 1970a; Pace and Price, 1974). Each train of action potentials was preceded by a slow depolarization of the resting potential and is reminiscent of the slow waves described in smooth muscle (Section IV,A). The analogy with smooth muscle can be carried further because a detailed analysis of the ionic basis of these β-cell action potentials revealed that they were predominantly due to calcium entry (Dean and Matthews, 1970b). The way in which this influx of calcium triggers the release of insulin granules is not known. Calcium may promote the movement of granules toward the plasma membrane by stimulating the contraction of the microfilament-microtubular complex which is thought to mediate granule release (Lacy, Howell, Young, and Fink, 1968; Malaisse et al., 1971; Orci, Gabbay, and Malaisse, 1972).

Having established the central role of calcium in the release mechanism, it remains to consider the mode of action of various other agents known to influence the release of insulin. The action of many of these agents appears to be linked with that of cyclic AMP. Before considering the action of these agents, it is necessary to establish that cyclic AMP plays no direct role in mediating the action of glucose. Glucose has no effect either on the intracellular cyclic AMP level of resting cells (Montague and Cook, 1971; Cooper, Ashcroft, and Randle, 1973) or on the activity of adenylyl cyclase in particulate fractions (Davis and Lazurus, 1972; Goldfine, Roth, and Birnbaumer, 1972). On the other hand, glucagon appears to stimulate insulin release by increasing the level of cyclic AMP in β cells (Fig. 3) by activating adenylyl cyclase (Sussman and Vaughan, 1967; Turtle and Kipnis, 1967). The importance of cyclic AMP in insulin release is evident from numerous studies showing that inhibition of phosphodiesterase with caffeine, theophylline, tolbutamide, or 3-isobutyl-1-methylxanthine (IBMX) can potentiate the action of glucose (Montague and Cook, 1971; Ashcroft, Randle, and Täljedal, 1972). During the action of these agents, there is an increase in the intracellular level of cyclic AMP (Turtle and Kipnis, 1967; Montague and Cook, 1971). However, an increase in cyclic AMP concentration is not, in itself, a sufficient stimulus to induce the release of insulin. Cyclic AMP is effective only when operating in the presence of a certain critical level of glucose. For example, glucagon does not stimulate insulin release in the presence of 3.3 mM glucose, but it becomes an effective stimulus when the glucose level is elevated to 16.5 mM (Hales and Milner, 1968). In a parallel experiment, Cooper et al. (1973) showed that IBMX produced a marked increase in the cyclic AMP concentration at both 3.3 mM and 30 mM glucose, but an increase in the release of insulin was observed only at the higher glucose concentration. Similar observations have been made by Montague and Cook (1971). At the low glucose concentration, therefore, there was an increase in the intracellular level of cyclic AMP but no release of insulin. These studies seem to indicate that cyclic AMP plays a permissive role in insulin secretion by potentiating the effects of glucose.

Cyclic AMP may sensitize the β cells to glucose by modulating calcium homeostasis (Fig. 3). It was argued earlier that insulin release was probably controlled by the intracellular level of calcium. Cyclic AMP may stimulate the release of calcium from intracellular reservoirs and thus increase the resting level of calcium which will sensitize the cell to glucose. A clue to the internal action of cyclic AMP was provided by Brisson, Malaisse-Lagae, and Malaisse (1972) and by Brisson and Malaisse (1973), who showed that theophylline increased the efflux of calcium from prelabeled β cells. The ability of intracellular cyclic AMP to mobilize intracellular calcium would certainly account for the ability of dibutyryl cyclic AMP or theophylline to restore the action of glucose in low calcium media (Brisson et al., 1972). Cyclic AMP by itself is not capable of stimulating the release of insulin but, by modulating calcium homeostasis, it is capable of sensitizing the β cell to glucose.

It is of interest to speculate on the possible role of cyclic AMP in the control of insulin release *in vivo*. Since the main hormone regulating the level of cyclic AMP in β cells is glucagon, its release from the neighboring α cells during periods of low glucose concentration may serve to prime the β cells to respond to the inevitable rise in glucose level induced by glucagon. This ability of glucagon to influence the responsiveness of the β cells to glucose may represent a fine adjustment of the major feedback mechanisms responsible for ensuring plasma glucose homeostasis.

E. Exocrine Pancreas

The main function of the exocrine pancreas is to synthesize and secrete a range of digestive enzymes. These enter the duodenum in a flow of pancreatic juice which is rich in bicarbonate. The bicarbonate is also important in that it neutralizes hydrogen ions entering from the stomach. The two functions of enzyme and fluid secretion are carried out by different cell types (Fig. 4). Acetylcholine or pancreozymin stimulates the acinar cells to release enzymes, whereas secretin stimulates the centroacinar cells to secrete fluid. In attempting to understand how the exocrine pancreas is regulated, we are faced with a duality of both function and control. The presence of two cell types which are under independent control has doubtless contributed to much of the confusion in the literature.

Another possible source of confusion may arise from the fact that enzyme secretion can be observed only if there is a parallel flow of fluid to flush the enzymes out of the pancreas. For example, when the pancreas is stimulated with secretin to initiate fluid secretion, there is always an initial but temporary appearance of enzyme which had accumulated in the ducts while the gland was at rest (Case and Scratcherd, 1972a; Case, Johnson, Scratcherd, and Sherratt, 1972). Particular care must be exercised in interpreting results, especially with tissue slices. Some agents may increase enzyme secretion indirectly by stimulating fluid secretion, which then helps to flush enzyme out of the slice. A further complicating factor in studies on the pancreas is the considerable differences which exist

EXOCRINE PANCREAS

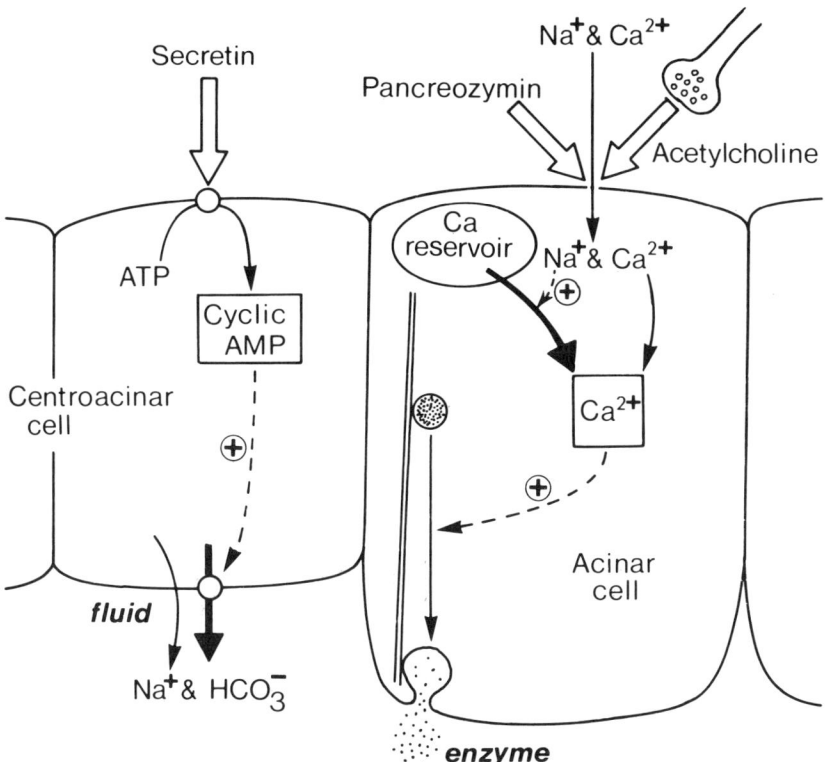

FIG. 4. Separate control of fluid and enzyme secretion by the exocrine pancreas. Secretin stimulates the centroacinar cells to form cyclic AMP, which is thought to increase fluid secretion. The mechanism of secretion is not fully understood, but may consist of an active secretion of bicarbonate ions with sodium following passively. Pancreozymin, or acetylcholine, acts on the acinar cells to induce an influx of sodium and calcium. Case and Clausen (1973) suggest that the sodium may stimulate a release of calcium from intracellular reservoirs. The increase in the intracellular level of calcium is then responsible for releasing the zymogen granules.

among species. If one takes all these complications into account, a consistent picture begins to emerge of the control of the exocrine pancreas. The ability of secretin to initiate fluid secretion is apparently mediated by cyclic AMP, whereas the release of enzymes, regulated by acetylcholine or pancreozymin, is mediated by a calcium-dependent process (Fig. 4).

After the pancreas is treated with secretin, there is a rapid increase in the intracellular level of cyclic AMP which precedes the onset of fluid secretion (Case et al., 1972). Benz, Eckstein, Matthews, and Williams (1972) have also observed that secretin can increase the intracellular level of cyclic AMP. Theophylline, which can increase cyclic AMP levels (Case et al., 1972), can stimulate fluid secretion directly and can also potentiate the effect of submaximal doses of secretin (Case and Scratcherd, 1972a). Prostaglandins, which are capable of activating adenylyl cyclase in many systems, can stimulate electrolyte secretion by the saline-perfused pancreas (Case and Scratcherd, 1972b). This stimulatory action of prostaglandins was potentiated by theophylline. There is thus considerable evidence linking an increase in cyclic AMP with the stimulatory action of secretin on fluid secretion (Fig. 4). Very little is known about how cyclic AMP increases the sodium and bicarbonate transport mechanisms responsible for the flow of pancreatic juice.

Although earlier studies pointed to a role for cyclic AMP in the secretion of enzymes (Kulka and Sternlicht, 1968; Ridderstrap and Bonting, 1969; Morisset and Webster, 1971; Case et al., 1972; Heisler, Fast, and Tenenhouse, 1972), subsequent studies have contradicted these reports by indicating that cyclic AMP is not a mediator (Benz et al., 1972; Case and Scratcherd, 1972a). On balance, the evidence against cyclic AMP is probably more convincing. Much of the evidence linking cyclic AMP with enzyme secretion was obtained by applying this nucleotide, or its dibutyryl derivative, directly to tissue slices, which may produce erroneous results as discussed earlier. Another important consideration is that pancreozymin can cross-react with the secretin receptor. Using purified hormones, Heatley (1968) found that secretin had no effect on enzyme secretion, but pancreozymin caused considerable fluid secretion. High doses of pancreozymin can stimulate the same adenylyl cyclase normally activated by secretin (Rutten, De Pont, and Bonting, 1972). Benz et al. (1972) found that, in the presence of 3 mM theophylline, pancreozymin had no effect on the level of cyclic AMP, but this was elevated after treatment with secretin. However, this increase in cyclic AMP level was not associated with any change in amylase secretion. This evidence, when taken in conjunction with the observation that direct application of dibutyryl cyclic AMP or cyclic AMP has no effect (Benz et al., 1972; Case and Scratcherd, 1972a), would seem to rule out any direct role for cyclic AMP in regulating enzyme secretion.

Calcium is a much more likely candidate as the mediator of the action of either acetylcholine or pancreozymin on enzyme release. The calcium ionophore A 23187 can stimulate the release of amylase (Williams and Lee, 1974). An external supply of calcium is essential for enzyme secretion (Hokin, 1966; Robberecht and Christophe, 1971; Benz et al., 1972; Heisler et al., 1972; Kanno, 1972; Case, 1973). The role of calcium in stimulus-secretion coupling has emerged from both electrical and calcium flux studies. The resting potential of guinea pig acinar cells is approximately −40 mV (Dean and Matthews, 1972; Matthews and Petersen, 1973), whereas that from rat pancreas is lower (−33 mV) (Kanno, 1972). Both cholinergic agents and pancreozymin induce rapid changes in the membrane potential, whereas secretin has no effect. The cells from guinea pig, which have the higher resting potential, are depolarized; the rat cells are hyperpolarized. This difference, which probably depends on the difference in the resting potentials, may not mean too much because other studies have shown that a change in membrane potential *per se* is not a sufficient stimulus for secretion. In the presence of atropine, which blocks the action of endogenous acetylcholine, depolarization of the acinar cell membrane with high potassium has no effect on enzyme release. This is in contrast to muscle and various other secretory cells (Section II,A), where high potassium stimulates cell activity. In the pancreas, therefore, the secretogogues have direct effects on ion fluxes, which then lead to the change in membrane potential and not vice versa. Indeed, Petersen (1973*a*) has shown that there are large decreases in resistance during the action of either acetylcholine or pancreozymin. The next problem to consider is how this increased permeability to ions stimulates the exocytosis of zymogen granules.

By studying the ionic requirements for the acetylcholine-induced membrane depolarization, Matthews and Petersen (1973) concluded that there was an increased influx of both sodium and calcium (Fig. 4). Although the sodium current predominates, they suggested that the influx of calcium was the significant event in mediating secretion. However, other studies have shown that the inhibitory effect of removing calcium takes time to develop. There is also little evidence for an influx of calcium during the action of acetylcholine or pancreozymin on enzyme secretion (Argent, Case, and Scratcherd, 1973; Case, 1973; Heisler and Grondin, 1973). The view is growing, therefore, that an intracellular reservoir of calcium may be more important than extracellular calcium in the regulation of enzyme secretion (Argent et al., 1973; Case, 1973; Case and Clausen, 1973). Support for this notion is provided by the large increases in calcium efflux observed during the action of either acetylcholine or pancreozymin (Case and Clausen, 1973; Heisler and Grondin, 1973; Matthews, Petersen, and Williams, 1973). The increase in calcium efflux preceded the release of amylase, and there also was a very close correlation between these two parameters. A curious feature of this increase in calcium flux was its temporary nature; i.e., the flux rate rapidly returned to the control level whereas the release of amylase remained high. However, if the gland was allowed to recover, another increase in calcium efflux could be elicited on further stimulation with acetylcholine (Matthews et al., 1973). The source of this calcium, which contributes to the increased efflux, has not been established. Matthews et al. (1973) speculate that it originates from the inner surface of the plasma membrane, whereas Case and Clausen (1973) suggest that it comes from an intracellular reservoir (Fig. 4). If the latter idea is correct, it will be necessary to explain how receptor activation on the plasma membrane leads to this release of calcium from the intracellular reservoir. In this case, there is no evidence that cyclic AMP functions in calcium mobilization as occurs in various other systems. Case and Clausen (1973) suggest that the influx of sodium may be responsible for releasing the calcium (Fig. 4). Removal of sodium from the bathing medium greatly reduces the effect of pancreozymin on both calcium and amylase release.

As in a number of other secretory cells, stimulus-secretion coupling in the pancreas has some remarkable similarities with stimulus-contraction coupling in muscle. This similarity may include the apparent utilization of an intracellular pool of calcium during cell activation.

F. Insect Salivary Gland

The salivary glands of the blowfly are an ideal model system for studying the interactions of cyclic AMP and calcium during cell activation. The secretory region of these insect salivary glands consists of a long thin tube which secretes isosmotic potassium chloride when stimulated with 5-hydroxytryptamine (5-HT) (Berridge and Prince, 1972a). The structure of the glands and the various techniques used to monitor salivary gland activity were summarized in the first volume of this series (Berridge and Prince, 1972c). One of the main advantages of this secretory system is that it is made up of a single layer of homogeneous cells (Oschman and Berridge, 1970). The mechanism of fluid secretion has been studied in some detail (Berridge, Lindley, and Prince, 1975a,b,c) and is summarized in Fig. 5. The driving force is a potassium pump located on the luminal membrane; chloride follows passively, and this movement of ions into the lumen creates the osmotic gradient to entrain a parallel flow of water. Potassium and chloride enter the cell passively across the basal plasma membrane to replace the ions which are being transported into the lumen. The movement of ions across the apical surface generates characteristic potentials which are easy to measure and provide a way of continuously monitoring cellular activity (Berridge and Prince, 1971, 1972a,b; Prince and Berridge, 1972, 1973). Since cyclic AMP stimulates the cation pump whereas calcium stimulates chloride movement, it is possible to continuously monitor the intracellular action of these two second messengers. These insect salivary glands provide an ideal opportunity to study how cyclic AMP and calcium interact during the action of a hormone.

There are several lines of evidence pointing to a role for cyclic AMP in the action of 5-HT. (1) During the action of 5-HT there is an increase in the intracellular level of cyclic AMP (Prince et al., 1972). (2) Exogenous cyclic AMP and some closely related derivatives are capable of stimulating fluid secretion (Berridge, 1970, 1973). (3) Theophylline can stimulate secretion and can also greatly potentiate the action of both 5-HT and exogenous cyclic AMP (Berridge, 1970). Electrophysiological studies suggested that cyclic AMP acts by stimulating a potassium pump. However, exogenous cyclic AMP was not able to duplicate all the potential changes seen during the action of 5-HT, and it became clear that there was a component of 5-HT action which was not mimicked by cyclic AMP. Further studies revealed that 5-HT can stimulate chloride movement using calcium as a second messenger (Fig. 5). In attempting to analyze how 5-HT influenced calcium homeostasis, it became clear that there is an interesting feedback relationship operating between cyclic AMP and calcium.

The relationship between these two second messengers first emerged from studying cell activation in calcium-free media (Prince et al., 1972; Prince and

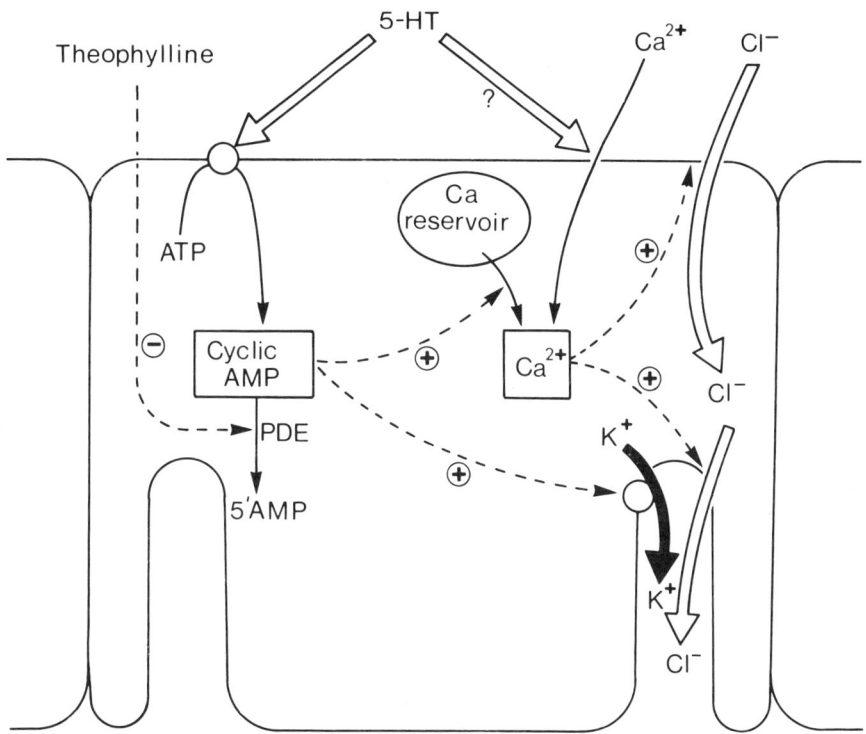

FIG. 5. Stimulation of fluid secretion by 5-HT in the blowfly salivary gland. 5-HT increases the synthesis of cyclic AMP and may also stimulate the influx of calcium. Cyclic AMP has two actions in the cell: it increases the release of intracellular calcium and may also stimulate the potassium pump on the apical membrane. Calcium acts to increase the chloride permeability of both the basal and apical membranes.

Berridge, 1973; Berridge et al., 1974). When salivary glands are stimulated in a calcium-free environment, there is an apparent independence of external calcium because normal rates of secretion are evident for approximately 10 min. However, after this initial period the rate gradually declines toward the resting rate of secretion. This temporary independence of external calcium is achieved by the mobilization and utilization of an internal reservoir of calcium. Once this store is depleted, the glands become totally dependent on external calcium; normal rates of secretion are seen immediately after the readdition of calcium to the depleted glands. Treatment of these insect salivary glands for extended periods in calcium-free media has no apparent deleterious effects. The effects produced in calcium-free media are reversible, and there is no change in transepithelial resistance—indicating that the cell junctions remain intact, contrary to what happens in toad bladder (Section III,H). Mobilization of the internal calcium reservoir seems to occur only during stimulation with 5-HT and not in resting glands (Berridge et al., 1974). Since both 5-HT and cyclic AMP can increase the efflux of ^{45}Ca from prelabeled glands (Prince et al., 1972), it appears as if one important intracellular action of cyclic AMP is to release calcium from intracellular reservoirs (Fig. 5). The problem of trying to decide whether, under normal conditions, a cell uses extra- or intracellular calcium is difficult to resolve, but may hold the key to our understanding of cell activation in many systems. Before considering this aspect further, it is important to demonstrate that calcium is indeed a second messenger. The fact that 5-HT action is calcium-dependent is a first step, but is by no means final proof that calcium functions as a second messenger.

Studies with the calcium ionophore A 23187 have provided more direct evidence that calcium has a second messenger role (Prince et al., 1973). Stimulation of secretion with A 23187 is dependent on calcium. However, there are indications that the ionophore may act not only by altering the influx of calcium from the outside, but may also release calcium from internal reservoirs *(unpublished observation)*. If added to salivary glands in a calcium-free medium, A 23187 can induce a temporary burst of secretion suggesting that, like 5-HT, it is capable of mobilizing intracellular calcium (Fig. 6). The fact that subsequent addition of 5-HT has no effect on the rate of secretion clearly indicates that the glands have been depleted of calcium. The glands are now totally dependent on external calcium, as shown by the marked increases in secretory rate when calcium is readmitted (Fig. 6). Calcium continues to stimulate secretion even after 5-HT and the ionophore have been removed by repeated washing, which suggests that, once the ionophore has entered the membrane, it reverses very slowly, if at all. Under these conditions, the addition of external calcium is a sufficient stimulus for secretion suggesting that calcium may well have a second messenger action in salivary glands. A similar dependence on external calcium is observed during stimulation of secretion with high potassium media (Berridge et al., 1975c). Stimulation with high potassium or the ionophore A 23187 both seem to rely on an increased influx of external calcium, and the obvious question arises whether there is a similar increased influx of calcium during stimulation with 5-HT. Studies with radiocalcium certainly indicate that there is an increased uptake of ^{45}Ca (Prince et al., 1972), but such studies are subject to the reservations outlined earlier concerning the uncertainty of influx measurements (Section II,A). If 5-HT does increase the influx of calcium, it will be interesting to know whether the two events, stimulation of adenylyl cyclase and a change in calcium permeability, are the result of interaction with a single receptor. Despite the uncertainty about the contribution of extracellular calcium during the normal action of 5-HT, there is considerable evidence to implicate calcium in the activation of these salivary glands.

In order to firmly establish a second messenger action for calcium, it is necessary to show that calcium has an effect on an intracellular effector system con-

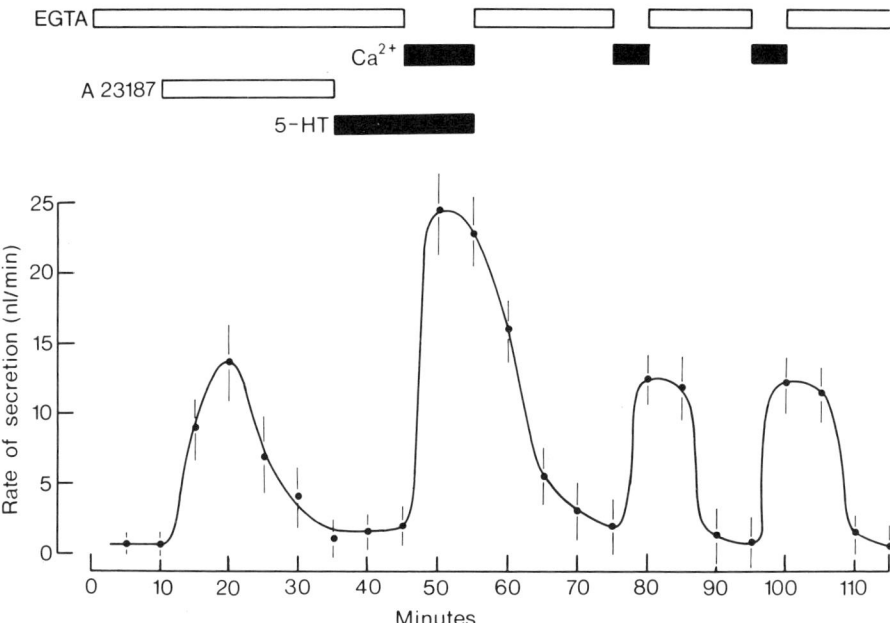

FIG. 6. Effect of the calcium ionophore A 23187 (2×10^{-5} M) on fluid secretion by isolated salivary glands of the blowfly. The ionophore was applied to the gland in a calcium-free medium containing 1 mM EGTA. Subsequent treatment with 1×10^{-8} M 5-HT had no effect unless calcium (2 mM) was present. After 5-HT and the ionophore were removed, the addition of calcium still caused marked increases in fluid secretion.

nected with secretion. Previous studies on potential changes during the action of 5-HT suggested that calcium may act by increasing the passive permeability of the membrane to chloride (Berridge and Prince, 1971, 1972b; Prince and Berridge, 1972, 1973). More direct evidence for such an action has emerged from subsequent studies on the effects of 5-HT on membrane resistance (Berridge et al., 1975a). During the action of 5-HT there are rapid and large decreases in resistance which occur simultaneously across both the apical and basal membranes. These changes in resistance are apparently mediated by calcium (Fig. 5). By regulating chloride permeability, calcium functions as an important link in the chain of events leading to an increase in fluid secretion (Fig. 5).

The exact role of cyclic AMP is still not clear. It certainly seems to mediate release of intracellular calcium as was described earlier. There are also indications, based on electrical recordings, that cyclic AMP stimulates the potassium pump on the apical (luminal) membrane (Fig. 5). It is not clear yet whether cyclic AMP acts directly on the pump mechanisms or indirectly by altering a related ionic event. It is conceivable that pump rate is also regulated by calcium and that there is a hierarchy of calcium-dependent events activated sequentially as the calcium level rises. The pump activity seems to require less calcium than the chloride permeability changes. This is well illustrated when both parameters are studied during calcium depletion (Berridge and Prince, 1972c; Prince and Berridge, 1973). The negative potential, which depends on the elevated chloride permeability, declines much sooner than the rate of secretion, which is a measure of the activity of the potassium pump.

G. Mammalian Salivary Gland

Salivary glands of mammals secrete both fluid and enzymes. The problem of control is complicated because these two different secretory processes take place in the same cell and yet they can be regulated independently of each other. The organization in salivary glands differs from that in the pancreas, where the processes of enzyme and fluid secretion are carried out by separate cells (Section III,E). The salivary gland has a racemose structure. The secretory mechanisms responsible for initiating secretion are located in the acinar cells which make up the bulbous acini (Berridge and Oschman, 1972).

A characteristic feature of salivary gland function is that long periods of quiescence are interrupted by short periods of intense secretory activity. These short bursts of secretion are regulated by both components of the autonomic nervous system (Fig. 7). Acetylcholine released from the parasympathetic nerves regulates fluid secretion, whereas norepinephrine released from the sympathetic nerve endings is primarily concerned with the secretion of enzymes. During sympathetic stimulation, the enzymes are carried out of the gland in a small flow of saliva so that there is a partial activation of fluid secretion during the action of norepinephrine (Fig. 7). Both calcium and cyclic nucleotides have been implicated in the control of salivary secretion, but there is no satisfactory explanation of how these two second messengers interact to permit enzyme and fluid secretion to be stimulated independently of each other. The following hypothesis sets out to coordinate current information on the role of cyclic nucleotides and calcium in the control of salivary gland secretion.

Attempts to understand how acetylcholine controls fluid secretion have been hampered by a lack of information on the mechanism of ion and water transport. In order to understand the secretory process, it is necessary to characterize events at both surfaces of the cell. Unfortunately, it is extremely difficult to monitor either the ionic or the electrical events taking place across the apical (luminal) surface of mammalian salivary glands. Most of our current information, therefore, has been derived from studying what happens across the basal surface. During the action of acetylcholine, there are marked changes in ion fluxes across this surface with concomitant changes in potential. The potential across the basal plasma membrane was relatively low (-20 to -30 mV), and was found to hyperpolarize markedly during the action of acetylcholine (Fritz and Botelho, 1969; Petersen and Poulsen, 1969; Petersen, 1970a). This hyperpolarization is thought to arise through an increase in potassium permeability. However, the potential failed to reach the potassium equilibrium potential, which has lead to the suggestion that this outward potassium current is short-circuited by a simulta-

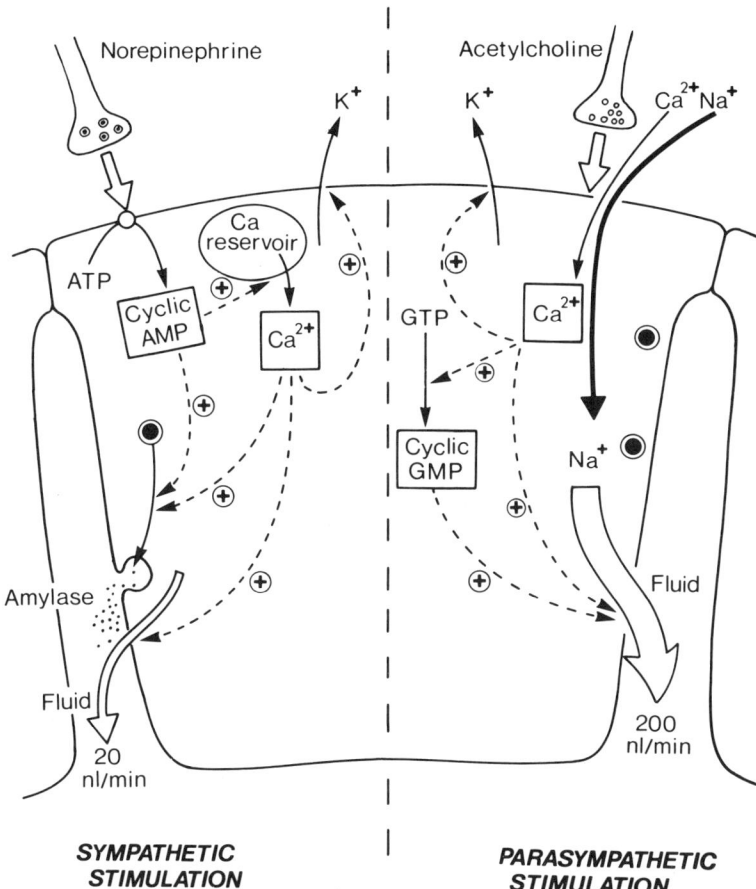

FIG. 7. A hypothesis to account for the independent control of enzyme and fluid secretion in mammalian salivary glands. Norepinephrine released from the sympathetic nerves stimulates the synthesis of cyclic AMP, which is probably responsible for initiating the release of amylase. Part of the action of cyclic AMP may involve a release of calcium from intracellular reservoirs. The calcium may assist in amylase release and may also be responsible for initiating the efflux of potassium and the small flow of fluid seen during sympathetic stimulation. Acetylcholine released from the parasympathetic nerves may stimulate an influx of both sodium and calcium. The latter may account for an increased efflux of potassium and the increased level of cyclic GMP. Both cyclic GMP and calcium may be responsible for switching on the secretory mechanisms located on the luminal surface. The intracellular level of calcium is probably much higher during parasympathetic than during sympathetic stimulation.

neous inward flow of sodium (Petersen, 1970a). These opposite movements of potassium out of the gland, and sodium into the gland, are certainly consistent with various physiological observations. It has been known for a long time that there is a transient loss of cellular potassium during salivary gland stimulation

(Burgen, 1956). The opposite influx of sodium into the gland would also be a necessary prerequisite for a cell which is secreting large volumes of saliva rich in sodium.

The possibility that the ionic changes which have just been described across the basal surface represent primary events during cell stimulation must take into account the long delay which exists between the arrival of acetylcholine at the cell surface and the initiation of the hyperpolarization. For example, after stimulating the parasympathetic nerves innervating cat submandibular glands, the onset of the secretory potential had a latency of 350 msec (Creed and Wilson, 1969). These authors calculated that no more than 60 msec could be accounted for by nerve conduction, ganglionic delay, release of acetylcholine, and diffusion of the transmitter to the gland surface. This very long latency seemed to rule out any direct action of acetylcholine on sodium and potassium permeability. However, recent electrical studies on salivary glands, which have higher resting potentials than those recorded earlier, reveal that acetylcholine produces a biphasic response (Nishiyama and Kagayama, 1973; Petersen, 1973b). There is an initial rapid depolarization, which is probably caused by an increase in sodium permeability, followed by a hyperpolarization with a time course resembling that recorded previously. It is conceivable that calcium enters the gland together with sodium during the initial depolarization and that the former is responsible for the subsequent increase in potassium permeability (Fig. 7).

Selinger, Batzri, Eimerl, and Schramm (1973) have demonstrated that external calcium is essential for the release of potassium from rat parotid slices during α-adrenergic stimulation. Apart from this requirement for potassium release, calcium is also essential for the maintenance of fluid secretion (Douglas and Poisner, 1963; Thorn, 1974). The ability of the calcium ionophore A 23187 to stimulate a large release of potassium from these parotid slices provides further evidence that an influx of calcium triggers the change in potassium permeability (Selinger et al., 1974). It would be instructive to know whether A 23187 is capable of stimulating fluid secretion as has been observed in insect salivary glands (Section III,F). Although potassium release is certainly associated with the normal response to acetylcholine, there is still no information on how sodium entry and potassium exit across the basal surface are related to the ion-transport processes which are responsible for generating the flow of saliva across the apical surface of the cell.

In another secretory system, the avian salt gland, Peaker (1971) has suggested that the influx of sodium may raise the intracellular level sufficiently to activate sodium pumps on the apical surface. However, Poulsen (1973) ruled out such a mechanism for salivary glands because he was not able to evoke secretion when he raised the intracellular level of sodium by inhibiting the sodium-potassium pump by lowering either the temperature or the external potassium. Another possibility is that the apical pump mechanism is regulated by calcium (Petersen, 1970b; Schneyer, Young, and Schneyer, 1972; Fig. 7).

There is no evidence that cyclic AMP plays any role in regulating fluid secretion, but an increase in the cyclic GMP level has been noted during the action of methacholine (Schultz et al., 1973). As in other systems, the increase in cyclic GMP is dependent on external calcium and further argues for an influx of calcium during the action of acetylcholine. Cyclic GMP must be included, together with calcium, as a possible candidate for regulating fluid secretion (Fig. 7).

Stimulation of the sympathetic nerves results in a large release of enzyme together with a small volume of fluid. The ability of norepinephrine to stimulate enzyme secretion independently of a large output of fluid is a β-adrenergic effect and uses cyclic AMP as a second messenger. Exogenous cyclic AMP can induce a large release of amylase (Bdolah and Schramm, 1965; Malamud, 1972). Adrenergic stimulation of parotid glands results in a marked increase in the intracellular cyclic AMP concentration (Rasmussen and Tenenhouse, 1968; Guidotti, Weiss, and Costa, 1972). The picture is complicated, however, by several observations which indicate that calcium is necessary for enzyme secretion (Rubin, 1970; Selinger and Naim, 1970). However, it is not clear yet whether this calcium requirement includes a function as a second messenger. The calcium ionophore A 23187 induces a twofold increase in enzyme release, which is insignificant when compared to the 20-fold increase in enzyme secretion produced by isoproterenol (Selinger et al., 1974). An increase in the intracellular level of calcium by itself is apparently not a sufficient stimulus for enzyme release. Cyclic AMP must also be present. However, an increase in the intracellular level of calcium may be essential to activate the small increase in fluid secretion which accompanies the release of enzyme. It looks as if this calcium may be derived from intracellular sources, because enzyme secretion is much less sensitive to removal of external calcium than is the potassium release mechanism described earlier (Batzri and Selinger, 1973; Selinger et al., 1973). Harfield and Tenenhouse (1973) found that the rat parotid gland continued to secrete amylase in a calcium-free medium for at least 40 min. Therefore, if calcium does have a second messenger role during sympathetic stimulation, it apparently can be derived from some internal reservoir (Fig. 7). Salivary glands possess an enormous intracellular store of calcium (Feinstein and Schramm, 1970). Selinger, Naim, and Lasser (1970) have isolated a microsomal preparation from rat parotid and submaxillary glands which displays energy-dependent calcium accumulation. This microsomal preparation also possesses a very specific cyclic AMP binding site (Salomon and Schramm, 1970), so it is conceivable that cyclic AMP may release calcium from this intracellular reservoir (Fig. 7). It is postulated that the small amount of calcium released during adrenergic stimulation accounts for the small increase in fluid secretion which occurs in parallel with the release of enzyme. During sympathetic stimulation, therefore, fluid secretion may be activated by the same mechanism as that used during parasympathetic stimulation. Such a view is substantiated by the observation that sympathetic stimulation induces very similar secretory potentials (Creed and Wilson, 1969) and losses of potassium (Petersen, 1970c) as occur during parasympathetic stimulation.

The cornerstone of the hypothesis summarized in Fig. 7 is that calcium is obligatory for both fluid and enzyme secretion. During the action of acetylcholine, there is a large influx of sodium and calcium. The latter may then activate potassium efflux and the fluid secretory mechanism on the apical surface. On the other hand, norepinephrine acts by increasing the intracellular level of cyclic AMP which, in turn, may release calcium from an intracellular reservoir. The small increase in calcium may then stimulate fluid secretion and may also function together with cyclic AMP to stimulate enzyme secretion. Rasmussen (1970) postulated that cyclic AMP may prime the system by phosphorylating the membrane of the secretory granule through the activation of a specific protein kinase. Granule movement to the periphery would then be sensitive to calcium. The granule membrane certainly appears to be very responsive to cyclic AMP, because Schramm, Selinger, Salomon, Eytan, and Batzri (1972) have shown that during stimulation with isoproterenol or dibutyryl cyclic AMP, the granule membrane is thrown into long pseudopodia orientated preferentially toward the lumen. These pseudopodia may subsequently fuse with the cell membrane to form a narrow tube through which the granule contents drain into the lumen.

H. Toad Bladder

The bladder of the toad responds to antidiuretic hormone (ADH) with a large uptake of both sodium and water. The rate-limiting step for the absorption of both ions and water is their entry across the mucosal (apical) surface (Civan and Frazier, 1968; Handler, Preston, and Orloff, 1972; DiBona, Civan, and Leaf, 1969). Since ADH is effective only when added from the serosal surface, we have to consider how the information is transferred across the cell to influence the permeability properties of the opposite surface.

Both cyclic AMP and calcium seem to be important in mediating the action of ADH. The evidence implicating cyclic AMP is impressive and is summarized in detail by Orloff and Handler (1967) and by Robison et al. (1971). The intracellular level of cyclic AMP increases during the action of ADH (Handler, Butcher, Sutherland, and Orloff, 1965). Theophylline, which is also capable of mimicking the action of ADH (Orloff and Handler, 1962), can similarly elevate the intracellular level of cyclic AMP. Exogenous cyclic AMP can mimic the action of ADH on both sodium and water transport (Orloff and Handler, 1967). Despite this strong evidence that cyclic AMP has a central role in the action of ADH, there is still some confusion concerning its exact site and mode of action. Much of the confusion seems to arise from the fact that there is an apparent interaction between cyclic AMP and calcium which is not fully understood.

The problem of studying calcium homeostasis in the toad bladder is complicated because there are several points where calcium has an important role, some of which may not be directly related to any second messenger function. For example, if calcium is removed from the bathing medium, the epithelial cell junctions loosen up. As the cells fall apart, there are large decreases in transepithelial resistance which are unrelated to any hormonal effect (Hays, Singer, and

Malamed, 1965). Apart from holding the epithelial cells together, a critical level of calcium is necessary for the transduction of information from the ADH receptor to the adenylyl cyclase (Bockaert, Roy, and Jard, 1972; Roy, Bockaert, Rajerison, and Jard, 1973). In the absence of calcium, or in the presence of excess calcium, this transduction step is inhibited and the action of ADH on water transport is blocked, but there is little effect on the increase in sodium transport (Petersen and Edelman, 1964; Argy, Handler, and Orloff, 1967). The effect of exogenous cyclic AMP on water transport is not affected by excess calcium. The two effects of the hormone can thus be dissociated and seem to indicate that cyclic AMP alone is incapable of regulating both water and sodium movement. However, this conclusion must be weighed against the observation that exogenous cyclic AMP is capable of stimulating both sodium and water transport (Orloff and Handler, 1967). The current literature contains at least two hypotheses which could account for this apparent paradox.

The first hypothesis, originally proposed by Petersen and Edelman (1964), is that the hormone stimulates the formation of cyclic AMP at two separate sites, one calcium-sensitive and responsible for regulating water movements, the other insensitive to calcium and responsible for increasing sodium transport. However, there are obvious problems in having two separate pools of cyclic AMP within the same cell. Perhaps it will be necessary to consider that these two separate sites may represent two separate cell types as suggested by Orloff and Handler (1967). On the basis of the structural organization of the toad bladder, it is conceivable that the mitochondria-rich cells are responsible for sodium transport whereas water moves via the granular cells. Both cell types would respond to ADH by synthesizing cyclic AMP, but would differ in their sensitivity to calcium. Perhaps the two separate receptor systems isolated by Bourguet and Morel (1967) are located on the two separate cell types.

The second hypothesis, which has not been formulated quite so explicitly as the separate site hypothesis, attempts to incorporate calcium as a second messenger in the action of ADH. There are numerous hints in the literature that calcium may function in concert with cyclic AMP. Thorn and Schwartz (1965) showed that the efflux of ^{45}Ca from prelabeled bladders increased during the action of vasopressin, and they suggested that the mobilization of calcium within the bladder is an important step in the action of this hormone. The possibility that calcium may mediate some of the actions of the hormone may help to explain some of the present paradoxes in the literature. For example, Wong, Bedwani, and Cuthbert (1972) have found little correlation between the level of cyclic AMP and the changes in water flow. They suggest that ADH may have additional effects not mediated by cyclic AMP. Perhaps this additional role is concerned with calcium homeostasis, as proposed by Snart and Dalton (1972). The calcium level of mitochondria seems to fall during hormonal stimulation; this release is mediated by cyclic AMP (Besley and Snart, 1971; Snart and Dalton, 1972). It is possible, therefore, that the paradox outlined earlier might be resolved if both cyclic AMP and calcium function as second messengers. Which of these two is ultimately responsible for regulating water and sodium transport is not clear. During the action of ADH or dibutyryl cyclic AMP, there is a substantial drop in the degree of phosphorylation of a specific membrane protein (DeLorenzo, Walton, Curran, and Greengard, 1973). This dephosphorylation, which appears to precede the permeability changes, seems to result from cyclic AMP activating a phosphatase which specifically removes the phosphate groups from the membrane proteins (DeLorenzo and Greengard, 1973). Oschman, Wall, and Gupta (1974) have suggested that this removal of phosphate groups from the membrane proteins may alter calcium binding and lead to an increase in water transport. There may be an increase in membrane fluidity which permits water molecules to pass more freely through the phospholipid bilayer.

I. Adrenal Cortex

One of the earliest recognized actions of cyclic AMP was to mediate the effect of adrenocorticotropic hormone (ACTH) on steroidogenesis (Haynes, 1958). The site and mode of action of cyclic AMP on steroidogenesis has been characterized in considerable detail (Gill, 1972; Garren, Gill, Masui, and Walton, 1971; see also review by Halkerston, *This Volume*). ACTH stimulates adenylyl cyclase causing a rapid increase in the intracellular level of cyclic AMP which is followed, after a lag of a few minutes, by an increase in steroid synthesis and release. A similar lag period is observed during the action of exogenous cyclic AMP (Beall and Sayers, 1972; Mackie, Richardson, and Schulster, 1972) and involves the time required for cyclic AMP to activate the synthesis of a protein which is essential for speeding up the rate-limiting step in steroidogenesis (conversion of cholesterol to pregnenolone). Cyclic AMP seems to act by stimulating a protein kinase which phosphorylates a component of the ribosomes, resulting in increased rates of translation (Fig. 8). A second messenger role for cyclic AMP is well established.

But there is evidence to suggest that calcium may also function as a second messenger during ACTH stimulation of steroidogenesis. However, the exact role of calcium is confused, and it is difficult to dissociate its effects as a second messenger from its other functions within the cell. A number of processes associated with the synthesis and release of corticosteroids require calcium, but it is still not absolutely clear whether these processes are sensitive to *changes* in calcium concentration. Farese (1971) considers that one action of calcium is to speed up the transfer of amino acids from amino acyl transfer RNA to the growing protein chain. In this respect, calcium would function in concert with cyclic AMP to speed up the synthesis of the rate-limiting protein necessary to initiate steroid synthesis (Fig. 8). Another possible site of calcium action is to speed up the hydroxylation reactions which occur later in the synthetic pathway (Péron, Guerra, and McCarthy, 1965; Hirshfield and Koritz, 1964). Finally, calcium may be important in the release of steroids from the cell. The mechanism of release is unknown, but the steroids are apparently not stored in granules and there is no evidence of exocytosis. At the beginning of ACTH action there is a

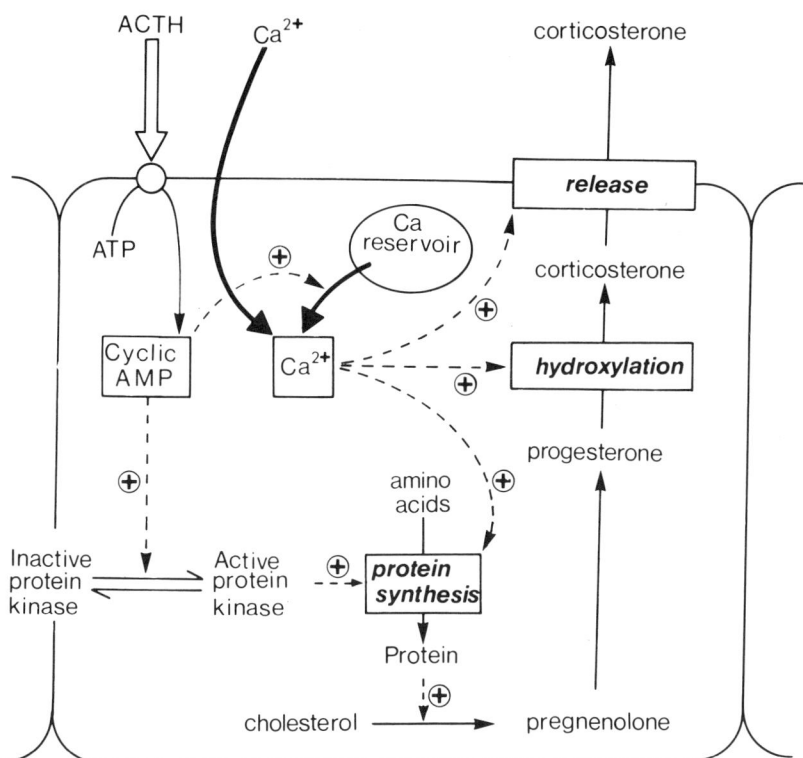

FIG. 8. Interaction of cyclic AMP and calcium in the control of steroidogenesis in the adrenal cortex. ACTH causes an increase in cyclic AMP, which acts by stimulating a protein kinase to phosphorylate a ribosomal component, which stimulates the synthesis of a protein essential for an early step in steroidogenesis. There may also be an increase in the intracellular level of calcium, which may stimulate steroidogenesis at several points.

rapid synthesis of steroid, which builds up within the cells before there is any release. It seems that release begins only when the steroids reach a critical level (Jaanus, Rosenstein, and Rubin, 1970). In the absence of calcium, the steroid level in the gland is found to rise to a much higher level, suggesting that steroid synthesis continues but that release is inhibited by the absence of calcium (Jaanus et al., 1970). Calcium may thus play an important part in regulating the release of steroid from the cell (Fig. 8). Since calcium has such a marked effect on both the synthesis and release of steroid, it may well function as a second messenger in the action of ACTH.

Even though there is some evidence pointing to a second messenger action for calcium, the exact source of this calcium is still in doubt. There are suggestions in the literature that it may either enter from the outside or be released from intracellular reservoirs. The problem of trying to establish a role for extracellular calcium is complicated by the fact that the action of ACTH at its membrane receptor complex appears to require calcium (Bär and Hechter, 1969; Lefkowitz, Roth, and Pastan, 1970). Calcium is not required for ACTH to bind to the receptor, but appears to be needed for the transduction step leading to the activation of adenylyl cyclase. The lack of external calcium can be partially alleviated by raising the concentration of ACTH (Sayers, Beall, and Seelig, 1972). Even though the extra ACTH can cause a considerable increase in the cyclic AMP level, which under normal conditions would cause near maximal rates of steroid synthesis, there is only a partial stimulation of biosynthesis. Other workers (Carchman, Jaanus, and Rubin, 1971; Rubin, Carchman, and Jaanus, 1972a,b) have also been able to dissociate an elevation of internal cyclic AMP from an increase in steroid production by removing external calcium. Removal of calcium caused a sixfold increase in the level of cyclic AMP, but there was no secretion of steroid. Subsequent addition of ACTH produced little further increase in the cyclic AMP level, but there was a large output of steroid which was abolished if ACTH was added in the presence of EGTA. These experiments suggest that, under normal conditions, ACTH may increase the influx of calcium into the adrenocortical cells (Fig. 8).

The effect of ACTH on the electrical properties of the adrenocortical cells provides further evidence for an increase in calcium influx (Matthews and Saffran, 1967, 1968, 1973). ACTH is capable of stimulating large increases in steroid output with no change in the membrane potential. When studied in potassium-free solution, however, ACTH depolarizes the cell, and at a critical depolarization (-46 mV) the membrane begins to display rapid fluctuations in potential, resembling action potentials (Matthews and Saffran, 1968, 1973). The spikes are of long duration (600 to 1,800 msec) and, since they are not blocked by tetrodotoxin (which blocks sodium channels), they may result from a prolonged influx of calcium. These action potentials are not simply the result of membrane depolarization, because they are not seen when cells are depolarized with ouabain or excess potassium; they seem to be specific to the action of ACTH. An effect of ACTH on calcium entry is also suggested by the fact that lanthanum and local anesthetics inhibit steroidogenesis (Matthews and Saffran, 1973).

Studies on the effect of ACTH on ^{45}Ca entry are somewhat contradictory. Jaanus and Rubin (1971) found little evidence for an increased influx of calcium, whereas Leier and Jungmann (1973) report the opposite. A similar contradiction is found with respect to the effect of ACTH on the net accumulation of calcium; Jaanus and Rubin (1971) recorded no increase, whereas Leier and Jungmann (1973) observed a net uptake. Clearly, further studies are necessary to resolve the problem of whether or not external calcium enters the cell during the increase in steroidogenesis.

Another possible source of calcium is the intracellular reservoirs. Although Jaanus and Rubin (1971) found little evidence for an effect of ACTH on calcium influx, they did show that there is a redistribution of internal calcium. The ability of dibutyryl cyclic AMP to stimulate steroidogenesis in the absence of calcium

(Birmingham and Bartová, 1973) might be explained by release of calcium from some intracellular pool (Fig. 8).

Much more work needs to be done on the exact role of calcium in the adrenal cortex, but the evidence so far seems to indicate that it does play a second messenger role and that it cooperates with cyclic AMP in mediating the action of ACTH on steroidogenesis.

J. Photoreceptors

The link between photon capture and the subsequent ionic permeability changes responsible for the generator potential has puzzled physiologists for a long time. The problem is particularly apparent in rods, where the photochemical reaction (bleaching of rhodopsin) occurs on the internal disk membranes, which have no membranous connection with the plasma membrane, where the change in sodium conductance takes place (Fig. 9).

Considerable excitement was generated by the claim that cyclic AMP might play an important role in photoreception (Bitensky, Gorman, and Miller, 1971; Bitensky, Miller, Gorman, Neufeld, and Robison, 1972). Light was found to reduce the cyclic AMP level in preparations of purified disk membranes, and there was a close correspondence between the change in cyclic AMP concentration and the amount of bleaching of rhodopsin. Although this relationship suggests a possible direct role in coupling photon capture and sodium conductance, there is no direct evidence that cyclic AMP has any effect on membrane permeabilities (Miller, 1973). If anything, the direct application of cyclic AMP inhibits the receptor potential produced by light stimulation in *Limulus* (Wulff, 1971). The relatively slow changes in cyclic nucleotide level would also appear to rule out any direct second messenger action in photoreception (Pannbacker, 1973). However, there is every reason to believe that cyclic nucleotides may modulate the responsiveness of the cells to light (Bitensky, Miki, Marcus, and Keirns, 1973). Before considering this modulator role of cyclic nucleotides, it is essential to return to the problem of how the bleaching of rhodopsin in the disk membranes is translated into a permeability change in the surface membrane.

There is growing evidence that calcium released from the disks may be the missing link in photoreception. Wald, Brown, and Gibbons (1963) originally suggested that the transduction mechanism may involve either activation of an enzyme or a change of membrane permeability. Mason, Fager, and Abrahamson (1974) have demonstrated that the latter may apply because light caused a release of calcium from the disk membranes. Cone (1972) has suggested that when activated by light the rhodopsin molecules, which normally lie on the surface of the disk membrane, tumble into the hydrophobic phase of the membrane where they rapidly rotate, thus functioning as a carrier to mediate the efflux of calcium. In the experiments of Mason et al. (1974), there were clear indications that the amount of calcium released was directly related to the amount of bleaching. The disk membranes also have a pump mechanism capable of rapidly sequestering

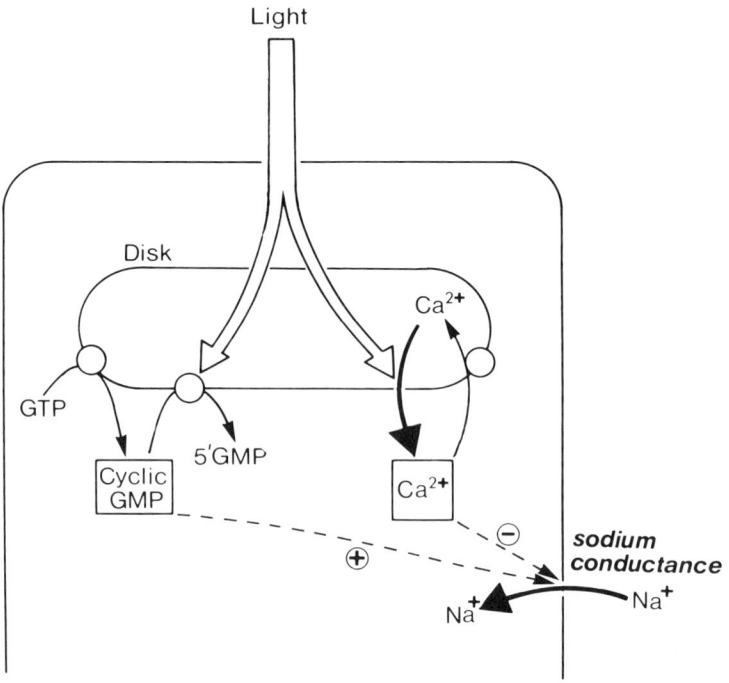

FIG. 9. Light interacting with rhodopsin in the disk membranes of photoreceptors may lead to a release of calcium as well as a stimulation of phosphodiesterase. In vertebrates, calcium reduces sodium conductance, leading to a hyperpolarization of the surface membrane. Cyclic GMP may sensitize the membrane to calcium by increasing the number of available sodium channels. Light has a similar effect on the intracellular level of cyclic AMP (not shown in this diagram).

calcium (Neufeld, Miller, and Bitensky, 1972; Mason et al., 1974). During the action of light, therefore, calcium may be released from the disk to interact with the plasma membrane to alter its permeability to sodium (Fig. 9). In vertebrates, the primary effect of light is to decrease the membrane permeability to sodium, causing the cell to hyperpolarize (Tomita, 1970a). (In invertebrates the response to light is exactly opposite in that the change in receptor potential is a depolarization.) The action of calcium in vertebrate receptors, therefore, is to plug up the sodium channels, thus increasing the membrane resistance leading to a hyperpolarization. The effect of light can be mimicked by raising the external calcium concentration, thus providing some evidence for the second messenger action of calcium (Brown and Pinto, 1974). More direct evidence for a second messenger role of calcium has been provided by Brown and Blinks (1972), who recorded transient increases in the intracellular level of calcium during light treatment.

Most of the light-dark adaptation which occurs in the eye takes place within the photoreceptor itself. Photoreceptors are remarkable in that they can function over a 10^6 range of light intensity. At one extreme, the dark-adapted rod can respond to a single photon. This exquisite sensitivity corresponds with a high level of cyclic nucleotides which are thought to sensitize the receptor. Bitensky et al. (1973) suggest that cyclic nucleotides may phosphorylate the sodium channels, thus making them sensitive to the inhibitory action of calcium. As the light intensity rises and the cyclic nucleotide level declines, the number of susceptible sodium channels decreases and much more calcium is necessary to produce an effect. The original idea that this was caused by a light-dependent inactivation of adenylyl cyclase is probably incorrect, because there is increasing evidence that the decrease in cyclic nucleotide concentration is brought about by a light-dependent activation of phosphodiesterase (Fig. 9; Chader, Bensinger, Johnson, and Fletcher, 1973; Miki, Keirns, Marcus, Freeman, and Bitensky, 1973; Goridis and Virmaux, 1974).

The phosphodiesterase of photoreceptors is much more active against cyclic GMP than against cyclic AMP. Changes in the cyclic GMP level certainly occur faster than those of cyclic AMP (Pannbacker, 1973). Pannbacker (1973) has raised the possibility that the decrease in cyclic GMP which occurs in the light may arise through an inhibition of guanylyl cyclase activity brought about by the light-induced release of calcium from the inside of the disk. Despite continuing uncertainty about how light influences the levels of cyclic GMP and cyclic AMP, it is clear that there is a marked decline in their levels with increasing light intensity. Whether or not they both act on the plasma membrane, as suggested earlier, remains to be seen. It is possible that one may act to modify the susceptibility of the membrane to calcium whereas the other may function as a feedback regulator of calcium movement across the disk membrane. Preliminary studies have shown that cyclic AMP has no effect on calcium binding by the disk membranes (Neufeld et al., 1972). The suggestion that these nucleotides modulate the action of calcium introduces yet another example of how cyclic nucleotides and calcium may interact to regulate cellular activity.

K. Summary of Monodirectional Systems

A characteristic feature of monodirectional systems is that cell activation is regulated in one direction only, and recovery is mediated simply by removing the stimulant. In cases where there are two different stimulants, these either mediate the same (e.g., acetylcholine and pancreozymin in the pancreas) or parallel (e.g., acetylcholine and catecholamines in skeletal muscle or mammalian salivary gland) cellular events. In many of the examples which have been analyzed, calcium appears to be the primary second messenger. It plays a central role in stimulus-contraction coupling in skeletal muscle and in stimulus-secretion coupling during the release of vesicles or granules from nerves, the anterior pituitary, chromaffin cells, β cells, and the exocrine pancreas. Although cyclic AMP has been implicated as a second messenger in many of these examples, careful analysis of the available evidence suggests that it acts indirectly by augmenting the action of calcium (see Fig. 14). A classic example of its role as a modulator of the calcium signal occurs in β cells during the stimulation of insulin secretion by glucagon (see Fig. 3).

Cyclic AMP does not always play a supporting role in monodirectional systems because there are examples, particularly in tissues specialized for various metabolic functions, where it has a more direct action on certain processes. It certainly regulates steroidogenesis in the adrenal cortex, and it also has direct effects on the thyroid which have not been dealt with in this chapter. In order to understand the action of ACTH on the adrenal cortex, however, it is also necessary to include the effects of calcium on various key events during both the synthesis and release of steroid. Cyclic AMP and calcium may also function in concert to alter certain membrane properties associated with the stimulation of toad bladder, insect salivary glands, or photoreceptors. It is not clear yet exactly how these two second messengers interact to influence these various membrane parameters. There are indications in the toad bladder, as well as in postganglionic neurons, that cyclic AMP may act by phosphorylating or dephosphorylating various key membrane components. These actions of cyclic AMP could also account for the increased pump rates seen in insect salivary gland and exocrine pancreas or the change in ionic permeability which is thought to occur in the eye and in presynaptic terminals.

The mammalian salivary gland provides an interesting example where both cyclic AMP and calcium regulate two separate processes within the same cell. Calcium seems to regulate fluid secretion, whereas cyclic AMP stimulates the release of enzyme. The latter is one of the few cases where cyclic AMP seems to play a direct role in mediating granule release by exocytosis. Skeletal muscle is another tissue in which cyclic AMP and calcium regulate two separate intracellular events. Calcium regulates contraction (Sandow, 1970), whereas cyclic AMP mediates the stimulatory effects of catecholamines on glycogenolysis (Robison et al., 1971).

IV. BIDIRECTIONAL SYSTEMS

A. Smooth Muscle

Smooth muscle is notorious for the diversity of its structure and control mechanisms. Despite the large variations in properties of different smooth muscles, a number of basic features are beginning to emerge concerning the intracellular second messengers concerned with contraction and relaxation. Smooth muscle is a classic bidirectional system in that its contractile activity is regulated by two opposing groups of neurotransmitters. In general, those agents which stimulate contraction act by raising the intracellular level of calcium, whereas the relaxing agents lower the level by a mechanism which, in some cases, is regulated by cyclic

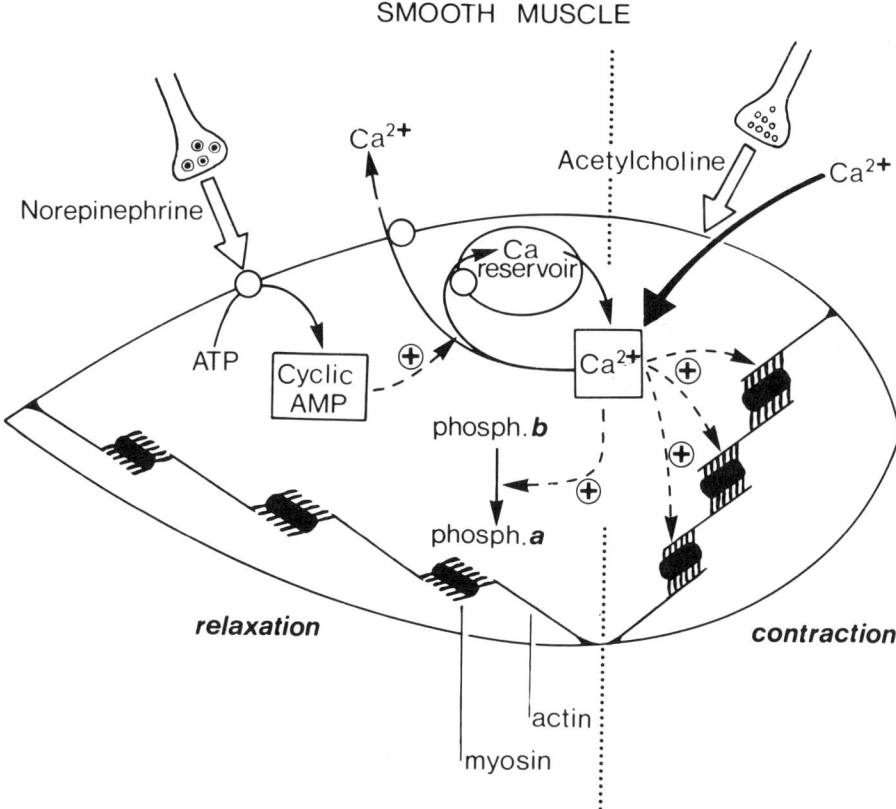

FIG. 10. Bidirectional control of smooth muscle contraction. Stimulants such as acetylcholine elevate the intracellular level of calcium, leading to contraction *(right)*. Energy metabolism may also be entrained to the level of calcium, which converts phosphorylase *b* to phosphorylase *a*. Relaxants such as norepinephrine act via cyclic AMP, which lowers the level of calcium and thus leads to relaxation *(left)*.

AMP (Fig. 10). An understanding of smooth muscle contractility thus revolves around the question of how the intracellular level of calcium is regulated.

Before trying to understand how contracting agents raise the intracellular level of calcium, it is necessary to know the source of this calcium. Smooth muscles use both extracellular and intracellular supplies of calcium (Greenberg, Long, and Diecke, 1973). Certain muscles, such as taenia coli, are very dependent on extracellular calcium and rapidly cease contracting when transferred to a calcium-free medium, whereas other muscles, in particular various vascular muscles, can function for a long time in the absence of extracellular calcium (Freeman and Daniel, 1973; Greenberg et al., 1973). Such variations in the susceptibility of different muscles to the removal of external calcium is well correlated with the relative volumes of the sarcoplasmic reticulum (Devine et al., 1972). Certain vascular muscles, which are relatively independent of external calcium, have a more profuse sarcoplasmic reticulum which lies very close (within 10 nM) to the plasma membrane. As in skeletal muscle, there is uncertainty about how a signal in the plasma membrane is transmitted across the 10-nM gap to the membrane of the sarcoplasmic membrane. Devine et al. (1972) noted periodic densities linking the two membranes which closely resemble those reported in skeletal muscle (Franzini-Armstrong, 1970). Perhaps these densities function to insulate a space connecting the two membranes which would prevent possible ionic signals from short-circuiting to the cytoplasmic fluid. It would be very interesting to know if various tracer molecules (e.g., lanthanum or peroxidase) can penetrate into the spaces delineated by these densities. Although the coupling mechanism is still unknown, there is considerable information concerning ionic events in the plasma membrane.

Most smooth muscle cells are excitable, and the action potential may play a direct role in mediating contraction. In taenia coli, for example, the charge carrier during the action potential is calcium (Bülbring and Tomita, 1970; Tomita, 1970b). Enough calcium enters during the action potential to induce a small, but significant, contraction. A train of action potentials is necessary to raise the calcium sufficiently to produce a full contraction (Goodford, 1970). In most multiunit muscles, the action potentials are initiated directly by the neurotransmitter, whereas in certain single-unit muscles the action potentials may arise spontaneously as part of an endogenous myogenic rhythm. In the latter case, the rhythm consists of "slow waves" of depolarization which, when they reach a critical level, give rise to a train of action potentials (Golenhofen, 1970). There is direct stimulus-contraction coupling across the surface membrane mediated by an influx of calcium. The action potential induces a momentary increase in calcium permeability which allows sufficient extracellular calcium to flood into the cell to partially activate the contractile machinery (Fig. 10). In those smooth muscles which derive calcium from the sarcoplasmic reticulum, the action potential may also be responsible for inducing release of this intracellular store. Muscles which utilize calcium in the sarcoplasmic reticulum respond to an action potential with a much larger contraction than those muscles which rely on extracellular calcium.

Relaxation of smooth muscles is achieved by lowering the intracellular level of calcium in the environment surrounding the contractile elements. Calcium is either pumped out across the plasma membrane or sequestered into the sarcoplasmic reticulum (Fig. 10). Extrusion of calcium from the cell is mediated by a specific calcium pump and not by an exchange reaction with sodium (Casteels, Goffin, Raeymaekers, and Wuytack, 1973). Microsomal preparations, which have membrane fragments derived from both the plasma membrane and the sarcoplasmic reticulum, have been isolated from various smooth muscles and are capable of sequestering calcium (Carsten, 1969; Andersson and Nilsson, 1972; Baudouin, Meyer, Fermandjian, and Morgat, 1972; Fitzpatrick, Landon, Debbas, and Hurwitz, 1972). Those muscles which utilize extracellular calcium appear to have the pump located on the plasma membrane, whereas muscles running on intracellular calcium have the pump located on the sarcoplasmic reticulum (Hurwitz et al., 1973). In order for these calcium pumps to bring about relaxation, they must remove calcium faster than it enters by the processes described earlier. Relaxant drugs can tip the balance in favor of calcium removal, either by inhibiting calcium entry or by stimulating the calcium pumps.

In certain smooth muscles, such as guinea pig taenia coli or rabbit colon, α-adrenergic agents cause relaxation by hyperpolarizing the plasma membrane by a mechanism involving an increase in potassium permeability (Jenkinson and Morton, 1967). The hyperpolarization would prevent the potential from dropping to the critical level where the ionic permeabilities responsible for stimulating contraction are induced. Inhibition of these calcium entry mechanisms allows the calcium pumps to lower the intracellular level of calcium sufficiently to cause relaxation.

The second method of favoring calcium removal is to stimulate the calcium pumps directly by a mechanism involving cyclic AMP (Fig. 10). Andersson (1972) and Bär (1974) have summarized the evidence implicating cyclic AMP in the relaxation of smooth muscle. In most smooth muscles, stimulation of β-adrenergic receptors results in an increase in cyclic AMP concentration which precedes relaxation. Exogenous cyclic AMP can relax smooth muscles directly, and a number of phosphodiesterase inhibitors (e.g., papavarine) relax smooth muscles through their ability to raise the intracellular level of cyclic AMP (Kukovetz and Pöch, 1970). Cyclic AMP has been reported to stimulate the uptake of calcium into the various microsomal fractions which have been prepared from different smooth muscles (Andersson, 1972; Andersson and Nilsson, 1972). The molecular basis for this stimulation has not been studied, but it may resemble that which has been described in the heart (Section II,D).

During the contraction-relaxation cycle of smooth muscle, there are changes in the cyclic GMP level (Schultz et al., 1973; Schultz, Hardman, and Sutherland, 1974). Acetylcholine produces a rapid and large increase in the concentration of cyclic GMP in the ductus deferens by a mechanism which is dependent on extracellular calcium. Preincubation of the muscle in a calcium-free medium reduces the control level of cyclic GMP and totally abolishes the stimulatory effect of acetylcholine. Potassium-induced contractures, which are dependent on an influx of external calcium, also cause a significant increase in the level of cyclic GMP. The increase in cyclic GMP thus appears to be a secondary event brought about by the increase in intracellular calcium concentration associated with contraction. An increase in cyclic GMP concentration, however, is not always associated with contraction, because Schultz et al. (1973) found that after treatment with the phosphodiesterase inhibitor 3-isobutyl-1-methylxanthine (SC-2964), the ductus deferens relaxed and both the cyclic AMP and cyclic GMP levels were elevated. These increases in cyclic GMP, under conditions where the muscle is either contracting or relaxing, seem to rule out any direct regulatory role at the level of the contractile machinery. However, as mentioned earlier (Section II,D), it is conceivable that cyclic GMP may function as a feedback signal to regulate the intracellular level of calcium (Schultz et al., 1973; Hardman et al., 1974).

In many smooth muscles, contractility and metabolism are tightly coupled. Energy is derived from glycogen, and phosphorylase is the rate-limiting enzyme in glycogen breakdown. In both arterial and uterine smooth muscle, conversion of inactive phosphorylase *b* to the active form phosphorylase *a* during contraction is apparently mediated not by cyclic AMP but by calcium (Namm, 1971; Diamond, 1973). Therefore, when calcium is elevated to activate contraction, there is a simultaneous activation of energy metabolism to provide the requisite ATP molecules. In other words, the mechanical and metabolic events are tightly coupled by being regulated by the common mediator calcium (Fig. 10).

B. Heart

Heart activity is regulated by action potentials which are relayed from the pacemaker centers. Each action potential leads to an increase in the intracellular level of calcium which activates contraction (Fig. 11). Relaxation follows as this calcium is removed by uptake and extrusion mechanisms. Superimposed upon this constant ebb and flow of calcium, which follows each action potential, there are subtle adjustments of calcium homeostasis which can significantly alter the nature of the calcium signal during each heart beat. The major modulation concerns both the rate and quantity of calcium delivered during the initiation of contraction and the rate at which calcium is removed during relaxation. Under normal conditions, this modulation of the basic calcium signal is primarily the concern of catecholamines using cyclic AMP as a second messenger (Robison, Butcher, Øye, Morgan, and Sutherland, 1965). In order to appreciate how cyclic AMP acts, it is necessary to understand the main features of the onset and recovery of the basic calcium signal during the contraction-relaxation cycle.

Both extracellular and intracellular calcium contribute to the increase in calcium level during the initiation of contraction (Langer, 1973). The origin of the calcium during stimulus-contraction coupling is complicated because much of the calcium which enters during an action potential does not immediately reach the contractile machinery but first enters the sarcoplasmic reticulum; it is during the next action potential that this calcium is released into the cytoplasm. The sarcoplasmic reticulum seems to act as a halfway house for calcium entering from the outside. This means that the quantity and the speed of delivery of the calcium signal during each action potential is determined by events during the previous beat. A critical factor is the quantity of calcium accumulated in the sarcoplasmic reticulum which is then available for release during the next action potential. There seem to be several ways of regulating the amount of calcium which accumulates in this reservoir. Either the entry of calcium from the outside is

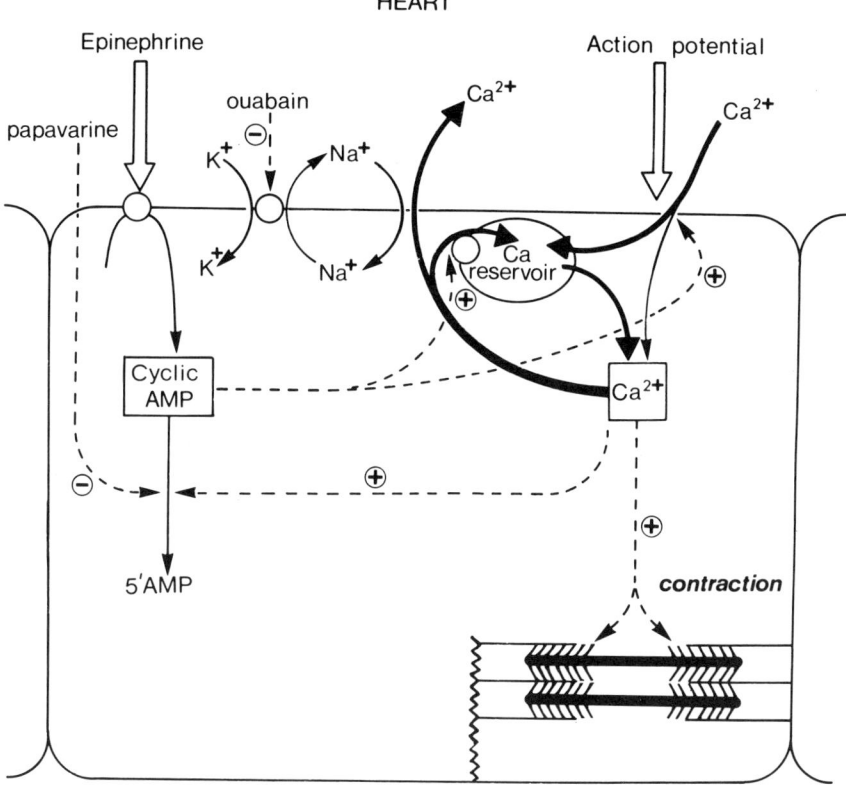

FIG. 11. Stimulus-contraction coupling in the heart. The action potential causes an influx of external calcium, most of which enters the calcium reservoir (sarcoplasmic reticulum) but some of which goes directly to the contractile elements. Calcium, which entered the sarcoplasmic reticulum during the previous beat, is also released and contributes to the build-up of free calcium responsible for initiating contraction. Relaxation occurs when this calcium level is reduced by being pumped into the sarcoplasmic reticulum or by extrusion across the surface membrane in exchange for sodium. Epinephrine exerts its positive inotropic effect via cyclic AMP, which modulates the ebb and flow of calcium during each heart beat. Cyclic AMP can either prolong the influx of calcium across the plasma membrane or enhance the uptake of calcium into the sarcoplasmic reticulum.

prolonged, or the rate of calcium transport into the sarcoplasmic reticulum is increased.

The action potential responsible for switching on contraction consists of two parts; there is a fast inward sodium current followed by a slow inward calcium current, which accounts for the long plateau phase of the action potential (Beeler and Reuter, 1970; Tritthart, Volkmann, Weiss, and Fleckstein, 1973). Not only does this action potential allow extracellular calcium to enter the cell, but it may also be responsible for stimulating the release of calcium from the sarcoplasmic reticulum (Fig. 11). One important action of the catecholamines is to prolong the slow inward current, and this effect can be mimicked by dibutyryl cyclic AMP. This ability of cyclic AMP to stimulate calcium entry may account for the increased uptake of ^{45}Ca seen during the action of epinephrine or dibutyryl cyclic AMP (Meinertz, Nawrath, and Scholz, 1973). By increasing calcium entry, cyclic AMP will increase the amount of calcium entering the sarcoplasmic reticulum, thus increasing the quantity available for release during the next action potential.

Cyclic AMP can also influence the quantity of calcium in the sarcoplasmic reticulum by stimulating the calcium pump (Fig. 11). There seem to be two ways of removing intracellular calcium. First, it can be removed from the cell in exchange for sodium (Reuter and Sietz, 1968; Langer, 1973). Such a mechanism ultimately depends on the sodium-potassium exchange pump maintaining a low internal level of sodium (Fig. 11). Cardiac glycosides, such as ouabain, may exert their inotropic effect by inhibiting the Na,K-activated ATPase, thus indirectly leading to an increased accumulation of calcium. Second, calcium can be pumped into the sarcoplasmic reticulum. The ability of cyclic AMP to stimulate this pump was described in detail in Section II,D.

Cyclic AMP seems to mediate the positive inotropic action of catecholamines by modulating the complex processes responsible for generating the calcium signal during stimulus-contraction coupling. However, the action of cyclic AMP may not be restricted to periods of catecholamine treatment, because Brooker (1973) has found marked oscillations in the intracellular level of cyclic AMP during a single contraction-relaxation cycle. Cyclic AMP may thus play an important role in modulating calcium homeostasis, and this function might be augmented during the action of catecholamines. Brooker considers that the oscillation in cyclic AMP which occurs during each myocardial contraction is caused by transient activation of adenylyl cyclase during each action potential. Another possibility is that the cyclic AMP level is somehow entrained to the fluctuating calcium level through a feedback effect on phosphodiesterase (Fig. 11) as suggested by Teo and Wang (1973).

C. Melanophores

A large number of cold-blooded vertebrates can adjust their coloration to suit their background (Bagnara and Hadley, 1973). This adaptive coloration is achieved through the dispersion or aggregation of melanin granules contained within stellate-shaped cells called melanophores. In a dispersed melanophore, the ovoid black melanin granules radiate out into the arms, thus darkening the skin, whereas in a lightened skin the granules are aggregated in the cell body. This bidirectional movement of melanin granules is regulated by opposing hormones: melanophore-stimulating hormone (α-MSH) or β-adrenergic agents induce dispersion, whereas melatonin and α-adrenergic agents mediate aggregation.

The mechanism moving the granules to and fro within the long extensions of the melanophore is not clear. Electron microscopic studies have revealed the presence of microtubules and microfilaments, both of which have been implicated in granule movement (Bikle, Tilney, and Porter, 1966; McGuire and Moellmann, 1972). However, just how this microtubule-microfilament system is organized to provide the bidirectional movement of granules has not been established. Green (1968) has made the interesting observation that movement in the two directions has different characteristics. Centripetal movement during aggregation is a smooth regular motion, with all the granules migrating together at a steady rate of 2 to 6 μ/sec. Centrifugal movement during dispersion is slower and less coordinated. Green has speculated that this dispersion might be an "energy-storage process," whereas the opposite process of aggregation may be an "energy-release process." This asymmetry in the nature of granule movement in the two directions, which appears to be analogous to the contraction-relaxation cycle in muscles, could have considerable bearing on our understanding of the underlying control mechanisms.

Not much is known about the intracellular mediators responsible for the two opposing effects of aggregation and dispersion except that cyclic AMP is intimately involved. There is considerable evidence that α-MSH and β-adrenergic agents increase the intracellular level of cyclic AMP, which then leads to a dispersion of the melanin granules. During the action of α-MSH, the cyclic AMP concentration of the skin increases in parallel with an increase in darkening (Abe, Butcher, Nicholson, Baird, Liddle, and Liddle, 1969a). Direct application of exogenous cyclic AMP, or dibutyryl cyclic AMP, could also induce darkening (Bitensky and Burstein, 1965; Novales and Davis, 1967; Abe, Robison, Liddle, Butcher, Nicholson, and Baird, 1969b; Van der Veerdonk and Konijn, 1970). Further evidence linking the skin-darkening activity of α-MSH with an increase in cyclic AMP has come from experiments with methylxanthines, which inhibit the intracellular breakdown of cyclic AMP. Theophylline and caffeine, which alone can significantly increase the intracellular concentration of cyclic AMP (Abe et al., 1969a), can mimic the action of α-MSH by inducing granule dispersion (Wright and Lerner, 1960; Bitensky and Burstein, 1965; Novales and Davis, 1967). There is considerable evidence, therefore, implicating cyclic AMP in pigment dispersion.

The way in which melatonin and α-adrenergic agents control pigment aggregation is not as clearly understood. There are reports that melatonin can reduce the intracellular level of cyclic AMP (Abe et al., 1969b). Both melatonin and norepinephrine block both the darkening and the increase in cyclic AMP levels induced by α-MSH. However, neither of these agents has an effect on the darkening induced by cyclic AMP, suggesting that their inhibitory action is exerted on adenylyl cyclase. It is not clear whether the decrease in cyclic AMP is directly responsible for aggregation or whether it occurs indirectly due to other changes induced by the aggregating agents. Since aggregation can be induced by α-adrenergic agents, we can suspect, by analogy with other systems, that it might be regulated by calcium. Unfortunately, there is little information on calcium homeostasis in melanophores. Calcium appears to be essential for the action of α-MSH (Novales, 1971; Vesely and Hadley, 1971). However, since theophylline or dibutyryl cyclic AMP can cause dispersion in the absence of calcium, the effect of calcium is apparently related to the interaction of α-MSH with its receptor.

As in other bidirectional systems, it is conceivable that granule movement in one direction is mediated by an increase in the level of calcium, whereas the opposite process depends on cyclic AMP acting to lower the calcium concentration.

D. Blood Platelets

Blood platelets are concerned with sealing damaged blood vessels. There are two important phases of hemostasis; blood platelets initiate the process by rapidly aggregating to form a plug which is subsequently stabilized by the coagulation of plasma around the adhering platelets. I shall concentrate on the early phases, when the normally nonadhesive disk-shaped circulating platelets are suddenly transformed into oval "sticky" structures which aggregate together to form a clot. Apart from this change in shape and surface properties, the aggregating platelet is also induced to release various granules which contain a range of compounds including 5-HT, ADP, and calcium. Some of these agents, in particular ADP, play an important role in transforming incoming platelets so that they become sticky before they reach the developing clot. The plasma membranes of the aggregated platelets provide a lipoprotein surface which catalyses the formation of thrombin from the intrinsic coagulation system. Like ADP, thrombin can initiate the aggregation and degranulation of incoming platelets, and it also plays a vital role in stabilizing the clot by converting fibrinogen to fibrin. Once the clot is formed and covered with fibrin, there is a process of clot retraction which seems to involve a concerted contraction of the platelets, which somehow pulls the fibrin fibers together to form a water-tight seal. During this complicated sequence of events we can recognize three main events: (1) a rapid change in shape and surface properties which initiates aggregation, (2) release of granules by exocytosis, and (3) contraction of the pseudopodia during clot retraction. The enormous interest in platelets derives partly from the fact that these processes leading to clot formation are reversible by increasing the cyclic AMP level within the platelet. We thus have a classic bidirectional system in that a variety of unrelated agents will induce aggregation and release, whereas a separate group of agents, which raise the internal level of cyclic AMP, will reverse clot formation (Fig. 12). Before considering how cyclic AMP acts, we must consider the aggregation-release-retraction phenomenon.

A wide range of compounds will initiate the aggregation sequence. Apart from various agents (collagen, thrombin, ADP) which occur normally during the clotting process, a wide range of other chemicals will induce aggregation and release, including a tumor-promoting phorbal ester (Zucker, Troll, and Belman, 1974), vasopressin (Haslam and Rosson, 1972), and the calcium ionophore A

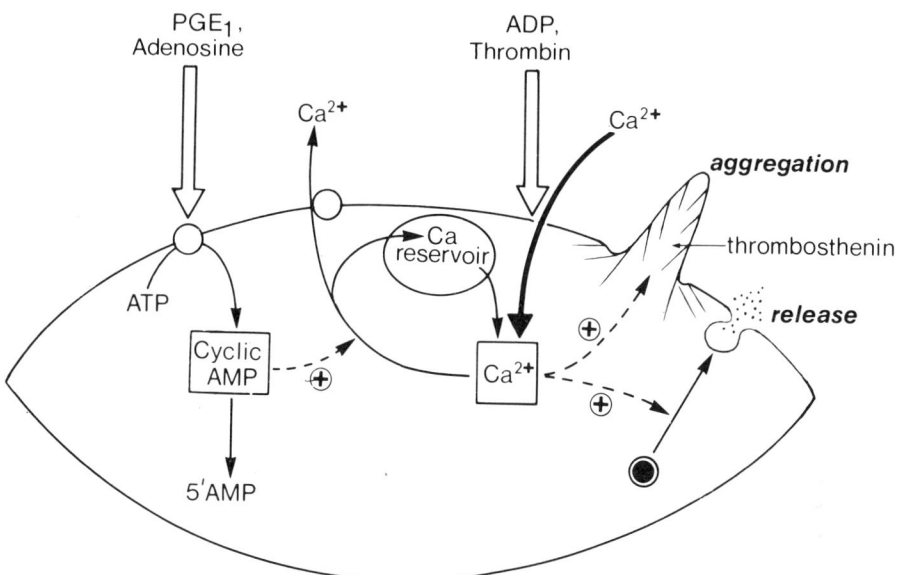

FIG. 12. Bidirectional control mechanism in blood platelets. ADP, or thrombin, induces an influx of calcium which is responsible not only for altering the surface leading to aggregation but also for stimulating the release reaction. Agents such as PGE_1, or adenosine, function to maintain the status quo by stimulating the formation of cyclic AMP which may act by facilitating the removal of calcium.

23187 (Feinman and Detwiler, 1974; Yuen and Macey, 1974). Although all these agents share few chemical similarities, there is one unifying feature—they all have effects on the cell membrane.

One possible common factor connecting the action of all these aggregating agents is cyclic AMP. As mentioned earlier, an increase in cyclic AMP can inhibit aggregation and the release reaction irrespective of the inducing agent. Since we are dealing with a bidirectional system, it is logical to assume that, if inhibition of aggregation is mediated by increasing cyclic AMP, aggregation would be associated with a fall in cyclic AMP; however, this has not been observed consistently (Salzman, Kensler, and Levine, 1972). Some authors (Marquis, Becker, and Vigdahl, 1970; Cole, Robison, and Hartmann, 1971; Moskowitz, Harwood, Reid, and Krishna, 1971; Haslam and Taylor, 1971) have reported that norepinephrine and epinephrine will decrease the internal cyclic AMP level in cells which already have elevated levels through the action of PGE_1 or the phosphodiesterase inhibitor papavarine. This effect of catecholamines is an α-adrenergic effect because it was blocked by phentolamine and not by propranolol. There are reports that adenylyl cyclase is inhibited by collagen, 5-HT, and thrombin but not by ADP (Salzman and Levine, 1971). On the other hand, Droller and Wolfe (1972) contend that there is an increase in the intracellular level of cyclic AMP during the action of thrombin. In the light of this apparent uncertainty concerning the role of cyclic AMP in the induction of platelet aggregation and release, it is important to consider the alternative possibility that platelet activation is mediated by an increase in the level of calcium rather than by a decrease in the level of cyclic AMP. Indeed, the latter may well occur indirectly through an increase in the level of calcium (Haslam and Taylor, 1971), through a possible feedback effect on phosphodiesterase as outlined in Section II,D and Fig. 1.

There are a number of reasons for implicating calcium as a regulator of platelet activity. Blood platelets have a large quantity (15% of the total protein) of thrombosthenin, which is very similar to the actomyosin of skeletal muscle (Adelstein, Pollard, and Kuehl, 1971; Bettex-Galland, Probst, and Behnke, 1972; Hanson et al., 1973). Furthermore, platelets possess a tropomyosin-like protein and there is reason to believe that calcium may regulate the activity of thrombosthenin by means of a troponin-tropomyosin complex as in skeletal muscle. In electron micrographs, the thrombosthenin is arranged in the form of microfilaments which lie very close to the plasma membrane, and White (1968) has speculated that these microfilaments may play an important role in mediating the surface changes which occur during aggregation. These microfilaments may also play an important role in clot retraction, and Cohen and de Vries (1973) have demonstrated that their contraction is regulated by calcium. There are also strong precedents from other cell systems for suggesting that the release of granules from blood platelets is mediated by calcium. There is growing evidence that calcium may regulate all phases of the aggregation-release-retraction sequence (Fig. 12).

As mentioned earlier, all the agents which stimulate aggregation and release share a common mode of action in that they all act on the surface and could lead to an influx of calcium or to a release of intracellular calcium (Fig. 12). Feinman and Detwiler (1974) have shown that the calcium ionophore A 23187 can stimulate release in the absence of external calcium, and they postulate that secretion is triggered by release of internal calcium. The ability of platelets to utilize internal calcium may have led to some of the confusion concerning the calcium requirement of the various platelet reactions.[1] Despite some uncertainties about calcium homeostasis in platelets, most authors agree that extracellular calcium is essential for both aggregation and release (Haslam, 1964; Vigdahl, Marquis, and Tavormina, 1969; Mustard, Kinlough-Rathbone, Jenkins, and Packham,

---

[1] When considering calcium homeostasis in platelets we should probably ignore the large pool of nonexchangeable calcium located in the granules, which probably plays no immediate role in cell activation. However, we must consider the possibility that this calcium will be extruded to the outside during the release reaction and may thus lead to a local build-up of extracellular calcium and thereby influence incoming platelets.

1972; Kinlough-Rathbone, Chahil, and Mustard, 1973; Haslam and Rosson, 1972; Zucker et al., 1974). The change in shape, which precedes aggregation, can occur in the absence of calcium (Haslam and Rosson, 1972; Mustard et al., 1972). A small proportion of the release reaction is also partly independent of calcium (Sneddon and Williams, 1973). Sneddon and Williams (1973) thus suggest that these processes, which are apparently independent of external calcium, may use calcium which is bound to the membrane. Increasing the concentration of thrombin can partially overcome the absence of calcium (Sneddon, 1972; Kinlough-Rathbone, Chahil, and Mustard, 1974). Sneddon and Williams (1973) have separated the release reaction into a calcium-independent induction phase, during which thrombin increases the sensitivity of the plasma membrane, and a calcium-dependent release phase. A similar separation has been achieved in mast cells as described in the next section. Thrombin is a proteolytic enzyme and, since sensitization of platelets can be achieved equally well with trypsin, it appears that removal of various surface components increases the calcium permeability (Sneddon and Williams, 1973). The ability of proteolytic enzymes to induce these dramatic changes in surface properties has some interesting analogies with what occurs in cultured cells (Section V,B). Further evidence that an influx of calcium stimulates blood platelets comes from the observation that the calcium ionophore A 23187 will induce both aggregation (Yuen and Macey, 1974) and release (Feinman and Detwiler, 1974).

A role for calcium in clot retraction has been firmly established by Cohen and de Vries (1973), who demonstrated that the tension developed within a clot was directly related to the level of calcium. The clot relaxes below 10^{-7} M calcium, but contracts above 10^{-6} M. All phases of the aggregation-release-retraction sequence may thus be regulated by the internal level of calcium. The available evidence suggests that this is elevated by increasing the membrane permeability to calcium, but the possibility of calcium being released from internal reservoirs cannot be ignored at this stage.

All phases of the aggregation-release-retraction sequence can be reversed by raising the intracellular concentration of cyclic AMP. One of the most potent stimulators of platelet adenylyl cyclase is PGE_1 (Fig. 12). Particulate fractions of adenylyl cyclase are stimulated by PGE_1 (Marquis, Vigdahl, and Tavormina, 1969; Moskowitz et al., 1971). PGE_1 increases the intracellular level of cyclic AMP simultaneously with the observed rate of disaggregation (Vigdahl et al., 1969; Cole et al., 1971; Harwood, Moskowitz, and Krishna, 1972). In the presence of phentolamine, epinephrine causes a marked increase in cyclic AMP concentration, and this β-effect is thus opposite to the lowering of cyclic AMP caused by α-stimulation (Haslam and Taylor, 1971). Adenosine is a powerful inhibitor of aggregation and was originally thought to act by competing with ADP. However, Haslam and Rosson (1972) found that adenosine inhibits aggregation induced by vasopressin more effectively than that induced by ADP. Adenosine may thus act through a direct stimulation of adenylyl cyclase. Phosphodiesterase inhibitors, which lead indirectly to an increase in the cyclic AMP concentration, can also inhibit aggregation (Mills and Smith, 1971). Finally, the direct application of cyclic AMP, especially the dibutyryl derivative, is capable of inhibiting aggregation (1965; Marquis et al., 1969; Salzman and Levine, 1971; Zucker et al., 1974). The inhibitory effect of cyclic AMP apparently extends to the final process of clot retraction, because Majno, Bouvier, Gabbiani, Ryan, and Statkov (1972) have shown that in an isotonic contractile system, a clot relaxes in the presence of the phosphodiesterase inhibitors papavarine and theophylline.

The way in which cyclic AMP inhibits these various platelet processes is unknown. There are many analogies with smooth muscle in that calcium activates the platelets whereas cyclic AMP attempts to maintain the status quo. As in smooth muscle, cyclic AMP may exert its inhibitory effect by speeding up the removal of the calcium signal (Fig. 12). Grette (1963) first demonstrated that platelets possess a "relaxing factor" similar to that in skeletal muscle. Statland, Heagan, and White (1969) extended these observations by showing that the relaxing factor consisted of membranous vesicles which possess a calcium-activated ATPase capable of sequestering calcium. They suggest that the enzyme is probably located on the plasma membrane and is responsible for extruding calcium from the cell. If calcium is responsible for mediating the aggregation-release-retraction sequence, stimulation of this calcium extrusion pump by cyclic AMP could explain how the latter can reverse or inhibit all aspects of this sequence.

In summary, the activity of platelets appears to be critically dependent on the intracellular level of calcium. Aggregating agents may act by increasing the permeability of the plasma membrane to calcium, but a release of internal calcium may also occur. Cyclic AMP is capable of reversing the effect of these aggregating agents, and it is postulated that cyclic AMP may act by lowering the level of calcium by speeding up membrane-bound calcium extrusion pumps on either the surface or the internal reservoirs (Fig. 12).

E. Mast Cells

Histamine release from mast cells and leukocytes is triggered by the interaction of an antigen with a membrane-bound antibody. This release process, which is regulated by calcium, resembles a bidirectional system in that the antigen-induced release can be reversed by agents which elevate the intracellular level of cyclic AMP. The action of cyclic AMP as a modulator of a wide range of inflammatory and immunological responses has been reviewed by Bourne, Lichtenstein, Melmon, Henney, Weinstein, and Shearer (1974) and by Kaliner and Austen (1974*a*). Histamine is stored within large granules and is released to the exterior by exocytosis (Bloom and Haegermark, 1967; Röhlich, Anderson, and Uvnäs, 1971). The release reaction in mast cells, like that just described in blood platelets, seems to depend on the balance between two opposing control systems which may operate by adjusting the intracellular level of calcium.

Under normal conditions, interaction of an antigen with a surface antibody apparently leads to an influx of calcium (Kaliner and Austen, 1974a). The antibody, which resides in the membrane, is analogous to a hormone receptor with the antigen acting as a hormone. Lichtenstein (1971) has divided the release process into two stages. The antigen first activates the cells by a mechanism which is independent of calcium; this is followed by histamine release, which is calcium-dependent. Histamine release can be induced by a range of other agents, such as the cationic compound 48/80 (Hogberg and Uvnäs, 1960), α-adrenergic and cholinergic agents (Kaliner, Orange, and Austen, 1972), and the calcium ionophore A 23187 (Foreman et al., 1973; Cochrane and Douglas, 1974). All these compounds probably act by increasing the calcium permeability of the plasma membrane and thus mimic the action of the antigen. In keeping with such a hypothesis, histamine release is almost totally dependent on extracellular calcium (Hogberg and Uvnäs, 1960; Foreman and Mongar, 1972a,b; Cochrane and Douglas, 1974). The antiasthmatic drug, disodium cromoglycate, may decrease the release of histamine by stabilizing the membrane and thus preventing the increase in calcium permeability (Orr, Hall, Gwilliam, and Cox, 1971).

The ability of the calcium ionophores to stimulate release provides strong evidence that an influx of extracellular calcium is the basis of the coupling mechanism (Foreman et al., 1973; Cochrane and Douglas, 1974). In the absence of calcium there is very little release, but when calcium is added release proceeds by the normal process of exocytosis. Foreman et al. (1973) demonstrated that the ionophores induced calcium uptake into the cells. Convincing evidence that an increase in the intracellular level of calcium elicits release was provided by injecting calcium directly into mast cells (Kanno, Cochrane, and Douglas, 1973). Mast cells, which had been impaled with micropipettes, released histamine within seconds of introducing calcium by iontophoresis.

Although there is substantial evidence that histamine release is driven by an influx of external calcium, there is some evidence suggesting that a mobilization of intracellular calcium may also be important. During antigen-induced histamine release from rat mast cells, there is an initial drop in the cellular content of calcium (Kaliner and Austen, 1974b). This release of calcium coincides with a sudden fall in the level of cyclic AMP. It is conceivable that cyclic AMP regulates the intracellular distribution of calcium by promoting the uptake of calcium into intracellular reservoirs. Such an explanation is certainly consistent with other observations showing that cyclic AMP can inhibit histamine release.

The calcium-induced release of histamine can be prevented by raising the intracellular level of cyclic AMP (Lichtenstein, Levy, and Ishizaka, 1970; Loeffler, Lovenberg, and Sjoerdsma, 1971). Both dibutyryl cyclic AMP and theophylline were capable of immediately stopping the release of histamine. Agents which elevate the intracellular level of cyclic AMP in mast cells and leukocytes by other mechanisms are also capable of inhibiting the release of histamine. There is a preliminary report by Roy and Warren (1974) that disodium cromoglycate may act by inhibiting phosphodiesterase, in addition to its postulated role as a membrane stabilizer as mentioned earlier. The ability of β-adrenergic agents, prostaglandins, and histamine to increase the intracellular level of cyclic AMP correlates well with their ability to inhibit antigen-induced release of histamine (Bourne, Lichtenstein, and Melmon, 1972; Assem and Richter, 1971; Orange, Kaliner, Laraia, and Austen, 1971). The well-known therapeutic role of epinephrine in certain allergic conditions (anaphylaxis and asthma) may depend not only on its bronchodilator and circulatory effects (as a smooth muscle relaxant; see Section IV,A), but also on its ability to reduce inflammation by inhibiting the release of histamine from leukocytes and mast cells (Bourne et al., 1972).

Exactly how cyclic AMP inhibits release has not been established, but Orange et al. (1971) have suggested that it may act by reducing the intracellular level of calcium. By analogy with smooth muscle, we can speculate that cyclic AMP may lower internal calcium by stimulating calcium removal by one of the pump mechanisms discussed earlier (Section II,A).

F. Liver

The liver responds to catabolic hormones such as glucagon and the catecholamines with a prompt release of glucose. The action of both hormones is mediated by cyclic AMP. Glucagon is much more effective than epinephrine in elevating the intracellular level of cyclic AMP (Exton, Robison, Sutherland, and Park, 1971), which can rise to concentrations far in excess of that required to initiate the various metabolic events outlined in Fig. 13. This excessive production of cyclic AMP may be important in stimulus-division coupling (Section V,E). Under normal conditions, cyclic AMP has at least three effects. (1) It stimulates the breakdown of glycogen to glucose; (2) it inhibits the synthesis of glycogen, thus making more glucose available for release; and (3) it stimulates the synthesis of glucose (gluconeogenesis) from various simple precursors (Fig. 13). The way in which cyclic AMP mediates these three processes, all of which contribute to the release of glucose, will not be described since there are excellent reviews on this subject (Exton, Mallette, Jefferson, Wong, Friedmann, Miller, and Park, 1970; Robison et al., 1971). These biochemical events initiated by glucagon or epinephrine are reasonably well understood. However, these catabolic hormones also induce marked changes in membrane potential and ionic fluxes. Although the significance of these membrane events is still unknown, they are worth studying in detail because they may provide clues about how insulin opposes the action of these catabolic hormones.

A characteristic effect of glucagon or catecholamines on the liver is to increase the efflux of potassium. This efflux of potassium has many similarities to that seen during stimulation of mammalian salivary glands with either catecholamines or acetylcholine (Section III,G). As in salivary glands, potassium efflux from the liver is associated with a marked hyperpolarization of the membrane potential (Friedmann, Somlyo, and Somlyo, 1971; Haylett and Jenkinson, 1972; Friedmann and Dambach, 1973; Lambotte, 1973). Haylett and Jenkinson (1972) have

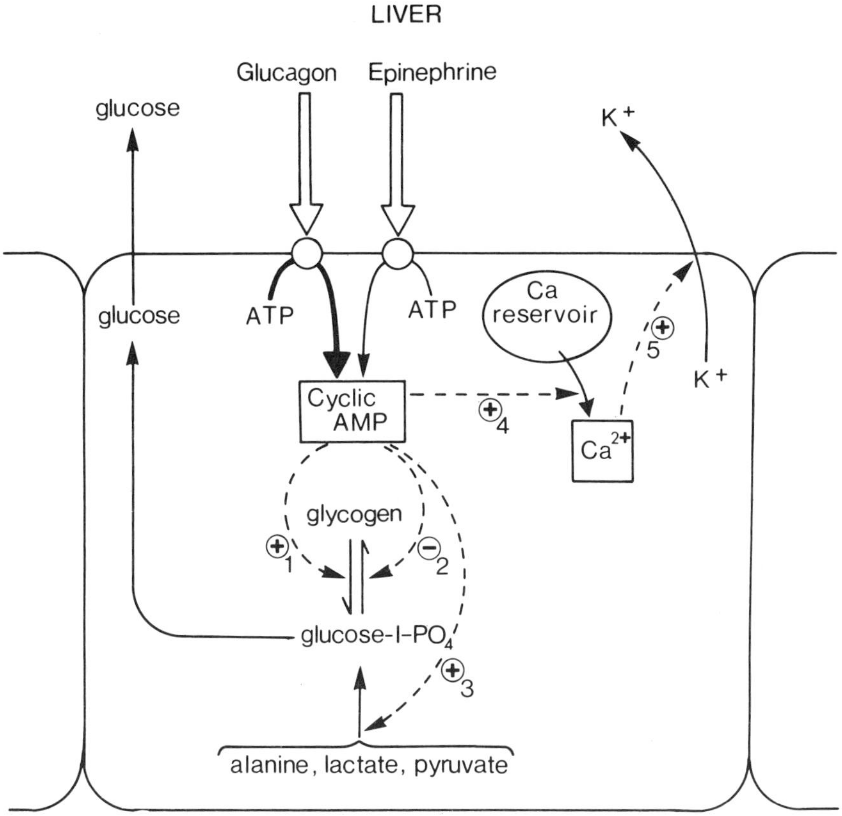

FIG. 13. Interactions of cyclic AMP and calcium in the control of liver metabolism. Glucagon and epinephrine stimulate the synthesis of cyclic AMP, which mediates the release of glucose by activating phosphorylase (1), by inhibiting the formation of glycogen (2), and by stimulating gluconeogenesis (3). Cyclic AMP may also stimulate the release of calcium from intracellular reservoirs (4) which, in turn, may stimulate the efflux of potassium (5).

shown that the increase in potassium efflux is caused by a marked decrease in membrane resistance. Potassium efflux and hyperpolarization of the membrane are not direct effects of the catabolic hormones, because these membrane events can be mimicked by either cyclic AMP or cyclic GMP (Friedmann et al., 1971; Friedmann and Dambach, 1973). The cyclic nucleotides may either alter membrane potential directly, or they may act by releasing calcium (Fig. 13). There is some evidence for the latter possibility: Borle (1974) has found that cyclic AMP can stimulate the release of calcium from isolated liver mitochondria, which are known to accumulate large quantities of calcium (Drahota et al., 1965). Such mobilization of calcium from the mitochondria could account for the increased efflux of calcium which occurs when prelabeled livers are treated with either glucagon or cyclic AMP (Friedmann and Park, 1968; Friedmann, 1972). This increase in calcium efflux precedes the efflux of potassium, suggesting that the two events may be related (Fig. 13). Calcium is known to increase potassium permeability in red blood cells (Gárdos, 1958; Romero and Whittam, 1971) and in nerve cells (Meech, 1974).

The relationship of these changes in ion fluxes to the changes in metabolism induced by the catabolic hormones remains to be established. The action of glucagon on gluconeogenesis was reduced in adrenalectomized animals, but its effect was restored by prior treatment with the glucocorticoid dexamethasone (Exton, Friedmann, Wong, Brineaux, Corbin, and Park, 1972). These experiments indicate that some change in ion distribution might be important in liver function. Friedmann and Rasmussen (1970) have suggested that calcium may play some role in regulating liver metabolism. The ability of cyclic AMP to stimulate glycogenolysis or gluconeogenesis was blocked by tetracaine, which also inhibited the increase in calcium efflux. On the other hand, Exton et al. (1972) found that removing calcium reduced the control rates of glycogenolysis, and gluconeogenesis but did not reduce the stimulatory effect of cyclic AMP. Clearly, much more needs to be done in order to establish the role of calcium in liver metabolism. In Section V,E, it will be argued that calcium may regulate cellular proliferation during liver regeneration.

The fact that cyclic AMP can induce the release of intracellular calcium indicates that the liver possesses certain features of a monodirectional system. However, the liver has been classified as a bidirectional system because the catabolic effects of glucagon and epinephrine are antagonized by insulin. Unfortunately, very little is known about the mode of action of insulin, so it is difficult to speculate about the role of second messengers in its antagonistic effects. Unlike the other bidirectional systems, however, the antagonism may be centered around cyclic AMP, which seems to be the primary second messenger in this system. Some of the effects of insulin might be mediated by lowering the cyclic AMP level (Butcher, Sneyd, Park, and Sutherland, 1966), either by inhibiting adenylyl cyclase (Illiano and Cuatrecasas, 1972) or by stimulating phosphodiesterase (Loten and Sneyd, 1970; Vaughan, 1972). There seems to be no information on the effects of insulin on calcium homeostasis in liver. Williams, Exton, Friedmann, and Park (1971) have reported that insulin inhibits the increase in potassium efflux induced by low doses of cyclic AMP. They favor the view that this action of insulin is explained by a decrease in the level of cyclic AMP, but they do not consider the possibility that insulin might inhibit potassium efflux by reducing the intracellular level of calcium. Such a reduction in the level of calcium may facilitate some of the anabolic events normally stimulated by insulin. There clearly is a need for more information on the effects of insulin on calcium homeostasis.

G. Summary of Bidirectional Systems

This analysis of a range of bidirectional systems indicates that cellular activity depends on the intracellular level of calcium. The antagonistic action of the opposing stimulants seems to depend on altering the various processes responsible for regulating the internal level of calcium. Those agents which increase cell activity in one direction seem to act by increasing the influx of calcium either from the outside or from the intracellular reservoirs. There is good evidence for this in heart, smooth muscle, blood platelets, and mast cells. Since melanophores are a typical bidirectional system, it seems reasonable to suppose that an increase in calcium mediates the aggregation of melanin granules.

The opposing hormones seem to exert their antagonistic effects by lowering the intracellular level of calcium. This is mediated by an increase in the intracellular level of cyclic AMP (Fig. 14). There are indications from some of the systems (smooth muscle and heart) that cyclic AMP acts by speeding up the removal of calcium. This mechanism may be applicable to the other bidirectional systems (melanophores, blood platelets, and mast cells).

Hepatic (and adipose) cells represent an exception in that cellular activity in one direction is regulated by cyclic AMP. The mode of action of insulin, which antagonizes the catabolic effects of glucagon and epinephrine, is unknown. In

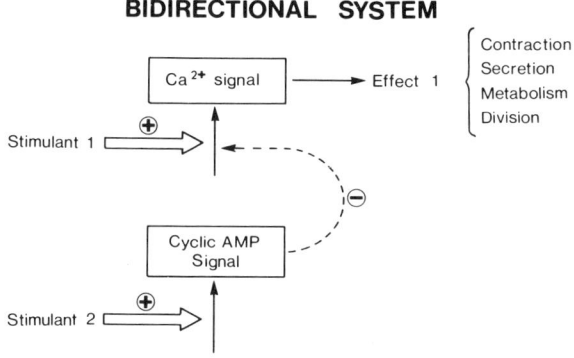

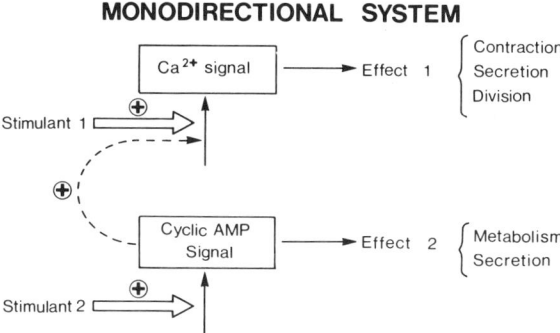

FIG. 14. Summary of the main features of mondirectional and bidirectional systems. See text for further details.

some respects, the liver resembles a monodirectional system in that cyclic AMP is capable of releasing intracellular calcium. A similar release of calcium from intracellular reservoirs may also occur during the action of lipolytic hormones on adipose cells (Hales et al., 1974). During the action of catabolic hormones, therefore, there appears to be an increase in the intracellular level of both cyclic AMP and calcium. If insulin exercizes its antagonistic action at the level of these second messengers, it would be expected to lower the intracellular concentration of both second messengers. Although there is some evidence that insulin may lower the level of cyclic AMP under certain circumstances, this is certainly not its primary mode of action. It would be interesting to know what effects insulin has on calcium homeostasis.

V. CONTROL OF CELL DIVISION

A. The Interactions of Cyclic Nucleotides and Calcium in Stimulus-Division Coupling

The knowledge which has been gained by studying the role of calcium and cyclic nucleotides in the control of differentiated cells can be applied to the problem of how cell division is regulated. Although the various cellular processes which unfold during the course of a cell cycle are numerous and complex, there is reason to believe that the signal which initiates these division processes may be relatively simple (Fig. 15). I have assumed that most of the events which occur during the cell cycle are part of a division program which, once set in motion, unfolds in an orderly and sequential fashion. Some of the major events of the division sequence are shown in Fig. 15, and more details on the cell cycle can be found in reviews by Baserga (1970) and Mitchison (1971). By adopting this simplifying assumption that most of the cell cycle is relatively independent of external factors and consists of running through a preexisting division program, we can concentrate our attention on how this program is switched on. When considered in this way, control of cell division can be analyzed in the same way as the mode of action of a hormone. This becomes all the more interesting in view of the evidence that program selection is made during a relatively short period early in G_1. Since the signal to divide usually comes from the outside, it is necessary to understand how the cell receives this signal and how this information is relayed to the nucleus where the division program is switched on. Different intracellular signals may be responsible for switching on the alternative programs which either maintain the cell in an undifferentiated resting phase (G_o) or cause it to differentiate (Fig. 15). It will be argued that the intracellular second messengers, in particular calcium, which are normally associated with the action of hormones, are also responsible for switching on division. As in hormonal responses, the cell membrane plays a crucial role in detecting the division signals and transducing them into intracellular messengers which will initiate division.

Considerable excitement resulted from the finding that cyclic nucleotides may

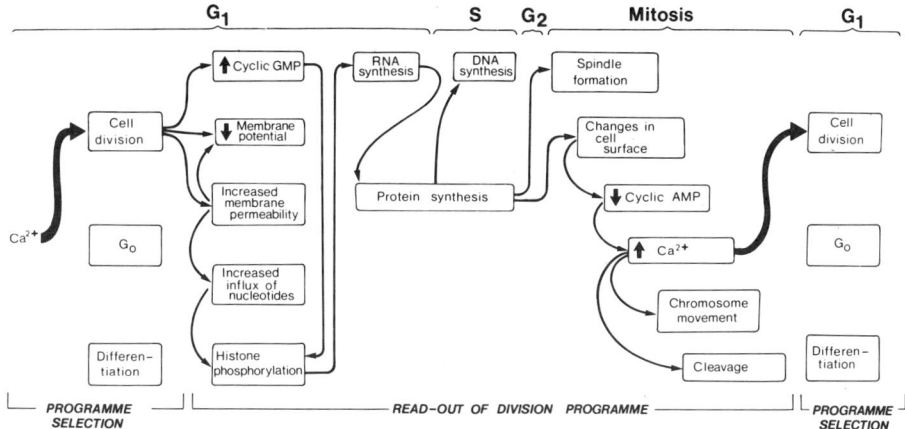

FIG. 15. Summary of the main events during a cell cycle. Early in G_1 there is a critical period during which the cell is faced with three possible alternatives. It can either differentiate, remain in a resting state (G_0), or proceed to divide. Each of these alternatives will have a specific program (only that for cell division is illustrated). If the program for cell division is selected, there is a sequential read-out of the division program terminating in cell cleavage. As the cells pass from mitosis into G_1, they must once again make a further program selection. The interaction between cyclic nucleotides and calcium may play a critical role in determining this program selection. A high level of calcium may initiate division, whereas a high level of cyclic AMP may switch on differentiation.

regulate cell division (Abell and Monahan, 1973; Ryan and Heidrick, 1974). The story began with the observation that cyclic AMP could inhibit division in a number of cell lines (Bürk, 1968; Ryan and Heidrick, 1968). Further details of this inhibitory effect of cyclic AMP will be given later. The main conclusion to emerge from these studies on cyclic AMP in cultured cells was that the stimulus to divide appeared to be a decrease in the intracellular level of cyclic AMP. Most cancer cells which divide in an uncontrolled manner have lower than normal levels of cyclic AMP (Ryan and Heidrick, 1974). Initiation of cell division in contact-inhibited 3T3 cells with either fresh serum or trypsin is associated with a precipitous fall in the intracellular level of cyclic AMP. Unfortunately, there are a number of other examples where initiation of cell division is associated with an increase in the intracellular level of cyclic AMP. Such contradictions begin to cast doubt on the possibility that changes in cyclic AMP level are directly involved with switching on cell division. However, since changes in cyclic AMP metabolism have such profound effects on cell growth, we cannot ignore the possibility that cyclic AMP plays a vital role as a modulator of the division signal.

In attempting to find out what the division signal might be, a number of examples of cell division were analyzed in the hope of identifying a common factor which might provide a clue to the signaling mechanism. A common denominator in all the examples chosen was an increase in the level of calcium during the short critical period when cell division is being switched on. Calcium may thus play a central role in stimulus-division coupling. This conclusion is of interest because it ties in nicely with another hypothesis which considers that cell division is regulated by cyclic GMP (Hadden, Hadden, Haddox, and Goldberg, 1972; Goldberg et al., 1974).

On the basis of the calcium hypothesis, the changes in cyclic GMP which occur during the initiation of cell division in lymphocytes (Hadden et al., 1972), fibroblasts (Goldberg et al., 1974), and epidermal cells (Voorhees, Duell, Stawiski, and Harrell, 1974) probably arise secondarily to the increase in the intracellular level of calcium. Indeed, the subsequent increase in cyclic GMP, with a concomitant phosphorylation of certain key nuclear and cytoplasmic proteins, may be an obligatory first step in the division program outlined earlier (Fig. 15). The primary switch to start it all off, however, is more likely to be a sudden increase in the intracellular level of calcium. Cyclic AMP can be brought into this hypothesis as a modulator of the calcium signal. Indeed, it is possible to explain the opposite effects cyclic AMP has on cell division in different cells on the basis of its positive or negative feedback effects on calcium homeostasis as described in Sections III and IV. In liver and parotid glands, cyclic AMP may switch on cell division through its ability to release calcium (Fig. 16). In other cells, such as various cultured cells, which are bidirectional in character, cyclic AMP may inhibit cell division through its ability to lower the level of calcium (Fig. 16).

The systems which are analyzed in more detail later include these cultured cells, lymphocytes, salivary gland, and liver. Before dealing with these selected examples, it is worthwhile to point out that an increase in the intracellular level of calcium may also be associated with cell division in several other systems.

Calcium may play an important role in initiating cell division after fertilization. There is associated with fertilization an increase in membrane permeability resulting in marked changes in membrane potential and ionic fluxes. Immediately after fertilization there is a temporary depolarization followed by a marked hyperpolarization due to an increase in potassium conductance (Steinhardt, Lundin, and Mazia, 1971; Ito and Yoshioka, 1973). This biphasic potential is very similar to that found in the mammalian salivary gland after stimulation with acetylcholine (Section III,G). The initial depolarization may be caused by an influx of sodium, whereas the subsequent hyperpolarization depends on the increase in potassium conductance (Steinhardt et al., 1971). As in the salivary gland, this increased potassium permeability may reflect an increase in the intracellular level of calcium which may enter the egg together with sodium. Clothier and Timourian (1972) have reported a marked increase in calcium uptake immediately after fertilization, and this could account for the increase in concentration of free calcium described by Nakamura and Yasumasu (1974). The importance of calcium as a stimulus has been substantiated by the observation that the calcium ionophore A 23187 can initiate most of the events normally associated with fertilization, including DNA synthesis and mitosis (Steinhardt and Epel, 1974).

Cyclic nucleotides and calcium may also play a role in regulating cell division in plants. The cytokinins, which are thought to promote cell division, were

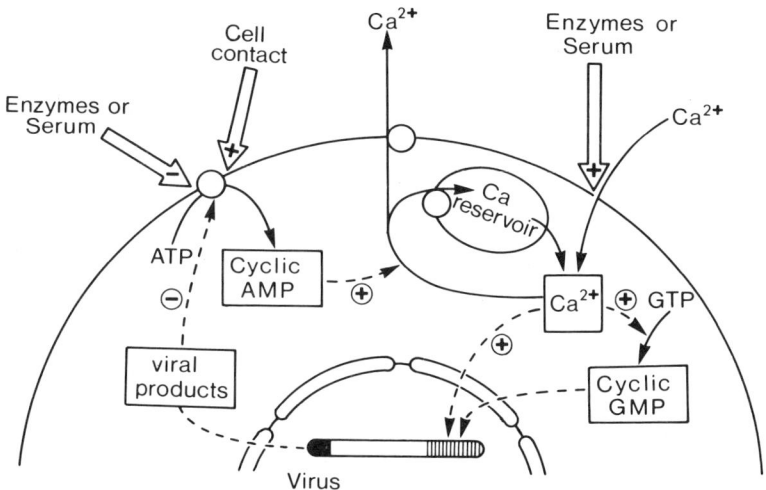

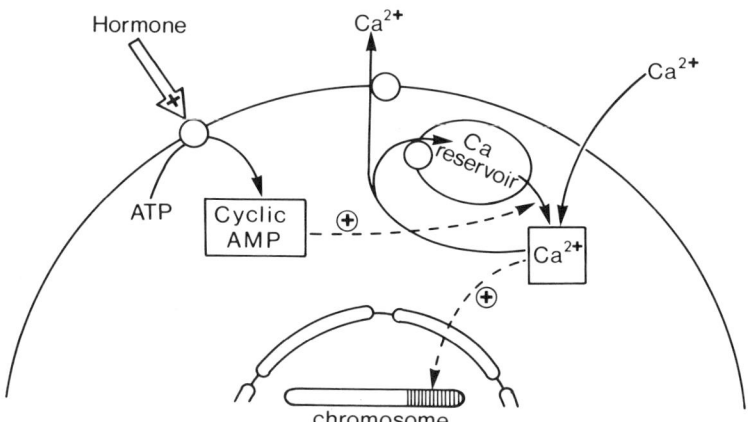

FIG. 16. The interactions of cyclic nucleotides and calcium during stimulus-division coupling. (a) In cultured fibroblasts, surface contact stimulates the synthesis of cyclic AMP, which then lowers the intracellular level of calcium by increasing its removal from the cell or its uptake into intracellular reservoirs. Alteration of surface properties either from within (by transforming viruses—the DNA viruses attach themselves to the host's DNA) or from the outside (absence of cell contact, treatment with fresh serum, or enzyme digestion), inhibits adenylyl cyclase activity, which leads to an accumulation of calcium since the removal mechanisms will slow down. Some of these treatments may also increase calcium influx. (b) In mammalian salivary gland and liver, hormonal stimulation leads to an increase in the intracellular level of cyclic AMP, which then releases calcium from the intracellular reservoirs. In both cases, the increase in the intracellular level of calcium may then switch on the division program located on the chromosomes. A rise in the level of calcium may secondarily elevate the level of cyclic GMP (illustrated in a), which could function to initiate some of the early events of the division program as outlined in Fig. 15.

reported to inhibit phosphodiesterase in crown-gall tumor cells (Woods, Lin, and Braun, 1972). Theophylline also inhibited this phosphodiesterase and stimulated cell division. The cytokinins may also have marked effects on calcium homeostasis in that they enhanced the uptake of calcium into fungal cells (LeJohn and Cameron, 1973a,b). The localization of cyclic GMP in the root tips of bean seedlings where proliferation is rapid (Haddox, Stephenson, and Goldberg, 1974) further suggests a role for these second messengers in the control of cell division in plants.

An understanding of the interactions between cyclic nucleotides and calcium may be crucial if we are to unravel the sequence of events which occurs during stimulus-division coupling. These interactions will be developed more fully by analyzing some specific examples in greater detail.

B. Cultured Cells

Studies on the growth of various cell lines in culture have provided many important insights into the control of cell division. There are several advantages to be gained from studying cells *in vitro*. First, it is possible to synchronize the cells so that they all pass through the various stages of the cell cycle in unison. Second, most of the cell lines can be transformed with various viruses; these transformed cells lack growth control and thus provide an interesting comparison with normal cells. Normal cells proliferate until they fill up the available space in the culture dish, at which point they cease growing. This contact inhibition of growth is not irreversible, because when the cells from a confluent culture are replated they begin further rounds of growth and division until confluency is reached again. This delicate control mechanism breaks down in transformed cells, which continue growing with little or no regard to cell density; i.e., they lose the ability to react appropriately to their neighbors at confluency.

The abnormal growth of transformed cells seems to be caused by a lesion of the surface glycoproteins. This surface component may resemble the glycoprotein which has been described in red blood cells (Bretscher, 1973; Marchesi, Jackson, Segrest, and Kahane, 1973). The protein chain spans across the membrane, and the end which protrudes from the outer surface of the membrane is encrusted with a number of short carbohydrate chains. The carbohydrate composition of the side chains is different in the transformed cells (Warren, Fuhrer, and Buck, 1972). Viral-induced alterations in the nature of this surface glycoprotein may cause small but subtle alterations in membrane conformation sufficient to alter the complex mechanisms responsible for regulating cellular growth. The discovery by Aub, Tieslau, and Lankester (1963) that a lipase preparation from wheat germ would specifically agglutinate malignant cells highlights the differences in membrane properties. Just how the agglutinin molecule causes the cancerous cells to clump together is not clear, but it does require the lectin molecule to bind to the surface of the cell. Normal and transformed cells bind equivalent amounts of the agglutinin molecule conconavalin A (Con A), but only the latter are agglutinated. It appears that the two cell types have the same number of agglutinin receptors, which may resemble the glycoproteins described earlier. It has been suggested that in normal cells these surface molecules are relatively fixed, whereas in transformed cells they are free to migrate laterally within the membrane owing to its greater fluidity (Nicolson, 1972; Noonan and Burger, 1973a,b; Rosenblith, Ukena, Yin, Berlin, and Karnovsky, 1973; Vlodavsky, Inbar, and Sachs, 1973).

The multivalent Con A molecule will bind to one surface glycoprotein and in time can draw in additional binding sites to form clusters (Noonan and Burger, 1973b). It is not yet clear whether agglutination is directly related to this clustering of the Con A-binding sites or depends upon some other property related to the enhanced membrane fluidity of transformed cells (de Petris, Raff, and Mallucci, 1973). Direct chemical analysis of membrane components, together with these indirect studies comparing the responses to specific plant agglutinins, clearly reveal that a surface lesion is present in transformed cells which may account for the defects in their growth control mechanisms.

One important consequence of the surface lesion in transformed cells appears to be a defective adenylyl cyclase. Bürk (1968) showed that hamster cells transformed with polyoma virus had lower levels of adenylyl cyclase. There is a striking inverse correlation between the rate of growth and the internal concentration of cyclic AMP in a number of different transformed cells (Otten, Johnson, and Pastan, 1971; Abell and Monahan, 1973). The fastest growing cells had the lowest cyclic AMP levels. Of even greater interest was the observation that at confluency, the normal contact-inhibited cells produced a dramatic increase in their intracellular concentration of cyclic AMP. On the contrary, transformed cell lines which lack growth control failed to show an increase in cyclic AMP concentration at confluency. It would appear that the information received during cell contact is transduced in normal cells into an increase in the intracellular level of cyclic AMP which then switches off the division process. A defective adenylyl cyclase system, which does not respond appropriately during contact with neighboring cells, allows the colliding cells to escape from the constraints normally imposed by an increase in cyclic AMP concentration. The increased membrane fluidity found in these transformed cells may prevent the subunits of the adenylyl cyclase system from aggregating appropriately to generate the increase in cyclic AMP which is necessary to prevent further division. This lack of endogenous cyclic AMP can be counteracted by adding cyclic AMP to the bathing medium, which then reverses many of the abnormal properties of transformed cells and strongly suggests that cyclic AMP plays a vital role in regulating cell division. The ability of exogenous cyclic AMP to normalize the growth, morphology, and motility of transformed cells has been summarized by Abell and Monahan (1973) and by Ryan and Heidrick (1974), and will not be considered further. The main points to emerge from all these studies is that low levels of cyclic AMP are associated with cell division, whereas an elevation of cyclic AMP inhibits growth.

It would seem, therefore, that a drop in the level of cyclic AMP may be a signal to divide. Some evidence for this idea has come from studies on temperature-sensitive mutants of the transforming viruses. When chick embryo fibroblasts were injected with a temperature-sensitive mutant of the Rous sarcoma virus, they grew normally at 41°C but were transformed at 37°C (Anderson, Johnson, and Pastan, 1973). When shifted to the lower temperature there was a rapid fall in the intracellular level of cyclic AMP.

Similar changes in the intracellular level of cyclic AMP are found when normal cells are stimulated to divide. If stationary cell cultures, which are contact-inhibited, are provided with fresh serum or are treated with various enzymes (trypsin, pronase), a further round of cell division will be initiated (Otten, Johnson, and Pastan, 1972; Froehlich and Rachmeler, 1972; Noonan and Burger, 1973c; Bombik and Burger, 1973; Oey, Vogel, and Pollack, 1974). These studies are particularly instructive because the enzymes specifically remove parts of the surface glycoproteins mentioned earlier and appear to induce a surface lesion not unlike that present in transformed cells. For example, Noonan and Burger (1973c) found that a 5-min treatment with pronase was enough to induce cell division. Immediately after the pronase treatment, the cells were susceptible to agglutination with Con A for approximately 3 hr, but thereafter the surface returned to normal as the damage inflicted by the enzyme was repaired. Of even greater interest was the observation that associated with this altered membrane configuration, which suggested an increase in fluidity, there was the familiar decline in cyclic AMP concentration (Otten et al., 1972; Sheppard, 1972; Burger, Bombik, Breckenridge, and Sheppard, 1972). The importance of this sudden drop in cyclic AMP as a signal to divide can be appreciated from the fact that the stimulatory effect of fresh serum, or enzyme digestion, can be prevented by adding cyclic AMP to the bathing medium (Froehlich and Rachmeler, 1972; Bombik and Burger, 1973; Noonan and Burger, 1973c). The critical period for switching on cell division can be analyzed by varying the periods which cyclic AMP is applied. If dibutyryl cyclic AMP is added 5 min after pronase treatment, there are no cell divisions. However, if the addition of dibutyryl cyclic AMP is delayed for 10 min after enzyme treatment, it does not inhibit division. The division program is switched on within minutes of enzyme treatment, and suggests that there is a narrow launch window during which the decision to divide is made. Once the decision to divide has been made and the initial processes of the division program have been switched on, the whole process is apparently irreversible, thus strongly arguing for the existence of an autonomous division program as discussed earlier (Fig. 15).

If a drop in the intracellular level of cyclic AMP is the signal to divide, we are left to consider how normal cells divide in subconfluent cultures but cease dividing at confluency. A possible answer to this problem has come from studies on how the intracellular level of cyclic AMP varies during the cell cycle (Burger et al., 1972; Millis, Forrest, and Pious, 1972; Sheppard and Prescott, 1972). The most pronounced and consistent change in all the cells studied was a large fall in the cyclic AMP level at the time of mitosis. The concentration returns to normal as the cells pass into G_1. Therefore, the cells have a low cyclic AMP level during the critical period early in G_1 when the decision to divide is made. Whether or not a cell decides to divide again will probably depend on how fast the cyclic AMP level returns to normal as the cells pass through the M-to-G_1 interface. Presumably, as the number of cell contacts increase, the adenylyl cyclase will be more active (Fig. 16a) and the cyclic AMP level will be returned to normal before the cells pass through this critical decision-making period early in G_1. If they pass through this period with a high level of cyclic AMP, there will be no further division and the cells will be contact-inhibited (further division can be initiated by fresh serum or treatment with enzymes as discussed above). A low level of cyclic AMP may thus be the signal to divide during normal growth.

The fall in cyclic AMP at mitosis may be part of the division program as outlined in Fig. 15. During mitosis, the cell membrane assumes a configuration which closely resembles that found in transformed cells. There are marked changes in the carbohydrate content of the cell coat (Glick, Gerner, and Warren, 1971), and they respond to agglutinin molecules in much the same way as transformed cells (Fox, Sheppard, and Burger, 1971; Noonan and Burger, 1973a; Noonan, Levine, and Burger, 1973). During normal mitosis, therefore, the cells assume both the surface configuration and the reduced level of cyclic AMP which seem to be associated with the signal to divide.

Although a drop in the intracellular level of cyclic AMP seems to be associated with an initiation of division, we cannot be sure that cyclic AMP is the primary signal. One important conclusion to emerge from analyzing a wide range of control systems is that cyclic AMP often plays a secondary role in cell activation by modulating the intracellular level of calcium. It is important, therefore, to consider whether or not calcium plays a role in regulating cell division. Not much is known about calcium homeostasis in dividing cells, but there are sufficient clues for us to predict that these cultured cells are examples of a bidirectional system. In other words, a high level of cyclic AMP will lower the intracellular level of calcium and vice versa. Indeed, it is conceivable that cyclic AMP may play a critical role in regulating the level of calcium during all phases of the cell cycle. Most fibroblasts are motile. Since the cells have contractile microfilaments, which are thought to provide the propulsive forces (Huxley, 1973), it is reasonable to suppose that motility is regulated by calcium (Perdue, 1971). Addition of dibutyryl cyclic AMP, or elevation of the internal level of cyclic AMP with PGE_1, rapidly reduces the motility of L-929 fibroblasts (Johnson, Morgan, and Pastan, 1972). Cyclic AMP may thus act by reducing the intracellular level of calcium, perhaps by stimulating the ATP-dependent calcium-transport mechanism present on the plasma membrane of fibroblasts (Perdue, 1971). A similar interaction between cyclic AMP and calcium may take place during mitosis. The movement of chromosomes during anaphase might be mediated by actin-like filaments (Forer, 1969; Gawadi, 1971; Hinkley and Telser, 1974). Likewise, Schroeder (1973) has speculated that cell cleavage at telophase is mediated by a ring of microfilaments which contract and assist in constricting the cell into two equal parts. If chromosome movement and cleavage are mediated by actin-like microfilaments, it seems reasonable to suspect that their contraction is regulated by calcium. Baker and Warner (1972) have obtained direct evidence that calcium is involved by showing that cleavage is blocked by injecting the calcium chelator EGTA into dividing cells from early embryos of *Xenopus laevis*. Under normal conditions, calcium probably comes from some intracellular pool, because cleavage continued even when embryos were immersed in a calcium-free medium. Some preliminary experiments using aequorin suggested that in similar calcium-free conditions there may be temporary increases in the intracellular calcium concentration at the time of mitosis. These experiments seem to indicate that the calcium level rises during mitosis and suggest further that these fluctuations in intracellular calcium concentration can occur independently of external calcium.

The proposed increase in calcium concentration necessary for chromosome movement and cleavage during mitosis thus occurs at a time when the cyclic AMP level is plunging to its lowest point in the cell cycle. Such an inverse relationship is characteristic of bidirectional cells and suggests that cyclic AMP may regulate the intracellular level of calcium. It is interesting to note that Smets (1972) has found that dibutyryl cyclic AMP inhibits cytokinesis, causing the accumulation of a large number of mitotic cells. If the argument about the interaction between cyclic AMP and calcium is correct, we can predict that transformed cells, which have low cyclic AMP levels, should have higher levels of calcium. It is of great interest, therefore, to find that normal chicken fibroblasts divide very slowly in a calcium-free medium, but after transformation with Schmidt-Ruppin Rous sarcoma virus, they become almost independent of external calcium and divide rapidly (Balk, 1971; Balk, Whitfield, Youdale, and Braun, 1973). Since the growth of 3T3 cells is totally dependent on external calcium, Boynton, Whitfield, Isaacs, and Morton (1974) have suggested that calcium may be an important division stimulus. A change in calcium homeostasis during transformation, or during the course of a normal cell division, could arise independently of cyclic AMP. For example, the increase in fluidity which is associated with the division signal may alter the permeability properties of the plasma membrane, allowing for a greater influx of calcium. However, since the administration of cyclic AMP can reverse many of the characteristics of transformed cells, it seems that a lesion in the cyclic AMP system leading indirectly to a rise in the level of calcium together with an increased membrane fluidity is a more likely possibility. Membrane fluidity, and the intracellular levels of cyclic AMP and calcium, are all closely interrelated and seem to be of central importance in regulating division.

In summary, tissue culture cells provide several opportunities for studying the control of cell division. In all cases a drop in the intracellular level of cyclic AMP or, more likely, an increase in the level of calcium, may be the signal to divide. During normal growth the decision to divide is probably taken early in G_1. As the cells come out of mitosis they will have a low cyclic AMP concentration, but a high level of calcium which may provide the signal for a further division. At high cell density, these mitogenic second messenger levels may be rapidly reversed after mitosis, and the cell will pass through its critical program selection period (Fig. 15) with high cyclic AMP and low calcium levels which are not mitogenic.[2] However, such contact-inhibited cells can be induced to divide again by procedures, such as adding fresh serum or brief enzyme treatment, which will lower the cyclic AMP level and probably raise the calcium level (Fig. 16). Hülser and Frank (1971) have found that within minutes of adding "growth-stimulating protein" to embryonic rat cells, the membrane potential depolarizes from -55 to -12 mV, which is entirely consistent with a sudden elevation in the intracellular level of calcium. In these cells, therefore, an increase in the intracellular level of calcium may be the primary signal responsible for switching on the division program. The change in calcium level during the initiation of division may elevate the level of cyclic GMP, which may play a role in initiating the early events of the division program as outlined earlier (Fig. 15).

During treatment with fresh serum the fall in cyclic AMP level is associated with an equally large and rapid increase in the level of cyclic GMP (Seifert and Rudland, 1974). When the level of cyclic AMP was increased by treatment with PGE_1, there was a simultaneous fall in the level of cyclic GMP (Rudland, Seeley, and Seifert, 1974). These parallel but reciprocal changes in cyclic AMP and cyclic GMP levels, which occur in many cell types in various stages of growth, are readily explicable if these two nucleotides are linked by the common intermediate calcium in a bidirectional system as shown in Fig. 16a.

C. Lymphocytes

Lymphocytes play an important role in the initiation of immunological reactions. The ability of lymphocytes to respond to a foreign molecule by dividing to form a population of immunologically active lymphocytes provides an intriguing model system for studying the control of cell division. Phytohemagglutinin (PHA) is a particularly effective mitogen.

Soon after the addition of PHA, it is possible to detect the early components of the division program (Fig. 15), such as the increased uptake of nucleosides (Hausen and Stein, 1968), inorganic phosphate (Cross and Ord, 1971), and potassium (Quastel and Kaplan, 1970). Associated with these surface alterations, there are also rapid changes in the cytoplasm and nucleus such as a rapid incorporation of ^{32}P into nuclear phosphoproteins and an increase in acetate incorporation into histones. These initial biochemical events are followed, after about 30 min, by an increase in protein and RNA synthesis, which is followed much later by DNA

---

[2] It is of interest that high cyclic AMP levels can stimulate differentiation in certain cells. A rapid increase in the level of cyclic AMP will not only inhibit cell division by lowering the level of calcium, but may also play an important role in selecting the differentiation program (Fig. 15).

synthesis (Cross and Ord, 1971; Averner, Brock, and Jost, 1972). Although these details of the division program are becoming clearer, there is still considerable confusion about how PHA switches the process on.

There is a strong possibility that calcium may play a direct role in stimulus-division coupling during the action of PHA. The calcium ionophore A 23187 was capable of mimicking the effect of PHA on thymidine uptake and incorporation into DNA (Maino et al., 1974). The stimulatory effect of PHA is totally dependent on external calcium (Alford, 1970; Whitney and Sutherland, 1972a). A stimulatory role for calcium is consistent with the observation that the calcium-requiring step occurs early in stimulation, because removal of calcium shortly after PHA stimulation is not inhibitory. Calcium is thus essential for the early activation process. Using ^{45}Ca, it is possible to demonstrate that PHA increases the uptake of calcium into the lymphocytes (Allwood, Asherson, Davey, and Goodford, 1971; Whitney and Sutherland, 1972b; Goldberg et al., 1974). The mitogenic action of PHA is thus associated with an increase in the intracellular level of calcium.

There are also marked changes in cyclic nucleotide levels during PHA stimulation. Smith, Steiner, Newberry, and Parker (1971a) showed that PHA caused the cyclic AMP level to rise rapidly, but after a few hours it returned to the resting level and by 16 to 20 hr it declined below the resting level. These changes in the cyclic AMP level during the early stages of the cell cycle may explain some of the contradictory results concerning the effect of exogenous cyclic AMP. On the basis of experiments by Smith and others (Smith, Steiner, and Parker, 1971b; Hirschhorn, Grossman, and Weissmann, 1970; Averner et al., 1972), we might expect cyclic AMP to stimulate mitosis, but the opposite was found. The effect of PHA on thymidine incorporation was suppressed if the cyclic AMP level of lymphocytes was increased by treating them with PGE_1 or cholera toxin (DeRubertis, Zenser, Adler, and Hudson, 1974). However, high cyclic AMP may stimulate early events in the cell cycle but may inhibit subsequent events which seem to depend upon the presence of low levels of cyclic AMP. There is also a complicated relationship between the doses of cyclic AMP and PHA (Webb, Stites, Perlman, Luong, and Fudenberg, 1973). In order for experiments of this kind to be meaningful, it is essential to restrict the application of nucleotides to specific times during the cell cycle, as was done in the experiments on fibroblasts discussed earlier. If purified PHA or Con A are used as mitogens, there is apparently no change in the intracellular level of cyclic AMP but there is a very marked increase in the intracellular level of cyclic GMP (Hadden et al., 1972). The ability of cyclic GMP and its 8-bromo-derivative to increase thymidine incorporation suggests an important role for this nucleotide. Dibutyryl cyclic AMP could inhibit thymidine incorporation stimulated by Con A, but it had no effect on that induced by 8-bromo-cyclic GMP (Weinstein, Chambers, Bourne, and Melmon, 1974). This last observation suggests that lymphocytes, like fibroblasts, are bidirectional systems (Fig. 16a). By speeding up the removal of calcium, dibutyryl cyclic AMP may prevent the formation of cyclic GMP which may be vital for initiating early events in the cell division program. These results lead to the idea that cyclic GMP is the intracellular signal regulating proliferation not only in lymphocytes but in other cells as well (Goldberg et al., 1974). However, as argued earlier, it is conceivable that the increase in cyclic GMP level develops indirectly through the increase in the level of calcium, and it is the latter which should be considered as the primary signal for mitosis.

D. Salivary Glands

It has been known for some time that β-adrenergic drugs such as isoproterenol can induce mitosis in mammalian submaxillary or parotid salivary glands (Selye, Veilleux, and Cantin, 1961). A single injection of isoproterenol can stimulate division in 80% of the cells of the parotid gland (Baserga, 1970). Mitosis can also be stimulated by supramaximal electrical stimulation of the preganglionic fibers innervating the gland (Muir, Pollock, and Turner, 1973). The timing of the different processes of the division program have been analyzed in detail, but the nature of the triggering mechanism is unknown (Baserga, 1970).

A possible clue to the division stimulus comes from the observation that, associated with the action of isoproterenol, the cells are stimulated to secrete and are rapidly depleted of stored enzyme. Control of enzyme secretion in salivary glands by β-adrenergic agents is described in detail in Section III,G. An excessive elevation of the intracellular second messengers which normally regulate secretion may provide the stimulus to divide. Schneyer (1974) has noted that the stimulus to divide is connected with the secretion of amylase. Stimulation of enzyme secretion with β-adrenergic drugs is associated with an increase in the intracellular level of cyclic AMP, and Guidotti et al. (1972) have suggested that this increase in the cyclic AMP level, which occurs within 1 min of administering isoproterenol, is responsible for stimulating division.

In these salivary glands, therefore, mitosis is associated with an increase in the intracellular level of cyclic AMP, whereas in the cultured cells described earlier such an increase in cyclic AMP appears to be inhibitory. This paradox disappears if we assume that the mitogenic signal is calcium and not cyclic AMP (Fig. 16b). Salivary glands are a classic monodirectional system and, as argued earlier (Section III,G), an increase in the level of cyclic AMP probably causes a parallel rise in the level of calcium. This increase in calcium may activate guanylyl cyclase to account for the parallel increase in the level of cyclic GMP which occurs when the parotid gland is stimulated with isoproterenol (Durham, Baserga, and Butcher, 1974). It is interesting to note that the increase in the level of cyclic GMP lags behind the rise in cyclic AMP level. Once again we find that the stimulation of mitosis is associated with an increase in the level of both calcium and cyclic GMP.

If calcium functions in stimulus-division coupling in salivary glands, one would expect parasympathetic stimulation to switch on mitosis because there is considerable evidence linking an influx of calcium with the action of acetylcholine

(Section III,G). Indeed, Schneyer (1974) has now shown that stimulation of the parasympathetic nervous system can stimulate division. Since parasympathetic stimulation is not usually associated with an increase in the cyclic AMP level, these salivary gland studies provide further evidence to implicate calcium as the primary internal mitogenic stimulus.

E. Liver Regeneration

Cells within the liver retain the ability to divide. If two-thirds of the liver is removed, the cells within the remaining fragment proliferate rapidly to restore the original size within a few days (Bucher, 1963). A fascinating feature of regeneration is its precise regulatory properties. The original size is always achieved, irrespective of the amount removed. Removal of part of the liver provides a division signal to the remaining cells, and this signal apparently wanes once the initial cell population has been restored. A possible explanation for this regulatory phenomenon, based on hormonal feedback mechanisms, will emerge from the following description of how liver regeneration might be controlled.

There is convincing evidence that the mitogenic signal is hormonal in nature. Nearly all the cells are stimulated and they divide in unison, suggesting that the stimulus to divide acts on all the cells simultaneously. Parabiosis experiments indicate that the division signal arrives from outside the liver (Moolten and Bucher, 1967; Sakai, 1970). If the blood system of a hepatectomized rat is connected to that of a normal rat, the liver cells of the latter will begin to divide. The amount of division seen in the normal partner depends on the amount of liver removed from the experimental rat. Total hepatectomy provides the largest stimulus (Fisher, Szuch, Levine, and Fisher, 1971). These experiments using totally hepatectomized rats confirm that the division stimulus arises from outside the liver. The effect of removing the liver in one rat generates a division signal which is capable of inducing cell division in the liver of the normal parabiotic partner. Fisher et al. (1971) have shown that by using a postcaval shunt (blood entering the partially removed liver is diverted into the systemic circulation and becomes available to the normal liver) the effect of 70% hepatectomy becomes equivalent to total hepatectomy. Such experiments have led Fisher et al. (1971) to speculate that removal of the liver alters the balance between "portal blood factors" and the available liver cell population. There are several reasons for speculating that these portal blood factors may be equated with the normal hormones such as glucagon, and to a lesser extent epinephrine, which are responsible for regulating liver metabolism (Section IV,F).

The liver is the effector organ for the feedback mechanisms responsible for regulating the level of blood glucose. An increase in the glucose concentration leads to a release of insulin (as described in Section III), whereas a drop in the glucose level initiates a release of glucagon which stimulates the various metabolic processes in the liver responsible for restoring the glucose level in the blood (Fig. 13). By removing the major source of blood sugar, hepatectomy will lower the blood glucose and activate the feedback mechanisms responsible for restoring the glucose level. A normal release of glucagon is unlikely to restore the glucose level because of the small size of the liver fragment, and the feedback mechanisms will probably overcompensate by producing excess glucagon in an effort to raise the glucose level. The remaining cells rapidly lose their glycogen immediately following hepatectomy (Verity, Travis, and Brown, 1972), suggesting that they are intensely active during this period. The liver cells are probably subjected to supramaximal hormonal treatment, which may provide the stimulus to divide.

There are a number of additional observations which further suggest that the stimulus to divide is caused by hyperstimulation of the cells as the remaining part of the liver attempts to fulfill the metabolic obligations of the whole organ. A complex mixture containing triiodothyronine, amino acids, glucagon, and heparin can induce DNA synthesis in normal livers with a time course similar to that in regenerating livers (Short, Brown, Husakova, Gilbertson, Zemel, and Lieberman, 1972). In the normal liver, the action of glucagon is associated with an elevation in the intracellular level of cyclic AMP (Section IV,F). During the infusion of this mixture there are large increases in the level of cyclic AMP within the liver. After hepatectomy there are similar increases in the intracellular level of cyclic AMP which precede the synthesis of DNA (MacManus, Franks, Youdale, and Braceland, 1972). As in the mammalian salivary gland, therefore, the stimulus to divide is associated with an increase in the intracellular level of cyclic AMP. Similar correlations between proliferation and the level of cyclic AMP can be found by studying either liver development or liver tumors. Shortly after birth there is a peak in the cyclic AMP level correlated with a peak in both basal and hormone-sensitive adenylyl cyclase (Christoffersen, Mørland, Osnes, and Øye, 1973). The postnatal elevation in cyclic AMP level occurs during a period of rapid cell division (Rohr, Wirz, Henning, Riede, and Bianchi, 1971). An elevation in the level of cyclic AMP, or in adenylyl cyclase activity, has been found for a variety of liver tumors (Brown, Chattopadhyay, Morris, and Pennington, 1970; Chayoth, Epstein, and Field, 1972; Christoffersen, Mørland, Osnes, and Elgjo, 1972). Thomas, Murad, Looney, and Morris (1973) studied a series of liver tumors of different growth rates and found that the levels of both cyclic AMP and cyclic GMP were consistently higher than those found in normal liver.

In all the above examples (liver regeneration, development, and tumors), the mitogenic signal is associated with an elevation of cyclic AMP. Under normal circumstances glucagon stimulates liver metabolism by elevating the cyclic AMP level without initiating division. If cyclic AMP is part of the mitogenic signal, we must assume that during the hyperstimulation necessary to promote division the cyclic AMP level rises considerably above the level normally required to stimulate metabolism. A characteristic feature of glucagon action is its ability to elevate the cyclic AMP level far in excess of that required to stimulate metabolism (Exton et al., 1971). The excessive production of cyclic AMP may be related to the control of other cell functions, and there may be a hierarchy of cellular processes activated sequentially as the level of cyclic AMP rises. Park, Lewis, and Exton (1972) arranged various processes in the following series of increasing insensitivity (the concentration of glucagon necessary to produce half-maximal stimulation is shown in parentheses): glycogenolysis (5×10^{-11} M) < gluconeogenesis (1×10^{-10} M) < ureogenesis (5×10^{-8} M). Presumably mitosis would come at the end of this hierarchy and would be activated at cyclic AMP levels above those necessary to stimulate ureogenesis.

Mitosis in the liver thus seems to be stimulated by an increase in the level of cyclic AMP which is opposite to the situation in various cultured cells described earlier. As in the salivary gland, this paradox can be resolved by postulating that the primary mitogenic signal is calcium and not cyclic AMP (Fig. 16b). Activation of liver metabolism seems to contain components of a typical monodirectional system in that cyclic AMP appears to elevate the level of calcium by mobilizing intracellular reservoirs (Section IV,F). The marked increase in the cyclic GMP level in various liver tumors (Thomas et al., 1973) is certainly consistent with an elevated intracellular level of calcium.

Liver regeneration may thus be closely linked to the homeostatic mechanisms responsible for maintaining a constant level of blood glucose. As long as the liver is incapable of supplying the body's requirement for glucose, its cells will be constantly stimulated with supramaximal levels of glucagon with concomitant increases in the intracellular levels of cyclic AMP and calcium. As soon as the cell population in the liver returns to normal and is capable of satisfying the body's demand for glucose, the feedback mechanisms will be dampened down and division will cease because the intracellular levels of cyclic AMP and calcium will return to normal.

VI. SUMMARY AND CONCLUSIONS

Several conclusions have emerged from this analysis of individual control systems. First, it is evident that calcium is a very important second messenger controlling a wide range of cellular processes. Notable exceptions are various metabolic tissues such as the liver, adipose tissue, and the adrenal cortex, where cyclic AMP seems to be the primary second messenger. Second, in those tissues which are regulated by calcium, cyclic AMP often plays an important role as a modulator of the calcium signal. It can act either as a negative or a positive feedback regulator. In monodirectional systems cyclic AMP augments, and in bidirectional systems it opposes the calcium signal (Fig. 14). This latter conclusion is different from that outlined in the Yin-Yang hypothesis (Goldberg et al., 1974), in which cyclic GMP is thought to oppose the action of cyclic AMP. In Sections II,B and II,D it was argued that cyclic GMP is not a primary second messenger in that it is probably dependent on an increase in intracellular calcium (Fig. 1). It is much more realistic, therefore, to consider that the antagonism between second messengers in bidirectional systems is between calcium and cyclic AMP, and not between cyclic GMP and cyclic AMP.

In most monodirectional systems, an increase in the intracellular level of calcium is responsible for stimulating cellular activity (Fig. 14). The cellular processes regulated include contraction (skeletal muscle) and secretion (adrenal medulla and cortex, β cells, fluid secretion in insects and mammalian salivary gland, amylase from the exocrine pancreas, nerve and neurosecretory terminals, anterior pituitary). In many of these systems, cyclic AMP may play a secondary role to augment the calcium signal or to facilitate its action, as occurs in the eye. In some systems, however, cyclic AMP plays a primary role in stimulating certain cell processes, particularly metabolism (adrenal cortex, thyroid, and skeletal muscle). It may also play a role in regulating amylase secretion in mammalian salivary gland and fluid secretion in insect salivary gland and the exocrine pancreas. However, the exact mode of action of cyclic AMP in these secretory events remains to be determined.

In bidirectional systems, the major second messenger stimulating cellular activity in one direction is again calcium, which mediates contraction (smooth and cardiac muscle), secretion (blood platelets, mast cells), and metabolism (smooth muscle). Stimulation of the cell in the opposite direction depends on a decline in the calcium level brought about by the negative feedback effect of cyclic AMP (Fig. 14). Major exceptions are the liver and adipose cells, which are clearly bidirectional but use cyclic AMP rather than calcium as the primary second messenger. Unfortunately, not much is known about how insulin acts, so it is premature to speculate about possible antagonistic reactions at the second messenger level in these metabolic tissues. Some possible effects of insulin are discussed briefly in Section IV,F.

Calcium may also play a central role in stimulus-division coupling. Previous studies on the role of cyclic nucleotides in cell division have been paradoxical in that high levels of cyclic AMP seem to be inhibitory in some systems but permissive in others. However, by analyzing these different examples using the concepts derived from mono- and bidirectional control systems, it is clear that an increase in intracellular calcium could be a constant feature of the stimulus to divide in all the systems studied (fibroblasts, lymphocytes, salivary gland, and liver). This finding clearly indicates that calcium may be an important intracellular mitogenic signal during stimulus-division coupling. Calcium may act on the nucleus directly, or it may have an indirect effect by altering the intracellular distribution of ions such as sodium.

ACKNOWLEDGMENTS

I am particularly grateful to Drs. A. M. Katz, N. A. Thorn, N. D. Goldberg, C. A. Schneyer, and G. Schultz for providing me with copies of their unpublished papers. I would also like to thank Mr. J. W. Rodford, who did the drawings, and Mrs. M. V. Clements for her care and patience in preparing the manuscript.

Much of the reading for this review was done during a visit to Woods Hole, and I would like to express my gratitude to Jim Oschman and Betty Wall for their help and hospitality.

REFERENCES

Abe, K., Butcher, R. W., Nicholson, W. E., Baird, C. E., Liddle, R. A., and Liddle, G. W. (1969a): Adenosine 3′,5′-monophosphate (cyclic AMP) as the mediator of the actions of melanocyte stimulating hormone (MSH) and norepinephrine on the frog skin. *Endocrinology*, 84:362–368.

Abe, K., Robison, G. A., Liddle, G. W., Butcher, R. W., Nicholson, W. E., and Baird, C. E. (1969b): Role of cyclic AMP in mediating the effects of MSH, norepinephrine, and melatonin on frog skin color. *Endocrinology*, 85:674–682.

Abell, C. W., and Monahan, T. M. (1973): The role of adenosine 3′,5′-cyclic monophosphate in the regulation of mammalian cell division. *Journal of Cell Biology*, 59:549–558.

Adelstein, R. S., Conti, M. A., Johnson, G. S., Pastan, I., and Pollard, T. D. (1972): Isolation and characterization of myosin from cloned mouse fibroblasts. *Proceedings of the National Academy of Sciences of the U.S.A.*, 69:3693–3697.

Adelstein, R. S., Pollard, T. D., and Kuehl, W. M. (1971): Isolation and characterization of myosin and two myosin fragments from human blood platelets. *Proceedings of the National Academy of Sciences of the U.S.A.*, 68:2703–2707.

Alford, R. H. (1970): Metal cation requirements for phytohemagglutinin-induced transformation of human peripheral blood lymphocytes. *Journal of Immunology*, 104:698–703.

Allwood, G., Asherson, G. L., Davey, M. J., and Goodford, P. J. (1971): The early uptake of radioactive calcium by human lymphocytes treated with phytohemagglutinin. *Immunology*, 21:509.

Anderson, W. B., Johnson, G. S., and Pastan, I. (1973): Transformation of chick-embryo fibroblasts by wild-type and temperature-sensitive Rous Sarcoma virus alters adenylate cyclase activity. *Proceedings of the National Academy of Sciences of the U.S.A.*, 70:1055–1059.

Andersson, R. G. G. (1972): Cyclic AMP and calcium ions in mechanical and metabolic responses of smooth muscles; influence of some hormones and drugs. *Acta Physiologica Scandinavica*, Supplement 382:1–59.

Andersson, R. G. G., and Nilsson, K. (1972): Cyclic AMP and calcium in relaxation in intestinal smooth muscle. *Nature, New Biology*, 238:119–120.

Appleman, M. M., Thompson, W. J., and Russell, T. R. (1973): Cyclic nucleotide phosphodiesterase. *Advances in Cyclic Nucleotide Research*, 3:65–98.

Argent, B. E., Case, R. M., and Scratcherd, T. (1973): Amylase secretion by the perfused cat pancreas in relation to the secretion of calcium and other electrolytes and as influenced by the external ionic environment. *Journal of Physiology*, 230:575–593.

Argy, W. P., Handler, J. S., and Orloff, J. (1967): Ca^{++} and Mg^{++} effects on toad bladder responses to cyclic AMP, theophylline, and ADH analogues. *American Journal of Physiology*. 213:803–808.

Ashcroft, S. J. H., Randle, P. J., and Täljedal, I.-B. (1972): Cyclic nucleotide phosphodiesterase activity in normal mouse pancreatic islets. *FEBS Letters*, 20:263–266.

Assem, E. S. K., and Richter, A. W. (1971): Comparison of *in vivo* and *in vitro* inhibition of the anaphylactic mechanism by β-adrenergic stimulants and disodium cromoglycate. *Immunology*, 21:729–739.

Aub, J. C., Tieslau, C., and Lankester, A. (1963): Reactions of normal and tumor cell surfaces to enzymes. I. Wheat-germ lipase and associated mucopolysaccharides. *Proceedings of the National Academy of Sciences of the U.S.A.*, 50:613–619.

Averner, M. J., Brock, M. L., and Jost, J.-P. (1972): Stimulation of ribonucleic acid synthesis in horse lymphocytes by exogenous cyclic adenosine 3′,5′-monophosphate. *Journal of Biological Chemistry*, 247:413–417.

Bagnara, J. T., and Hadley, M. E. (1973): *Chromatophores and Color Change*. Prentice-Hall, Englewood Cliffs, N.J.

Baker, P. F. (1970): Sodium-calcium exchange across the nerve cell membrane. In: *Calcium and Cellular Function*, edited by A. W. Cuthbert, pp. 96–107. Macmillan, London.

Baker, P. F., Hodgkin, A. L., and Ridgway, E. B. (1971): Depolarization and calcium entry in squid giant axons. *Journal of Physiology*, 218:709–755.

Baker, P. F., Meves, H., and Ridgway, E. B. (1973): Calcium entry in response to maintained depolarization of squid axons. *Journal of Physiology*, 231:527–548.

Baker, P. F., and Warner, A. E. (1972): Intracellular calcium and cell cleavage in early embryos of *Xenopus laevis*. *Journal of Cell Biology*, 53:579–581.

Balk, S. D. (1971): Calcium as a regulator of the proliferation of normal but not of transformed,

chicken fibroblasts in a plasma-containing medium. *Proceedings of the National Academy of Sciences of the U.S.A.,* 68:271–275.
Balk, S. D., Whitfield, J. F., Youdale, T., and Braun, A. C. (1973): Roles of calcium, serum, plasma, and folic acid in the control of proliferation of normal and Rous Sarcoma virus-infected chicken fibroblasts. *Proceedings of the National Academy of Sciences of the U.S.A.,* 70:675–679.
Banks, P., Biggins, R., Bishop, R., Christian, B., and Currie, N. (1969): Sodium ions and the secretion of catecholamines. *Journal of Physiology,* 200:797–805.
Bär, H.-P. (1974): Cyclic nucleotides and smooth muscle. *Advances in Cyclic Nucleotide Research,* 4:195–237.
Bär, H.-P., and Hechter, O. (1969): Adenyl cyclase and hormone action. I. Effects of adrenocorticotropic hormone, glucagon, and epinephrine on the plasma membrane of rat fat cells. *Proceedings of the National Academy of Sciences of the U.S.A.,* 63:350–356.
Baserga, R. (1970): Induction of DNA synthesis by a purified compound. *Federation Proceedings,* 29:1443–1446.
Batzri, S., and Selinger, Z. (1973): Enzyme secretion mediated by the epinephrine β-receptor in rat parotid slices. Factors governing efficiency of the process. *Journal of Biological Chemistry,* 248:356–360.
Baudouin, M., Meyer, P., Fermandjian, S., and Morgat, J.-L. (1972): Calcium release induced by interaction of angiotensin with its receptors in smooth muscle cell microsomes. *Nature,* 235:336–338.
Bdolah, A., and Schramm, M. (1965): The function of 3',5'-cyclic AMP in enzyme secretion. *Biochemical and Biophysical Research Communications,* 18:452–454.
Beall, R. J., and Sayers, G. (1972): Isolated adrenal cells: steroidogenesis and cyclic AMP accumulation in response to ACTH. *Archives of Biochemistry and Biophysics,* 148:70–76.
Beeler, G. W., and Reuter, H. (1970): Membrane calcium current in ventricular myocardial fibres. *Journal of Physiology,* 207:191–209.
Benz, L., Eckstein, B., Matthews, E. K., and Williams, J. A. (1972): Control of pancreatic amylase release *in vitro:* Effect of ions, cyclic AMP, and colchicine. *British Journal of Pharmacology,* 46:66–77.
Berl, S., Puszkin, S., and Nicklas, W. J. (1973): Actomyosin-like protein in brain. *Science,* 179:441–446.
Berridge, M. J. (1970): The role of 5-hydroxytryptamine and cyclic AMP in the control of fluid secretion by isolated salivary glands. *Journal of Experimental Biology,* 53:171–186.
Berridge, M. J. (1973): The effects of derivatives of adenosine 3',5'-monophosphate on fluid secretion by the salivary glands of *Calliphora. Journal of Experimental Biology,* 59:595–606.
Berridge, M. J., Lindley, B. D., and Prince, W. T. (1975a): Membrane permeability changes during stimulation of isolated salivary glands by 5-hydroxytryptamine. *Journal of Physiology,* 244:549–567.
Berridge, M. J., Lindley, B. D., and Prince, W. T. (1975b): Studies on the mechanism of fluid secretion by isolated salivary glands of *Calliphora. Journal of Experimental Biology (in preparation).*
Berridge, M. J., Lindley, B. D., and Prince, W. T. (1975c): Stimulus-secretion coupling in an insect salivary gland: Cell activation by elevated potassium levels. *Journal of Experimental Biology (in press).*
Berridge, M. J., Lindley, B. D., and Prince, W. T. (1974): Role of calcium and cyclic AMP in controlling fly salivary gland secretion. In: *Secretory Mechanisms of Exocrine Glands,* Alfred Benzon Symposium VII, edited by N. A. Thorn and O. H. Petersen, pp. 2101–2109. Munksgaard, Copenhagen.
Berridge, M. J., and Oschman, J. L. (1972): *Transporting Epithelia.* Academic Press, New York.
Berridge, M. J., and Prince W. T. (1971): The electrical response of isolated salivary glands during stimulation with 5-hydroxytryptamine and cyclic AMP. *Philosophical Transactions of the Royal Society,* 262:111–120.
Berridge, M. J., and Prince, W. T. (1972a): The role of cyclic AMP and calcium in hormone action. In: *Advances in Insect Physiology,* Vol. 9, edited by J. E. Treherne, M. J. Berridge, and V. B. Wigglesworth, pp. 1–49. Academic Press, London.
Berridge, M. J., and Prince, W. T. (1972b): Transepithelial potential changes during stimulation of isolated salivary glands with 5-hydroxytryptamine and cyclic AMP. *Journal of Experimental Biology,* 56:139–153.
Berridge, M. J., and Prince, W. T. (1972c): The role of cyclic AMP in the control of fluid secretion. *Advances in Cyclic Nucleotide Research,* 1:137–147.

Besley, G. T. N., and Snart, R. N. (1971): Effect of vasopressin on the uptake of calcium ions by kidney mitochondria and on the concentration of adenosine 3',5'-cyclic monophosphate in toad bladder. *Biochemical Journal,* 125:60P.

Bettex-Galland, M., Probst, E., and Behnke, O. (1972): Complex formation with heavy meromyosin of the isolated actin-like component of thrombosthenin, the contractile protein from blood platelets. *Journal of Molecular Biology,* 68:533–535.

Bikle, D., Tilney, L. G., and Porter, K. R. (1966): Microtubules and pigment migration in the melanophores of *Fundulus heteroclitus* L. *Protoplasma,* 61:322–345.

Birmingham, M. K., and Bartová, A. (1973): Effects of calcium and theophylline on ACTH- and dibutyryl cyclic AMP-stimulated steroidogenesis and glycolysis by intact mouse adrenal glands *in vitro. Endocrinology,* 92:743–749.

Bitensky, M. W., and Burstein, S. R. (1965): Effects of cyclic adenosine monophosphate and melanocyte-stimulating hormone on frog skin *in vitro. Nature,* 208:1282–1284.

Bitensky, M. W., Gorman, R. E., and Miller, W. H. (1971): Adenyl cyclase as a link between photon capture and changes in membrane permeability of frog photoreceptors. *Proceedings of the National Academy of Sciences of the U.S.A.,* 68:561–562.

Bitensky, M. W., Miller, W. H., Gorman, R. E., Neufeld. A. H., and Robison, G. A. (1972): Digitonin effects on photoreceptor adenylate cyclase. *Science,* 175:1363–1364.

Bitensky, M. W., Miki, N., Marcus, F. R., and Keirns, J. J. (1973): The role of cyclic nucleotides in visual excitation. *Life Sciences,* 13:1451–1472.

Blaustein, M. P., and Hodgkin, A. L. (1969): The effect of cyanide on the efflux of calcium from squid axons. *Journal of Physiology,* 200:497–527.

Bloom, G. D., and Haegermark, Ö. (1967): Studies on morphological changes and histamine release induced by bee venom, *n*-decylamine and hypotonic solutions in rat peritoneal mast cells. *Acta Phyiologica Scandinavica,* 71:257–269.

Bockaert, J., Roy, C., and Jard, S. (1972): Oxytocin-sensitive adenylate cyclase in frog bladder epithelial cells. *Journal of Biological Chemistry,* 247:7073–7081.

Bombik, B. M., and Burger, M. M. (1973): c-AMP and the cell cycle: Inhibition of growth stimulation. *Experimental Cell Research,* 80:88–94.

Borgeat, P., Chavancy, G., Dupont, A., Labrie, F., Arimura, A., and Schally, A. V. (1972): Stimulation of adenosine 3',5'-cyclic monophosphate accumulation in anterior pituitary gland *in vitro* by synthetic luteinizing hormone-releasing hormone. *Proceedings of the National Academy of Sciences of the U.S.A.,* 69:2677–2681.

Borle, A. B. (1973): Calcium metabolism at the cellular level. *Federation Proceedings,* 32:1944–1950.

Borle, A. B. (1974): Cyclic AMP stimulation of calcium efflux from kidney, liver and heart mitochondria. *Journal of Membrane Biology,* 16:221–236.

Borowitz, J. L., Fuwa, K., and Weiner, N. (1965): Distribution of metals and catecholamines in bovine adrenal medulla subcellular fractions. *Nature,* 205:42–43.

Bourguet, J., and Morel, F. (1967): Independence des variations de permeabilite a l'eau et au sodium produites par les hormones neurohypophysaires sur la vessie de grenoville. *Biochimica et Biophysica Acta,* 135:693–700.

Bourne, H. R., Lichtenstein, L. M., and Melmon, K. L. (1972): Pharmacologic control of allergic histamine release *in vitro:* Evidence for an inhibitory role of 3',5'-adenosine monophosphate in human leukocytes. *Journal of Immunology,* 108:695–705.

Bourne, H. R., Lichtenstein, L. M., Melmon, K. L., Henney, C. S., Weinstein, Y., and Shearer, G. M. (1974): Modulation of inflammation and immunity by cyclic AMP. *Science,* 184:19–28.

Boynton, A. L., Whitfield, J. F., Isaacs, R. J., and Morton, H. J. (1974): Control of 3T3 cell proliferation by calcium. *In vitro,* 10:12–17.

Bray, D. (1972): Cytoplasmic actin: A comparative study. *Cold Spring Harbor Symposia on Quantitative Biology,* 37:567–571.

Breckenridge, B. McL., Burn, J. H., and Matschinsky, F. M. (1967): Theophylline, epinephrine and neostigmine facilitation of neuromuscular transmission. *Proceedings of the National Academy of Sciences of the U.S.A.,* 57:1893–1897.

Bretscher, M. S. (1973): Membrane structure: Some general principles. *Science,* 181:622–629.

Brierley, G. P., Murer, E., and Bachmann, E. (1964): Studies on ion transport. III. The accumulation of calcium and inorganic phosphate by heart mitochondria. *Archives of Biochemistry and Biophysics,* 105:89–102.

Brinley, F. J. (1973): Calcium and magnesium transport in single cells. *Federation Proceedings,* 32:1735–1739.
Brisson, G. R., Camu, F., Malaisse-Lagae, F., and Malaisse, W. J. (1971): Effect of a local anesthetic upon calcium uptake and insulin secretion by isolated islets of Langerhans. *Life Sciences,* 10(I):445–448.
Brisson, G. R., and Malaisse, W. J. (1973): The stimulus-secretion coupling of glucose-induced insulin release. XI. Effects of theophylline and epinephrine on ^{45}Ca efflux from perifused islets. *Metabolism,* 22:455–465.
Brisson, G. R., Malaisse-Lagae, F., and Malaisse, W. J. (1972): The stimulus-secretion coupling of glucose-induced insulin release. VII. A proposed site of action for adenosine-3′,5′-cyclic monophosphate. *Journal of Clinical Investigation,* 51:232–241.
Brooker, G. (1973): Oscillation of cyclic adenosine monophosphate concentration during the myocardial contraction cycle. *Science,* 182:933–934.
Brown, H. D., Chattopadhyay, S. K., Morris, H. P., and Pennington, S. N. (1970): Adenyl cyclase activity in Morris Hepatomas 7777, 7794A and 9618A. *Cancer Research,* 30:123–126.
Brown, J. E., and Blinks, J. R. (1972): Changes in [Ca^{++}] of *Limulus* ventral photoreceptors measured with aequorin. *Biological Bulletin,* 143:456.
Brown, J. E., and Pinto, L. H. (1974): Ionic mechanism for the photoreceptor potential of the retina of *Bufo marinus. Journal of Physiology,* 236:575–591.
Bucher, N. L. R. (1963): Regeneration of mammalian liver. *International Review of Cytology,* 15:245–300.
Bülbring, E., and Tomita, T. (1970): Calcium and the action potential in smooth muscle. In: *Calcium and Cellular Function,* edited by A. W. Cuthbert, pp. 249–260. Macmillan, London.
Burgen, A. S. V. (1956): The secretion of potassium in saliva. *Journal of Physiology,* 132:20–39.
Burger, M. M., Bombik, B. M., Breckenridge, B. McL., and Sheppard, J. R. (1972): Growth control and cyclic alterations of cyclic AMP in the cell cycle. *Nature, New Biology,* 239:161–163.
Bürk, R. R. (1968): Reduced adenyl cyclase activity in a polyoma virus transformed cell line. *Nature,* 219:1272–1275.
Butcher, R. W., Sneyd, J. G. T., Park, C. R., and Sutherland, E. W. (1966): Effect of insulin on adenosine 3′,5′-monophosphate in the rat epididymal fat pad. *Journal of Biological Chemistry,* 241:1651–1653.
Carchman, R. A., Jaanus, S. D., and Rubin, R. P. (1971): The role of adrenocorticotropin and calcium in adenosine cyclic 3′,5′-phosphate production and steroid release from isolated, perfused cat adrenal gland. *Molecular Pharmacology,* 7:491–499.
Carsten, M. E. (1969): Role of calcium binding by sarcoplasmic reticulum in the contraction and relaxation of uterine smooth muscle. *Journal of General Physiology,* 55:414–426.
Case, R. M. (1973): Calcium and gastrointestinal secretion. *Digestion,* 8:269–288.
Case, R. M., and Clausen, T. (1973): The relationship between calcium exchange and enzyme secretion in the isolated rat pancreas. *Journal of Physiology,* 235:75–102.
Case, R. M., Johnson, M., Scratcherd, T., and Sherratt, H. S. A. (1972): Cyclic adenosine 3′,5′-monophosphate concentration in the pancreas following stimulation by secretin, cholecystokinin-pancreozymin and acetylcholine. *Journal of Physiology,* 223:669–684.
Case, R. M., and Scratcherd, T. (1972a): The actions of dibutyryl cyclic adenosine 3′,5′-monophosphate and methyl xanthines on pancreatic exocrine pancreas. *Journal of Physiology,* 223:649–667.
Case, R. M., and Scratcherd, T. (1972b): Prostaglandin action on pancreatic blood flow and on electrolyte and enzyme secretion by exocrine pancreas *in vivo* and *in vitro. Journal of Physiology,* 226:393–405.
Casteels, R., Goffin, J., Raeymaekers, L., and Wuytack, F. (1973): Calcium pumping in the smooth muscle cells of the taenia coli. *Journal of Physiology,* 231:19P.
Caswell, A. H., and Pressman, B. C. (1972): Kinetics of transport of divalent cations across sarcoplasmic reticulum vesicles induced by ionophores. *Biochemical and Biophysical Research Communications,* 49:292–298.
Chader, G. J., Bensinger, R., Johnson, M., and Fletcher, R. T. (1973): Phosphodiesterase: An important role in cyclic nucleotide regulation in the retina. *Experimental Eye Research,* 17:483–486.
Chayoth, R., Epstein, S., and Field, J. B. (1972): Increased cyclic AMP levels in malignant hepatic nodules of ethionine treated rats. *Biochemical and Biophysical Research Communications,* 49:1663–1670.

Christoffersen, T., Mørland, J., Osnes, J.-B., and Elgjo, K. (1972): Hepatic adenyl cyclase: Alterations in hormone response during treatment with a chemical carcinogen. *Biochimica et Biophysica Acta,* 279:363–366.
Christoffersen, T., Mørland, J., Osnes, J.-B., and Øye, I. (1973): Development of cyclic AMP metabolism in rat liver. *Biochimica et Biophysica Acta,* 313:338–349.
Cittadini, A., Scarpa, A., and Chance, B. (1973): Calcium transport in intact Ehrlich ascites tumor cells. *Biochimica et Biophysica Acta,* 291:246–259.
Civan, M. M., and Frazier, H. S. (1968): The site of the stimulatory action of vasopressin on sodium transport in toad bladder. *Journal of General Physiology,* 51:589–605.
Clothier, G., and Timourian, H. (1972): Calcium uptake and release by dividing sea urchin eggs. *Experimental Cell Research,* 75:105–110.
Cochrane, D. E., and Douglas, W. W. (1974): Calcium-induced extrusion of secretory granules (exocytosis) in mast cells exposed to 48/80 or the ionophores A 23187 and X-537A. *Proceedings of the National Academy of Sciences of the U.S.A.,* 71:408–412.
Cohen, I., and de Vries, A. (1973): Platelet contractile regulation in an isometric system. *Nature,* 246:36–37.
Cole, B., Robison, G. A., and Hartmann, R. C. (1971): Studies on the role of cyclic AMP in platelet function. *Annals of the New York Academy of Science,* 185:477–487.
Cone, R. A. (1972): Rotational diffusion of rhodopsin in the visual receptor membrane. *Nature, New Biology,* 236:39–43.
Cooper, R. H., Ashcroft, S. J. H., and Randle, P. J. (1973): Concentration of adenosine 3',5'-cyclic monophosphate in mouse pancreatic islets measured by a protein-binding radioassay. *Biochemical Journal,* 134:599–605.
Creed, K. E., and Wilson, J. A. F. (1969): The latency of response of secretory acinar cells to nerve stimulation in the submandibular gland of the cat. *Australian Journal of Experimental Biology and Medical Science,* 47:135–144.
Cross, M. E., and Ord, M. G. (1971): Changes in histone phosphorylation and associated early metabolic events in pig lymphocyte cultures transformed by phytohaemagglutinin or 6-N,2'-O-dibutyryladenosine 3',5'-monophosphate. *Biochemical Journal,* 124:241–248.
Davis, B., and Lazurus, N. R. (1972): Insulin release from mouse islets. Effect of glucose and hormones on adenylate cyclase. *Biochemical Journal,* 129:373–379.
Dean, P. M., and Matthews, E. K. (1970a): Glucose-induced electrical activity in pancreatic islet cells. *Journal of Physiology,* 210:255–264.
Dean, P. M., and Matthews, E. K. (1970b): Electrical activity in pancreatic islet cells: Effect of ions. *Journal of Physiology,* 210:265–275.
Dean, P. M., and Matthews, E. K. (1972): Pancreatic acinar cells: Measurement of membrane potential and miniature depolarization potentials. *Journal of Physiology,* 225:1–13.
Deery, D. J., and Howell, S. L. (1973): Rat anterior pituitary adenyl cyclase activity: GTP requireof prostaglandin E_1 and E_2 and synthetic luteinizing hormone-releasing hormone activation. *Biochimica et Biophysica Acta,* 329:17–22.
DeLorenzo, R. J., and Greengard, P. (1973): Activation by adenosine 3',5'-monophosphate of a membrane-bound phosphoprotein phosphatase from toad bladder. *Proceedings of the National Academy of Sciences of the U.S.A.,* 70:1831–1835.
DeLorenzo, R. J., Walton, K. G., Curran, P. F., and Greengard, P. (1973): Regulation of phosphorylation of a specific protein in toad-bladder membrane by antidiuretic hormone and cyclic AMP, and its possible relationship to membrane permeability changes. *Proceedings of the National Academy of Sciences of the U.S.A.,* 70:880–884.
de Petris, S., Raff, M. C., and Mallucci, L. (1973): Ligand-induced redistribution of concanavalin A receptors on normal, trypsinized and transformed fibroblasts. *Nature, New Biology,* 244:275–278.
DeRubertis, F. R., Zenser, T. V., Adler, W. H., and Hudson, T. (1974): Role of cyclic adenosine-3',5'-monophosphate in lymphocyte mitogenesis. *Journal of Immunology,* 113:151–161.
Devine, C. E., Somlyo, A. V., and Somlyo, A. P. (1972): Sarcoplasmic reticulum and excitation-contraction coupling in mammalian smooth muscles. *Journal of Cell Biology,* 52:690–718.
Diamond, J. (1973): Phosphorylase, calcium, and cyclic AMP in smooth muscle contraction. *American Journal of Physiology,* 225:930–937.
DiBona, D. R., Civan, M. M., and Leaf, A. (1969): The cellular specificity of the effect of vasopressin on toad urinary bladder. *Journal of Membrane Biology,* 1:79–91.

Douglas, W. W. (1968): Stimulus-secretion coupling: The concept and clues from chromaffin and other cells. *British Journal of Pharmacology,* 34:451–474.
Douglas, W. W., Kanno, T., and Sampson, S. R. (1966): Intracellular recording from adrenal chromaffin cells: Effects of acetylcholine, hexamethonium and potassium on membrane potentials. *Journal of Physiology,* 186:125P.
Douglas, W. W., Kanno, T., and Sampson, S. R. (1967): Influence of the ionic environment on the membrane potential of adrenal chromaffin cells and on the depolarizing effect of acetylcholine. *Journal of Physiology,* 191:107–121.
Douglas, W. W., and Poisner, A. M. (1962): On the mode of action of acetylcholine in evoking adrenal medullary secretion: Increased uptake of calcium during the secretory response. *Journal of Physiology,* 162:385–392.
Douglas, W. W., and Poisner, A. M. (1963): The influence of calcium on the secretory response of the submaxillary gland to acetylcholine or to noradrenaline. *Journal of Physiology,* 165:528–541.
Douglas, W. W., and Poisner, A. M. (1964): Stimulus-secretion coupling in a neurosecretory organ: The role of calcium in the release of vasopressin from the neurohypophysis. *Journal of Physiology,* 172:1–18.
Douglas, W. W., and Rubin, R. P. (1961): The role of calcium in the secretory response of the adrenal medulla to acetylcholine. *Journal of Physiology,* 159:40–57.
Douglas, W. W., and Rubin, R. P. (1963): The mechanism of catecholamine release from the adrenal medulla and the role of calcium in stimulus-secretion coupling. *Journal of Physiology,* 167:288–310.
Drahota, Z., Carafoli, E., Rossi, C. S., Gamble, R. L., and Lehninger, A. L. (1965): The steady state maintenance of accumulated Ca^{++} in rat liver mitochondria. *Journal of Biological Chemistry,* 240:2712–2720.
Dransfeld, H., Greeff, K., Hess, D., and Schorn, A. (1967): Die Abhängigkeit der Ca^{++} Aufnahme isolierter Mitochondrien des Herzmuskels von der Na- and K-Konzentration als mögliche Ursache der inotropen Digitaliswirkung. *Experientia,* 23:375–377.
Dreifuss, J. J., Grau, J. D., and Nordmann, J. J. (1973): Effects on the isolated neurohypophysis of agents which affect the membrane permeability to calcium. *Journal of Physiology,* 231:96P.
Droller, M. J., and Wolfe, S. M. (1972): Thrombin-induced increase in intracellular cyclic 3′,5′-adenosine monophosphate in human platelets. *Journal of Clinical Investigation,* 51:3094–3103.
Drummond, G. I., and Duncan, L. (1970): Adenyl cyclase in cardiac tissue. *Journal of Biological Chemistry,* 245:976–983.
Durbin, R. P., and Jenkinson, D. H. (1961): The calcium dependence of tension development in depolarized smooth muscle. *Journal of Physiology,* 157:90–96.
Durham, J. P., Baserga, R., and Butcher, F. R. (1974): The effect of isoproterenol and its analogs upon adenosine 3′,5′-monophosphate and guanosine 3′,5′-monophosphate levels in mouse parotid gland *in vivo.* Relationship to the stimulation of DNA synthesis. *Biochimica et Biophysica Acta,* 327:196–217.
Endo, M., Tanaka, M., and Ogawa, Y. (1970): Calcium induced release of calcium from the sarcoplasmic reticulum of skinned skeletal muscle fibres. *Nature,* 228:34–36.
Entman, M. L., Levey, G. S., and Epstein, S. E. (1969): Mechanism of action of epinephrine and glucagon on the canine heart. *Circulation Research,* 25:429–438.
Exton, J. H., Friedmann, N., Wong, E. H. A., Brineaux, J. P., Corbin, J. D., and Park, C. R. (1972): Interaction of glucocorticoids with glucagon and epinephrine in the control of gluconeogenesis and glycogenolysis in liver and of lipolysis in adipose tissue. *Journal of Biological Chemistry,* 247:3579–3588.
Exton, J. H., Mallette, L. E., Jefferson, L. S., Wong, E. H. A., Friedmann, N., Miller, T. B., and Park, C. R. (1970): The hormonal control of hepatic gluconeogenesis. *Recent Progress in Hormone Research,* 26:411–457.
Exton, J. H., Robison, G. A., Sutherland, E. W., and Park, C. R. (1971): Studies on the role of adenosine 3′,5′-monophosphate in the hepatic actions of glucagon and catecholamines. *Journal of Biological Chemistry,* 246:6166–6177.
Farese, R. V. (1971): Calcium as a mediator of adrenocorticotrophic hormone action on adrenal protein synthesis. *Science,* 173:447–450.
Feinman, R. D., and Detwiler, T. C. (1974): Platelet secretion induced by divalent cation ionophores. *Nature,* 249:172–173.
Feinstein, H., and Schramm, M. (1970): Energy production in rat parotid gland. Relation to enzyme secretion and effects of calcium. *European Journal of Biochemistry,* 13:158–163.

Ferrendelli, J. A., Kinscherf, D. A., and Chang, M. M. (1973): Regulation of levels of guanosine cyclic 3′,5′-monophosphate in the central nervous system: Effects of depolarizing agents. *Molecular Pharmacology,* 9:445–454.

Fisher, B., Szuch, P., Levine, M., and Fisher, E. R. (1971): A portal blood factor as the humoral agent in liver regeneration. *Science,* 171:575–577.

Fitzpatrick, D. F., Landon, E. J., Debbas, G., and Hurwitz, L. (1972): A calcium pump in vascular smooth muscle. *Science,* 176:305–306.

Fleischer, N., Donald, R. A., and Butcher, R. W. (1969): Involvement of adenosine 3′,5′-monophosphate in release of ACTH. *American Journal of Physiology,* 217:1287–1291.

Ford, L. E., and Podolsky, R. J. (1970): Regenerative calcium release within muscle cells. *Science,* 167:58–59.

Foreman, J. C., and Mongar, J. L. (1972a): Dual effect of lanthanum on histamine release from mast cells. *Nature, New Biology,* 240:255–256.

Foreman, J. C., and Mongar, J. L. (1972b): The role of the alkaline earth ions in anaphylactic histamine secretion. *Journal of Physiology,* 224:753–769.

Foreman, J. C., Mongar, J. L., and Gomperts, B. D. (1973): Calcium ionophores and movement of calcium ions following the physiological stimulus to a secretory process. *Nature,* 245:249–251.

Forer, A. (1969): Chromosome movements during cell division. In: *Handbook of Molecular Cytology,* edited by A. Lima-de-Faria, pp. 553–601. North-Holland, Amsterdam.

Fox, T. O., Sheppard, J. R., and Burger, M. M. (1971): Cyclic membrane changes in animal cells: Transformed cells permanently display a surface architecture detected in normal cells only during mitosis. *Proceedings of the National Academy of Sciences of the U.S.A.,* 68:244–247.

Franzini-Armstrong, C. (1970): Studies on the triad. I. Structure of the junction in frog twitch fibres. *Journal of Cell Biology,* 47:488–499.

Freeman, D. J., and Daniel, E. E. (1973): Calcium movement in vascular smooth muscle and its detection using lanthanum as a tool. *Canadian Journal of Physiology and Pharmacology,* 51:900–913.

Friedmann, N. (1972): Effects of glucagon and cyclic AMP on ion fluxes in the perfused liver. *Biochimica et Biophysica Acta,* 274:214–225.

Friedmann, N., and Dambach, G. (1973): Effects of glucagon, 3′,5′-AMP and 3′,5′-GMP on ion fluxes and transmembrane potential in perfused livers of normal and adrenalectomized rats. *Biochimica et Biophysica Acta,* 307:399–403.

Friedmann, N., and Park, C. R. (1968): Early effects of 3′,5′-adenosine monophosphate on the fluxes of calcium and potassium in the perfused liver of normal and adrenalectomized rats. *Proceedings of the National Academy of Sciences of the U.S.A.,* 61:504–508.

Friedmann, N., and Rasmussen, H. (1970): Calcium, manganese and hepatic gluconeogenesis. *Biochimica et Biophysica Acta,* 222:41–52.

Friedmann, N., Somlyo, A. V., and Somlyo, A. (1971): Cyclic adenosine and guanosine monophosphates and glucagon: Effect on liver membrane potentials. *Science,* 171:400–402.

Fritz, M. E., and Botelho, S. Y. (1969): Role of autonomic nerve impulses in secretion by the parotid gland of the cat. *American Journal of Physiology,* 216:1392–1398.

Froehlich, J. E., and Rachmeler, M. (1972): Effect of adenosine 3′,5′-cyclic monophosphate on cell proliferation. *Journal of Cell Biology,* 55:19–31.

Garcia, A. G., Kirpekar, S. M., and Prat, J. C. (1975): A calcium ionophore stimulating the secretion of catecholamines from the cat adrenal. *Journal of Physiology,* 244:253–262.

Gárdos, G. (1958): The function of calcium in the potassium permeability of human erythrocytes. *Biochimica et Biophysica Acta,* 30:653–654.

Garren, L. D., Gill, G. N., Masui, H., and Walton, G. M. (1971): On the mechanism of action of ACTH. *Recent Progress in Hormone Research,* 27:433–474.

Gawadi, N. (1971): Actin in the mitotic spindle. *Nature,* 234:410.

George, W. J., Wilkerson, R. D., and Kodowitz, P. J. (1973): Influence of acetylcholine on contractile force and cyclic nucleotide levels in the isolated perfused rat heart. *Journal of Pharmacology and Experimental Therapeutics,* 184:228–235.

Gill, G. N. (1972): Mechanism of ACTH action. *Metabolism,* 21:571–588.

Glick, M. C., Gerner, E. W., and Warren, L. (1971): Changes in the carbohydrate content of the KB cell during the growth cycle. *Journal of Cellular Physiology,* 77:1–6.

Goldberg, A. L., and Singer, J. J. (1969): Evidence for the role of cyclic AMP in neuromuscular transmission. *Proceedings of the National Academy of Sciences of the U.S.A.,* 64:134–141.

Goldberg, N. D., Haddox, M. K., Dunham, E., Lopez, C., and Hadden, J. W. (1974): The Yin Yang hypothesis of biological control: Opposing influences of cyclic GMP and cyclic AMP in the regulation of cell proliferation and other biological processes. In: *Control of Proliferation in Animal Cells,* edited by B. Clarkson and R. Baserga, pp. 609–625. Cold Spring Harbor Laboratory.

Goldberg, N. D., O'Dea, R. F., and Haddox, M. K. (1973): Cyclic GMP. *Advances in Cyclic Nucleotide Research,* 3:155–223.

Goldfine, I. D., Roth, J., and Birnbaumer, L. (1972): Glucagon receptors in β-cells. Binding of ^{125}I-glucagon and activation of adenylate cyclase. *Journal of Biological Chemistry,* 247:1211–1218.

Golenhofen, K. (1970): Slow rhythms in smooth muscle (minute-rhythm). In: *Smooth Muscle,* edited by E. Bülbring, A. F., Brading, A. W. Jones, and T. Tomita, pp. 316–342. Edward Arnold, London.

Goodford, P. J. (1970): Ionic interactions in smooth muscle. In; *Smooth Muscle,* edited by E. Bülbring, A. F. Brading, A. W. Jones, and T. Tomita, pp. 100–121. Edward Arnold, London.

Goridis, C., and Virmaux, N. (1974): Light-regulated guanosine 3',5'-monophosphate phosphodiesterase of bovine retina. *Nature,* 248:57–58.

Green, L. (1968): Mechanism of movements of granules in melanocytes of *Fundulus heteroclitus. Proceedings of the National Academy of Sciences of the U.S.A.,* 59:1179–1186.

Greenberg, S., Long, J. P., and Diecke, F. P. J. (1973): Differentiation of calcium pools utilized in the contractile response of canine arterial and venous smooth muscle to norepinephrine. *Journal of Pharmacology and Experimental Therapeutics,* 185:493–504.

Greengard, P. and Kebabian, J. W. (1974): Role of cyclic AMP in synaptic transmission in the mammalian peripheral nervous system. *Federation Proceedings,* 33:1059–1067.

Grette, K. (1963): Relaxing factor in extracts of blood platelets and its function in the cells. *Nature,* 198:488–489.

Grodsky, G. M., and Bennett, L. L. (1966): Cation requirements for insulin secretion in the isolated perfused pancreas. *Diabetes,* 15:910–913.

Guidotti, A., and Costa, E. (1973): Involvement of adenosine 3',5'-monophosphate in the activation of tyrosine hydroxylase elicited by drugs. *Science,* 179:902–904.

Guidotti, A., Weiss, B., and Costa, E. (1972): Adenosine 3',5'-monophosphate concentrations and isoproterenol induced synthesis of deoxyribonucleic acid in mouse parotid gland. *Molecular Pharmacology,* 8:521–530.

Hadden, J. W., Hadden, E. M., Haddox, M. K., and Goldberg, N. D. (1972): Guanosine 3',5'-cyclic monophosphate: A possible intracellular mediator of mitogenic influence in lymphocytes. *Proceedings of the National Academy of Sciences of the U.S.A.,* 69:3024–3027.

Haddox, M. K., Stephenson, J. H., and Goldberg, N. D. (1974): Cyclic GMP in meristematic and elongating regions of bean root. *Federation Proceedings,* 33:1755Abs.

Hales, C. N., Luzio, J. P., Chandler, J. A., and Herman, L. (1974): Localization of calcium in the smooth endoplasmic reticulum of rat isolated fat cells. *Journal of Cell Science,* 15:1–15.

Hales, C. N., and Miller, R. D. G. (1968): Cations and secretion of insulin from rabbit pancreas *in vitro. Journal of Physiology,* 199:177–187.

Handler, J. S., Butcher, R. W., Sutherland, E. W., and Orloff, J. (1965): The effect of vasopressin and of theophylline on the concentration of adenosine 3',5'-phosphate in the urinary bladder of the toad. *Journal of Biological Chemistry,* 240:4524–4526

Handler, J. S., Preston, A. S., and Orloff, J. (1972): Effect of ADH, aldosterone, ouabain, and amiloride on toad bladder epithelial cells. *American Journal of Physiology,* 222:1071–1074.

Hanson, J. P., Repke, D. I., Katz, A. M., and Aledort, L. M. (1973): Calcium ion control of platelet thrombosthenin ATPase activity. *Biochimica et Biophysica Acta,* 314:382–389.

Hardman, J. G., Schultz, G., and Sutherland, E. W. (1974): Cyclic GMP: Vestige or another intracellular messenger? In: *Cyclic AMP, Cell Growth, and the Immune Response,* edited by W. Braun, L. M. Lichtenstein, and C. W. Parker, pp. 223–226. Springer-Verlag, New York.

Harfield, B., and Tenenhouse, A. (1973): Effect of EGTA on protein release and cyclic AMP accumulation in rat parotid gland. *Canadian Journal of Physiology and Pharmacology,* 51:997–1001.

Harwood, J. P., Moskowitz, J., and Krishna, G. (1972): Dynamic interaction of prostaglandin and norepinephrine in the formation of adenosine 3',5'-monophosphate in human and rabbit platelets. *Biochimica et Biophysica Acta,* 261:444–456.

Haslam, R. J. (1964): Role of adenosine diphosphate in the aggregation of human blood-platelets by thrombin and by fatty acids. *Nature,* 202:765–768.

Haslam, R. J., and Rosson, G. M. (1972): Aggregation of human blood platelets by vasopressin. *American Journal of Physiology,* 223:958–967.
Haslam, R. J., and Taylor, A. (1971): Effects of catecholamines on the formation of adenosine 3',5'-cyclic monophosphate in human blood platelets. *Biochemical Journal,* 125:377–379.
Hausen, P., and Stein, H. (1968): On the synthesis of RNA in lymphocytes stimulated by phytohemagglutinin. I. Induction of uridine-kinase and the conversion of uridine to UTP. *European Journal of Biochemistry,* 4:401–406.
Haylett, D. G., and Jenkinson, D. H. (1972): Effects of noradrenaline on potassium efflux, membrane potential and electrolyte levels in tissue slices prepared from guinea-pig liver. *Journal of Physiology,* 225:721–750.
Haynes, R. C. (1958): The activation of adrenal phosphorylase by the adrenocorticotropic hormone. *Journal of Biological Chemistry,* 233:1220–1222.
Hays, R. M., Singer, B., and Malamed, S. (1965): The effect of calcium withdrawal on the structure and function of the toad bladder. *Journal of Cell Biology,* 25(Part 2):195–208.
Heatley, N. G. (1968): The assay of secretin in the rat. *Journal of Endocrinology,* 42:535–547.
Heisler, S., Fast, D., and Tenenhouse, A. (1972): Role of Ca^{2+} and cyclic AMP in protein secretion from rat exocrine pancreas. *Biochimica et Biophysica Acta,* 279:561–572.
Heisler, S., and Grondin, G. (1973): Effect of lanthanum on ^{45}Ca flux and secretion of protein from rat exocrine pancreas. *Life Sciences,* 13:783–794.
Hinkley, R., and Telser, A. (1974): Heavy meromyosin-binding filaments in the mitotic apparatus of mammalian cells. *Experimental Cell Research,* 86:161–164.
Hirschfield, I. N., and Koritz, S. B. (1964): The stimulation of pregnenolone synthesis in the large particles from rat adrenal by some agents which cause mitochondrial swelling. *Biochemistry,* 3:1994–1998.
Hirschhorn, R., Grossman, J., and Weissmann, G. (1970): Effect of cyclic 3',5'-adenosine monophosphate and theophylline on lymphocyte transformation. *Proceedings of the Society for Experimental Biology and Medicine,* 133:1361–1365.
Hitchcock, S. E., Huxley, H. E., and Szent-Györgyi, A. G. (1973): Calcium sensitive binding of troponin to actin-tropomyosin: A two-site model for troponin action. *Journal of Molecular Biology,* 80:825–836.
Hodgkin, A. L., and Keynes, R. D. (1957): Movements of labelled calcium in squid giant axons. *Journal of Physiology,* 138:253–281.
Hogberg, B., and Uvnäs, B. (1960): Further observations on the disruption of rat mesentery mast cells caused by compound 48/80, antigen-antibody reaction, lecithinase A and decylamine. *Acta Phyiologica Scandinavica,* 48:133–145.
Hokin, L. E. (1966): Effects of calcium omission on acetylcholine-stimulated amylase secretion and phospholipid synthesis in pigeon pancreas slices. *Biochimica et Biophysica Acta,* 115:219–221.
Hülser, D. F., and Frank, W. (1971): Stimulierung von Kulturen embryonaler Rattenzellen durch eine Proteinfraktion aus fötalem Kälberserum. *Zeitschrift für Naturforschung,* 26b:1045–1048.
Hurko, O., Elster, P., and Wurtman, R. J. (1974): Adenylate cyclase activity in bovine adrenal medulla. *Endocrinology,* 94:591–593.
Hurwitz, L. Fitzpatrick, D. F., Debbas, G., and Landon, E. J. (1973): Localization of calcium pump activity in smooth muscle. *Science,* 179:384–386.
Huxley, H. E. (1972): Structural changes in the actin- and myosin-containing filaments during contraction. *Cold Spring Harbor Symposia on Quantitative Biology,* 37:361–376.
Huxley, H. E. (1973): Muscular contraction and cell motility. *Nature,* 243:445–449.
Illiano, G., and Cuatrecasas, P. (1972): Modulation of adenylate cyclase activity in liver and fat cell membranes by insulin. *Science,* 175:906–908.
Ito, S., and Yoshioka, K. (1973): Effect of various ionic compositions upon the membrane potentials during activation of sea urchin eggs. *Experimental Cell Research,* 78:191–200.
Jaanus, S. D., Rosenstein, M. J., and Rubin, R. P. (1970): On the mode of action of ACTH on the isolated perfused adrenal gland. *Journal of Physiology,* 209:539–556.
Jaanus, S. D., and Rubin, R. P. (1971): The effect of ACTH on calcium distribution in the perfused cat adrenal gland. *Journal of Physiology,* 213:581–598.
Jaanus, S. D., and Rubin, R. P. (1974): Analysis of the role of cyclic adenosine 3',5'-monophosphate in catecholamine release. *Journal of Physiology,* 237:465–476.
Jenkinson, D. H., and Morton, I. K. M. (1967): Adrenergic blocking drugs as tools in the study of the actions of catecholamines on the smooth muscle membrane. *Annals of the New York Academy of Sciences,* 139:762–771.

Jenkinson, D. H., Stamenović, B. A., and Whitaker, B. D. L. (1968): The effect of noradrenaline on the end-plate potential in twitch fibres of the frog. *Journal of Physiology,* 195:743–754.

Johnson, G. S., Morgan, W. D., and Pastan, I. (1972): Regulation of cell motility by cyclic AMP. *Nature,* 235:54–56.

Jutisz, M., and de la Llosa, M. P. (1970): Requirement of Ca^{++} and Mg^{++} ions for the *in vitro* release of follicle-stimulating hormone from rat pituitary glands and its subsequent biosynthesis. *Endocrinology,* 86:761–768.

Kakiuchi, S., and Yamazaki, R. (1970): Calcium dependent phosphodiesterase activity and its activating factor (PAF) from brain. *Biochemical and Biophysical Research Communications,* 41:1104–1110.

Kakiuchi, S., Yamazaki, R., and Teshima, Y. (1971): Cyclic 3′,5′-nucleotide phosphodiesterase. IV. Two enzymes with different properties from brain. *Biochemical and Biophysical Research Communications,* 42:968–974.

Kakiuchi, S., Yamazaki, R., Teshima, Y., and Uenishi, K. (1973): Regulation of nucleoside cyclic 3′,5′-monophosphate phosphodiesterase activity from rat brain by a modulator and Ca^{2+}. *Proceedings of the National Academy of Sciences of the U.S.A.,* 70:3526–3530.

Kaliner, M., and Austen. K. F. (1974a): Cyclic nucleotides and modulation of effector systems in inflammation. *Biochemical Pharmacology,* 23:763–771.

Kaliner, M., and Austen, K. F. (1974b): Cyclic AMP, ATP, and reversed anaphylactic histamine release from rat mast cells. *Journal of Immunology,* 112:664–674.

Kaliner, M., Orange, R. P. and Austen, K. F. (1972): Immunological release of histamine and slow reacting substance of anaphylaxis from human lung. IV. Enhancement by cholinergic and alpha adrenergic stimulation. *Journal of Experimental Medicine,* 136:556–567.

Kanno, T. (1972): Calcium-dependent amylase release and electrophysiological measurements in cells of the pancreas. *Journal of Physiology,* 226:353–371.

Kanno, T., Cochrane, D. E., and Douglas, W. W. (1973): Exocytosis (secretory granule extrusion) induced by injection of calcium into mast cells. *Canadian Journal of Physiology and Pharmacology,* 51:1001–1004.

Katsumi, W., Kamberi, I. A., and McCann, S. M. (1969): *In vitro* response of the rat pituitary to gonadotrophin-releasing factors and to ions. *Endocrinology,* 85:1046–1056.

Kinlough-Rathbone, R. L., Chahil, A., and Mustard, J. F. (1973): Effect of external calcium and magnesium on thrombin-induced changes in calcium and magnesium of pig platelets. *American Journal of Physiology,* 224:941–945.

Kinlough-Rathbone, R. L., Chahil, A., and Mustard, J. F. (1974): Divalent cations and the release reaction of pig platelets. *American Journal of Physiology,* 226:235–239.

Kirchberger, M. A., Tada, M., Repke, D. I., and Katz, A. M. (1972): Cyclic adenosine 3′,5′-monophosphate-dependent protein kinase stimulation of calcium uptake by canine cardiac microsomes. *Journal of Molecular and Cellular Cardiology,* 4:673–680.

Krnjević, K., and Miledi, R. (1958): Some effects produced by adrenaline upon neuromuscular propagation in rats. *Journal of Physiology,* 141:291–304.

Kukovetz, W. R., and Pöch, G. (1970): Inhibition of cyclic-3′,5′-nucleotide-phosphodiesterase as a possible mode of action of papaverine and similarly acting drugs. *Naunyn Schmiedeberg's Archives of Pharmacology,* 267:189–194.

Kulka, R. G., and Sternlicht, E. (1968): Enzyme secretion in mouse pancreas mediated by adenosine-3′,5′-cyclic phosphate and inhibited by adenosine-3′-phosphate. *Proceedings of the National Academy of Sciences of the U.S.A.,* 61:1123–1128.

Kuo, J. F. (1974): Guanosine 3′,5′-monophosphate-dependent protein kinases in mammalian tissues. *Proceedings of the National Academy of Sciences of the U.S.A.,* 71:4037–4041.

Lacy, P. E., Howell, S. L., Young, D. A., and Fink, C. J. (1968): New hypothesis of insulin secretion. *Nature,* 219:1177–1179.

Lamb, J. F., and Lindsay, R. (1971): Effect of Na, metabolic inhibitors and ATP on Ca movements in L cells. *Journal of Physiology,* 218:691–708.

Lambotte, L. (1973): Effect of activation of α and β adrenergic receptors on the hepatic cell membrane potential in perfused dog liver. *Journal of Physiology,* 232:181–192.

Langan, T. A. (1973): Protein kinases and protein kinase substrates. *Advances in Cyclic Nucleotide Research,* 3:99–153.

Langer, G. A. (1973): Heart: Excitation-contraction coupling. *Annual Review of Physiology,* 35:55–86.

Lastowecka, A., and Trifaró, J. M. (1974): The effect of sodium and calcium ions on the release of catecholamines from the adrenal medulla: Sodium deprivation induces release by exocytosis in the absence of extracellular calcium. *Journal of Physiology,* 236:681–705.

Lee, K. S., and Shin, B. C. (1969): Studies on the active transport of calcium in human red cells. *Journal of General Physiology,* 54:713–729.

Lefkowitz, R. J., Roth, J., and Pastan, I. (1970): Effects of calcium on ACTH stimulation of the adrenal: Separation of hormone binding from adenyl cyclase activation. *Nature,* 228:864–866.

Leier, D. J., and Jungmann, R. A. (1973): Adrenocorticotropic hormone and dibutyryl adenosine cyclic monophosphate-mediated Ca^{2+} uptake by rat adrenal glands. *Biochimica et Biophysica Acta,* 329:196–210.

LeJohn, H. B., and Cameron, L. E. (1973a): Cytokinins regulate calcium binding to a glycoprotein from fungal cells. *Biochemical and Biophysical Research Communications,* 54:1053–1060.

LeJohn, H. B., and Cameron, L. E. (1973b): Cytokinins and magnesium ions may control the flow of metabolites and calcium ions through fungal cell membranes. *Biochemical and Biophysical Research Communications,* 54:1061–1066.

Lemaire, S., Pelletier, G., and Labrie, F. (1971): Adenosine 3′,5′-monophosphate-dependent protein kinase from bovine anterior pituitary gland. II. Subcellular distribution. *Journal of Biological Chemistry,* 246:7303–7310.

Lemay, A., and Labrie, F. (1972): Calcium-dependent stimulation of prolactin release in rat anterior pituitary *in vitro* by N^6-monobutyryl adenosine 3′,5′-monophosphate. *FEBS Letters,* 20:7–10.

Levy, J. V., Cohen, J. A., and Inesi, G. (1973): Contractile effects of a calcium ionophore. *Nature,* 242:461–463.

Lichtenstein, L. M. (1971): The immediate allergic response: *In vitro* separation of antigen activation, decay and histamine release. *Journal of Immunology,* 107:1122–1130.

Lichtenstein, L. M., Levy, D. A., and Ishizaka, K. (1970): *In vitro* reversed anaphalaxis: Characteristics of anti-IgE mediated histamine release. *Immunology,* 19:831–842.

Loeffler, L. J., Lovenberg, W., and Sjoerdsma, A. (1971): Effects of dibutyryl-3′,5′-cyclic adenosine monophosphate, phosphodiesterase inhibitors and prostaglandin E_1 on compound 48/80-induced histamine release from rat peritoneal mast cells *in vitro*. *Biochemical Pharmacology,* 20:2287–2297.

Loten, E. G., and Sneyd, J. G. T. (1970): An effect of insulin on adipose-tissue adenosine 3′,5′-cyclic monophosphate phosphodiesterase. *Biochemical Journal,* 120:187–193.

Luxoro, M., and Yañez, E. (1968): Permeability of the giant axon of *Dosidicus gigas* to calcium ions. *Journal of General Physiology,* 51:115–122S.

Mackie, C., Richardson, M. C., and Schulster, D. (1972): Kinetics and dose-response characteristics of adenosine 3′,5′-monophosphate production by isolated rat adrenal cells stimulated with adrenocorticotrophic hormone. *FEBS Letters,* 23:345–348.

MacLeod, R. M., and Lehmeyer, J. E. (1970): Release of pituitary growth hormone by prostaglandins and dibutyryl cyclic 3′,5′-monophosphate in the absence of protein synthesis. *Proceedings of the National Academy of Sciences of the U.S.A.,* 67:1172–1179.

MacManus, J. P., Franks, D. J., Youdale, T., and Braceland, B. M. (1972): Increases in rat liver cyclic AMP concentrations prior to the initiation of DNA synthesis following partial hepatectomy or hormone infusion. *Biochemical and Biophysical Research Communications,* 49:1201–1207.

Maddrell, S. H. P., and Gee, J. D. (1974): Potassium-induced release of the diuretic hormones of *Rhodnies prolixus* and *Glossina austeri:* Ca dependence, time course and localization of neurohaemal areas. *Journal of Experimental Biology,* 61:155–171.

Maino, V. C., Green, H. M., and Crumpton, M. J. (1974): The role of calcium ions in initiating transformation of lymphocytes. *Nature,* 251:324–327.

Majno, G., Bouvier, C. A., Gabbiani, G., Ryan, G. B., and Statkov, P. (1972): Kymographic recording of clot retraction: Effects of papavarine, theophylline and cytocholasin B. *Thrombosis et Diathesis Haemorrhagica,* 28:49–53.

Malaisse, W. J., Malaisse-Lagae, F., Walker, M. O., and Lacy, P. E. (1971): The stimulus-secretion coupling of glucose-induced insulin release. V. The participation of a microtubular-microfilamentous system. *Diabetes,* 20:257–265.

Malaisse-Lagae, F., and Malaisse, W. J. (1971): Stimulus-secretion coupling of glucose-induced insulin release. III. Uptake of ^{45}calcium by isolated islets of Langerhans. *Endocrinology,* 88:72–80.

Malamud, D. (1972): Amylase secretion from mouse parotid and pancreas: Role of cyclic AMP and isoproterenol. *Biochimica et Biophysica Acta,* 279:373–376.

Marchesi, V. T., Jackson, R. L., Segrest, J. P., and Kahane, I. (1973): Molecular features of the major glycoprotein of the human erythrocyte membrane. *Federation Proceedings,* 32:1833–1837.

Marquis, N. R., Becker, J. A., and Vigdahl, R. L. (1970): Platelet aggregation. III. An epinephrine induced decrease in cyclic AMP synthesis. *Biochemical and Biophysical Research Communications,* 39:783–789.

Marquis, N. R., Vigdahl, R. L., and Tavormina, P. A. (1969): Platelet aggregation. I. Regulation by cyclic AMP and prostaglandin E_1. *Biochemical and Biophysical Research Communications,* 36:965–972.

Martin, S., York, D. H., and Kraicer, J. (1973): Alterations in transmembrane potential of adenohypophysial cells in elevated potassium and calcium-free media. *Endocrinology,* 92:1084–1088.

Mason, W. T., Fager, R. S., and Abrahamson, E. W. (1974): Ion fluxes in disk membranes of retinal rod outer segments. *Nature,* 247:562–563.

Matthews, E. K. (1970): Calcium and hormone release. In: *Calcium and Cellular Function,* edited by A. W. Cuthbert, pp. 163–182. Macmillan, London.

Matthews, E. K., and Petersen, O. H. (1973): Pancreatic acinar cells: Ionic dependence of the membrane potential and acetylcholine-induced depolarization. *Journal of Physiology,* 231:283–295.

Matthews, E. K., Petersen, O. H., and Williams, J. A. (1973): Pancreatic acinar cells: Acetylcholine-induced membrane depolarization, calcium efflux and amylase release. *Journal of Physiology,* 234:689–701.

Matthews, E. K., and Saffran, M. (1967): Steroid production and membrane potential measurement in cells of the adrenal cortex. *Journal of Physiology,* 189:149–161.

Matthews, E. K., and Saffran, M. (1968): Effect of ACTH on the electrical properties of adrenocortical cells. *Nature (London),* 219:1369–1370.

Matthews, E. K., and Saffran, M. (1973): Ionic dependence of adrenal steroidogenesis and ACTH-induced changes in the membrane potential of adrenocortical cells. *Journal of Physiology,* 234:43–64.

McAfee, D. A., and Greengard, P. (1972): Adenosine 3',5'-monophosphate: Electrophysiological evidence for a role in synaptic transmission. *Science,* 178:310–312.

McGuire, J., and Moellmann, G. (1972): Cytocholasin B: Effects on microfilaments and movement of melanin granules within melanocytes. *Science,* 175:642–644.

Meech, R. W. (1974): The sensitivity of *Helix aspersa* neurons to injected calcium ions. *Journal of Physiology,* 237:259–277.

Meinertz, T., Nawrath, H., and Scholz, H. (1973): Dibutyryl cyclic AMP and adrenaline increase contractile force and ^{45}Ca uptake in mammalian cardiac muscle. *Naunyn Schmiedeberg's Archives of Pharmacology,* 277:107–112.

Miki, N., Keirns, J. J., Marcus, F. R., Freeman, J., and Bitensky, M. W. (1973): Regulation of cyclic nucleotide concentrations in photoreceptors and ATP-dependent stimulation of cyclic nucleotide phosphodiesterase by light. *Proceedings of the National Academy of Sciences of the U.S.A.,* 70:3820–3824.

Miledi, R. (1973): Transmitter release induced by injection of calcium ions into nerve terminals. *Proceedings of the Royal Society of London, Series B,* 183:421–425.

Miller, W. H. (1973): Cyclic nucleotides and photoreception. *Experimental Eye Research,* 16:357–363.

Milligan, J. V., and Kraicer, J. (1969): Calcium-45 uptake and potassium-induced release of ACTH from the adenohypophysis. *The Physiologist,* 12:303.

Milligan, J. V., and Kraicer, J. (1974): Physical characteristics of the Ca^{++} compartments associated with *in vitro* ACTH release. *Endocrinology,* 94:435–443.

Milligan, J. V., Kraicer, J., Fawcett, C. P., and Illner, P. (1972): Purified growth hormone releasing factor increased ^{45}Ca uptake into pituitary cells. *Canadian Journal of Physiology and Pharmacology,* 50:613–617.

Millis, A. J. T., Forrest, G., and Pious, D. A. (1972): Cyclic AMP in cultured human lymphoid cells: Relationship to mitosis. *Biochemical and Biophysical Research Communications,* 49:1645–1649.

Mills, D. C. B., and Smith, J. B. (1971): The influence on platelet aggregation of drugs that affect the accumulation of adenosine 3',5'-cyclic monophosphate in platelets. *Biochemical Journal,* 121:185–196.

Mitchison, J. M. (1971): *The Biology of the Cell Cycle.* Cambridge University Press, Cambridge.

Miyamoto, M. D., and Breckenridge, B. McL. (1974): A cyclic adenosine monophosphate link in the catecholamine enhancement of transmitter release at the neuromuscular junction. *Journal of General Physiology,* 63:609–624.

Montague, W., and Cook, J. R. (1971): The role of adenosine 3′,5′-cyclic monophosphate in the regulation of insulin release by isolated rat islets of Langerhans. *Biochemical Journal,* 122:115–120.
Moolten, F. L., and Bucher, N. L. R. (1967): Regeneration of rat liver: Transfer of humoral agent by cross circulation. *Science,* 158:272–274.
Morisset, J. A., and Webster, P. D. (1971): *In vitro* and *in vivo* effects of pancreozymin, urecholine, and cyclic AMP on rat pancreas. *American Journal of Physiology,* 230:202–208.
Moskowitz, J., Harwood, J. P., Reid, W. D., and Krishna, G. (1971): The interaction of norepinephrine and prostaglandin E_1 on the adenyl cyclase system of human and rabbit blood platelets. *Biochimica et Biophysica Acta,* 230:279–285.
Muir, T. C., Pollock, D., and Turner, C. J. (1973): The effects of electrical stimulation of the sympathetic nerves on the size and mitotic index of rat salivary glands. *Journal of Physiology,* 232:43P.
Muller, R. U., and Finkelstein, A. (1974): The electrostatic basis of Mg^{++} inhibition of transmitter release. *Proceedings of the National Academy of Sciences of the U.S.A.,* 71:923–926.
Mustard, J. E., Kinlough-Rathbone, R. L., Jenkins, C. S. P., and Packham, M. A. (1972): Modification of platelet function. *Annals of the New York Academy of Sciences,* 201:343–359.
Nagata, N., and Rasmussen, H. (1970): Parathyroid hormone, 3′,5′-AMP, Ca^{++} and renal gluconeogenesis. *Proceedings of the National Academy of Sciences of the U.S.A.,* 65:368–374.
Nakamura, M., and Yasumasu, I. (1974): Mechanism for increase in intracellular concentration of free calcium in fertilized sea urchin egg. *Journal of General Physiology,* 63:374–388.
Nakazato, Y., and Douglas, W. W. (1974): Vasopressin release from the isolated neurohypophysis induced by a calcium ionophore, X-537 A. *Nature,* 249:479–481.
Namm, D. H. (1971): The activation of glycogen phosphorylase in arterial smooth muscle. *Journal of Pharmacology and Experimental Therapeutics,* 178:299–310.
Neufeld, A. H., Miller, W. H., and Bitensky, M. W. (1972): Calcium binding to retinal rod disk membranes. *Biochimica et Biophysica Acta,* 266:67–71.
Nicolson, G. L. (1972): Topography of membrane concanavalin A sites modified by proteolysis. *Nature, New Biology,* 239:193–197.
Nishiyama, A., and Kagayama, M. (1973): Biphasic secretory potentials in cat and rabbit submaxillary glands. *Experientia,* 29:161–163.
Noonan, K. D., and Burger, M. M. (1973*a*): Binding of [^{3}H] concanavalin A to normal and transformed cells. *Journal of Biological Chemistry,* 248:4286–4292.
Noonan, K. D., and Burger, M. M. (1973*b*): The relationship of concanavalin A binding to lectin-initiated cell agglutination. *Journal of Cell Biology,* 59:134–142.
Noonan, K. D., and Burger, M. M. (1973*c*): Induction of 3T3 cell division at the monolayer stage. *Experimental Cell Research,* 80:405–414.
Noonan, K. D., Levine, A. J., and Burger, M. M. (1973): Cell cycle-dependent changes in the surface membrane as detected with [^{3}H] concanavalin A. *Journal of Cell Biology,* 58:491–497.
Novales, R. R. (1971): On the role of cyclic AMP in the function of skin melanophores. *Annals of the New York Academy of Sciences,* 185:494–506.
Novales, R. R., and Davis, W. J. (1967): Melanin-dispersing effect of adenosine 3′,5′-monophosphate on amphibian melanophores. *Endocrinology,* 81:283–290.
Oey, J., Vogel, A., and Pollack, R. (1974): Intracellular cyclic AMP concentration responds specifically to growth regulation by serum. *Proceedings of the National Academy of Sciences of the U.S.A,* 71:694–698.
Orange, R. P., Kaliner, M. A., Laraia, P. J., and Austen, K. F. (1971): Immunological release of histamine and slow reacting substance of anaphylaxis from human lung. II. Influence of cellular levels of cyclic AMP. *Federation Proceedings,* 30:1725–1729.
Orci, L., Gabbay, K. H., and Malaisse, W. J. (1972): Pancreatic beta-cell web; its possible role in insulin secretion. *Science,* 175:1128–1130.
Orloff, J., and Handler, J. S. (1962): The similarity of effects of vasopressin, 3′,5′-AMP (cyclic AMP) and theophylline on the toad bladder. *Journal of Clinical Investigation,* 41:702–709.
Orloff, J., and Handler, J. S. (1967): The role of adenosine 3′,5′-phosphate in the action of antidiuretic hormone. *American Journal of Medicine,* 42:757–768.
Orr, T. S. C., Hall, D. E., and Allison, A. C. (1972): Role of contractile microfilaments in the release of histamine from mast cells. *Nature,* 236:350–351.
Orr, T. S. C., Hall, D. E., Gwilliam, J. M., and Cox, J. S. G. (1971): The effect of disodium cromoglycate on the release of histamine and degranulation of rat mast cells induced by compound 48/80. *Life Sciences,* 10(I):805–812.

Oschman, J. L., and Berridge, M. J. (1970): Structural and functional aspects of salivary fluid secretion in *Calliphora. Tissue and Cell,* 2:281–310.
Oschman, J. L., Wall, B. J., and Gupta, B. L. (1974): Cellular basis of water transport. *Symposia of the Society for Experimental Biology,* 28:305–350.
Otten, J., Johnson, G. S., and Pastan, I. (1971): Cyclic AMP levels in fibroblasts: Relationship to growth rate and contact inhibition of growth. *Biochemical and Biophysical Research Communications,* 44:1192–1198.
Otten, J., Johnson, G. S., and Pastan, I. (1972): Regulation of cell growth by cyclic adenosine 3′,5′-monophosphate. *Journal of Biological Chemistry,* 247:7082–7087.
Pace, C. S., and Price, S. (1974): Bioelectric effects of hexoses on pancreatic islet cells. *Endocrinology,* 94:142–147.
Pannbacker, R. G. (1973): Control of guanylate cyclase activity in the rod outer segment. *Science,* 182:1138–1140.
Pappano, A. J. (1970): Calcium-dependent action potentials produced by catecholamines in guinea pig atrial muscle fibers depolarized by potassium. *Circulation Research,* 27:379–390.
Park, C. R., Lewis, S. B., and Exton, J. H. (1972): Relationship of some hepatic actions of insulin to the intracellular level of cyclic adenylate. *Diabetes,* 21:439–446.
Parry, D. A. D., and Squire, J. M. (1973): Structural role of tropomyosin in muscle regulation: Analysis of the X-ray diffraction patterns from relaxed and contracting muscles. *Journal of Molecular Biology,* 75:33–55.
Peach, M. J. (1972): Stimulation of release of adrenal catecholamine by adenosine 3′,5′-cyclic monophosphate and theophylline in the absence of extracellular Ca^{2+}. *Proceedings of the National Academy of Sciences of the U.S.A.,* 69:834–836.
Peake, G. T., Steiner, A. L., and Daughaday, W. H. (1972): Guanosine 3′,5′-cyclic monophosphate is a potent pituitary growth hormone secretogogue. *Endocrinology,* 90:212–216.
Peaker, M. (1971): Intracellular concentrations of sodium, potassium and chloride in the salt-gland of the domestic goose and their relation to the secretory mechanism. *Journal of Physiology,* 213:399–410.
Perdue, J. F. (1971): The isolation and characterization of plasma membranes from cultured cells. III. The adenosine triphosphate-dependent accumulation of Ca^{2+} by chick embryo fibroblasts. *Journal of Biological Chemistry,* 246:6750–6759.
Perkins, J. P. (1973): Adenyl cyclase. *Advances in Cyclic Nucleotide Research,* 3:1–64.
Péron, F. G., Guerra, F., and McCarthy, J. L. (1965): Further studies on the effect of calcium ions and corticosteroidogenesis. *Biochimica et Biophysica Acta,* 110:277–289.
Petersen, M. J., and Edelman, I. S. (1964): Calcium inhibition of the action of vasopressin on the urinary bladder of the toad. *Journal of Clinical Investigation,* 43:583–594.
Petersen, O. H. (1970a): The dependence of the transmembrane salivary secretory potential on the external potassium and sodium concentration. *Journal of Physiology,* 210:205–215.
Petersen, O. H. (1970b): The ionic transports involved in the acetylcholine-induced change in membrane potential in acinar cells from salivary glands and their importance in the salivary secretion process. In: *Electrophysiology of Epithelial Cells,* edited by G. Giebisch, pp. 207–221. Schattauer Verlag, Stuttgart.
Petersen, O. H. (1970c): Some factors influencing stimulation-induced release of potassium from the cat submandibular gland to fluid perfused through the gland. *Journal of Physiology,* 208:431–447.
Petersen, O. H. (1973a): Mechanism of action of pancreozymin and acetylcholine on pancreatic acinar cells. *Nature, New Biology,* 244:73.
Petersen, O. H. (1973b): Membrane potential measurement in mouse salivary gland cells. *Experientia,* 29:160–161.
Petersen, O. H., and Poulsen, J. H. (1969): Secretory transmembrane potentials and electrolyte transients in salivary glands. In: *Exocrine Glands,* edited by S. Y. Botelho, F. P. Brooks, and W. B. Shelley, pp. 3–20. University of Pennsylvania Press, Philadelphia.
Poisner, A. M. (1973a): Direct stimulant effect of aminophylline on catecholamine release from the adrenal medulla. *Biochemical Pharmacology,* 22:469–476.
Poisner, A. M. (1973b): Caffeine-induced catecholamine secretion: Similarity to caffeine-induced muscle contraction. *Proceedings of the Society for Experimental Biology and Medicine,* 142:103–105.
Portzehl, H., Caldwell, P. C., and Ruegg, J. C. (1964): The dependence of contraction and relaxation of muscle fibres from the crab *Maia squinidado* on the internal concentration of free calcium ions. *Biochimica et Biophysica Acta,* 79:581–591.
Poulsen, J. H. (1973): An attempt to elicit salivary secretion by changing the intracellular sodium

and potassium concentrations without applying neurotransmitters. *Acta Physiologica Scandinavica,* 89:51A.

Prince, W. T., and Berridge, M. J. (1972): The effects of 5-hydroxytryptamine and cyclic AMP on the potential profile across isolated salivary glands. *Journal of Experimental Biology,* 56:323–333.

Prince, W. T., and Berridge, M. J. (1973): The role of calcium in the action of 5-hydroxytryptamine and cyclic AMP on salivary glands. *Journal of Experimental Biology,* 58:367–384.

Prince, W. T., Berridge, M. J., and Rasmussen, H. (1972): Role of calcium and adenosine-3',5'-cyclic monophosphate in controlling fly salivary gland secretion. *Proceedings of the National Academy of Sciences of the U.S.A.,* 69:553–557.

Prince, W. T., Rasmussen, H., and Berridge, M. J. (1973): The role of calcium in fly salivary gland secretion analyzed with the ionophore A 23187. *Biochimica et Biophysica Acta,* 329:98–107.

Quastel, M. R., and Kaplan, J. G. (1970): Early stimulation of potassium uptake in lymphocytes treated with PHA. *Experimental Cell Research,* 63:230–233.

Rahwan, R. G., and Borowitz, J. L. (1973): Mechanisms of stimulus-secretion coupling in adrenal medulla. *Journal of Pharmaceutical Sciences,* 62:1911–1923.

Rahwan, R. G., Borowitz, J. L., and Miya, T. S. (1973): The role of intracellular calcium in catecholamine secretion from the bovine adrenal medulla. *Journal of Pharmacology and Experimental Therapeutics,* 184:106–118.

Rasmussen, H. (1970): Cell communication, calcium ion, and cyclic adenosine monophosphate. *Science,* 170:404–412.

Rasmussen, H. (1971): Ionic and hormonal control of calcium homeostasis. *American Journal of Medicine,* 50:567–588.

Rasmussen, H., Kurokawa, K., Mason, J., and Goodman, D. B. P. (1971): Cyclic AMP, calcium and cell activation. In: *Calcium, Parathyroid Hormone and the Calcitonins,* pp. 492–501. International Congress Series No. 243. Excerpta Medica, Amsterdam.

Rasmussen, H., and Nagata, N. (1970): Hormones, cell calcium and cyclic AMP. In: *Calcium and Cellular Function,* edited by A. W. Cuthbert, pp. 198–213. Macmillan, London.

Rasmussen, H., and Tenenhouse, A. (1968): Cyclic adenosine monophosphate, Ca^{++}, and membranes. *Proceedings of the National Academy of Sciences of the U.S.A.,* 59:1364–1370.

Rebhun, L. I. (1972): Polarized intracellular particle transport: Saltatory movements and cytoplasmic streaming. *International Review of Cytology,* 32:93–137.

Reed, P. W., and Lardy, H. A. (1972): A 23187: A divalent cation ionophore. *Journal of Biological Chemistry,* 247:6970–6977.

Reuter, H., and Seitz, H. (1968): The dependence of calcium efflux from cardiac muscle on temperature and external ion composition. *Journal of Physiology,* 195:451–470.

Ridderstap, A. S., and Bonting, S. L. (1969): Cyclic AMP and enzyme secretion by the isolated rabbit pancreas. *Pflügers Archiv,* 313:62–70.

Robberecht, P., and Christophe, J. (1971): Secretion of hydrolases by perfused fragments of rat pancreas: Effect of calcium. *American Journal of Physiology,* 220:911–917.

Robison, G. A., Butcher, R. W., Øye, I., Morgan, M. E., and Sutherland, E. W. (1965): The effects of epinephrine on adenosine 3',5'-phosphate levels in the isolated perfused rat heart. *Molecular Pharmacology,* 1:168–177.

Robison, G. A., Butcher, R. W., and Sutherland, E. W. (1967): Adenyl cyclase as an adrenergic receptor. *Annals of the New York Academy of Sciences,* 139:703–723.

Robison, G. A., Butcher, R. W., and Sutherland, E. W. (1971): *Cyclic AMP.* Academic Press, New York and London.

Röhlich, P., Anderson, P., and Uvnäs, B. (1971): Electron microscope observations on compound 48/80-induced degranulation in rat mast cells. *Journal of Cell Biology,* 51:465–483.

Rohr, H. P., Wirz, A., Henning, L. C., Riede, U. N., and Bianchi, L. (1971): Morphometric analysis of the rat liver cells in the perinatal period. *Laboratory Investigation,* 24:128–139.

Romero, P. J., and Whittam, R. (1971): The control by internal calcium of membrane permeability to sodium and potassium. *Journal of Physiology,* 214:481–507.

Rosenblith, J. Z., Ukena, T. E., Yin, H. H., Berlin, R. D., and Karnovsky, M. J. (1973): A comparative evaluation of the distribution of concanavalin A-binding sites on the surfaces of normal, virally-transformed, and protease-treated fibroblasts. *Proceedings of the National Academy of Sciences of the U.S.A.,* 70:1625–1629.

Roy, A. C., and Warren, B. T. (1974): Inhibition of cAMP phosphodiesterase by disodium cromoglycate. *Biochemical Pharmacology,* 23:917–920.

Roy, C., Bockaert, J., Rajerison, R., and Jard, S. (1973): Oxytocin receptor in frog bladder epithelial cells. Relationship of [^{3}H] oxytocin binding to adenylate cyclase activation. *FEBS Letters,* 30: 329–334.

Rubin, R. P. (1970): The role of calcium in the release of neurotransmitter substances and hormones. *Pharmacological Reviews,* 22:389–428.

Rubin, R. P., Carchman, R. A., and Jaanus, S. D. (1972a): Role of cyclic 3′,5′-adenosine monophosphate on corticosteroid synthesis and release from the intact adrenal gland. *Biochemical and Biophysical Research Communications,* 47:1492–1497.

Rubin, R. P., Carchman, R. A., and Jaanus, S. D. (1972b): Role of calcium and adenosine cyclic 3′,5′-phosphate in action of adrenocorticotropin. *Nature, New Biology,* 240:150–152.

Rubin, R. P., Feinstein, M. B., Jaanus, S. D., and Paimre, M. (1967): Inhibition of catecholamine secretion and calcium exchange in perfused cat adrenal glands by tetracaine and magnesium. *Journal of Pharmacology and Experimental Therapeutics,* 155:463–471.

Rudland, P. S., Seeley, M., and Seifert, W. (1974): Cyclic GMP and cyclic AMP levels in normal and transformed fibroblasts. *Nature,* 251:417–419.

Russell, J. T., Hansen, E. L., and Thorn, N. A. (1974): Calcium and stimulus secretion coupling in the neurohypophysis. III. Calcium ionophore (A 23187)-induced release of vasopressin from isolated rat neurohypophyses. *Acta Endocrinologica,* 77:443–450.

Rutten, W. J., De Pont, J. J. H. H. M., and Bonting, S. L. (1972): Adenylate cyclase in the rat pancreas properties and stimulation by hormones. *Biochimica et Biophysica Acta,* 274:201–213.

Ryan, W. L., and Heidrick, M. L. (1968): Inhibition of cell growth by adenosine 3′,5′monophosphate. *Science,* 162:1484–1485.

Ryan, W. L. and Heidrick, M. L. (1974): Role of cyclic nucleotides in cancer. *Advances in Cyclic Nucleotide Research,* 4:81–116.

Sakai, A. (1970): Humoral factor triggering DNA synthesis after partial hepatectomy in the rat. *Nature,* 228:1186–1187.

Salomon, Y., and Schramm, M. (1970): A specific binding site for 3′,5′-cyclic AMP in rat parotid microsomes. *Biochemical and Biophysical Research Communications,* 38:106–111.

Salzman, E. W., Kensler, P. C., and Levine, L. (1972): Cyclic 3′,5′-adenosine monophosphate in human blood platelets. IV. Regulatory role of cyclic AMP in platelet function. *Annals of the New York Academy of Sciences,* 201:61–71.

Salzman, E. W., and Levine, L. (1971): Cyclic 3′,5′-adenosine monophosphate in human blood platelets. II. Effect of N^6-2′-O-dibutyryl cyclic 3′,5′-adenosine monophosphate on platelet function. *Journal of Clinical Investigation,* 50:131–141.

Samli, M. H., and Geschwind, I. I. (1968): Some effects of energy-transfer inhibitors and of Ca^{++}-free or K$^+$-enhanced media on the release of luteinizing hormone (LH) from the rat pituitary gland in vitro. *Endocrinology,* 82:225–231.

Sandow, A. (1970): Skeletal muscle. *Annual Review of Physiology,* 32:87–138.

Sayers, G., Beall, R. J., and Seelig, S. (1972): Isolated adrenal cells: Adrenocorticotropic hormone, calcium, steroidogenesis, and cyclic adenosine monophosphate. *Science,* 175:1131–1133.

Scarpa, A., and Inesi, G. (1972): Ionophore mediated equilibration of calcium ion gradients in fragmented sarcoplasmic reticulum. *FEBS Letters,* 22:273–276.

Schatzmann, H. J., and Vincenzi, F. F. (1969): Calcium movements across the membrane of human red cells. *Journal of Physiology,* 201: 369–395.

Schneider, M. F., and Chandler, W. K. (1973): Voltage dependent charge movement in skeletal muscle: A possible step in excitation-contraction coupling. *Nature,* 242:244–246.

Schneyer, C. A. (1974): Autonomic regulation of secretory activity and growth responses of rat parotid gland. In: *Secretory Mechanisms of Exocrine Glands,* Alfred Benzon Symposium VII, edited by N. A. Thorn and O. H. Petersen, pp. 42–55. Munksgaard, Copenhagen.

Schneyer, L. H., Young, J. A., and Schneyer, C. A. (1972): Salivary secretion of electrolytes. *Physiological Reviews,* 52:720–777.

Schofield, J. G. (1967): Role of cyclic 3′,5′-adenosine monosphosphate in the release of growth hormone in vitro. *Nature,* 215:1382–1383.

Schramm, M., Selinger, Z., Salomon, Y., Eytan, E., and Batzri, S. (1972): Pseudopodia formation by secretory granules. *Nature, New Biology,* 240:203–205.

Schroeder, T. E. (1973): Actin in dividing cells: Contractile ring filaments bind heavy meromyosin. *Proceedings of the National Academy of Sciences of the U.S.A.,* 70:1688–1692.

Schultz, G., Hardman, J. G., Schultz, K., Baird, C. E., and Sutherland, E. W. (1973): The importance of calcium ions for the regulation of guanosine 3',5'-cyclic monophosphate levels. *Proceedings of the National Academy of Sciences of the U.S.A.,* 70:3889–3893.

Schultz, G., Hardman, J. G., and Sutherland, E. W. (1974): Cyclic nucleotides and smooth muscle function. In: *Asthma; Pathology, Immunopharmacology and Treatment,* edited by K. F. Austen and L. M. Lichtenstein, pp. 123–138. Academic Press, New York and London.

Seifert, W. E. and Rudland, P. S. (1974): Possible involvement of cyclic AMP in growth control of cultured mouse cells. *Nature,* 248:138–140.

Selinger, Z., Batzri, S., Eimerl, S., and Schramm, M. (1973): Calcium and energy requirements for K^+ release mediated by the epinephrine α-receptor in rat parotid slices. *Journal of Biological Chemistry,* 248:369–372.

Selinger, Z., Eimerl, S., and Schramm, M. (1974): A calcium ionophore simulating the action of epinephrine on the α-adrenergic receptor. *Proceedings of the National Academy of Sciences of the U.S.A.,* 71:128–131.

Selinger, Z., and Naim, E. (1970): The effect of calcium on amylase secretion by rat parotid slices. *Biochimica et Biophysica Acta,* 203:335–337.

Selinger, Z., Naim, E., and Lasser, M. (1970): ATP-dependent calcium uptake by microsomal preparations from rat parotid and submaxillary glands. *Biochimica et Biophysica Acta,* 203:326–334.

Selye, H., Veilleux, R., and Cantin, M. (1961): Excessive stimulation of salivary gland growth by isoproterenol. *Science,* 133:44–45.

Sheppard, J. R. (1972): Difference in the cyclic adenosine 3',5'-monophosphate levels in normal and transformed cells. *Nature, New Biology,* 236:14–16.

Sheppard, J. R., and Prescott, R. R. (1972): Cyclic AMP levels in synchronized mammalian cells. *Experimental Cell Research,* 75:293–296.

Short, J., Brown, R. F., Husakova, A., Gilbertson, J. R., Zemel, R., and Lieberman, I. (1972): Induction of deoxyribonucleic acid synthesis in the liver of the intact animal. *Journal of Biological Chemistry,* 247:1757–1766.

Siggins, G. R., Battenberg, E. F., Hoffer, B. J., Bloom, F. E., and Steiner, A. L. (1973): Noradrenergic stimulation of cyclic adenosine monophosphate in rat Purkinje neurons: An immunocytochemical study. *Science,* 179:585–588.

Siggins, G. R., Hoffer, B. J., and Bloom, F. E. (1971a): Studies on norepinephrine-containing afferents to Purkinje cells of rat cerebellum. III. Evidence for mediation of norepinephrine effects by cyclic 3',5'-adenosine monophosphate. *Brain Research,* 25:535–553.

Siggins, G. R., Oliver, A. P., Hoffer, B. J., and Bloom, F. E. (1971b): Cyclic adenosine monophosphate and norepinephrine: Effects on transmembrane properties of cerebellar Purkinje cells. *Science,* 171:192–194.

Simpson, L. L. (1968): The role of calcium in neurohumoral and neurohormonal extrusion processes. *Journal of Pharmacy and Pharmacology,* 20:889–910.

Singer, J. J., and Goldberg, A. L. (1970): Cyclic AMP and transmission at the neuromuscular junction. In: *Role of Cyclic AMP in Cell Function, Advances in Biochemical Psychopharmacology, Vol. 3,* edited by P. Greengard and E. Costa. Raven Press, New York.

Smets, L. A. (1972): Contact inhibition of transformed cells incompletely restored by dibutyryl cyclic AMP. *Nature, New Biology,* 239:123–124.

Smith, J. W., Steiner, A. L., Newberry, W. M., and Parker, C. W. (1971a): Cyclic adenosine 3',5'-monophosphate in human lymphocytes. Alterations after phytohemagglutinin stimulation. *Journal of Clinical Investigation,* 50:432–441.

Smith, J. W., Steiner, A. L., and Parker, C. W. (1971b): Human lymphocyte metabolism. Effect of cyclic and non-cyclic nucleotides on stimulation by phytohemagglutinin. *Journal of Clinical Investigation,* 50:442–448.

Snart, R. S., and Dalton, T. (1972): Thermal activation of the cyclic AMP stimulated sodium transport across isolated toad bladder and Na^+/K^+ ATPase in toad bladder homogenates. *Experientia,* 28:1028–1029.

Sneddon, J. M. (1972): Divalent cations and the blood platelet release reaction. *Nature, New Biology,* 236:103–104.

Sneddon, J. M., and Williams, K. I. (1973): Effect of cations on the blood platelet release reaction. *Journal of Physiology,* 235:625–637.

Spencer, T., and Bygrave, F. L. (1972): Modification by calcium ions of adenine nucleotide translocation in rat liver mitochondria. *Biochemical Journal,* 129:355–365.

Statland, B. E., Heagan, B. M., and White, J. G. (1969): Uptake of calcium by platelet relaxing factor. *Nature,* 223:521–522.

Steinhardt, R. A., and Epel, D. (1974): Activation of sea-urchin eggs by a calcium ionophore. *Proceedings of the National Academy of Sciences of the U.S.A.,* 71:1915–1919.

Steinhardt, R. A., Lundin, L., and Mazia, D. (1971): Bioelectric responses of the echinoderm egg to fertilization. *Proceedings of the National Academy of Sciences of the U.S.A.,* 68:2426–2430.

Stormorken, H. (1969): The release reaction of secretion. *Scandanavian Journal of Haematology,* Suppl. 9:1–24.

Streeto, J. M. (1969): Renal cortical adenyl cyclase: Effect of parathyroid hormone and calcium. *Metabolism,* 18:968–973.

Sussman, K. E., and Vaughan, G. D. (1967): Insulin release after ACTH, glucagon and adenosine-3',5'-phosphate (cyclic AMP) in the perfused isolated rat pancreas. *Diabetes,* 16:449–454.

Tada, M., Kirchberger, M. A., Iorio, J.-A. M., and Katz, A. M. (1974a): Effects of a cardiac adenosine 3',5'-monophosphate-dependent protein kinase on the cardiac sarcoplasmic reticulum. III. Role of a 22,000 dalton protein component. *Journal of Biological Chemistry (in press).*

Tada, M., Kirchberger, M. A., Repke, D. I., and Katz, A. M. (1974b): The stimulation of calcium transport in cardiac sarcoplasmic reticulum by adenosine 3',5'-monophosphate-dependent protein kinase. *Journal of Biological Chemistry,* 249:6174–6180.

Teo, T. S., and Wang, J. H. (1973): Mechanism of activation of a cyclic adenosine 3',5'-monophosphate phosphodiesterase from bovine heart by calcium ions. *Journal of Biological Chemistry,* 248:5950–5955.

Thomas, E. W., Murad, F., Looney, W. B., and Morris, H. P. (1973): Adenosine 3',5'-monophosphate and guanosine 3',5'-monophosphate: Concentrations in Morris hepatomas of different growth rates. *Biochimica et Biophysica Acta,* 297:564–567.

Thorn, N. A. (1974): Role of calcium in secretory processes. In: *Secretory Mechanisms of Exocrine Glands,* Alfred Benzon Symposium VII, edited by N. A. Thorn, and O. H. Petersen, pp. 305–326. Munksgaard, Copenhagen.

Thorn, N. A., and Schwartz, I. L. (1965): Effect of antidiuretic hormone on washout curves of radiocalcium from isolated toad bladder tissue. *General and Comparative Endocrinology,* 5:710.

Tomita, T. (1970a): Electrical activity of vertebrate photoreceptors. *Quarterly Review of Biophysics,* 3:179–222.

Tomita, T. (1970b): Electrical properties of mammalian smooth muscle. In: *Smooth Muscle,* edited by E. Bülbring, A. F. Brading, A. W. Jones, and T. Tomita, pp. 197–243. Edward Arnold, London.

Tritthart, H., Volkmann, R., Weiss, R., and Fleckstein, A. (1973): Calcium-mediated action potentials in mammalian myocardium. *Naunyn Schmiedeberg's Archives of Pharmacology,* 280:239–252.

Turtle, J. R., and Kipnis, D. M. (1967): An adrenergic receptor mechanism for the control of cyclic 3',5'-adenosine monophosphate synthesis in tissues. *Biochemical and Biophysical Research Communications,* 28:797–802.

Vale, W., Burgus, R., and Guillemin, R. (1967): Presence of calcium ions as a requisite for the *in vitro* stimulation of TSH-release by hypothalamic TRF. *Experientia,* 23:853–855.

Vale, W., and Guillemin, R. (1967): Potassium-induced stimulation of thyrotropin release *in vitro.* Requirement for presence of calcium and inhibition by thyroxine. *Experientia,* 23:855–857.

Van de Veerdonk, F. C. G., and Konijn, Th. M. (1970): The role of adenosine 3',5'-cyclic monophosphate and catecholamines in the pigment migration process in *Xenopus laevis. Acta Endocrinologica,* 64:364–376.

Vaughan, M. (1972): The role of insulin in regulation of cyclic AMP metabolism. In: *Insulin Action,* edited by I. B. Fritz. Academic Press, New York.

Verity, M. A., Travis, G., and Brown, W. J. (1972): Glycogen mobilization after partial hepatectomy. *Laboratory Investigation,* 27:108–114.

Vesely, D. L., and Hadley, M. E. (1971): Calcium requirement for melanophore-stimulating hormone action on melanophores. *Science,* 173:923–925.

Vigdahl, R. L., Marquis, N. R., and Tavormina, P. A. (1969): Platelet aggregation. II. Adenyl cyclase, prostaglandin E_1, and calcium. *Biochemical and Biophysical Research Communications,* 37:409–415.

Vlodavsky, I., Inbar, M., and Sachs, L. (1973): Membrane changes and adenosine triphosphate content in normal and malignant transformed cells. *Proceedings of the National Academy of Sciences of the U.S.A.,* 70:1780–1784.

Voorhees, J. J., Duell, E. A., Stawiski, M., and Harrell, E. R. (1974): Cyclic nucleotide metabolism in normal and proliferating epidermis. *Advances in Cyclic Nucleotide Research,* 4:117–162.

Wald, G., Brown, P. K., and Gibbons, I. R. (1963): The problem of visual excitation. *Journal of the Optical Society of America,* 53:20–35.

Walsh, D. A., and Ashby, C. D. (1973): Protein kinases: Aspects of their regulation and diversity. *Recent Progress in Hormone Research,* 29:329–353.

Warren, G. B., Toon, P. A., Birdsall, N. J. M., Lee, A. G., and Metcalfe, J. C. (1974): Reconstitution of a calcium pump using defined membrane components. *Proceedings of the National Academy of Sciences of the U.S.A.,* 71:622–626.

Warren, L., Fuhrer, J. P., and Buck, C. A. (1972): Surface glycoproteins of normal and transformed cells: A difference determined by sialic acid and a growth-dependent sialyl transferase. *Proceedings of the National Academy of Sciences of the U.S.A.,* 69:1838–1842.

Webb, D. R., Stites, D. P., Perlman, J. D., Luong, D., and Fudenberg, H. H. (1973): Lymphocyte activation: The dualistic effect of cAMP. *Biochemical and Biophysical Research Communications,* 53:1002–1008.

Weinstein, Y., Chambers, D. A., Bourne, H. R., and Melmon, K. L. (1974): Cyclic GMP stimulates lymphocyte nucleic acid synthesis. *Nature,* 251:352–353.

White, J. G. (1968): Fine structural alterations induced in platelets by adenosine diphosphate. *Blood,* 31:604–622.

Whitney, R. B., and Sutherland, R. M. (1972a): Requirement for calcium ions in lymphocyte transformation stimulated by phytohemagglutinin. *Journal of Cellular Physiology,* 80:329–338.

Whitney, R. B., and Sutherland, R. M. (1972b): Enhanced uptake of calcium by transforming lymphocytes. *Cellular Immunology,* 5:137–147.

Wilber, J. F., Peake, G. T., and Utiger, R. D. (1969): Thyrotropin release *in vitro:* Stimulation by cyclic 3′,5′-adenosine monophosphate. *Endocrinology,* 84:758–760.

Williams, J. A., and Lee, M. (1974): Pancreatic acinar cells: Use of a Ca^{2+} ionophore to separate enzyme release from the earlier steps in stimulus-secretion coupling. *Biochemical and Biophysical Research Communication,* 60:542–548.

Williams, T. F., Exton, J. H., Friedmann, N., and Park, C. R. (1971): Effects of insulin and adenosine 3′,5′-monophosphate on K^+ flux and glucose output in perfused rat liver. *American Journal of Physiology,* 221:1645–1651.

Wilson, D. F. (1974): The effects of dibutyryl cyclic adenosine 3′,5′-monophosphate, theophylline and aminophylline on neuromuscular transmission in the rat. *Journal of Pharmacology and Experimental Therapeutics,* 188:447–452.

Wong, P. Y. D., Bedwani, J. R., and Cuthbert, A. W. (1972): Hormone action and the levels of cyclic AMP and prostaglandins in the toad bladder. *Nature, New Biology,* 238:27–31.

Woods, H. N., Lin, M. C., and Braun, A. C. (1972): The inhibition of plant and animal adenosine 3′,5′-cyclic monophosphate phosphodiesterase by a cell-division-promoting substance from tissues of higher plant species. *Proceedings of the National Academy of Sciences of the U.S.A.,* 69:403–406.

Wright, M. R., and Lerner, A. B. (1960): On the movement of pigment granules in frog melanocytes. *Endocrinology,* 66:599–609.

Wulff, V. J. (1971): The effect of cyclic AMP on *Limulus* lateral eye retinular cells. *Vision Research,* 11:1493–1495.

Yang, Y., and Perdue, J. F. (1972): Contractile proteins of cultured cells. I. The isolation and characterization of an actin-like protein from cultured chick embryo fibroblasts. *Journal of Biological Chemistry,* 247:4503–4509.

Yuen, M., and Macey, R. (1974): Platelet aggregation induced by a calcium ionophore. *Federation Proceedings,* 33:269Abs.

Zor, U., Kaneko, T., Schneider, H. P. G., McCann, S. M., and Field, J. B. (1970): Further studies of stimulation of anterior pituitary cyclic adenosine 3′,5′-monophosphate formation by hypothalamic extract and prostaglandins. *Journal of Biological Chemistry,* 245:2883–2888.

Zor, U., Kaneko, T., Schneider, H. P. G., McCann, S. M., Lowe, I. P., Bloom, G., Borland, B., and Field, J. B. (1969): Stimulation of anterior pituitary adenyl cyclase activity and adenosine 3′,5′-cyclic phosphate by hypothalmic extract and prostaglandin E_1. *Proceedings of the National Academy of Sciences of the U.S.A.,* 63:918–925.

Zucker, M. B., Troll, W., and Belman, S. (1974): The tumor-promoting phorbol ester (12-O-tetradecanoylphorbol-13-acetate), a potent aggregating agent for blood platelets. *Journal of Cell Biology,* 60:325–336.

Cyclic AMP and Adrenocortical Function

Ian D. K. Halkerston

Department of Biochemistry, University of Massachusetts Medical School, 55 Lake Avenue North, Worcester, Massachusetts 01605

CONTENTS

I.	Introduction	100
II.	Hormonal Activation of Adrenocortical Cells	102
	A. The Initiating Hormone-Receptor Interaction	102
	1. ACTH Receptors in the Adrenal Cortex	102
	2. Cholera Enterotoxin	104
	3. Angiotensin Receptors	104
	4. Prostaglandin Receptors	106
	B. Adrenocortical Adenylyl Cyclase	106
	C. Calcium, Cyclic AMP, and Adrenal Steroidogenesis . . .	108
	D. Prostaglandins, Cyclic AMP, and Adrenal Steroidogenesis .	111
III.	Hormonal Stimulation of Adrenal Cortex and Cyclic Nucleotide Levels	113
IV.	Adrenal Steroidogenesis and Inhibitors of Cyclic Nucleotide Phosphodiesterase	116
V.	Effect of Exogenous Cyclic Nucleotides on Adrenal Steroidogenesis	117
	A. Cyclic AMP and Dibutyryl Cyclic AMP	117
	B. On the Specificity of Cyclic AMP Stimulation of Adrenal Steroidogenesis	118
VI.	On the Mechanism of Action of Cyclic AMP in the Adrenal Cortex	120
VII.	Concluding Remarks	126
VIII.	Acknowledgments	126
IX.	References	126

I. INTRODUCTION

The cortical region of the adrenal gland atrophies following hypophysectomy, although the outermost cells (zona glomerulosa) do not. Pituitary adrenocorticotropin (ACTH) maintains the cortical cell population and in addition acutely regulates the secretory function of the cells of the zona fasiculata-reticularis. This ability of ACTH to rapidly increase steroid synthesis and output is gradually lost after removal of the pituitary, so that in the rat by 6 days postoperation a single injection of ACTH is no longer capable of inducing an increased rate of steroid synthesis. Thus the interaction of ACTH with adrenocortical tissue can result in both adrenal hyperplasia and enhanced glucocorticoid secretion, with both of these responses dependent on "chronic" ACTH stimulation. In addition, both responses can be elicited in the hypophysectomized animal by a naturally occurring peptide of 39 residues chain length and by a synthetic peptide comprised of the first 24 residues of natural ACTH ($ACTH_{1-24}$).

The involvement of adenosine 3',5'-monophosphate (cyclic AMP) in the sequence of events triggered by ACTH in the adrenal cortex was discovered by Haynes and his associates (Haynes and Berthet, 1957; Haynes, 1958; Haynes, Koritz and Péron, 1959; Haynes, Sutherland, and Rall, 1960). These early studies closely paralleled those being carried out by Sutherland and his group on the action of epinephrine on liver glycogenolysis (Sutherland and Rall, 1960; Rall and Sutherland, 1961), and formed part of the experimental data upon which the second messenger hypothesis of cyclic AMP involvement in hormone action was based (Sutherland, Oye, and Butcher, 1965; Robison, Butcher, and Sutherland, 1971). This concept of the regulation of adrenocortical function envisages an interaction between circulating ACTH and a receptor situated on the outer surface of the adrenal cell plasma membrane. The interaction leads to an activation of adenylyl cyclase (situated on the inner surface of the membrane), which catalyzes the formation of cyclic AMP and pyrophosphate from ATP. The increased intracellular levels of cyclic AMP are believed to modulate the activity of a variety of systems involved in adrenocortical cell function (biosynthesis and secretion of steroid hormones) and in hypertrophy and hyperplasia of the adrenal cortex when the need arises. This is the second messenger role of the cyclic nucleotide, i.e., the translation of the extrinsic message into the cellular response. Destruction of the intracellular cyclic AMP can be accomplished by the action of 3',5'-cyclic nucleotide phosphodiesterase, which converts cyclic AMP to 5'-AMP (Butcher and Sutherland, 1962).

A great deal of experimental data has been obtained which "fits" within this skeletal picture of the mechanism of action of a wide variety of humoral agents, including most polypeptide hormones, secretagogues, biogenic amines, and in certain cases prostaglandins (Robison et al., 1971). In most cases of hormonal action the picture is far from complete; in particular, there is no certain knowledge of the mode of transmission of message between excited receptor and the catalytic site of adenylyl cyclase. Furthermore, the mechanism(s) involved in the induction of the cellular response to hormone by increased cellular levels of cyclic AMP, while known in some detail for some tissues, is quite obscure in others. At the present time the adrenal cortex definitely belongs to the second category.

Studies on the role of cyclic AMP in hormone action have been materially aided by a set of guidelines laid down by Sutherland's group (Robison et al., 1971), which may be quoted directly here:

(1) The hormone should be capable of stimulating adenylyl cyclase in broken cell preparations from the appropriate cells, while hormones which do not produce the response should not stimulate adenylyl cyclase.

(2) The hormone should be capable of increasing the intracellular level of cyclic AMP in intact cells, while inactive hormones should not increase cyclic AMP levels. It should be demonstrated that the effect on the level of cyclic AMP occurs at dose levels of the hormone which are at least as small as the smallest levels which are capable of producing a physiological response. The increase in the level of cyclic AMP should precede or at least not follow the physiological response.

(3) It should be possible to potentiate the hormone (i.e., increase the magnitude of the physiological response) by administering the hormone together with theophylline or other phosphodiesterase inhibitors. The hormone and the phosphodiesterase inhibitor should act synergistically.

(4) It should be possible to mimic the physiological effect of the hormone by the addition of exogenous cyclic AMP.

In the case of ACTH action in the adrenal cortex most of the above criteria have been met by experimental data, although recent studies have focused upon apparent discrepancies in the induction of increased cellular cyclic AMP levels and the induction of steroidogenesis. As well as their usefulness in guiding past experimentation in this area, the four criteria also form a useful framework upon which a discussion of the role of cyclic AMP may be constructed. It is not the reviewer's intent to provide a detailed historical account of the development of the second messenger hypothesis of cyclic AMP in the adrenal cortex, as most of the early work has been covered by a number of excellent reviews, treated either from the point of view of the mechanism of action of ACTH (Kowal, 1970a; Garren, Gill, Masui, and Walton, 1971a; Ferguson, 1972; Mulrow, 1972; Gill, 1972) or of cyclic nucleotide involvement in cell regulation (Robison et al., 1971; Hardman, Robison, and Sutherland, 1971; Jost and Rickenberg, 1971; Bitensky and Gorman, 1972; Wicks, 1974). Rather, it appeared to the reviewer to be more useful to emphasize those aspects of recent research not' covered by earlier reviews, and to dwell particularly upon those areas where, according to some investigators, some modification of the second messenger hypothesis of cyclic AMP involvement in ACTH action may be required.

II. HORMONAL ACTIVATION OF ADRENOCORTICAL CELLS

A. The Initiating Hormone-Receptor Interaction

1. ACTH Receptors in the Adrenal Cortex

Brief exposure of adrenal tissue to ACTH leads to a persistent effect in terms of increased steroid production (Birmingham, Kurlents, Lane, Muhlstock, and Traikov, 1960; Taunton, Roth, and Pastan, 1967). With an exposure of less than 5 min, subsequent treatment with trypsin obliterates the effect, but the activation becomes progressively less reversible with increasing exposure time. Anti-ACTH antibody acts in a manner similar to trypsin (Taunton et al., 1967).

Several groups of investigators have presented evidence that ACTH does not have to enter the cell in order to stimulate steroidogenesis. Schimmer, Ueda, and Sato (1968) showed that $ACTH_{1-20}$ diazotized to p-aminobenzolyl cellulose could stimulate steroidogenesis in cultured cells derived from a mouse adrenal tumor (Buonassisi, Sato, and Cohen, 1962), although the insoluble fibers were visible in the microscope and appeared to be too large to enter the cell. Further, the choice of the eicosapeptide reduced the possibility that stimulation of the cells was due to peptides cleaved from the complex, as any chains shorter than 20 amino acids in length would have greatly reduced biological activity. Richardson and Schulster (1972) pointed out that cleavage of the diazo-linkage had not been ruled out in the studies of Schimmer et al. (1968), and in their own investigations showed that $ACTH_{1-24}$ diazotized to beads of polyacrylamide stimulated corticosterone production in suspensions of isolated rat adrenal cells. They provided evidence that the ACTH-polyacrylamide complex was not releasing free biologically active peptides by examination of supernatants (from incubations of adrenal cells with the complex) for solubilized ACTH activity. Similar studies were carried out by Selinger and Civen (1971), who used agarose beads as the insoluble carrier. The beads had a diameter about three times that of the adrenal cells, but were effective in stimulating corticosterone production in suspensions of isolated rat adrenal cells. The authors made the interesting observation that the beads and adrenal cells did not appear to become specifically adherent.

Studies of the interaction of ACTH with isolated adrenal "receptors" were initiated by Lefkowitz, Roth, and Pastan (1970b), who investigated the binding of biologically active monoiodo ^{125}I-ACTH to a subcellular fraction of an ACTH-responsive mouse adrenal tumor. The adrenal extract contained small fragments of membrane which were nevertheless included in agarose gels (Bio A 15M). Extracts that contained an ACTH-sensitive adenylyl cyclase activity were capable of binding ^{125}I-ACTH; extracts that lacked this activity did not bind the labeled hormone. Unlabeled ACTH inhibited the binding of ^{125}I-ACTH, and ACTH derivatives inhibited the binding of the labeled hormone in direct proportion to their biological activity. Further study of the interaction of ACTH and the adrenal fraction by this group suggested that there might be two distinct orders of ACTH-binding sites: a relatively small number of "high-order" binding sites with an association constant, K, of 9×10^{11} M^{-1} and a much larger number of "low-order" sites with a K of 3×10^7 M^{-1} (Lefkowitz, Roth, and Pastan, 1971).

The binding of synthetic (^{14}C-Phe) (Gln^5) β-corticotropin$_{1-20}$ amide to a particulate preparation from bovine adrenal cortex has been examined and a correlation made of adrenocorticotropic activity of ACTH analogues with degree of binding to the particulate preparation (Hofmann, Wingender, and Finn, 1970b). The preparation was extensively studied in terms of its original location in the cell, with the conclusion that the binding and adenylyl cyclase activities were associated with the plasma membrane (Finn, Widnell, and Hofmann, 1972). Using a series of synthetic homogeneous nonradioactive analogues or fragments of β-corticotropin amide to displace (^{14}C-Phe) (Gln^5) β-corticotropin$_{1-20}$ amide from the particulate fraction, a significant correlation between binding and *in vivo* adrenocorticotropic activity was established (Hofmann et al., 1970b). Correlations of hormone binding and biological activity have been made on the basis of adenylyl cyclase activation in the binding preparation (e.g., Grahame-Smith, Butcher, Ney, and Sutherland, 1967; Lefkowitz et al., 1970b; Finn et al., 1972; Ide, Tanaka, Nakamura, and Okabayashi, 1972); but frequently the degree of binding has been correlated with biological activity found *in vivo* in terms of ascorbic acid depletion (for review see Schwyzer, 1964) or corticosteroid output in the adrenal vein of rats, or *in vitro* corticosterone formation by rat adrenal sections (for reviews see Ramachandran, Chung, and Li, 1965; Desaulles, Barthe, Schär, and Staehlin, 1966). Different portions of the ACTH molecule appear to be involved in binding to the receptor from those concerned with excitation, be this adenylyl cyclase activation and/or some "other event." The sequence Lys-Lys-Arg-Arg (residues 15–18) appears to be particularly significant in relation to binding to the receptor (Hofmann, 1960; Tesser and Schwyzer, 1966; Hofmann, Andreatta, Bohn, and Moroder, 1970a), while the functionally important or excitatory site of the ACTH molecule appears to be the sequence -His-Phe-Arg-Trp-gly (residues 6–10) (for reviews see Hofmann, 1962; Hofmann et al., 1970a).

Despite considerable interest in the relationships between structure and activity among ACTH analogues, very few studies have been reported regarding the chemical nature of the ACTH receptor. Haksar, Baniukiewicz, and Péron (1973) and also Haksar, Maudsley, Kimmel, and Péron (1974a) have reported that treatment of isolated rat adrenal cells with neuraminidase diminished the steroidogenic response to ACTH without affecting the cellular response to exogenous cyclic AMP or N^6-2'-0 dibutyryl adenosine 3',5'-monophosphate (dibutyryl cyclic AMP). In attempts to clarify a possible role for sialic acid in the interaction of ACTH and adrenal cell receptor, these authors examined the effect of neuraminidase treatment of the cells on the steroidogenic action of three peptides, ACTH$_{1-39}$, ACTH$_{1-24}$, and β-ACTH$_{1-10}$. The latter peptide is believed to contain the sequence that excites the receptor, and although it lacks the binding sequence (Arg-Arg-Lys-Lys) it shows the same V_{max} for corticosterone production (Sayers, Beall, and Seelig, 1972) as $ACTH_{1-39}$ or $ACTH_{1-24}$ (Schwyzer, Schiller, Seelig, and Sayers, 1971). Since neuraminidase treatment reduced the activity of all three peptides, Haksar et al. (1974a) suggest that either the separation of excitatory and binding sites is not complete, or, if involved, sialic acid residues may be concerned with the transmission of the signal from the ACTH-receptor interaction to adenylyl cyclase, as has been suggested by Cuatrecasas and Illiano (1971) for the action of insulin on fat cells.

2. Cholera Enterotoxin

Cholera enterotoxin has been shown to stimulate adenylyl cyclase in the intestine (Sharp and Hynie, 1971; Kimberg, Field, Johnson, Henderson, and Gershon, 1971) and elevated cyclic AMP levels are believed to be responsible for the increased water loss and dehydration associated with infection (Schaefer, Lust, Sircar, and Goldberg, 1970; Pierce, Carpenter, Elliot, and Greenough, 1971). Cholera enterotoxin also stimulates adenylyl cyclase in other tissues, including adrenal tumor cells (Wolff, Temple, and Cook, 1973; Donta, King, and Sloper, 1973) and cells isolated from normal rat adrenals (Haksar, Maudsley, and Péron, 1974b). In contrast to the very rapid response of adrenal adenylyl cyclase found on stimulation with ACTH, the response to enterotoxin is characterized by a lag period of the order of 30 min. Also, in contrast to the effect of ACTH on isolated adrenal cell corticosteroidogenesis and cyclic AMP accumulation, treatment with neuraminidase *enhances* the effect of enterotoxin on these parameters of cellular response (Haksar, et al., 1974b). Other studies have indicated that the receptor for enterotoxin may be a ganglioside (van Heyningen, Carpenter, Pierce, and Greenough, 1971; Holmgren, Lönnroth, and Svennerholm, 1973; King and van Heyningen, 1973; Cuatrecasas, 1973a,b), the addition of neuraminidase-resistant monosialosylganglioside with enterotoxin preventing the stimulation of intestinal adenylyl cyclase by enterotoxin (King and van Heyningen, 1973). Haksar et al. (1974b) suggest that the increased sensitivity of the neuraminidase-treated cells to enterotoxin may be due to the conversion of complex di- and trisialosylgangliosides to monosialosylganglioside in the cell membrane.

3. Angiotensin Receptors

The octapeptide angiotensin II is a potent stimulator of aldosterone secretion by the adrenal cortex in man (Laragh, Angers, Kelly, and Lieberman, 1960; Biron, Koiw, Nowaczynski, Brouillet, and Genest, 1961) and sheep adrenal transplants (Blair-West, Coghlan, Denton, Goding, Munro, Peterson, and Wintour, 1962) and of both cortisol and aldosterone secretion in the hypophysectomized dog (Carpenter, Davis, and Ayers, 1961; Slater, Barbour, Henderson, Casper, and Bartter, 1963) and bovine adrenal slices (Kaplan and Bartter, 1962; Kaplan, 1965).

A selective uptake of ^{125}I-labeled angiotensin I and II by target tissue was described by Bumpus, Smeby, Page, and Khairallah (1964) and examined in more detail by Goodfriend and Lin (1970). The latter group found saturable, specific binding of radioactive angiotensin I and II by intact cell and broken cell preparations from uterus, kidney, and adrenal cortex, with uterine and renal binding of angiotensin II more pronounced than the binding of angiotensin I. On the other hand, bovine adrenal cortex showed a greater capacity to bind angiotensin I than angiotensin II. More recent studies have been reported in which the binding of tritiated and monoiodinated angiotensin II by homogenates and subcellular fractions of bovine and rat adrenal cortex has been examined (Glossman, Baukal, and Catt, 1974a). The angiotensin II binding sites of bovine adrenal cortex were enriched severalfold in microsomal fractions, and results from electron microscopic examination and studies of marker enzymes suggested that the binding sites were located on the plasma membrane. In contrast to the results of Goodfriend and Lin (1970), the decapeptide angiotensin I exhibited relatively low affinity for angiotensin II binding sites in the system used by Glossman et al. (1974a), and the uptake of ^{125}I-labeled angiotensin I was lower than that of labeled angiotensin II. Binding studies performed under steady-state conditions at 22°C, with degradation of angiotensin II minimized by addition of glucagon and dithiothreitol, indicated the presence of two kinds of binding sites with different affinities for the octapeptide. The equilibrium binding constant for the high-affinity site was calculated to be 2×10^9 M^{-1} using either (^{3}H)- or (^{125}I)-labeled angiotensin II, and was relatively constant from one preparation to another. The number of low-affinity sites, and their affinity constants, showed more marked variation, apparently due to changes which occurred in the receptor preparation during isolation and storage. Freshly isolated receptor preparations did not always display the low-affinity sites.

In continuation of their studies on the binding of angiotensin II to bovine adrenal cortex preparations, Glossman, Baukal, and Catt (1974b) found that angiotensin II binding is inhibited with high specificity by GTP, 5'-guanilylimido-phosphate (Gpp(NH)p), and ITP. Half-maximal inhibition of angiotensin II binding was achieved with 1.6×10^{-7} M GTP in the presence of a nucleotide regenerating system, whereas Gpp(NH)p was equally effective in the absence of the regenerating system. Cyclic GMP, 5'-GMP, ATP, and UTP at concentrations 1,000 to 10,000-fold greater than those of GTP or Gpp(NH)p showed little or no inhibition of angiotensin II binding. The authors suggest that the actions of the guanyl nucleotides are based on conformational change at the receptor level.

The relationship of angiotensin II binding with adrenocortical receptors to activation of adenylyl cyclase is not clear at the present time. Peytreman, Nicholson, Brown, Liddle, and Hardman (1973a) compared the activity of ACTH and angiotensin II with respect to cyclic AMP formation and biosynthesis of cortisol and corticosterone in suspensions of cells isolated from the fasiculata region of calf adrenals. Angiotensin II produced a rapid but transient increase in cyclic AMP levels, although steroidogenesis continued as long as that induced by ACTH. Angiotensin II and ACTH were synergistic, not merely additive, in their effect on cyclic AMP levels, a finding which could not be explained by inhibition of phosphodiesterase activity by angiotensin II. Further, angiotensin II did not activate adenylyl cyclase activity of subcellular particulate fractions from the adrenal cells under conditions where ACTH and NaF were effective. Failure to obtain a stimulation of adenylyl cyclase by the direct addition of angiotensin II had previously been reported (Goodfriend and Lin, 1970; Schorr and Ney, 1971), although an activation of adenylyl cyclase in bovine adrenal cortex homogenates has been briefly reported on (Glossman, Baukal, and Catt, 1973).

4. Prostaglandin Receptors

Dazord, Morera, Bertrand, and Saez (1974) have recently described some properties of a prostaglandin-binding preparation obtained from human and sheep adrenal glands. The preparation from sheep glands was highly purified and the binding activity ascribed by the authors to plasma membrane. The binding of prostaglandins PGE_1 and PGE_2 was found to be specific, with a binding constant of the order of 10^8 M^{-1}. ACTH did not inhibit the binding of PGE_1 or PGE_2, nor was the binding inhibited by calcium ions (1 to 10 mM) or the calcium chelating agent, ethylene glycol-bis-(β-amino-ethyl ether)-N,N^1 tetraacetic acid (EGTA). Adenylyl cyclase activity in these preparations was stimulated by either PGE_1 or PGE_2, and both exhibited additive effects on cyclase activation when added with ACTH. EGTA, which inhibits the stimulation of adrenal adenylyl cyclase by ACTH (Bär and Hechter, 1969), had no effect on the stimulation induced by these prostaglandins. Calcium ions (0.04 to 10 mM) inhibited basal activity but did not affect the PGE stimulation. The authors believe that the prostaglandin receptors are distinct from the ACTH receptors, and that the prostaglandins are not obligatory intermediates in the action of ACTH.

Despite the evidence (see Section II,D) that prostaglandins are steroidogenic in the adrenal, the possibility should be considered that the adrenal prostaglandin receptors isolated by Dazord et al. (1974) do not in fact come from the plasma membranes of steroidogenic cells, but perhaps from other cellular elements such as those connected with the vascular structure of the gland. It may be recalled that the action of prostaglandins on adipose tissue cyclic AMP levels appeared confusing until the response of adipocytes and stromal elements were examined separately (Butcher and Baird, 1968, 1969).

B. Adrenocortical Adenylyl Cyclase

The demonstration by Haynes (1958) that ACTH added to incubated slices from bovine adrenal cortex increased the content of cyclic AMP could have been

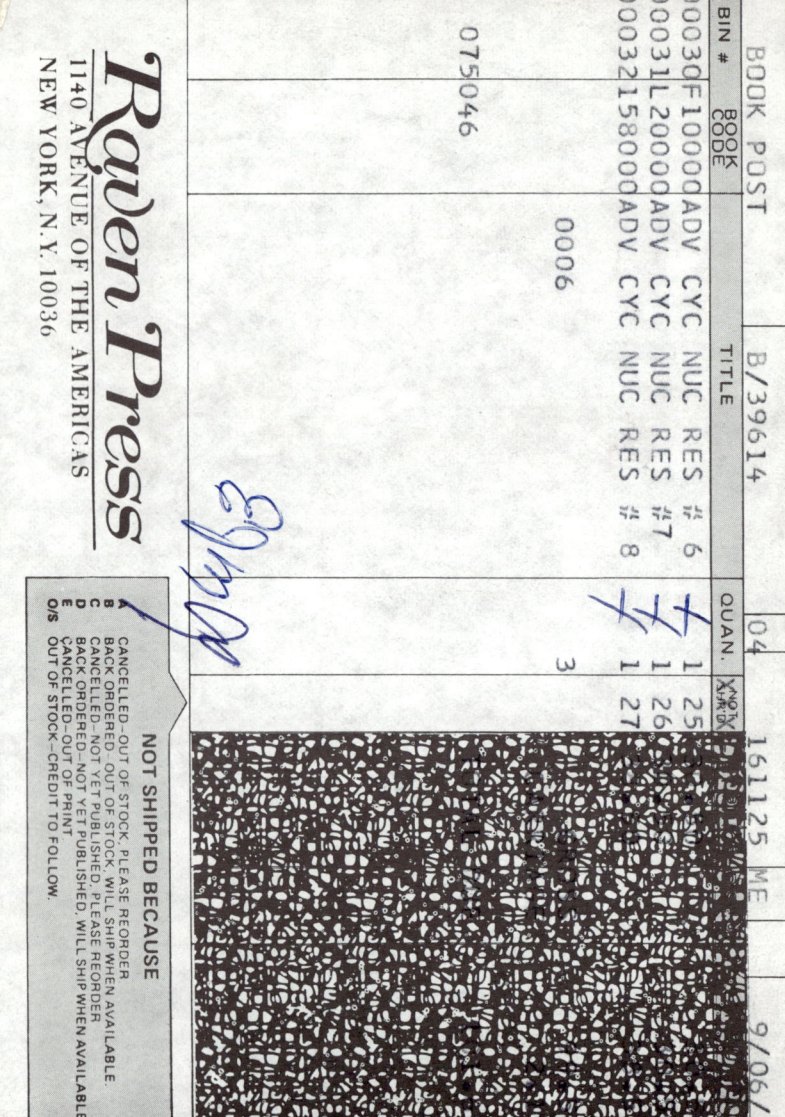

due either to stimulation of adenylyl cyclase or to an inhibition of phosphodiesterase activity. A direct stimulation of adenylyl cyclase activity by ACTH in whole homogenates of rat adrenal glands was obtained by Grahame-Smith et al. (1967). The enzyme activity was strongly stimulated by NaF as in other tissues (Sutherland, Rall, and Menon, 1962).

Taunton, Roth, and Pastan (1969) obtained a subcellular particulate fraction from an ACTH-sensitive mouse adrenal tumor which showed ACTH-sensitive adenylyl cyclase activity. Epinephrine, arginine vasopressin, parathyroid hormone, insulin, glucagon, and thyrotrophin were without stimulatory effect. Prostaglandin E_1 was also without effect on the ACTH- or fluoride-induced increase in adenylyl cyclase activity. Further purification of this preparation by tissue disintegration in the presence of a phospholipid and fluoride ion resulted in a population of vesicular profiles varying in size from 300 to 800 Å, studded with granules of an average diameter of 90 Å. The molecular weight was estimated to be 3 to 7 million daltons. Adenylyl cyclase activity could be stimulated by ACTH after removal of the fluoride by dialysis (Pastan, Pricer, and Blanchette-Mackie, 1970).

In a study of adenylyl cyclase activity in subcellular fractions of bovine adrenal cortex, Bär and Hechter (1969) showed that the stimulatory effect of ACTH could be abolished by treatment with EGTA and restored by addition of calcium. Further, they reported that basal levels of adenylyl cyclase activity were increased on removal of calcium. Although a low concentration of calcium was required for ACTH stimulation of adenylyl cyclase activity, addition of calcium above 1 mM reduced both basal and hormone-stimulated activity. In a parallel study of ACTH binding to "receptors" and activation of adenylyl cyclase, Lefkowitz, Roth, and Pastan (1970a) found that a subcellular preparation from a line of steroid-producing ACTH-sensitive adrenal tumors bound ^{125}I-ACTH by a process independent of calcium, as addition of EGTA had no effect upon the binding. In contrast, although the adenylyl cyclase of the preparation could be activated by both ACTH and NaF in the absence of added calcium, addition of 1 mM EGTA slightly inhibited activation of the enzyme by fluoride but completely eliminated cyclase activation by ACTH. Addition of calcium at increasing concentrations at low level restored the activation by ACTH, but by 2 mM Ca^{2+} the inhibitory effect reappeared. These authors suggested that Ca^{2+} is required for a step between hormone binding and adenylyl cyclase activation. A study of adenylyl cyclase activity and its response to ACTH was made in a subcellular fraction of bovine adrenal tissue by Kelly and Koritz (1971). Washing of the particulate preparation with ethylenediaminetetra-acetic acid (EDTA) resulted in a decreased stimulation by ACTH; although addition of Ca^{2+} (1 mM) to the EDTA-washed preparation reduced basal cyclase activity, the percent-stimulation by ACTH increased. The authors suggested that their results could be interpreted on the basis of an ACTH-mediated release of Ca^{2+} inhibition, or alternatively, on the basis of ACTH-sensitive and -insensitive adenylyl cyclase molecules with Ca^{2+} preferentially inhibiting the insensitive enzyme.

C. Calcium, Cyclic AMP, and Adrenal Steroidogenesis

A requirement for calcium in the action of ACTH on steroidogenesis by rat adrenal tissue *in vitro* was first reported by Birmingham, Elliott, and Valere (1953), later studies by Péron and Koritz (1958) showing that enhanced levels of potassium in the incubation medium could substitute for calcium. When the process of steroidogenesis by rat adrenal tissue homogenates was examined by Koritz and Péron (1959), they found that the addition of calcium had a strong stimulatory effect if the homogenates were fortified with glucose-6-phosphate and nicotinamide adenine dinucleotide phosphate (NADP). Thus it appeared at the time that the calcium requirement might be involved in mitochondrial and/or microsomal steroid transformations. However, although the interaction of ACTH with adrenal receptors does not require calcium (Birmingham et al., 1960; Bär and Hechter, 1969; Lefkowitz, Roth, and Pastan, 1970a; Kelly and Koritz, 1971), an optimal concentration of calcium is required for activation of adenylyl cyclase by ACTH (Kelly and Koritz, 1971). A requirement for calcium at more than one step in the sequence linking the ACTH-receptor interaction and steroidogenesis can be inferred from the finding that cyclic AMP, while capable of stimulating steroidogenesis in rat adrenals in the absence of extracellular Ca^{2+}, was less effective than when Ca^{2+} was present (Birmingham et al., 1960).

Further information on the role of calcium in the process of steroidogenesis by rat adrenal tissue *in vitro* has been provided by Farese (1971a,b), who showed that deletion of Ca^{2+} from the incubation medium decreased the steroidogenic action of ACTH and exogenous cyclic AMP to approximately the same extent (40 to 50%). Furthermore, the incorporation of label from ^{3}H-leucine into adrenal protein was also reduced to a comparable degree, whereas incorporation of ^{14}C-orotic acid into adrenal RNA was increased. Farese (1971a) suggested that a major reason for the calcium requirement during ACTH and cyclic AMP action might be the maintenance of optimal protein synthesis thought to be required for the steroidogenic effect of both substances (Ferguson, 1963; Garren, Ney, and Davis, 1965). Results from more detailed studies on the action of Ca^{2+} on adrenal protein synthesis suggest that the mechanism of the Ca^{2+} effect might be to speed up the transfer of amino acids from amino acyl tRNA to the growing peptide chain (Farese, 1971b).

Sayers et al. (1972) studied the calcium requirement for ACTH stimulation of corticosterone production and cyclic AMP accumulation by isolated rat adrenal cells. They found a parallel reduction in corticosterone and cyclic AMP formation in response to ACTH when calcium was omitted from the incubation medium, and showed that increasing levels of calcium ranging from 0 to 7.65 mM increased the stimulation of both parameters progressively. As the highest concentration of calcium (7.65 mM) supporting an enhanced response to ACTH was considerably in excess of that shown to inhibit adenylyl cyclase activity in subcellular fractions of adrenocortical tissue (Bär and Hechter, 1969; Kelly and

Koritz, 1971; Lefkowitz et al., 1971), Sayers et al. (1972) proposed that adenylyl cyclase in the intact plasma membrane is located within a compartment wherein Ca^{2+} concentration is low and remains unaffected by the concentration in the extracellular space. As the concentration of Ca^{2+} in the incubation medium increases from 0 to 7.65 mM, the strength of the signal generated by the interaction of ACTH with its receptor and transmitted to the adenylyl cyclase compartment is proportionately increased.

Haksar and Péron (1972a, 1973) have compared the requirement for calcium for ACTH stimulation of steroidogenesis by isolated rat adrenal cells with the requirement shown by dibutyryl cyclic AMP. At low levels of ACTH the calcium requirement was absolute, but at higher levels of ACTH the calcium requirement diminished. In contrast, the calcium requirement was about the same for all levels of dibutyryl cyclic AMP tested. In the presence of EGTA even very high concentrations of ACTH could not stimulate steroidogenesis, whereas the stimulatory effect of dibutyryl cyclic AMP was still observed even in the presence of 5 mM EGTA. From these studies it appears that the requirement for calcium in ACTH action, while involved in more than one step, is greater for events preceding the formation of cyclic AMP than for those that follow. A similar conclusion was reached by Birmingham and Bartova (1973), who showed that exogenous calcium is required for the stimulation of both steroidogenesis and glycolysis by ACTH in the intact mouse adrenal *in vitro*, whereas both dibutyryl cyclic AMP-evoked steroidogenesis and glycolysis proceeded efficiently in a calcium-free medium. They suggest that exogenous calcium might act by enhancing ACTH-evoked synthesis of cyclic AMP or by preventing its destruction, favoring the former view as theophylline at 10 mM stimulated steroid production less effectively than calcium in the absence of ACTH.

There is a great deal of evidence supporting the concept that calcium might act as a "second messenger" in the response of secretory cells to stimuli (Berridge, *This Volume*). Certain data accumulated from studies of ACTH action on steroid release from cat adrenals perfused *in situ* may be interpreted as contributing to this view (Rubin, Jaanus, and Miele, 1969; Jaanus, Rosenstein, and Rubin, 1970; Carchman, Jaanus, and Rubin, 1971; Jaanus and Rubin, 1971; Rubin, Carchman, and Jaanus, 1972a,b; Jaanus, Carchman, and Rubin, 1972). The cat adrenal perfused *in situ* with Locke's solution is sensitively responsive to ACTH, and allows the measurement of steroid release as opposed to synthesis. Calcium is required in the perfusion medium for ACTH stimulation of steroidogenesis, which is directly related quantitatively to the extracellular concentration of calcium up to 0.5 mM (Rubin, Jaanus, Carchman, and Puig, 1973). The requirements for ACTH stimulation of steroid release appear to be similar to those for acetylcholine stimulation of medullary catecholamine release, despite the fact that the latter hormones are stored within membrane-bound granules in contrast to the lack of sequestered stores of corticosteroids in the cortex (Rubin et al., 1969). Although steroid secretion is generally thought not to involve exocytosis, results from a recent study (Rubin, Shield, McCauley, and Laycock, 1974) suggest that release of protein (exocytosis?) accompanies ACTH stimulation of adrenocortical cells.

Prolonged perfusion of cat adrenals with Na-free (sucrose replacement) or K-free solutions does not markedly inhibit steroid output in response to ACTH, and neither does excess K (56 mM) produce consistent or marked increases in basal steroid output or affect the response to ACTH (Jaanus et al., 1970). Examination of the response of the perfused cat adrenal in terms of both steroid release and cyclic AMP accumulation in adrenal tissue allows a dissociation of cyclic AMP levels and steroid release (Carchman et al., 1971). Perfusion for 40 min with ACTH leads to a sustained rise in both tissue cyclic AMP and corticosteroid release. However, after brief exposure (5 min) to ACTH, cyclic AMP levels fall to control values within 30 min, while corticosteroid release is still near maximum. This pattern has been repeatedly observed in other systems, where stimulation of a cell results in a rapid elevation of cyclic AMP levels (cf. Robison et al., 1971). However, theophylline was found to elevate both basal and ACTH-stimulated cyclic AMP levels in cat adrenals, without augmenting ACTH-induced steroid release. Likewise, perfusion with Ca^{2+}-free (or K^+-free) media increases cyclic AMP levels three- to sixfold but does not augment steroid release. Addition of EGTA to the perfusate further increases cyclic AMP levels in the tissue, without increasing steroid release, and under these conditions ACTH has no further stimulatory effect on the levels of cyclic AMP found. When attempting to correlate the data obtained from these studies on the perfused cat adrenal, it may be important to bear in mind certain differences between the behavior of cat adrenals and that of rat adrenals from which so much data have been obtained. The concentrations of cyclic AMP found in cat adrenals perfused with EGTA (600 nmoles/g of tissue) are greater than those reported for the maximum response of rat adrenals to ACTH (Grahame-Smith et al., 1967). According to Carchman et al. (1971), the perfused cat adrenal preparation has very high basal levels of cyclic AMP compared to rat adrenal tissue and a high variation from one preparation to another (5.3 to 74.5 nmoles/g), although agreement between right and left adrenals is acceptable. In addition, perfusion of cat adrenals with cyclic AMP or its dibutyryl derivative (0.5 to 1.0 mM) does not increase the rate of steroid release into the perfusate (Carchman et al., 1971), which is in contrast to perfusion systems in other species (Hilton, Kruesi, Nedeljkovic, and Scian, 1961; Cushman, Alter, and Hilton, 1966) or results from *in vivo* administration of dibutyryl cyclic AMP (Imura, Matsukura, Matsuyama, Setsuda, and Miyake, 1965).

The steroid *content* of cat adrenals perfused with Locke's solution plus ACTH was found to be only 10% of the amount secreted, but glands perfused with Ca^{2+}-free medium + ACTH contained much higher amounts of steroid despite negligible steroid release (Jaanus et al., 1970). Investigation of steroid synthesis (as opposed to release) in perfused cat adrenals by Jaanus et al. (1972) revealed that the incorporation of ^3H-acetate into adrenal steroids was diminished, rather than increased, in glands perfused with Ca^{2+}-free media, despite elevated cyclic AMP levels, a finding consistent with the reduced steroid release. However, in glands perfused with K^+-free media, where cyclic AMP levels are also elevated but steroid release in response to ACTH decreased, the synthesis of steroid hormones is increased.

From these interesting studies of the response of the perfused cat adrenal, it appears that the mechanisms involved in both synthesis and release are highly dependent on Ca^{2+}, whereas activation of adenylyl cyclase is less affected. One is left with an impression that the difference between data collected from rat adrenals compared to the perfused cat adrenal preparation may be more quantitative than qualitative in nature. It would seem premature to conclude that increased intracellular levels of cyclic AMP cannot induce increased steroidogenesis unless an ACTH-induced "other event" occurs, when the manipulation used to increase levels of cyclic AMP involves creating a deficiency of factors involved in the synthesis and release of steroids.

Some data are available on calcium flux in adrenals, although different pictures are obtained from rat as compared to cat adrenal systems. Jaanus and Rubin (1971) have investigated the radiocalcium (^{45}Ca) space and content of perfused cat adrenals and the influence of ACTH on these parameters. The rate of efflux of ^{45}Ca was measured by perfusing glands with ^{45}Ca-containing medium followed by perfusion with calcium-free medium. Addition of ACTH to the calcium-free perfusion medium slowed the rate of ^{45}Ca-efflux from the glands during the first 20 min of washout, but had no effect on total calcium content of the cortex. The authors concluded that a translocation of calcium occurs during stimulation of the cortex by ACTH, but the source of the calcium ions was thought not to be the extracellular fluid; rather, it was suggested that ACTH might shift calcium from a rapidly exchanging to a more slowly exchanging intracellular pool. Leier and Jungmann (1973) found that ACTH and cyclic AMP caused a significant accumulation of $^{45}Ca^{2+}$ by rat adrenals incubated in the presence of $^{45}Ca^{2+}$-containing media. They provided evidence indicating that the accumulation of $^{45}Ca^{2+}$ was due to a net increase of calcium in the tissue and not only to an increased rate of exchange of extracellular ^{45}Ca with the intracellular calcium pool.

D. Prostaglandins, Cyclic AMP, and Adrenal Steroidogenesis

Cyclic AMP accumulation has been shown to be affected by prostaglandins in a number of tissues (cf. Butcher and Baird, 1968; Robison et al., 1971). The rat adrenal gland contains prostaglandins, as do most tissues, and in addition contains relatively high concentrations of prostaglandin precursor in the form of cholesterol arachidonate (Flack, Jessup, and Ramwell, 1969; Goodman, 1965). Flack, Jessup, and Ramwell (1968) observed that ACTH reduced the prostaglandin content of superfused rat adrenals, detecting more activity in the effluent than in the glands themselves. Subsequently this group showed that corticosterone release was increased by addition of prostaglandin E_2 to the perfusate of adrenal bisects from acutely hypophysectomized rats. The initial rates of corticosteroid synthesis during prostaglandin E_2 and ACTH stimulation were similar, both showing a peak output at 60 min. However, the response to prostaglandin E_2 decayed much more rapidly than the response to ACTH, indicating that the overall corticosteroidogenic response to prostaglandin E_2 is different from that to ACTH (Flack et al., 1969; Flack and Ramwell, 1972). These authors noted that the transient response of the superfused rat adrenal to prostaglandins was similar to the response of the perfused dog adrenal to L-vasopressin (Hilton, Scian, Westermann, Nakano, and Knuesi, 1960), although the prostaglandin effect was more sustained than that of L-vasopressin in the superfused rat adrenal. The stimulation of steroidogenesis by prostaglandin E_2 (28 μM) and a submaximal concentration (10 mμ/ml) of ACTH were additive, with a peak at 60 to 90 min. Prostaglandin E_2, when added to the perfusate alone, induced its maximal steroidogenesis at this time. The authors point out that since 2 hr after superfusion the prostaglandin E_2 effect has decayed, and since the response to PGE_2 and ACTH was greater than the response to ACTH alone, it is possible that PGE_2 may in fact potentiate ACTH action. It might be pointed out that in the superfused rat adrenal system it is steroid release that is monitored even though this type of preparation is far less sensitive to ACTH than "intact" gland perfusion systems. To what extent PGE_2 stimulation of steroid release is due to neosynthesis has not been determined. PGE_2 does not increase cyclic AMP levels in rat adrenal glands *in vitro* (Zor, Kaneko, Lowe, Bloom, and Field, 1969; Flack and Ramwell, 1972), even though corticosterone release is increased.

Peng, Six, and Munson (1970) found that prostaglandin E_1 depleted adrenal ascorbic acid and cholesterol and increased plasma and renal corticosterone in intact rats. However, PGE_1 did not affect adrenal ascorbic acid in cortisol-pretreated or in 24-hr hypophysectomized rats, indicating that it has no direct ACTH-like effect on the adrenal cortex. PGE_1 infused directly to the autotransplanted adrenal gland of Na-deficient sheep did not increase cortisol and corticosterone rates significantly (Blair-West, Coghlan, Denton, Funder, Scoggins, and Wright, 1971). However, Saruta and Kaplan (1972) found both PGE_1 and PGE_2 to significantly increase the synthesis of aldosterone, corticosterone, and to a lesser degree cortisol in outer slices of beef adrenal tissue. The stimulation by PGE_1 was found to require calcium, to be inhibited by puromycin but not actinomycin D, to increase cyclic AMP levels, and to show no additive effect with exogenous cyclic AMP. The authors suggest that these results are in keeping with a hypothesis that PGE_1 shares a receptor site on the plasma membrane with ACTH.

In an attempt to evaluate the relationship between prostaglandins and adrenal steroidogenesis, Gallant and Brownie (1973) administered an inhibitor of prostaglandin synthesis (indomethacin) to hypophysectomized rats and subsequently measured the *in vivo* production of corticosterone in response to ACTH, dibutyryl cyclic AMP, and PGE_2. The indomethacin-treated rats had a significantly reduced response to ACTH in terms of plasma corticosterone levels, whereas the response to dibutyryl cyclic AMP was not affected. Intraperitoneal injection of PGE_2 to indomethacin-treated rats restored the normal response to ACTH stimulation. In contrast to the report by Flack et al. (1969), PGE_2 itself had no significant effect on plasma corticosterone levels. Gallant and Brownie (1973) suggest that the endogenous levels of prostaglandins in the adrenal cortex of the hypophysectomized rats are sufficient to allow maximal stimulation of corticosteroidogenesis by ACTH, and when these endogenous levels are reduced by indomethacin treatment there is a decreased response to ACTH. They suggest also that PGE_2, even though devoid of activity by itself, may regulate the action of ACTH, possible sites of action including the binding of ACTH to its membrane "receptor," transmission of the signal arising from the ACTH receptor to the catalytic moiety of adenylyl cyclase, or modification of membrane-bound enzymes involved in the metabolism of cyclic AMP.

III. HORMONAL STIMULATION OF ADRENAL CORTEX AND CYCLIC NUCLEOTIDE LEVELS

Following the demonstration by Haynes (1958) that ACTH added to the incubation medium caused the accumulation of cyclic AMP by bovine adrenocortical slices, Grahame-Smith et al. (1967) examined this effect in rat adrenals in more detail. They showed that the increases in adrenal cyclic AMP induced by ACTH occurred before increases in the rate of adrenal steroidogenesis, that increasing doses of ACTH produced increasing concentrations of adrenal cyclic AMP as steroidogenesis was progressively stimulated, and that adrenal concentrations of cyclic AMP remained elevated while the rate of steroidogenesis was maintained. These particular experiments were carried out using quartered rat adrenals with high concentrations of ACTH (50 to 100 mU/ml), reflecting the relative insensitivity of this preparation to ACTH. When the effect of graded doses of ACTH (0 to 1,000 mU) on adrenal cyclic AMP (at 3 min) was examined in the rat *in vivo,* they found that as the dose of ACTH was increased up to 1 mU the adrenal cyclic AMP concentration gradually increased, accompanied by a gradual increase in corticosterone output in the adrenal venous blood. Above this dose of ACTH, however, the adrenal cyclic AMP concentration continued to rise, reaching a maximum at a dose of 50 mU of ACTH, despite the fact that steroid release reached a maximum at an ACTH dose of 2 mU. This type of study has been repeated by a number of laboratories using *in vitro* adrenal preparations of different kinds. In particular, studies carried out using the sensitive isolated rat adrenal cell preparations (Beall and Sayers, 1972; Mackie, Richardson, and Schulster, 1972; Nakamura, Ide, Okabayashi, and Tanaka, 1972; Kong, Moyle, and Ramachandran, 1972; Albano, Brown, Ekins, Price, Tait, and Tait, 1973; Haksar et al., 1974a) or the perfused cat adrenal (Carchman et al., 1971) have provided a more detailed picture of the relationship of cyclic AMP levels to steroidogenesis. The perfusion of the isolated cat adrenal for 40 min with Locke's solution containing 0.4 to 400 μU/ml or $ACTH_{1-24}$ produced a sustained rise in both tissue cyclic AMP levels and corticosteriod release. After perfusion with $ACTH_{1-24}$ for 5 min, tissue cyclic AMP levels fell to control values within 30 min, while corticosteroid release was still near maximum (Carchman et al., 1971). The studies with the isolated adrenal cell systems have clearly shown that low concentrations of ACTH stimulate steroidogenesis without causing detectable changes in the concentration of cyclic AMP (Beall and Sayers, 1972; Mackie et al., 1972; Nakamura et al., 1972). According to Beall and Sayers, low concentrations of ACTH (5 to 25 μU/ml) stimulate steroidogenesis in trypsin-dissociated rat adrenal cells without causing detectable changes in the concentration of cyclic AMP, while concentrations of ACTH of 50 to 250 μU/ml cause parallel increases in cyclic AMP and corticosterone production. High concentrations of ACTH (250 to 10,000 μU/ml) cause additional increases in the concentration of cyclic AMP without causing further increases in corticosterone accumulation.

Studies on the effect of the o-nitrophenyl sulphenyl derivative of $ACTH_{1-39}$ (in which the single tryptophan residue is modified) on corticosterone production and cyclic AMP levels have provided intriguing data. Ramachandran and Lee (1970a) found that $(Trp(NPS)^9)$ $ACTH_{1-39}$ was inactive in stimulating lipolysis by isolated fat cells of the rat, but inhibitory to the lipolytic action of $ACTH_{1-39}$ at a ratio of ACTH:inhibitor of 1:25. Further study showed that it inhibited ACTH activation of adenylyl cyclase in rat fat cell ghosts (Ramachandran and Lee, 1970b). When tested in an isolated rat adrenal cell system, it produced only a small increment in cyclic AMP formation compared to that induced by ACTH, but inhibited ACTH-induced cyclic AMP accumulation very effectively (Kong et al., 1972). In other studies (Seelig, Kumar, and Sayers, 1972) $(Trp(NPS)^9)$ $ACTH_{1-39}$ was shown to stimulate corticosterone production in isolated rat adrenal cells but with a maximal response less than that of unmodified ACTH. At appropriate combinations of doses the derivative inhibited the steroidogenic action of $ACTH_{1-39}$. From these properties the authors characterized the derivative as a partial agonist.

Both groups of investigators referred to above have made direct comparisons of the effect of $ACTH_{1-39}$ and $(Trp(NPS)^9)$ $ACTH_{1-39}$ on both corticosteroidogenesis and cyclic AMP accumulation in isolated rat adrenal cells (Seelig and Sayers, 1973; Moyle, Kong, and Ramachandran, 1973). Although in many respects the data obtained by these two groups are very similar, there appear to be differences and their interpretation certainly differs. Moyle and his associates regard $(Trp(NPS)^9)$ $ACTH_{1-39}$ as being capable of inducing the same maximal response in terms of corticosterone production as $ACTH_{1-39}$, thus characterizing it as an agonist, although the concentration of the analogue required is some 70 times that of unmodified ACTH. The analogue was found to be capable of stimulating cyclic AMP accumulation but to a level of only one-thirtieth to one-hundredth of that produced by ACTH. They believe that this discrepancy could not be explained on the basis of different kinetics, since both peptides elevated cyclic AMP levels throughout the time of measurement. The same investigators also reported that (Trp(NPS)^9) ACTH$_{1-39}$ appeared to competitively inhibit the effect of ACTH on cyclic AMP formation but not its effect on corticosteroidogenesis. The relationship between steroid synthesis and cyclic AMP accumulation was different for ACTH and NPS-ACTH, much less cyclic AMP being accumulated (either intracellularly or totally) when NPS-ACTH stimulated steroid synthesis to 75% of the maximum level than when ACTH enhanced steroidogenesis half-maximally. The authors contend that their results indicate that there may be two receptors for ACTH in the adrenal cell population (in different cell types or in the same cell), NPS-ACTH stimulating one type of receptor but inhibiting the other. In addition, they conclude that only very small amounts of cyclic AMP are required for the stimulation of steroidogenesis, or a factor(s) other than cyclic AMP may be involved in mediating the steroidogenic function of ACTH.

Seelig and Sayers (1973) have also studied the effect of ACTH and ACTH analogues on both corticosterone production and cyclic AMP accumulation in isolated rat adrenal cells. From the log-dose response curves obtained, they have expressed the biological characteristics of ACTH analogues in terms of the affinity of the peptides for the receptor, as reflected in displacement of the log-dose response curves along the abscissa, and the capacity of the peptides to activate the receptor, as reflected in the value of the maximum rate of production of corticosterone ("intrinsic activity") (Rudinger and Krejcí, 1962; Ariens and Simonis, 1964). They repeated their previous finding that NPS-ACTH acts as a partial agonist, inducing only 77% of maximum corticosterone production and less than 1% of maximum cyclic AMP accumulation. However, they point out that ACTH$_{1-39}$ at a concentration which induces 77% of maximal corticosterone production effects very little elevation of cyclic AMP levels. They also differ from Moyle et al. (1973) in that they found (Trp(NPS)^9) ACTH$_{1-39}$ to be a competitive inhibitor of both corticosterone production and cyclic AMP accumulation induced by ACTH$_{1-39}$. Seelig and Sayers (1973) conclude that the apparent discrepancy in the correlation of cyclic AMP and corticosterone production is not at variance with the second messenger hypothesis. They also conclude that (Trp(NPS)^9) ACTH$_{1-39}$ and ACTH$_{1-39}$ are competing for the same receptor, as judged by the decrease in both cyclic AMP and corticosterone formation seen when varying concentrations of derivative were added to adrenal cells in combination with a fixed concentration of ACTH$_{1-39}$ and by the increase in both parameters when varying concentrations of ACTH$_{1-39}$ were added together with a fixed concentration of the derivative. It is clear from the studies of both Moyle et al. (1973) and Seelig and Sayers (1973) that ACTH$_{1-39}$ can increase cyclic AMP levels in adrenal cells greatly in excess of that needed for maximal steroidogenesis, a finding previously reported by others (Grahame-Smith et al., 1967; Beall and Sayers, 1972). It is also possible that the failure of low but nevertheless steroidogenic concentrations of ACTH to induce detectable increases in the level of cyclic AMP might be due to inadequacy of analytic techniques.

IV. ADRENAL STEROIDOGENESIS AND INHIBITORS OF CYCLIC NUCLEOTIDE PHOSPHODIESTERASE

From the results of a study of the effect of theophylline in concentrations ranging from 10^{-4} to 10^{-2} M, Halkerston, Feinstein, and Hechter (1966) concluded that the phosphodiesterase inhibitor had two antagonistic actions in rat adrenal bisects: (1) a small potentiation of ACTH stimulation of steroidogenesis seen only at certain concentrations of activator; and (2) a reduction of the stimulatory action of ACTH which was accompanied by a partial inhibition of the incorporation of radioactive amino acid into adrenal protein. The antagonistic effect of theophylline on ACTH action on adrenal steroidogenesis was also reported by Bieck, Stock, and Westermann (1969), who found that theophylline antagonized the steroidogenic action of both cyclic AMP and dibutyryl cyclic AMP. Sayers, Ma, and Giordano (1971) found that theophylline (10^{-3} M), added to the medium with isolated adrenal cells from the rat, potentiated the steroidogenic action of cyclic AMP (10^{-4} to 10^{-3} M) but did not augment ACTH-induced steroidogenesis. Kitabchi, Wilson, and Sharma (1971), who also used isolated rat adrenal cells, found caffeine (10^{-3} to 10^{-2} M) to inhibit steroidogenesis induced by either ACTH or dibutyryl cyclic AMP. These latter studies suggested that the inhibition was unrelated to the phosphodiesterase-inhibitory activity of caffeine, as no phosphodiesterase activity could be detected in their isolated rat adrenal cell preparation, although hydrolysis of the cyclic 3',5'-phosphates of adenosine, guanosine, and inosine was carried out effectively by adrenal tissue homogenates. On the other hand, Mackie and Schulster (1973) found that isolated rat adrenal cells prepared by collagenase digestion still showed 67% of the phosphodiesterase activity of intact glands, and trypsin-dissociated cells in their hands retained 37% of the phosphodiesterase activity. In their studies theophylline at 10^{-3} M potentiated the steroidogenic effect of submaximal concentrations of ACTH, this concentration of theophylline having no inhibitory action on the incorporation of ^{3}H-leucine into adrenal protein.

In an interesting study of the effect of theophylline on rat adrenocortical tissue *in vitro,* Leier and Jungmann (1971) found that a short incubation (5 to 15 min) with theophylline increased cholesterol side-chain cleavage activity without measurable effect on protein synthesis. Longer incubation times resulted in reduction of ^{14}C-glycine incorporation into protein but, more important, resulted in a marked reduction of free and esterified cholesterol stores. They suggest that it is the depletion of precursor stores during preincubation with theophylline rather than decreased protein synthesis that brings about the inhibition of ACTH-induced steroidogenesis.

Evidence that theophylline could potentiate ACTH-induced steroidogenesis *in vivo* was obtained by Marton, Stark, Katalin, and Varga (1971), who measured plasma corticosterone levels in hypophysectomized rats receiving ACTH alone or in combination with theophylline. However, in the perfused cat adrenal, Carchman et al. (1971) found an augmentation by theophylline of ACTH-induced cyclic AMP levels without further increase in corticosteroidogenesis.

More recently Peytreman, Nicholson, Liddle, Hardman, and Sutherland (1973b) have studied the effect of an inhibitor of phosphodiesterase (1-methyl, 3-isobutylxanthine) which is 10 times as potent as theophylline, and found clear evidence for a potentiation of ACTH action on rat adrenal cyclic AMP concentrations and corticosterone production both *in vitro* and *in vivo*.

V. EFFECT OF EXOGENOUS CYCLIC NUCLEOTIDES ON ADRENAL STEROIDOGENESIS

A. Cyclic AMP and Dibutyryl Cyclic AMP

Haynes et al. (1959) first showed that exogenously added cyclic AMP could stimulate steroidogenesis in rat adrenals *in vitro*. Since then the steroidogenic effect of cyclic AMP on adrenal tissue has been demonstrated in a wide variety of systems, ranging from adrenals perfused *in situ* to suspensions of isolated adrenocortical cells. The dibutyryl derivative of cyclic AMP has been shown in most systems used to be a more potent stimulator of steroidogenesis (Butcher, Ho, Meng, and Sutherland, 1965; Henion, Sutherland, and Posternak, 1967; Blecher, 1971; Kitabchi et al., 1971; Kitabchi and Sharma, 1971) and is also an effective stimulator *in vivo* (Imura et al., 1965). Dibutyryl cyclic AMP has also been shown to partially prevent the reduction in adrenal weight and responsiveness to ACTH in terms of steroidogenesis which follows hypophysectomy (Ney, 1969) and to stimulate adrenal ornithine decarboxylase in a dose-related manner similar to ACTH (Richman, Dobbins, Voina, Underwood, Mahaffee, Gitelman, Van Wyk, and Ney, 1973).

The dynamics of the steroidogenic response of quartered rat adrenals to ACTH, cyclic AMP, and dibutyryl cyclic AMP have been examined in a continuous-flow superfusion system (Pearlmutter, Rapino, and Saffran, 1971, 1973). Release of steroids into the perfusion medium following addition of ACTH or cyclic AMP was observed after a lag of 3 min (corrected for analytical lag time), but the delay following addition of dibutyryl cyclic AMP was 6 min. The longer lag time with dibutyryl cyclic AMP is consistent with the view that the derivative has to be transformed to an active compound on entering the cell, possibly the N^6-monobutyryl derivative (Kaukel and Hilz, 1972; Kaukel, Mundhenk, and Hilz, 1972). Working with HeLa cells, Kaukel et al. (1972) found that dibutyryl cyclic AMP and the $O^{2'}$-mono derivative do not bind to the kinase regulatory unit in the Gilman (1970) assay system, but the N^6-mono derivative binds as well as cyclic AMP. Further, the N^6-mono derivative was the only cyclic nucleotide found to accumulate in cells incubated with dibutyryl cyclic AMP. The greater potency of dibutyryl cyclic AMP in stimulating steroidogenesis compared to cyclic AMP itself cannot be explained on the basis of a faster rate of penetration, because in the superfused adrenal system ^{3}H-label from ^{3}H cyclic AMP enters as readily as that from ^{3}H-dibutyryl cyclic AMP (Pearlmutter et al., 1973). The

N^6-monobutyryl derivative is active in stimulating steroidogenesis in the superfusion system, but at 10 times the dose required for stimulation by dibutyryl cyclic AMP, which may reflect a slower rate of entry of the mono-derivative (Pearlmutter et al., 1973). However, Pearlmutter et al. (1973) observed that when dibutyryl cyclic AMP was added to superfused adrenals already maximally stimulated by ACTH (or cyclic AMP), a pronounced inhibition of steroidogenesis occurred which was not mimicked by addition of butyric acid to the perfusion medium.

In a qualitative sense all known effects of ACTH upon adrenal tissue *in vitro*, and in many cases *in vivo*, are reproducible by administration of cyclic AMP or its dibutyryl derivative. Thus dibutyryl cyclic AMP infused into hypophysectomized rats reduces the adrenal ascorbic acid concentration (Earp, Watson, and Ney, 1970), a response long used to assay ACTH (Sayers, Sayers, and Woodbury, 1948; Munson and Toepel, 1958).

B. On the Specificity of Cyclic AMP Stimulation of Adrenal Steroidogenesis

Working with monolayer cultures of functional mouse adrenal tumors, Stollar, Buonassisi, and Sato (1964) found that the cells showed a definite steroidogenic response to 5′-AMP as well as cyclic AMP. Kowal and Fiedler (1969), also using monolayer cultures of mouse adrenal tumor cells, reported that steroidogenesis was stimulted by adenosine, AMP, ADP, and ATP, in addition to cyclic AMP. Partial responses were seen with UMP and CTP. This apparent lack of specificity was not confined to cells derived from tumors, as Glinsman, Hern, Linarelli, and Farese (1969) obtained a stimulation of steroidogenesis by rat adrenal quarters with cyclic GMP, the magnitude of the response being at least as great as that induced by cyclic AMP. Further studies using the 3′,5′-cyclic nucleotides of uridine, cytidine, guanosine, and inosine showed that all were capable of increasing steroidogenesis in rat adrenals *in vitro*, with 3′,5′-cyclic IMP being at least as effective as cyclic AMP (Mahaffee, Watson, and Ney, 1970; Mahaffee and Ney, 1970). These authors also noted that the deoxyribose derivative of cyclic AMP was as effective in stimulating steroidogenesis as the parent compound. Using suspensions of rat adrenal cells, Rivkin and Chasin (1971) found 3′,5′-IMP to be about half as active as cyclic AMP at concentrations of 3 mM. All of the 3′,5′-cyclic nucleotides tested in their system showed synergism with 5.0 mM theophylline, the most striking effect being that shown with 3′,5′-cyclic GMP, and, in addition, 2′,5′-cyclic nucleotides of guanosine and uridine were potent inhibitors of ACTH and cyclic AMP-stimulated steroidogenesis. Further evidence that other 3′,5′-cyclic nucleotides were effective stimulators of adrenal steroidogenesis has been provided by the studies of Vapaatalo, Bieck, and Westermann (1972), who found that 3′,5′-cyclic GMP and 3′,5′-cyclic IMP could substitute for cyclic AMP in the stimulation of steroidogenesis by rat adrenal slices, although they were less potent than cyclic AMP. Similarly, Kitabchi and Sharma

(1971) found 3′,5′-cyclic GMP and 3′,5′-cyclic IMP to be effective in stimulating steroidogenesis in isolated rat adrenal cells.

A series of 8-substituted derivatives of cyclic AMP were examined for their ability to activate steroidogenesis in isolated rat adrenal cells, and most of them were found to be as active as the parent compound. Twelve of the 8-substituted derivatives were more potent activators than cyclic AMP, and three compounds, the 8-methylthio, 8-bromo, and 8-hydroxy derivatives were more potent than $N^6,2'$-O-dibutyryl cyclic AMP (Free, Chasin, Paik, and Hess, 1971). Comparison of structure-activity relationships in this latter study, together with those in other reports (e.g., Muneyama, Bauer, Shuman, Robins, and Simon, 1971), led the authors to suggest that the steroidogenic activities of the 8-substituted derivatives result from their ability to function as activators of cyclic AMP-dependent protein kinase. Kowal (1973) has recently presented data indicating that the steroidogenic activity of 3′,5′-cyclic CMP in monolayer cultures of mouse adrenal cells is due to the greater permeability of the cell membrane to this nucleotide than to the other 3′,5′-cyclic nucleotides including cyclic AMP.

Considerable interest has been expressed with regard to the possibility that 3′,5′-cyclic GMP might act as an intracellular second messenger (cf. Goldberg, O'Dea, and Haddox, 1973). Cyclic GMP does stimulate steroidogenesis in adrenal preparations, although generally it is less potent than cyclic AMP. Recently Whitley, Stowe, Ong, Ney, and Steiner (1974) have briefly reported values for adrenal cyclic AMP and cyclic GMP in hypophysectomized rats following ACTH administration. Cyclic GMP levels increased 150 to 300% within 1 hr of hypophysectomy, and fell to baseline levels (or below) within 15 min following ACTH administration, at which time cyclic AMP levels were increased at least 50-fold. The authors suggest that rat adrenal cyclic GMP is regulated by ACTH, either directly or secondarily to alterations in cyclic AMP levels. They also report that intracellular localization of the two cyclic nucleotides by immunofluorescent techniques showed a prominent nuclear location for cyclic GMP and cytoplasmic location for cyclic AMP.

Kitabchi, Nathans, James, Bower, Wilson, and Kitchell (1974) have also briefly reported on changes in cyclic AMP and cyclic GMP levels during ACTH stimulation of steroidogenesis in isolated rat adrenal cells. Physiologic concentrations of ACTH (1 to 10 μU/ml) did not increase cyclic AMP levels measurably, but did increase cyclic GMP concentrations. With higher concentrations of ACTH (>50 μU/ml), cyclic AMP levels rose but cyclic GMP levels did not. The authors suggest that there is an inverse relationship between the levels of the two cyclic nucleotides, a low concentration of ACTH eliciting cyclic GMP formation while high ACTH concentrations inhibit GMP formation. Cyclic AMP formation is thus elicited only with high (supraphysiologic) concentrations of ACTH. In view of suggestions from other studies of ACTH action that cyclic AMP may not be the sole translator of the initiating action of ACTH (e.g., Moyle et al., 1973; Rubin et al., 1973), confirmation of the interesting studies of Whitley et al. (1974) and Kitabchi et al. (1974) will be most valuable.

VI. ON THE MECHANISM OF ACTION OF CYCLIC AMP IN THE ADRENAL CORTEX

By far the greater part of the information available is compatible with the concept that the steroidogenic effect of ACTH on the adrenal cortex is mediated via increased intracellular levels of cyclic AMP, although the results of some studies have been interpreted as supporting the idea that there may be an obligatory ACTH-mediated event that is separable from adenylyl cyclase activation (Rubin, Jaanus, Carchman, and Duig, 1973; Moyle et al., 1973). Despite considerable effort, however, the crucial question of how cyclic AMP "triggers" the steroidogenic mechanism is still not answered, and there is little insight into the mechanism(s) whereby the trophic effect of ACTH is translated.

The enzymatic pathways involved in the biosynthesis of corticosteroids have been extensively studied during the past 20 years (for reviews see Hechter, 1958; Garren, 1968; Tchen, 1968; Koritz, 1968; Boyd and Simpson, 1968; Harding, Bell, Oldham, and Wilson, 1968; Cooper, Narasimhulu, Rosenthal, and Estabrook, 1968; Cammer, Cooper, and Estabrook, 1968; Kimura, 1968; Purvis, Battu, and Péron, 1968; Simpson, Cooper, and Estabrook, 1969; Ferguson, 1972). Cholesterol, either synthesized from acetate within the adrenocortical cell or taken up from the plasma, is stored in esterified form in cytoplasmic droplets (Moses, Davis, Rosenthal, and Garren, 1969) and following hydrolysis to free cholesterol serves as the precursor for steroid hormones (Morris and Chaikoff, 1959; Davis and Garren, 1966; Dexter, Fishman, Ney, and Liddle, 1967b). Cholesterol is transformed to corticosteroids by a series of hydroxylations catalyzed by mixed-function oxidase requiring the participation of cytochrome P_{450}, a nonheme iron protein (adrenodoxin), NADPH, and molecular oxygen (for review see Simpson et al., 1969) and also involves the participation of an NAD^+-linked dehydrogenase and Δ^5-3 ketosteroid isomerase (Beyer and Samuels, 1956). The first step in the sequence, and the rate-limiting step, is the scission of the cholesterol side chain to yield pregnenolone and isocaproaldehyde which is rapidly converted to isocaproic acid (Stone and Hechter, 1954; Karaboyas and Koritz, 1965; Constantopoulos and Tchen, 1961). The enzyme system, which is located in the mitochondrial fraction (Hayano, Saba, Dorfman, and Hechter, 1956; Halkerston, Eichhorn, and Hechter, 1961), catalyzes a complex process involving probably the formation of 20α-hydroxy cholesterol and $20\alpha,22\epsilon$-dihydroxy cholesterol, but unless high concentrations of trapping agents are present in *in vitro* systems these intermediates do not accumulate but probably remain complexed to the enzyme (Tchen, 1968). The overall rate of the cleavage reaction is controlled by the rate of hydroxylation at C-20 (Shimizu, Hayano, Gut, and Dorfman, 1961; Koritz, 1962; Tchen, 1968). The rate-limiting cholesterol side-chain cleavage step is stimulated in the intact cell by ACTH or cyclic AMP (Stone and Hechter, 1954; Karaboyas and Koritz, 1965), but no unequivocal evidence has been obtained of a stimulation of this event by cyclic AMP in isolated adrenal mitochondria. An increased incorporation of radioactivity from ^{3}H-cholesterol into pregnenolone in the presence of cyclic AMP by isolated rat adrenal mitochondria has been described (Roberts, McCune, Creange, and Young, 1967), but the increased level of radioactive pregnenolone might have been due to an effect of cyclic AMP to block further metabolism of pregnenolone (Koritz, Yun, and Ferguson, 1968; McCune, Roberts, and Young, 1970). The intracellular location of the NAD^+-linked dehydrogenase and Δ^5-3 ketosteroid isomerase activities involved in the conversion of pregnenolone to progesterone is not entirely clear. Studies by Beyer and Samuels (1956) indicated an endoplasmic reticulum (microsomal) site, but more recently this activity has been described as occurring in both microsomal and mitochondrial fractions (McKune et al., 1970). Unfortunately, these studies do not clearly locate the site of pregnenolone-to-progesterone conversion in relation to the site of pregnenolone formation. Koritz and Hall (1964a) found pregnenolone (but not progesterone) to inhibit cholesterol sidechain cleavage, possibly by an allosteric mechanism (Koritz and Hall, 1964b), and in proposing that the rate of pregnenolone efflux from the mitochondria might regulate cholesterol side-chain cleavage (Koritz and Kumar, 1970) they were clearly invoking a membrane barrier between pregnenolone formation and its conversion to progesterone. Following hydroxylation of progesterone at C-21 or C-21 and C-17 (depending on species) by endoplasmic reticulum enzyme systems, the hydroxylated product has to return to the mitochondria for hydroxylation at C-11 (and 18 hydroxylation in the zona glomerulosa). The mode of secretion of corticosteroids from the adrenal cells is really unknown, and may not be simply a case of free diffusion, especially since the studies of Rubin and his associates indicate a role for calcium, and possibly exocytosis, in this process (Jaanus et al., 1970; Rubin et al., 1974).

In the intact adrenocortical cell the process of steroidogenesis is restrained in the absence of ACTH, although in homogenized adrenal tissue steroidogenesis proceeds rapidly as long as high levels of NADPH are maintained (Koritz and Péron, 1959). That the releasing effect of cell breakage might involve changes in mitochondrial permeability is supported by the finding of further stimulation of steroidogenesis in homogenates by high Ca^{2+} or freeze-thawing (Péron and Koritz, 1960). Upon complete withdrawal of ACTH *in vivo* (hypophysectomy), a number of enzyme activities of the steroidogenic pathway decrease markedly; e.g., cholesterol side-chain cleavage activity declines with a half-life of 3 to 4 days (Kimura, 1969; Doering and Clayton, 1969), and the activity of the cytochrome P_{450}-linked electron-transport system disappears with a half-life of 3 to 5 days (Purvis, Canick, Mason, McCarthy, and Estabrook, 1973).

Two inhibitors have been effectively used in studying control of steroidogenesis: aminoglutethimide, which inhibits cholesterol side-chain cleavage activity (Kahnt and Neher, 1966; Dexter, Fishman, Ney, and Liddle, 1967a,b), and cyanoketone (2α-cyano, $4,4,17\alpha$-trimethylandrost-5-en, 17β ol-3one), which blocks the conversion of pregnenolone to progesterone (Ferrari and Arnold, 1963; Goldman, Yakovac, and Bongiovani, 1965; Neville and Engel, 1968; Farese, 1971c). The report by Ferguson (1963) that an inhibitor of protein synthesis, puromycin, blocked the steroidogenic effects of ACTH and cyclic AMP *in vitro* provided a new dimension to studies on the regulation of steroidogenesis. Ferguson found that puromycin was not inhibitory to the enzymes of steroidogenesis (as it had no effect on this process in adrenal homogenates fortified with a NADPH-generating system) and concluded that ongoing protein synthesis was required for the stimulation of steroidogenesis by cyclic AMP. The role of protein synthesis in the action of cyclic AMP, including control of steroidogenesis, has recently been reviewed by Wicks (1974).

Intrigued by the suggestion that stimulation of steroidogenesis (an event which occurs within minutes and decays rapidly on withdrawal of hormone) (Liddle, Island, and Meador, 1962) might involve the synthesis of new protein as an obligatory event, Garren and his associates (Garren et al., 1965, 1971a) examined the effect of inhibitors of protein and RNA synthesis on ACTH-induced steroidogenesis in the rat *in vivo*. They demonstrated in elegant fashion that cycloheximide administered only 3 min prior to ACTH inhibited the rise in adrenal vein plasma corticosterone levels, which in the absence of inhibitor reached a maximum new steady state at about 20 min after hormone administration, at all doses of ACTH tested. Further, administration of cycloheximide at the new steady state promoted by ACTH resulted in a rapid decay of corticosterone output with a half-life of 7 to 8 min. The calculated half-life for the induction of steroidogenesis by ACTH was close to that calculated for the decay upon addition of cycloheximide to ACTH-stimulated glands, which is consistent with an event involving the induction of synthesis of a rapidly turning over protein (Schimke and Doyle, 1970; Wicks, 1974). In contrast, actinomycin D, an inhibitor of DNA-dependent RNA synthesis, did not prevent ACTH-induced steroidogenesis when given 2 hr before hormone, despite effective inhibition of adrenal RNA synthesis. This latter finding suggested that the messenger RNA involved in the synthesis of the protein was relatively stable and that control was at the level of translation. Similar conclusions were reached by other investigators using a variety of adrenal preparations, including superfused adrenal bisects (Schulster, Tait, Tait, and Mrotek, 1970), isolated adrenal cells (Schulster, Richardson, and Mackie, 1972), and adrenal tumor cells (Kowal, 1970a,b). The estimate of the half-life of the rapidly turning over protein varies from 2 to 4 min in isolated rat adrenal cells (Schulster et al., 1972) to 20 to 30 min in tumor cells (Kowal, 1970b) and 45 to 49 min in the superfused gland (Schulster et al., 1970). As pointed out by Wicks (1974), the different estimates reflect the differences in the rate of achievement of the new steady state reached after addition of ACTH, and may result from variations in the ability of ACTH to reach the adrenal cells in the different preparations.

The rapidly turning over protein appears to be involved at a step, or steps, subsequent to the generation of cyclic AMP, as the latter event is not blocked by inhibitors of protein or RNA synthesis (Grahame-Smith et al., 1967). The results of a number of studies show that the inhibition of protein synthesis affects the activation of the rate-limiting step of steroidogenesis, the conversion of cholesterol to pregnenolone, and not subsequent steps (Garren, 1968; Davis and Garren, 1968; Kowal, 1970a; Koritz and Kumar, 1970). Thus, cycloheximide does not block the transformation of pregnenolone to corticosterone in rat adrenals (Davis and Garren, 1968; Hall and Young, 1968) or the conversion of 20α-hydroxy cholesterol to corticosterone (Hall and Young, 1968). However, a recent unconfirmed report suggests that ACTH stimulates the conversion of ^{3}H-20α-cholesterol to ^{3}H-pregnenolone by isolated rat adrenal cells via a cycloheximide-insensitive step, but one which is not activated by cyclic AMP (Sharma, 1973).

The first proposal put forward to account for the steroidogenic action of cyclic AMP suggested that its primary effect was to increase the supply of reducing equivalents for the hydroxylations involved in the rate-limiting step of steroidogenesis (Haynes et al., 1960). Thus, increased glycogenolysis due to an ACTH-induced increase in cyclic AMP levels would provide glucose-6-phosphate for the hexose monophosphate pathway and result in increased formation of NADPH. However, it is now generally accepted that the reducing equivalents for the rate-limiting step of steroidogenesis are generated intramitochondrially, results of investigations in a number of laboratories emphasizing the importance of Krebs cycle intermediates and pyruvate in supporting steroidogenesis (Brownie and Grant, 1954; Péron, McCarthy, and Guerra, 1966; Harding et al., 1968; Lin, Haksar, and Péron, 1974). If, as has been suggested (Péron and Tsang, 1969; Haksar and Péron, 1972b; Lin et al., 1974), pyruvate may be the physiological substrate utilized by rat adrenal mitochondria for NADPH production for steroid hydroxylations, then an action of cyclic AMP to make additional pyruvate available could be one facet of the cyclic nucleotide effect. Despite the fact that glycolysis is stimulated by cyclic AMP in mouse adrenals (Bartova and Birmingham, 1971; Birmingham and Bartova, 1973), adrenal tumor cells (Kowal, 1970a), or rat adrenal gland sections (Tsang and Péron, 1971), and might be expected to be an important determinant for the continuation of the process of steroidogenesis, this action of cyclic AMP is not dependent on the rapidly turning over protein as it is not inhibited by cycloheximide (Kowal, 1970a).

The view that the supply of steroid precursors to the rate-limiting step in steroidogenesis might be the critical point of control has received a good deal of attention. The hydrolysis of cholesterol esters stored in cytoplasmic lipid droplets is stimulated by ACTH and dibutyryl cyclic AMP by a mechanism which is not blocked by cycloheximide (Davis and Garren, 1966, 1968; Davis, 1969). At this time it is not clear if cholesterol esterase activity is increased by a cyclic AMP-dependent protein kinase such as occurs with adipocyte lipase, but an adrenal cyclic AMP-dependent protein kinase can phosphorylate cholesterol esterase (Trzeciak and Boyd, 1974). When cholesterol esterase activity is stimulated by ACTH in the presence of cycloheximide, free cholesterol accumulates in the cytoplasmic lipid droplets (Garren et al., 1971a) rather than in the mitochondria (Garren et al., 1971a; Simpson, Jefcoate, Brownie, and Boyd, 1972), although some increase in mitochondrial cholesterol under similar conditions has been reported (Mahaffee, Reitz, and Ney, 1974). The accumulation of cholesterol in the lipid droplets when cycloheximide is present has led to the suggestion that the controlling step might involve a cholesterol-transporting protein, which might translocate free cholesterol to the mitochondria (Garren et al., 1971a; Mahaffee et al., 1974). A cholesterol-binding protein, which can enhance cholesterol sidechain cleavage activity of acetone-dried adrenal mitochondrial preparations, has been isolated from rat adrenals (Kan, Ritter, Ungar, and Dempsey, 1972), but because its level of activity in the rat adrenal does not appear to change upon administration of ACTH or after hypophysectomy (Kan and Ungar, 1973; Ungar, Kan, and McCoy, 1973), its possible relation to the rapidly turning over protein is obscure. The penetration of cholesterol into the mitochondria, if this can be considered separately from transport to the mitochondria, has long been a favored site of control (Hechter, 1955; Hayano et al., 1956; Jefcoate, Simpson, and Boyd, 1974; Bell and Harding, 1974). Recently Kahnt, Milani, Steffen, and Neher (1974) have concluded, from a study of the kinetics of conversion *in vitro* of doubly labeled cholesterol by intact and ruptured bovine adrenocortical mitochondria, that cyclic AMP has no direct effect on the transport process for cholesterol, the intramitochondrial enzymatic conversion, or the utilization of endogenous steroid precursor. The authors add that if it is involved, cyclic AMP seems to have an action at a premitochondrial site and to be related only indirectly to a transport mechanism or a precursor supply process.

Deeper insight into the regulation of cholesterol side-chain cleavage has been provided by studies of the interaction of cytochrome P_{450} with its substrates (Simpson et al., 1972; Jefcoate, Simpson, and Boyd, 1974; Bell and Harding, 1974). By optical difference spectra and electron paramagnetic resonance measurements, the degree of cytochrome P_{450} binding to cholesterol or deoxycorticosterone can be estimated, the active complex with cholesterol being predominantly in a high-spin state at pH 7 (Jefcoate, Hume, and Boyd, 1970; Jefcoate et al., 1974). Further, a different cytochrome P_{450} entity appears to be involved in cholesterol side-chain cleavage from that associated with 11β-hydroxylation (Whysner, Ramseyer, Kazmi, and Harding, 1969; Jefcoate and Gaylor, 1970; Jefcoate et al., 1974). In the adrenal *in vivo* only a small proportion of the cholesterol side-chain cleavage cytochrome P_{450} is bound to cholesterol, and if care is taken to prevent anoxia of the glands during removal and preparation of mitochondria, the latter likewise contain only a small proportion of the bound form of cytochrome P_{450} (Bell and Harding, 1974). Anoxia rapidly causes an accumulation of cytochrome P_{450}-cholesterol complex, which is enhanced by prior treatment with ACTH and blocked by cycloheximide (Jefcoate et al., 1974).

A reevaluation of cholesterol side-chain cleavage by isolated rat adrenal mitochondria has been recently reported by Bell and Harding (1974), who used a soluble cholesterol-lecithin complex as substrate for steroidogenesis. The rate of side-chain cleavage of cholesterol was found to be similar to that for 20α-hydroxycholesterol and approximately four times greater than the rate of 11β hydroxylation. Studies of the *in vivo* effect of ACTH on the levels of cytochrome P_{450} bound to cholesterol and to deoxycorticosterone showed that the concentrations of both complexes are increased by prior treatment with ACTH, and that cholesterol accumulates at its enzyme site only when hydroxylation is inhibited by anaerobiosis. The results of these studies appear to contradict two previously postulated mechanisms for the limitation of adrenal steroidogenesis. First, a point of control at the level of pregnenolone removal from the mitochondria (Koritz and Kumar, 1970) is not supported by the finding of rapid rates of pregnenolone conversion to products upon the addition of adrenal microsomes to the mitochondrial preparation, nor by the failure of pregnenolone to inhibit cholesterol side-chain cleavage except under abnormal conditions where the cytochrome P_{450}-cholesterol complex accumulates. Second, the studies indicate that the activity of the cholesterol side-chain cleavage system is neither inherently low nor affected by prior treatment with ACTH. Bell and Harding (1974) conclude that ACTH may stimulate steroidogenesis by transporting cholesterol to the cholesterol side-chain cleavage cytochrome P_{450} enzyme, and that the transported cholesterol is rapidly metabolized. Similar conclusions have been reached by Jefcoate et al. (1974).

In those few instances where there is a detailed understanding of the mechanisms involved in cyclic AMP mediation of hormonal effects, as with epinephrine action on glycogenolysis or lipolysis, the primary effect of cyclic AMP is the activation of protein kinase (see Robison et al., 1971). A cyclic AMP-dependent protein kinase isolated from adrenocortical tissue appears to be similar to other mammalian protein kinases (Gill and Garren, 1969, 1970, 1971; Walton and Garren, 1970; Garren et al., 1971b; Langan, 1973). Cyclic AMP-dependent protein kinase phosphorylation of adrenal ribosomal proteins has been described (for reviews see Garren et al., 1971b; Gill, 1972; Wicks, 1974) and, as mentioned before, adrenal cholesterol esterase will act as a substrate (Trzeciak and Boyd, 1974). Richardson and Schulster (1973) have studied the activation of protein kinase by ACTH in isolated rat adrenal cells and have found a rapid and complete activation of the kinase within 2 min of addition of ACTH to the cells. In response to a range of ACTH concentrations, a sigmoid log dose-response curve for protein kinase activation was obtained, with half-maximal stimulation at a concentration of about 1×10^{-3} IU ACTH/ml. As in the studies of ACTH stimulation of cyclic AMP levels (Beall and Sayers, 1972; Mackie et al., 1972; Moyle et al., 1973), certain low doses of ACTH which are clearly steroidogenic fail to cause a clear stimulation of protein kinase activity.

Garren et al. (1971a,b) have suggested that phosphorylation of ribosomes may be involved in the synthesis of the rapidly turning over protein required for cyclic AMP activation of steroidogenesis. Wicks (1974) points out that the degree of phosphorylation is rather extensive, and the number of proteins phosphorylated substantial, and suggests that this phenomenon is more likely to be associated with the trophic effect of cyclic AMP. The only direct evidence for the involvement of cyclic AMP in the trophic effect of ACTH is the finding by Ney (1969) that dibutyryl cyclic AMP given to hypophysectomized rats will cause a partial maintenance of adrenal weight and function. It is very difficult to maintain adequate prolonged stimulation of the adrenal *in vivo* by systemic administration of dibutyryl cyclic AMP, and it is not clear whether the partial effect is due to failure to maintain adequate blood levels of the cyclic nucleotide or whether other factors are involved in the trophic response.

VII. CONCLUDING REMARKS

There is such a wealth of data supporting the concept that cyclic AMP is the sole second messenger mediating the primary activation of adrenal cells by ACTH that any summary statement has to be confined to data that do not seem to fit this hypothesis. If the data are examined with respect to the criteria listed in the Introduction, there is but one facet where we have to be concerned with the possibility that cyclic AMP is not the *sole* second messenger. Clearly, steroidogenesis in adrenal cells can be measurably increased by ACTH under conditions where no increase in cyclic AMP is measurable. It is possible that available techniques of measuring cellular levels of cyclic AMP are inadequate to the task, and it is also possible that it is the concentration of cyclic AMP in a restricted compartment that is important, and currently immeasurable. The last two or three years have produced a considerable sharpening of our understanding of the enzymology of steroidogenesis and the importance of the processes involved in making cholesterol available to mitochondrial enzyme systems. It is true, however, that the link between cyclic AMP formation and cholesterol translocation is still something of an enigma.

VIII. ACKNOWLEDGMENTS

I would like to thank Dr. R. W. Butcher for his constant encouragement and advice during the writing of this review. I would also like to thank Drs. F. G. Péron and A. Haksar, of the Worcester Foundation for Experimental Biology, for providing me with copies of unpublished papers and for many most helpful discussions.

IX. REFERENCES

Albano, J. D. M., Brown, B. L., Ekins, R. P., Price, I., Tait, S. A. S., and Tait, J. F. (1973): The effect of increased K^+ concentration and serotonin on cyclic AMP and corticosterone output by dispersed adrenal zona glomerulosa cells. *Journal of Endocrinology,* 58:xi.

Ariens, E. J., and Simonis, A. M. (1964): A molecular basis for drug action. *Journal of Pharmacy and Pharmacology,* 16:137–157.

Bär, H. P., and Hechter, O. (1969): Adenyl cyclase and hormone action III. Calcium requirement for ACTH stimulation of adenyl cyclase. *Biochemical and Biophysical Research Communications,* 35:681–686.

Bartova, A., and Birmingham, M. K. (1971): Stimulation *in vitro* of lactic acid production by rodent adrenal glands. Effects of glucose, ACTH, 3′,5′-cyclic AMP, ouabain, potassium, anaerobiosis and corticosterone. *Endocrinology,* 88:845–856.

Beall, R. J., and Sayers, G. (1972): Isolated adrenal cells: Steroidogenesis and cyclic AMP accumulation in response to ACTH. *Archives of Biochemistry and Biophysics,* 148:70–76.

Bell, J. J., and Harding, B. W. (1974): The acute action of adrenocorticotropic hormone on adrenal steroidogenesis. *Biochimica et Biophysica Acta,* 348:285–298.

Beyer, K. F., and Samuels, L. T. (1956): Distribution of steroid-3β-ol-dehydrogenase in cellular structures of the adrenal gland. *Journal of Biological Chemistry,* 219:69–76.

Bieck, P., Stock, K., and Westermann, E. (1969): Wirkung von cyclischem Adenosin-3',5'-monophosphat (3',5'-AMP) und seinem Dibutyrylderivat (DBA) auf Lipolyse, Glycogenolyse und Corticosteronsynthese. *Naunyn-Schimiedeberg's Archiv Pharmakologie,* 263:387–405.

Birmingham, M. K., and Bartova, A. (1973): Effects of calcium and theophylline on ACTH- and dibutyryl cyclic AMP-stimulated steroidogenesis and glycolysis by intact mouse adrenal glands in vitro. *Endocrinology,* 92:743–749.

Birmingham, M. K., Elliot, F. H., and Valere, P. H. L. (1953): The need for the presence of calcium for the stimulation in vitro of rat adrenal glands by adrenocorticotrophic hormone. *Endocrinology,* 53:687–689.

Birmingham, M. K., Kurlents, E., Lane, R., Muhlstock, B., and Traikov, H. (1960): Effects of calcium on the potassium and sodium content of rat adrenal glands, on the stimulation of steroid production by adenosine 3',5'-monophosphate, and on the response of the adrenal to short contact with ACTH. *Canadian Journal of Biochemistry,* 38:1077–1085.

Biron, P., Koiw, E., Nowaczynski, W., Brouillet, J., and Genest, J. (1961): The effects of intravenous infusions of valine-5 angiotensin II and other pressor agents on urinary electrolytes and corticosteroids, including aldosterone. *Journal of Clinical Investigation,* 40:338–347.

Bitensky, M. W., and Gorman, R. E. (1972): Chemical mediation of hormone action. *Annual Review of Medicine,* 23:263–284.

Blair-West, J. R., Coghlan, J. P., Denton, D. A., Funder, J. W., Scoggins, B. A., and Wright, R. D. (1971): Effects of prostaglandin E_1 upon the steroid secretion of the adrenal of the sodium deficient sheep. *Endocrinology,* 88:367–371.

Blair-West, J. R., Coghlan, J. P., Denton, D. A., Goding, J. R., Munro, J. A., Peterson, R. E., and Wintour, M. (1962): Humoral stimulation of adrenal cortical secretion. *Journal of Clinical Investigation,* 41:1606–1627.

Blecher, M. (1971): Biological effects and catabolic metabolism of 3',5'-cyclic nucleotides and derivatives in rat adipose tissue and liver. *Metabolism,* 20:63–77.

Boyd, G. S., and Simpson, E. R. (1968): Studies on the conversion of cholesterol to pregnenolone in bovine adrenal mitochondria. In: *Functions of the Adrenal Cortex,* Vol. 1, edited by K. W. McKerns, pp. 49–76. Appleton-Century-Crofts, New York.

Brownie, A. C., and Grant, J. K. (1954): The in vitro hydroxylation of steroid hormones I. Factors influencing the enzymic 11β-hydroxylation of 11-deoxycorticosterone. *Biochemical Journal,* 57: 255–263.

Bumpus, F. M., Smeby, R. R., Page, I. H., and Khairallah, P. A. (1964): Distribution and metabolic fate of angiotensin II and various derivatives. *Canadian Medical Association Journal,* 90:190–193.

Buonassisi, V., Sato, G., and Cohen, A. I. (1962): Hormone-producing cultures of adrenal and pituitary tumor origin. *Proceedings of the National Academy of Sciences (U.S.),* 48:1184–1190.

Butcher, R. W., and Baird, C. E. (1968): Effects of prostaglandins on adenosine 3',5'-monophosphate levels in fat and other tissue. *Journal of Biological Chemistry,* 243:1713–1717.

Butcher, R. W., and Baird, C. E. (1969): The regulation of cyclic AMP and lipolysis in adipose tissue by hormones and other agents. In: *Drugs Affecting Lipid Metabolism,* edited by W. L. Holmes, L. A. Carlson, and R. Paoletti, pp. 5–23. Plenum Press, New York.

Butcher, R. W., Ho, P. J., Meng, H. C., and Sutherland, E. W. (1965): Adenosine 3',5'-monophosphate in biological materials II. The measurement of adenosine 3',5'-monophosphate in tissues and the role of the cyclic nucleotide in the lipolytic response of fat to epinephrine. *Journal of Biological Chemistry,* 240:4515–4523.

Butcher, R. W., and Sutherland, E. W. (1962): Adenosine 3',5'-phosphate in biological materials I. Purification and properties of cyclic 3',5'-nucleotide phosphodiesterase and use of this enzyme to characterize adenosine 3',5'-phosphate in human urine. *Journal of Biological Chemistry,* 237: 1244–1250.

Cammer, W., Cooper, D. Y., and Estabrook, R. W. (1968): Electron transport reactions for steroid hydroxylation by adrenal cortex mitochondria. In: *Functions of the Adrenal Cortex,* Vol. 2, edited by K. W. McKerns, pp. 943–992. Appleton-Century-Crofts, New York.

Carchman, R. A., Jaanus, S. D., and Rubin, R. P. (1971): The role of adrenocorticotropin and calcium in adenosine cyclic 3′,5′-phosphate production and steroid release from the isolated, perfused cat adrenal gland. *Molecular Pharmacology,* 7:491–499.

Carpenter, C. C. J., Davis, J. O., and Ayers, C. R. (1961): Relation of renin, angiotensin II, and experimental hypertension to aldosterone secretion. *Journal of Clinical Investigation,* 40:2026–2042.

Constantopoulos, G., and Tchen, T. T. (1961): Cleavage of cholesterol side chain by adrenal cortex: Cofactor requirement and product of cleavage. *Journal of Biological Chemistry,* 236:65–67.

Cooper, D. Y., Narasimhulu, S., Rosenthal, O., and Estabrook, R. W. (1968): In: *Functions of the Adrenal Cortex, Vol. 2,* edited by K. W. McKerns, pp. 897–942. Appleton-Century-Crofts, New York.

Cuatrecasas, P. (1973a): Interaction of *Vibrio cholerae* enterotoxin with cell membranes. *Biochemistry,* 12:3547–3558.

Cuatrecasas, P. (1973b): Gangliosides and membrane receptors for cholera toxin. *Biochemistry,* 12:3558–3577.

Cuatrecasas, P., and Illiano, G. (1971): Membrane sialic acid and the mechanism of insulin action in adipose tissue cells. *Journal of Biological Chemistry,* 246:4938–4946.

Cushman, P., Alter, S., and Hilton, J. G. (1966): Cortisol secretion by the dog adrenal: Effects of cyclic adenosine monophosphate, dichloroisoproterenol, dihydroergotamine and adrenaline. *Journal of Endocrinology,* 34:271–272.

Davis, W. W. (1969): Stimulation of adrenal cholesterol ester hydrolysis by dibutyryl cyclic AMP. *Federation Proceedings,* 28:701.

Davis, W. W., and Garren, L. D. (1966): Evidence for the stimulation by adrenocorticotropic hormone of the conversion of cholesterol esters to cholesterol in the adrenal, *in vivo. Biochemical and Biophysical Research Communications,* 24:805–810.

Davis, W. W., and Garren, L. D. (1968): On the mechanism of action of adrenocorticotropic hormone. The inhibitory site of cyclohexamide in the pathway of steroid biosynthesis. *Journal of Biological Chemistry,* 243:5153–5157.

Dazord, A., Morera, M., Bertrand, J., and Saez, J. M. (1974): Prostaglandin receptors in human and ovine adrenal glands: Binding and stimulation of adenyl cyclase in subcellular preparations. *Endocrinology,* 95:352–359.

Desaulles, P. A., Barthe, P., Schär, B., and Staehlin, M. (1966): The adrenocorticotrophic properties of α-melanophone-stimulating hormone. *Acta Endocrinologica,* 51:609–618.

Dexter, R. N., Fishman, L. M., Ney, R. L., and Liddle, G. W. (1967a): Inhibition of adrenal corticosteroid synthesis by aminoglutethimide: Studies of the mechanism of action. *Journal of Clinical Endocrinology,* 27:473–480.

Dexter, R. N., Fishman, L. M., Ney, R. L., and Liddle, G. W. (1967b): An effect of adrenocorticotrophic hormone on adrenal cholesterol accumulation. *Endocrinology,* 81:1185–1187.

Doering, C. H., and Clayton, R. B. (1969): Cholesterol side chain cleavage activity in the adrenal gland of the young rat: Development and responsiveness to adrenocorticotropic hormone. *Endocrinology,* 85:500–511.

Donta, S. T., King, M., and Sloper, K. (1973): Induction of steroidogenesis in tissue culture by cholera enterotoxin. *Nature New Biology,* 243:246–247.

Earp, H. S., Watson, B. S., and Ney, R. L. (1970): Adenosine 3′,5′-monophosphate as the mediator of ACTH-induced ascorbic acid depletion in the rat adrenal. *Endocrinology,* 87:118–123.

Farese, R. V. (1971a): On the requirement for calcium during the steroidogenic effect of ACTH. *Endocrinology,* 89:1057–1063.

Farese, R. V. (1971b): Calcium as a mediator of adrenocorticotrophic hormone action on adrenal protein synthesis. *Science,* 173:447–450.

Farese, R. V. (1971c): Stimulation of pregnenolone synthesis by ACTH in rat adrenal sections. *Endocrinology,* 89:958–962.

Ferguson, J. J., Jr. (1963): Protein synthesis and adrenocorticotrophin responsiveness. *Journal of Biological Chemistry,* 238:2754–2759.

Ferguson, J. J., Jr. (1972): The mechanism of action of adrenocorticotropic hormone. In: *Biochemical Actions of Hormones,* edited by G. Litwack, pp. 317–335. Academic Press, New York.

Ferrari, R. A., and Arnold, A. (1963): Inhibition of β-hydroxysteroid dehydrogenase. I. Structural characteristics of some steroidal inhibitors. *Biochimica et Biophysica Acta,* 77:349–356.

Finn, F. M., Widnell, C. C., and Hofmann, K. (1972): Localization of an adrenocorticotropic

hormone receptor on bovine adrenal cortical membranes. *Journal of Biological Chemistry,* 247: 5695–5702.
Flack, J. D., Jessup, R., and Ramwell, P. W. (1968): Effect of prostaglandins on the pituitary-adrenal axis in the intact and hypophysectomized rat. *Proceedings of the 24th International Union of Physiological Sciences,* Washington, D.C., 7:137.
Flack, J. D., Jessup, R., and Ramwell, P. W. (1969): Prostaglandin stimulation of rat corticosteroidogenesis. *Science,* 163:691–692.
Flack, J. D., and Ramwell, P. W. (1972): A comparison of the effects of ACTH, cyclic AMP, dibutyryl cyclic AMP, and PGE_2 on corticosteroidogenesis in vitro. *Endocrinology,* 90:371–377.
Free, C. A., Chasin, M., Paik, V. S., and Hess, S. M. (1971): Steroidogenesis and lipolytic activities of 8-substituted derivatives of cyclic 3',5'-adenosine monophosphate. *Biochemistry,* 10:3785–3789.
Gallant, S., and Brownie, A. C. (1973): The in vivo effect of indomethacin and prostaglandin E_2 on ACTH and DBCAMP-induced steroidogenesis in hypophysectomized rats. *Biochemical and Biophysical Research Communications,* 55:831–836.
Garren, L. D. (1968): The mechanism of action of adrenocorticotropic hormone. *Vitamins and Hormones,* 26:119–145.
Garren, L. D., Gill, G. N., Masui, H., and Walton, G. M. (1971a): On the mechanism of action of ACTH. *Recent Progress in Hormone Research,* 27:433–478.
Garren, L. D., Gill, G. N., and Walton, G. M. (1971b): The isolation of a receptor for adenosine-3',5'-cyclic monophosphate (cAMP) from adrenal cortex: The role of the receptor in the mechanism of action of cAMP. *Annals of the New York Academy of Sciences,* 185:210–226.
Garren, L. D., Ney, R. L., and Davis, W. W. (1965): Studies on the role of protein synthesis in the regulation of corticosterone production by adrenocorticotrophic hormone in vivo. *Proceedings of the National Academy of Sciences (U.S.),* 53:1443–1450.
Gill, G. N. (1972): Mechanism of ACTH action. *Metabolism,* 21:571–588.
Gill, G. N., and Garren, L. D. (1969): On the mechanism of action of adrenocorticotropic hormone: The binding of cyclic-3',5'-adenosine monophosphate to an adrenal cortical protein. *Proceedings of the National Academy of Sciences (U.S.),* 63:512–519.
Gill, G. N., and Garren, L. D. (1970): A cyclic-3',5'-adenosine monophosphate dependent protein kinase from the adrenal cortex: Comparison with a cyclic AMP binding protein. *Biochemical and Biophysical Research Communications,* 39:335–343.
Gill, G. N., and Garren, L. D. (1971): Role of the receptor in the mechanism of action of adenosine 3',5'-cyclic monophosphate. *Proceedings of the National Academy of Sciences (U.S.),* 68:786–790.
Gilman, A. G. (1970): A protein binding assay for adenosine 3':5'-monophosphate. *Proceedings of the National Academy of Sciences (U.S.),* 67:305–312.
Glinsman, W. H., Hern, E. P., Linarelli, L. G., and Farese, R. V. (1969): Similarities between effects of adenosine 3',5'-monophosphate and guanosine 3',5'-monophosphate on liver and adrenal metabolism. *Endocrinology,* 85:711–719.
Glossman, H., Baukal, A. J., and Catt, K. J. (1973): Angiotensin II receptors of the adrenal cortex. *Clinical Research,* 21:492.
Glossman, H., Baukal, A. J., and Catt, K. J. (1974a): Properties of angiotensin II receptors in the bovine and rat adrenal cortex. *Journal of Biological Chemistry,* 249:825–834.
Glossman, H., Baukal, A. J., and Catt, K. J. (1974b): Angiotensin II receptors in bovine adrenal cortex. Modification of angiotensin II binding by guanyl nucleotides. *Journal of Biological Chemistry,* 249:664–666.
Goldberg, N. D., O'Dea, R. F., and Haddox, M. K. (1973): Cyclic GMP. *Advances in Cyclic Nucleotide Research,* 3:155–233.
Goldman, A. S., Yakovac, W. C., and Bongiovanni, A. M. (1965): Persistent effects of a synthetic androstene derivative on activities of 3β-hydroxysteroid dehydrogenase and glucose-6-phosphate dehydrogenase in rats. *Endocrinology,* 77:1105–1118.
Goodfriend, T. L., and Lin, S. Y. (1970): Receptors for angiotensin I and II. *Circulation Research,* Supplement I to Vols XXVI and XXVII, I:163–170.
Goodman, D. S. (1965): Cholesterol ester metabolism. *Physiological Review,* 45:747–839.
Grahame-Smith, D. G., Butcher, R. W., Ney, R. L., and Sutherland, E. W. (1967): Adenosine 3',5'-monophosphate as the intracellular mediator of the action of adrenocorticotropic hormone on the adrenal cortex. *Journal of Biological Chemistry,* 242:5535–5541.
Haksar, A., Baniukiewicz, S., and Péron, F. G. (1973): Inhibition of ACTH-stimulated steroidogene-

sis in isolated rat adrenal cells treated with neuraminidase. *Biochemical and Biophysical Research Communications,* 52:959–966.
Haksar, A., Maudsley, D. V., Kimmel, G. L., and Péron, F. G. (1974a): ACTH-stimulation of cyclic adenosine 3′,5′-monophosphate formation in isolated rat adrenal cells. The role of membrane sialic acid. *Biochimica et Biophysica Acta,* 362:356–365.
Haksar, A., Maudsley, D. V., and Péron, F. G. (1974b): Neuraminidase treatment of adrenal cells increases their response to cholera enterotoxin. *Nature,* 251:514–515.
Haksar, A., and Péron, F. G. (1972a): Comparison of the Ca^{2+} requirement for the steroidogenic effect of ACTH and dibutyryl cyclic AMP in rat adrenal cell suspensions. *Biochemical and Biophysical Research Communications,* 47:445–450.
Haksar, A., and Péron, F. G. (1972b): The possible function of pyruvate in the steroidogenic response of rat adrenal cell suspensions to ACTH, cyclic-3′,5′-AMP and dibutyryl cyclic-3′,5′-AMP. *Journal of Steroid Biochemistry,* 3:847–857.
Haksar, A., and Péron, F. G. (1973): The role of calcium in the steroidogenic response of rat adrenal cells to adrenocorticotropic hormone. *Biochimica et Biophysica Acta,* 313:363–371.
Halkerston, I. D. K., Eichhorn, J., and Hechter, O. (1961): A requirement for reduced triphosphopyridine nucleotide for cholesterol side-chain cleavage by mitochondrial fractions of bovine adrenal cortex. *Journal of Biological Chemistry,* 236:374–380.
Halkerston, I. D. K., Feinstein, M., and Hechter, O. (1966): An anomalous effect of theophylline on ACTH and adenosine 3′,5′-monophosphate stimulation. *Proceedings of the Society for Experimental Biology and Medicine,* 122:896–900.
Hall, P. F., and Young, D. G. (1968): Site of action of tropic hormones upon the biosynthetic pathways to steroid hormones. *Endocrinology,* 82:559–568.
Harding, B. W., Bell, J. J., Oldham, S. B., and Wilson, L. D. (1968): Corticosteroid biosynthesis in adrenal cortical mitochondria. In: *Functions of the Adrenal Cortex, Vol. 2,* edited by K. W. McKerns, pp. 831–896. Appleton-Century-Crofts, New York.
Hardman, J. G., Robison, G. A., and Sutherland, E. W. (1971): Cyclic nucleotides. *Annual Review of Physiology,* 33:311–336.
Hayano, M., Saba, N., Dorfman, R. I., and Hechter, O. (1956): Some aspects of the biogenesis of adrenal steroid hormones. *Recent Progress in Hormone Research,* 12:79–123.
Haynes, R. C., Jr. (1958): The activation of adrenal phosphorylase by the adrenocorticotropic hormone. *Journal of Biological Chemistry,* 233:1220–1222.
Haynes, R. C., Jr., and Berthet, L. (1957): Studies on the mechanism of action of the adrenocorticotropic hormone. *Journal of Biological Chemistry,* 225:115–124.
Haynes, R. C., Jr., Koritz, S. B., and Péron, F. G. (1959): Influence of adenosine 3′,5′-monophosphate on corticoid production by rat adrenal glands. *Journal of Biological Chemistry,* 234:1421–1423.
Haynes, R. C., Jr., Sutherland, E. W., and Rall, T. W. (1960): The role of cyclic adenylic acid in hormone action. *Recent Progress in Hormone Research,* 16:121–138.
Hechter, O. (1955): Concerning possible mechanisms of hormone action. *Vitamins and Hormones,* 13:293–346.
Hechter, O. (1958): Conversion of cholesterol to steroid hormones. In: *Cholesterol,* edited by R. P. Cook, p. 309. Academic Press, New York.
Henion, W. F., Sutherland, E. W., and Posternak, T. (1967): Effects of derivatives of adenosine 3′,5′-phosphate on liver slices and intact animals. *Biochimica et Biophysica Acta,* 148:106–113.
Hilton, J. G., Kruesi, O. R., Nedeljkovic, R. I., and Scian, L. F. (1961): Adreno-cortical and medullary responses to adenosine 3′,5′-monophosphate. *Endocrinology,* 68:908–913.
Hilton, J. G., Scian, L. F., Westermann, C. D., Nakano, J., and Kruesi, O. R. (1960): Vasopressin stimulation of the isolated adrenal glands. Nature and mechanism of hydrocortisone secretion. *Endocrinology,* 67:298–310.
Hofmann, K. (1960): Preliminary observations relating structure and function in some pituitary hormones. In: *Protein, Structure and Function, Brookhaven Symposia in Biology,* 13:184–202.
Hofmann, K. (1962): Chemistry and function of polypeptide hormones. *Annual Review of Biochemistry,* 31:213–246.
Hofmann, K., Andreatta, R., Bohn, H., and Moroder, L. (1970a): Studies on polypeptides. XLV. Structure-function studies in the β-corticotropin series. *Journal of Medicinal Chemistry,* 13:339–345.
Hofmann, K., Wingender, W., and Finn, F. M. (1970b): Correlation of adrenocorticotropic activity

of ACTH analogs with degree of binding to an adrenal cortical particulate preparation. *Proceedings of the National Academy of Sciences (U.S.),* 67:829–836.

Holmgren, J., Lönnroth, I., and Svennerholm, L. (1973): Tissue receptor for cholera enterotoxin: Postulated structure from studies with G_{M_1} ganglioside and related glycolipids. *Infection and Immunity,* 8:208–214.

Ide, M., Tanaka, A., Nakamura, N., and Okabayashi, T. (1972): Stimulation by ACTH analogs of rat adrenal adenyl cyclase activity: Correlation with steroidogenic activity. *Archives of Biochemistry and Biophysics,* 149:189–196.

Imura, H., Matsukura, S., Matsuyama, H., Setsuda, T., and Miyake, T. (1965): Adrenal steroidogenic effect of adenosine 3',5'-monophosphate and its derivatives in vivo. *Endocrinology,* 76:933–937.

Jaanus, S. D., Carchman, R. A., and Rubin, R. P. (1972): Further studies on the relationship between cyclic AMP levels and adrenocortical activity. *Endocrinology,* 91:887–895.

Jaanus, S. D., Rosenstein, M. J., and Rubin, R. P. (1970): On the mode of action of ACTH on the isolated perfused adrenal gland. *Journal of Physiology,* 209:539–556.

Jaanus, S. D., and Rubin, R. P. (1971): The effect of ACTH on calcium distribution in the perfused cat adrenal gland. *Journal of Physiology,* 213:581–598.

Jefcoate, C. R., and Gaylor, J. L. (1970): Ligand interactions with hemoprotein P-450. Equilibria between high- and low-spin forms of P-450 in bovine adrenal mitochondria. *Biochemistry,* 9:3816–3823.

Jefcoate, C. R., Hume, R., and Boyd, G. S. (1970): Separation of two forms of cytochrome P450. Adrenal cortex mitochondria. *FEBS Letters,* 9:41–44.

Jefcoate, C. R., Simpson, E. R., and Boyd, G. S. (1974): Spectral properties of rat adrenal-mitochondrial cytochrome P-450. *European Journal of Biochemistry,* 42:539–551.

Jost, J. P., and Rickenberg, H. V. (1971): Cyclic AMP. *Annual Review of Biochemistry,* 40:741–774.

Kahnt, F. W., Milani, A., Steffen, H., and Neher, R. (1974): The rate-limiting step of adrenal steroidogenesis and adenosine 3':5'-monophosphate. *European Journal of Biochemistry,* 44:243–250.

Kahnt, F. W., and Neher, R. (1966): Über die adrenale Steroid-biosynthese in Vitro. III. Selektive Hemmung der Nebennierenrindenfunktion. *Helvetica Chimica Acta,* 49:725–732.

Kan, K. W., Ritter, M. C., Ungar, F., and Dempsey, M. E. (1972): The role of a carrier protein in cholesterol and steroid hormone synthesis by adrenal enzymes. *Biochemical and Biophysical Research Communications,* 48:423–429.

Kan, K. W., and Ungar, F. (1973): Characterization of an adrenal activator for cholesterol side chain cleavage. *Journal of Biological Chemistry,* 248:2868–2875.

Kaplan, N. M. (1965): The biosynthesis of adrenal steroids: Effects of angiotensin II, adrenocorticotropin, and potassium. *Journal of Clinical Investigation,* 44:2029–2039.

Kaplan, N. M., and Bartter, F. C. (1962): The effect of ACTH, renin, angiotensin II, and various precursors on biosynthesis of aldosterone by adrenal slices. *Journal of Clinical Investigation,* 41:715–724.

Karaboyas, G. C., and Koritz, S. B. (1965): Identity of the site of action of 3',5'-adenosine monophosphate and adrenocorticotropic hormone in corticosteroidogenesis in rat adrenal and beef adrenal cortex slices. *Biochemistry,* 4:462–468.

Kaukel, E., and Hilz, H. (1972): Permeation of dibutyryl cAMP into HeLa cells and its conversion to monobutyryl cAMP. *Biochemical and Biophysical Research Communications,* 46:1011–1018.

Kaukel, E., Mundhenk, K., and Hilz, H. (1972): N^6-monobutyryl adenosine 3':5'-monophosphate as the biologically active derivate of dibutyryl adenosine 3':5'-monophosphate in HeLa cells. *European Journal of Biochemistry,* 27:197–200.

Kelly, L. A., and Koritz, S. B. (1971): Bovine adrenal cortical adenyl cyclase and its stimulation by adrenocorticotropic hormone and NaF. *Biochimica et Biophysica Acta,* 237:141–155.

Kimberg, D. V., Field, M., Johnson, J., Henderson, A., and Gershon, E. (1971): Stimulation of intestinal mucosal adenyl cyclase by cholera enterotoxin and prostaglandins. *Journal of Clinical Investigation,* 50:1218–1230.

Kimura, T. (1968): Electron transfer system of steroid hydroxylases in adrenal mitochondria. In: *Function of the Adrenal Cortex, Vol. 2,* edited by K. W. McKerns, pp. 993–1006. Appleton-Century-Crofts, New York.

Kimura, T. (1969): Effects of hypophysectomy and ACTH administration on the level of adrenal cholesterol side-chain desmolase. *Endocrinology,* 85:492–499.

King, C. A., and van Heyningen, W. E. (1973): Deactivation of cholera toxin by a sialidase resistant monosialosylganglioside. *Journal of Infectious Diseases,* 127:639–647.

Kitabchi, A. E., Nathans, A. H., James, P., Bower, F., Wilson, D. B., and Kitchell, L. C. (1974): Cyclic 3',5'-GMP (cGMP) as possible mediator of steroidogenic action of ACTH at physiologic concentrations. *56th Annual Meeting, Endocrine Society, Atlanta, Ga.* Abstract #210. Supplement to *Endocrinology,* Vol. 94.

Kitabchi, A. E., and Sharma, R. K. (1971): Corticosteroidogenesis in isolated adrenal cells of rats. I. Effect of corticotropins and 3',5'-cyclic nucleotides on corticosterone production. *Endocrinology,* 88:1109–1116.

Kitabchi, A. E., Wilson, D. B., and Sharma, P. K. (1971): Steroidogenesis in isolated cells of rat. II. Effect of caffeine on ACTH and cyclic nucleotide-induced steroidogenesis and its relation to cyclic nucleotide phosphodiesterase. *Biochemical and Biophysical Research Communications,* 44: 898–904.

Kong, Y. C., Moyle, W. R., and Ramachandran, J. (1972): Inhibition of ACTH induced cyclic AMP synthesis in isolated rat adrenal cells by NPS-ACTH. *Proceedings of the Society for Experimental Biology and Medicine,* 141:350–352.

Koritz, S. B. (1962): The effect of calcium ions and freezing on the in vitro synthesis of pregnenolone by rat adrenal preparations. *Biochimica et Biophysica Acta,* 56:63–75.

Koritz, S. B. (1968): On the regulation of pregnenolone synthesis. In: *Functions of the Adrenal Cortex, Vol. 1,* edited by K. W. McKerns, pp. 27–48. Appleton-Century-Crofts, New York.

Koritz, S. B., and Hall, P. F. (1964a): End-product inhibition of the conversion of cholesterol to pregnenolone in an adrenal extract. *Biochemistry,* 3:1298–1304.

Koritz, S. B., and Hall, P. F. (1964b): Feedback inhibition by pregnenolone: A possible mechanism. *Biochimica et Biophysica Acta,* 93:215–217.

Koritz, S. B., and Kumar, A. M. (1970): On the mechanism of action of the adrenocorticotrophic hormone. The stimulation of the activity of enzymes involved in pregnenolone synthesis. *Journal of Biological Chemistry,* 245:152–159.

Koritz, S. B., and Péron, F. G. (1959): The stimulation in vitro by Ca^{2+}, freezing, and proteolysis of corticoid production by rat adrenal tissue. *Journal of Biological Chemistry,* 234:3122–3128.

Koritz, S. B., Yun, J., Ferguson, J. J., Jr. (1968): Inhibition of adrenal progesterone biosynthesis by 3',5'-AMP. *Endocrinology,* 82:620–622.

Kowal, J. (1970a): ACTH and the metabolism of adrenal cell cultures. *Recent Progress in Hormone Research,* 26:623–676.

Kowal, J. (1970b): Adrenal cells in tissue culture. VII. Effect of inhibitors of protein synthesis on steroidogenesis and glycolysis. *Endocrinology,* 87:951–965.

Kowal, J. (1973): Adrenal cells in tissue culture. X. On the mechanism of the stimulation of steroidogenesis by cyclic cytidine monophosphate. *Endocrinology,* 93:461–468.

Kowal, J., and Fiedler, R. M. (1969): Studies of adrenal cells in tissue culture. II. Steroidogenic responses to nucleosides and nucleotides. *Endocrinology,* 84:1113–1117.

Langan, T. A. (1973): Protein kinase and protein kinase substrates. *Advances in Cyclic Nucleotide Research,* 3:99–153.

Laragh, J. H., Angers, M., Kelly, W. G., and Lieberman, S. (1960): Hypotensive agents and pressor substances. *Journal of the American Medical Association,* 174:234–240.

Lefkowitz, R. J., Roth, J., and Pastan, I. (1970a): Effects of calcium on ACTH stimulation of the adrenal: Separation of hormone binding from adenyl cyclase activation. *Nature,* 228:864–866.

Lefkowitz, R. J., Roth, J., and Pastan, I. (1971): ACTH-receptor interaction in the adrenal: A model for the initial step in the action of hormones that stimulate adenyl cyclase. *Annals of the New York Academy of Sciences,* 185:195–209.

Lefkowitz, R. J., Roth, J., Pricer, W., and Pastan, I. (1970b): ACTH receptors in the adrenal: Specific binding of ACTH-^{125}I and its relation to adenyl cyclase. *Proceedings of the National Academy of Sciences (U.S.),* 65:745–752.

Leier, D. J., and Jungmann, R. A. (1971): Stimulation of adrenal cholesterol side chain cleavage activity by theophylline. *Biochimica et Biophysica Acta,* 239:320–328.

Leier, D. J., and Jungmann, R. A. (1973): Adrenocorticotropic hormone and dibutyryl adenosine cyclic monophosphate-mediated Ca^{2+} uptake by rat adrenal glands. *Biochimica et Biophysica Acta,* 329:196–210.

Liddle, G. W., Island, D., and Meador, C. K. (1962): Normal and abnormal regulation of corticotropin secretion in man. *Recent Progress in Hormone Research.* 18:125–166.

Lin, M., Haksar, A., and Péron, F. G. (1974): The role of the Krebs cycle in the generation of intramitochondrial reducing equivalents for the 11β-hydroxylation of deoxycorticosterone in isolated rat adrenal cells. *Archives of Biochemistry and Biophysics,* 164:429–439.

Mackie, C., Richardson, M. C., and Schulster, D. (1972): Kinetics and dose response characteristics of adenosine 3′,5′-monophosphate production by isolated rat adrenal cells stimulated with adrenocorticotrophic hormone. *FEBS Letters,* 23:345–348.

Mackie, C., and Schulster, D. (1973): Phosphodiesterase activity and the potentiation by theophylline of adrenocorticotrophin stimulated steroidogenesis and adenosine 3′,5′-monophosphate levels in isolated rat adrenal cells. *Biochemical and Biophysical Research Communications,* 53:545–551.

Mahaffee, D., and Ney, R. L. (1970): Effects of nucleotides possessing a 3′,5′-cyclic monophosphate on adrenal steroidogenesis. *Metabolism,* 19:1104–1108.

Mahaffee, D., Reitz, R. C., and Ney, R. L. (1974): The mechanism of action of adrenocorticotropic hormone. The role of mitochondrial cholesterol accumulation in the regulation of steroidogenesis. *Journal of Biological Chemistry,* 249:227–233.

Mahaffee, D., Watson, B., and Ney, R. L. (1970): The relationship between nucleotide structure and the stimulation of adrenal steroidogenesis. *Clinical Research,* 18:73.

Martou, J., Stark, E., Katalin, M., and Varga, B. (1971): Effects of theophylline on corticosterone secretory and adrenal growth response to ACTH in vivo. *Acta Physiologica Academiae Scientiarum Hungaricae,* 40:229–233.

McKune, R. W., Roberts, S., and Young, P. L. (1970): Competitive inhibition of adrenal Δ^5-3β-hydroxy steroid dehydrogenase and Δ^5-3 keto-steroid isomerase activities by adenosine 3′,5′-monophosphate. *Journal of Biological Chemistry,* 245:3859–3867.

Morris, M. D., and Chaikoff, I. L. (1959): The origin of cholesterol in liver, small intestine, adrenal gland, and testis of the rat: Dietary versus endogenous contributions. *Journal of Biological Chemistry,* 234:1095–1097.

Moses, H. L., Davis, W. W., Rosenthal, A. S., and Garren, L. D. (1969): Adrenal cholesterol: Localization by electron-microscope autoradiography. *Science,* 163:1203–1205.

Moyle, W. R., Kong, Y. C., and Ramachandran, J. (1973): Steroidogenesis and cyclic adenosine 3′,5′-monophosphate accumulation in rat adrenal cells. Divergent effects of adrenocorticotropin and its o-nitrophenyl sulfenyl derivative. *Journal of Biological Chemistry,* 248:2409–2417.

Mulrow, P. J. (1972): The adrenal cortex. *Annual Review of Physiology,* 34:409–424.

Muneyama, K., Bauer, R. J., Shuman, D. A., Robins, R. K., and Simon, L. (1971): Chemical synthesis and biological activity of 8-substituted adenosine 3′,5′-cyclic monophosphate derivatives. *Biochemistry,* 10:2390–2395.

Munson, P. L., and Toepel, W. (1958): Detection of minute amounts of adrenocorticotropic hormone by the effect on adrenal venous ascorbic acid. *Endocrinology,* 63:785–793.

Nakamura, M., Ide, M., Okabayashi, T., and Tanaka, A. (1972): Relation between steroidogenesis and 3′,5′-cyclic AMP production in isolated adrenal cells. *Endocrinologica Japonica,* 19:443–448.

Neville, A. M., and Engel, L. L. (1968): Inhibition of α- and β-hydroxy steroid dehydrogenases and steroid Δ-isomerase by substrate analogues. *Journal of Clinical Endocrinology,* 28:49–60.

Ney, R. L. (1969): Effects of dibutyryl cyclic AMP on adrenal growth and steroidogenic capacity. *Endocrinology,* 84:168–170.

Pastan, I., Pricer, W., and Blanchette-Mackie, J. (1970): Studies of an ACTH-activated adenyl cyclase from a mouse adrenal tumor. *Metabolism,* 19:809–817.

Pearlmutter, A. F., Rapino, E., and Saffran, M. (1971): ACTH and cyclic adenine nucleotides do not provoke identical adrenocortical responses. *Endocrinology,* 89:963–968.

Pearlmutter, A. F., Rapino, E., and Saffran, M. (1973): Comparison of steroidogenic effects of cAMP and dbcAMP in the rat adrenal gland. *Endocrinology,* 92:679–686.

Peng, T. C., Six, K. M., and Munson, P. L. (1970): Effects of prostaglandin E_1 on the hypothalamo-hypophyseal-adrenocortical axis in rats. *Endocrinology,* 86:202–206.

Péron, F. G., and Koritz, S. B. (1958): On the exogenous requirements for the action of ACTH in vitro on rat adrenal glands. *Journal of Biological Chemistry,* 233:256–259.

Péron, F. G., and Koritz, S. B. (1960): On the location of the stimulation in vitro by Ca^{++} and freezing of corticoid production by rat adrenal homogenates. *Journal of Biological Chemistry,* 235:1625–1628.

Péron, F. G., McCarthy, J. L., and Guerra, F. (1966): Inhibition of utilization of biological substrates for corticoid synthesis by high calcium concentrations. Possible role of transhydrogenase in corticosteroidogenesis. *Biochimica et Biophysica Acta,* 117:450–469.

Péron, F. G., and Tsang, C. P. W. (1969): Further studies on steroidogenesis. VI. Pyruvate and malate supported steroid 11β-hydroxylation in rat adrenal gland mitochondria. *Biochimica et Biophysica Acta,* 180:445–458.
Peytreman, A., Nicholson, W. E., Brown, R. D., Liddle, G. W., and Hardman, J. (1973a): Comparative effects of angiotensin and ACTH on cyclic AMP and steroidogenesis in isolated bovine adrenal cells. *Journal of Clinical Investigation,* 52:835–842.
Peytreman, A., Nicholson, W. E., Liddle, G. W., Hardman, J. G., and Sutherland, E. W. (1973b): Effects of methylxanthines on adenosine 3',5'-monophosphate and corticosterone in the rat adrenal. *Endocrinology,* 92:525–530.
Pierce, N. F., Carpenter, C. C. J., Elliott, H. L., and Greenough, W. B. (1971): Effects of prostaglandins, theophylline, and cholera exotoxin upon transmucosal water and electrolyte movement in canine jegunum. *Gastroenterology,* 60:22–32.
Purvis, J. L., Battu, R. G., and Péron, F. G. (1968): Generation and utilization of reducing power in the conversion of 11-deoxycorticosterone to corticosterone in rat adrenal mitochondria. In: *Functions of the Adrenal Cortex, Vol. 2,* edited by K. W. McKerns, pp. 1007–1055. Appleton-Century-Crofts, New York.
Purvis, J. L., Canick, J. I., Mason, J. I., McCarthy, J. L., and Estabrook, R. W. (1973): Life time of adrenal cytochrome-P450 as influenced by ACTH. *Annals of the New York Academy of Sciences,* 212:319–343.
Rall, T. W., and Sutherland, E. W. (1961): The regulatory role of adenosine-3',5'-phosphate. *Cold Spring Harbor Symposia on Quantitative Biology,* 26:347–354.
Ramachandran, J., Chung, D., and Li, C. H. (1965): Adrenocorticotropins. XXXIV. Aspects of structure-activity relationship of the ACTH molecule. Synthesis of a heptadecapeptide amide, an octadecapeptide amide, and a nonadecapeptide amide possessing high biological activities. *Journal of the American Chemical Society,* 87:2696–2708.
Ramachandran, J., and Lee, V. (1970a): Preparation and properties of the o-nitrophenyl sulfenyl derivative of ACTH: An inhibitor of the lipolytic action of the hormone. *Biochemical and Biophysical Research Communications,* 38:507–512.
Ramachandran, J., and Lee, V. (1970b): Divergent effects of o-nitrophenyl sulfenyl ACTH on rat and rabbit fat cell adenyl cyclases. *Biochemical and Biophysical Research Communications,* 41:-358–366.
Richardson, M. C., and Schulster, D. (1972): Corticosteroidogenesis in isolated adrenal cells: Effect of adrenocorticotrophic hormone, adenosine 3',5'-monophosphate and β^{1-24} adrenocorticotrophic hormone diazotized to polyacrylamide. *Journal of Endocrinology,* 55:127–139.
Richardson, M. C., and Schulster, D. (1973): The role of protein kinase activation in the control of steroidogenesis by adrenocorticotrophic hormone in the adrenal cortex. *Biochemical Journal,* 136: 993–998.
Richman, R., Dobbins, C., Voina, S., Underwood, L., Mahaffee, D., Gitelman, H. J., Van Wyk, J., and Ney, R. L. (1973): Regulation of ornithine decarboxylase by adrenocorticotropic hormone and cyclic AMP. *Journal of Clinical Investigation,* 52:2007–2015.
Rivkin, I., and Chasin, M. (1971): Nucleotide specificity of the steroidogenic response of rat adrenal cell suspensions prepared by collagenase digestion. *Endocrinology,* 88:664–670.
Roberts, S., McCune, R. W., Creange, J. E., and Young, P. L. (1967): Adenosine-3',5'-cyclic phosphate stimulation of steroidogenesis in sonically disrupted adrenal mitochondria. *Science,* 158: 372–374.
Robison, G. A., Butcher, R. W., and Sutherland, E. W. (1971): *Cyclic AMP.* Academic Press, New York.
Rubin, R. P., Carchman, R. A., and Jaanus, S. D. (1972a): Role of calcium and adenosine cyclic 3',5'-phosphate in action of adrenocorticotropin. *Nature New Biology,* 240:150–152.
Rubin, R. P., Carchman, R. A., and Jaanus, S. D. (1972b): Role of cyclic 3',5'-adenosine monophosphate on corticosteroid synthesis and release from the intact adrenal gland. *Biochemical and Biophysical Research Communications,* 47:1492–1497.
Rubin, R. P., Jaanus, S. D., Carchman, R. A., and Puig, M. (1973): Reversible inhibition of ACTH-induced corticosteroid release by cyclohexamide: Evidence for an unidentified cellular messenger. *Endocrinology,* 93:575–580.
Rubin, R. P., Jaanus, S. D., and Miele, E. (1969): Calcium dependent corticosteroid release from the perfused cat adrenal gland. *Experentia,* 25:1327–1328.
Rubin, R. P., Shield, B., McCauley, R., and Laycock, S. G. (1974): ACTH-induced protein release from the perfused cat adrenal gland: Evidence for exocytosis? *Endocrinology,* 95:370–378.

Rudinger, J., and Krejcí, I. (1962): Dose-response relations for some synthetic analogues of oxytocin, and the mode of action of oxytocin on the isolated uterus. *Experentia,* 18:585–588.
Saruta, T., and Kaplan, N. M. (1972): Adrenocortical steroidogenesis: The effects of prostaglandins. *Journal of Clinical Investigation,* 51:2246–2251.
Sayers, G., Beall, R. J., and Seelig, S. (1972): Isolated adrenal cells: Adrenocorticotropic hormone, calcium, steroidogenesis, and cyclic adenosine monophosphate. *Science,* 175:1131–1133.
Sayers, G., Ma, R. M., and Giordano, N. D. (1971): Isolated adrenal cells: Corticosterone production in response to cyclic AMP (adenosine 3',5'-monophosphate). *Proceedings of the Society for Experimental Biology and Medicine,* 136:619–622.
Sayers, M. S., Sayers, G., and Woodbury, L. A. (1948): The assay of adrenocorticotrophic hormone by the adrenal ascorbic acid-depletion method. *Endocrinology,* 42:379–393.
Schaefer, D. E., Lust, W. D., Sircar, B., and Goldberg, N. D. (1970): Elevated concentration of adenosine 3':5'-cyclic monophosphate in intestinal mucosa after treatment with cholera toxin. *Proceedings of the National Academy of Sciences (U.S.),* 67:851–856.
Schimke, R. T., and Doyle, D. (1970): Control of enzyme levels in animal tissues. *Annual Review of Biochemistry,* 39:929–976.
Schimmer, B. P., Ueda, K., and Sato, G. H. (1968): Site of action of adrenocorticotropic hormone. *Biochemical and Biophysical Research Communications,* 32:806–810.
Schorr, I., and Ney, R. L. (1971): Abnormal hormone responses of an adrenocortical cancer adenyl cyclase. *Journal of Clinical Investigation,* 50:1295–1300.
Schulster, D., Richardson, M. C., and Mackie, C. (1972): The mode of ACTH action in stimulating steroidogenesis; the obligatory role of adenosine 3':5'-cyclic monophosphate and the involvement of a rapidly turning over protein. *Biochemical Journal,* 129:8p–9p.
Schulster, D., Tait, S. A. S., Tait, J. F., and Mrotek, J. (1970): Production of steroids by in vitro superfusion of endocrine tissue. III. Corticosterone output from rat adrenals stimulated by adrenocorticotropin or cyclic 3',5'-adenosine monophosphate and the inhibitory effect of cyclohexamide. *Endocrinology,* 86:487–502.
Schwyzer, R. (1964): Chemistry and metabolic action of nonsteroid hormones. *Annual Review of Biochemistry,* 33:259–286.
Schwyzer, R., Schiller, P., Seelig, S., and Sayers, G. (1971): Isolated adrenal cells: Log dose response curves for steroidogenesis induced by $ACTH_{1-24}$, $ACTH_{4-10}$, and $ACTH_{5-10}$. *FEBS Letters,* 19:229–231.
Seelig, S., Kumar, S., and Sayers, G. (1972): Isolated adrenal cells: The partial agonists [Trp(Nps)^9] $ACTH_{1-39}$ and [Trp(Nps)^9] $ACTH_{1-24}$ (nitrophenyl sulfenyl derivatives of ACTH). *Proceedings of the Society for Experimental Biology and Medicine,* 139:1217–1219.
Seelig, S., and Sayers, G. (1973): Isolated adrenal cortex cells: ACTH agonists, partial agonists, antagonists; cyclic AMP and corticosterone production. *Archives of Biochemistry and Biophysics,* 154:230–239.
Selinger, R. C. L., and Civen, M. (1971): ACTH diazotized to agarose: Effects on isolated adrenal cells. *Biochemical and Biophysical Research Communications,* 43:793–799.
Sharma, R. K. (1973): Regulation of steroidogenesis by adrenocorticotropic hormone in isolated adrenal cells of rat. *Journal of Biological Chemistry,* 248:5473–5476.
Sharp, G. W. G., and Hynie, S. (1971): Stimulation of intestinal adenyl cyclase by cholera toxin. *Nature,* 229:266–269.
Shimizu, K., Hayano, M., Gut, M., and Dorfman, R. I. (1961): The transformation of 20α-hydroxy cholesterol to isocaproic acid and C_{21} steroids. *Journal of Biological Chemistry,* 236:695–699.
Simpson, E. R., Cooper, D. Y., and Estabrook, R. W. (1969): Metabolic events associated with steroid hydroxylation by the adrenal cortex. *Recent Progress in Hormone Research,* 25:523–562.
Simpson, E. R., Jefcoate, C. R., Brownie, A. C., and Boyd, G. S. (1972): The effect of ether anaesthesia stress on cholesterol-side-chain cleavage and cytochrome P-450 in rat-adrenal mitochondria. *European Journal of Biochemistry,* 28:442–450.
Slater, J. D. H., Barbour, B. H., Henderson, H. H., Casper, A. G. T., and Bartter, F. C. (1963); Influence of the pituitary and the renin-angiotensin system on the secretion of aldosterone, cortisol, and corticosterone. *Journal of Clinical Investigation,* 42:1504–1520.
Stollar, V., Buonassisi, V., and Sato, G. (1964): Studies on hormone secreting adrenocortical tumor in tissue culture. *Experimental Cell Research,* 35:608–616.
Stone, D., and Hechter, O. (1954): Studies on ACTH action in perfused bovine adrenals: The site of action of ACTH in corticosteroidogenesis. *Archives of Biochemistry and Biophysics,* 51:457–469.
Sutherland, E. W., Øye, I., and Butcher, R. W. (1965): The action of epinephrine and the role of

the adenyl cyclase system in hormone action. *Recent Progress in Hormone Research,* 21:263–642.

Sutherland, E. W., and Rall, T. W. (1960): The relation of adenosine-3',5'-phosphate and phosphorylase to the actions of catecholamines and other hormones. *Pharmacological Reviews,* 12:265–299.

Sutherland, E. W., Rall, T. W., and Menon, T. (1962): Adenyl cyclase I. Distribution, preparation and properties. *Journal of Biological Chemistry,* 237:1220–1227.

Taunton, O. D., Roth, J., and Pastan, I. (1967): The first step in ACTH action: Binding to tissue. *Journal of Clinical Investigation,* 46:1122.

Taunton, O. D., Roth, J., and Pastan, I. (1969): Studies on the adrenocorticotropic hormone-activated adenyl cyclase of a functional adrenal tumor. *Journal of Biological Chemistry,* 244:247–253.

Tchen, T. T. (1968): Conversion of cholesterol to pregnenolone in the adrenal cortex: Enzymology and regulation. In: *Functions of the Adrenal Cortex, Vol. 1,* edited by K. W. McKerns, pp. 3–26. Appleton-Century-Crofts, New York.

Tesser, G. I., and Schwyzer, R. (1966): Synthese des 17,18-Diornithin-β-corticotropin-(1-24)-tetracosapeptides, eines biologisch aktiven Analogons des adrenocorticotropen Hormons. *Helvetica Chimica Acta,* 49:1013.

Trzeciak, W. H., and Boyd, G. S. (1974): Activation of cholesteryl esterase in bovine adrenal cortex. *European Journal of Biochemistry,* 46:201–207.

Tsang, C. P. W., and Péron, F. G. (1971): Effects of adenosine-3',5'-monophosphate on steroidogenesis and glycolysis in the rat adrenal gland incubated in vitro. *Steroids,* 17:453–469.

Ungar, F., Kan, K. W., and McKoy, K. E. (1973): Activator and inhibitor factors in cholesterol side-chain cleavage. *Annals of the New York Academy of Sciences,* 212:276–289.

van Heyningen, W. E., Carpenter, C. C., Jr., Pierce, N. F., and Greenough, W. B. (1971): Deactivation of cholera toxin by a ganglioside. *Journal of Infectious Diseases,* 124:415–418.

Vapaatalo, H., Bieck, P., and Westermann, E. (1972): Actions of various cyclic nucleotides and purine bases on the synthesis of corticosterone in vitro. *Naunyn-Schmiedeberg's Archiv für Pharmakologie,* 275:435–443.

Walton, G. M., and Garren, L. D. (1970): An assay for adenosine 3',5'-cyclic monophosphate based on the association of the nucleotide with a partially purified binding protein. *Biochemistry,* 9:4223–4229.

Whitley, T. H., Stowe, N. W., Ong, S. H., Ney, R. L., and Steiner, A. L. (1974): Control and intracellular localization of adrenal cyclic GMP; comparison with cyclic AMP. *56th Annual Meeting, Endocrine Society, Atlanta, Ga.* Abstract #209, Supplement to *Endocrinology,* Vol. 94.

Whysner, J. A., Ramseyer, J., Kazmi, G. A., and Harding, B. W. (1969): Substrate induced spin state changes in cytochrome P450. *Biochemical and Biophysical Research Communications,* 36:795–801.

Wicks, W. D. (1974): Regulation of protein synthesis by cyclic AMP. *Advances in Cyclic Nucleotide Research,* 4:335–438.

Wolff, J., Temple, R., and Cook, G. H. (1973): Stimulation of steroid secretion in adrenal tumor cells by choleragen. *Proceedings of the National Academy of Sciences (U.S.),* 70:2741–2744.

Zor, U. T., Kaneko, T., Lowe, J. P., Bloom, G., and Field, J. P. (1969): Effect of thyroid-stimulating hormone and prostaglandins on thyroid adenyl cyclase activation and cyclic adenosine 3',5'-monophosphate. *Journal of Biological Chemistry,* 244:5189–5195.

The Role of Cyclic AMP in Gonadal Function

John M. Marsh

The Endocrine Laboratory, University of Miami School of Medicine, Miami, Florida 33152

CONTENTS

I.	Introduction	138
II.	Role of Cyclic AMP in the Stimulation of Steroidogenesis	139
	A. Ovarian Tissues	139
	1. Corpus Luteum	139
	a. Effect of Exogenous Cyclic AMP	140
	b. Endogenous Cyclic AMP	141
	c. Adenylyl Cyclase and Cyclic Nucleotide Phosphodiesterase	142
	2. Graafian Follicle	146
	a. Effect of Exogenous Cyclic AMP	146
	b. Endogenous Cyclic AMP	146
	3. Interstitial Tissue	148
	a. Effect of Exogenous Cyclic AMP	149
	b. Endogenous Cyclic AMP	149
	4. Whole Ovarian Preparations	150
	a. Effect of Exogenous Cyclic AMP	150
	b. Endogenous Cyclic AMP	150
	5. Summary	152
	B. Testis	153
	1. Effects of Exogenous Cyclic AMP	153
	2. Adenylyl Cyclase	154
	3. Measurement of Endogenous Cyclic AMP	156
	4. Reevaluation of the Role of Cyclic AMP in Testicular Steroidogenesis	158
	a. Evidence in Favor of a Role for Cyclic AMP	159
	b. Possible Methodologic Problems in Detecting Changes in Cyclic AMP and Suggestions for Further Studies	160
	c. The Possibility of Another Second Messenger	162
	d. Summary	162
	C. Possible Modes of Action of Cyclic AMP on Steroidogenesis	163
	1. Action via Increased Cofactors	163
	2. Action via Increased Substrate	167

        3. Action via Increased Transport of Cholesterol 169
        4. Action via Increased Side-Chain Cleavage Activity . . . 172
        5. Action via Increased Efflux of Pregnenolone from Mitochondria 173
        6. The Role of Protein Synthesis in the Stimulation of Steroidogenesis 174
        7. Summary 175

III. Other Roles for Cyclic AMP in Gonadal Functions 175
    A. Ovulation 176
    B. Ovum Maturation 177
    C. Luteinization 179
    D. Spermatogenesis 181
    E. Effects on Nucleic Acid Synthesis 184

IV. Conclusions 185

V. Acknowledgments 186

VI. References 186

I. INTRODUCTION

Although a great deal of work has been done implicating cyclic AMP in several gonadal functions, a complete understanding of its role in any one of these functions is still lacking. The possible mediatory role of cyclic AMP in the gonadotropic stimulation of steroidogenesis in ovaries and testes has been studied the longest and probably the most thoroughly, but even this action is far from being completely elucidated. The first indication that cyclic AMP was involved with the control of steroidogenesis came from the studies of Haynes and co-workers, who in a series of experiments (Haynes and Berthet, 1957; Haynes, Koritz, and Peron, 1959; Haynes, Sutherland, and Rall, 1960) showed that ACTH increased the endogenous level of cyclic AMP in the adrenal cortex, and that exogenous cyclic AMP mimicked the effect of ACTH on steroidogenesis. Guided by this work, other investigators began to study the role of cyclic AMP in the gonadotropic stimulation of steroidogenesis in the testis (Hall and Eik-Nes, 1962) and the ovary (Marsh and Savard, 1964a,b). From these early studies on gonadal steroidogenesis the field has expanded to where cyclic AMP has been implicated to some extent in nearly all actions of luteinizing hormone (LH) and some actions of follicle-stimulating hormone (FSH) in the ovary and the testis. I will attempt to review this work and try to describe the status of this field through Spring 1974.

II. ROLE OF CYCLIC AMP IN THE STIMULATION OF STEROIDOGENESIS

A. Ovarian Tissues

In order to organize the review of the literature on this topic, the studies are grouped according to the components of the ovary under investigation: the corpus luteum, the Graafian follicle, and the interstitial tissue. This allows the association of the endogenous changes in cyclic AMP or the effects of exogenous cyclic AMP with a particular ovarian function. In addition, there have been numerous studies carried out with whole ovarian preparations which have implicated cyclic AMP as part of a gonadotropic effect; these studies are discussed at the end of the section.

1. Corpus Luteum

The main function of the corpus luteum is the synthesis and secretion of progestins, and this has been shown in most species by *in vitro* and *in vivo* investigations to be, at least to some extent, controlled by LH. In some species, such as the cow, the human, and the monkey, LH appears to be the only controlling hormone, while in others, such as the hamster and the rat, LH is probably only one of several controlling hormones (Greep, 1971). In a series of studies (reviewed by Savard, Marsh, and Rice, 1965) an *in vitro* model system was developed in which the effect of LH on progesterone synthesis was determined in incubating slices of corpora lutea obtained from cows in early pregnancy. The slices of corpora lutea tissue were incubated in Krebs-Ringer bicarbonate buffer in the presence of LH or other test substances, and the amounts of progesterone and 20β hydroxypregn-4-en-3-one synthesized were measured. It was found that LH produced a marked increase in progestin synthesis in these incubating slices and that this effect was specific for hormones with LH activity (Mason and Savard, 1964*a*).

These incubating slices were also quite sensitive to LH, responding to concentrations as low as 1 to 2 ng/ml (Marsh and Savard, 1966*a*), which is about the same concentration of LH present in the plasma of the cow during pregnancy (Schams, Hoffman, Fischer, Marz, and Karg, 1972) or in the luteal phase of the estrous cycle (Karg, Hoffmann, and Schams, 1971). After the initial studies on the effect of LH on steroidogenesis in incubating bovine corpora lutea slices were completed, the investigation logically divided into two parts: (1) a study of the site of LH action on the steroidogenic pathway; and (2) an investigation of the mechanism of action. The site of the major effect of LH on the steroidogenic pathway has been shown by numerous investigators to be between cholesterol and pregnenolone (Ichii, Forchielli, and Dorfman, 1963; Hall and Koritz, 1964; Koritz and Hall, 1965; Hall and Young, 1968; Armstrong, Lee, and Miller, 1970). The study of the mechanism of LH action, on the other hand, has been largely a study of the possible role of cyclic AMP in this gonadotropic effect, and this is still under investigation.

a. Effect of exogenous cyclic AMP

Marsh and Savard (1964a,b, 1966b) found that the addition of exogenous cyclic AMP to incubating slices of cow corpora lutea caused a significant stimulation of progesterone synthesis in terms of mass, [1-^{14}C]acetate, and [7-^{3}H]cholesterol incorporation. At a concentration of 0.02 M the magnitude of the cyclic AMP effect was about equal to that produced by a saturating amount of LH, and the pattern of [1-^{14}C]acetate and [7-^{3}H]cholesterol utilization closely resembled that produced by LH. Other nucleotides, such as 3'-AMP, 5'-AMP, or ATP, had no effect in the system, which served to establish a certain degree of specificity to the action of cyclic AMP. The effect of LH was also not additive to that produced by a maximal amount of cyclic AMP, indicating that cyclic AMP might be a mediator of this action of LH. The concentration of cyclic AMP needed for this effect, however, far exceeded the endogenous level in tissues, but this could have been due to an inability of the exogenous nucleotide to readily penetrate cells or to its destruction by cyclic 3',5'-nucleotide phosphodiesterase. In this regard, the dibutyryl derivative of cyclic AMP, which has been reported to be resistant to phosphodiesterase degradation (Posternak, Sutherland, and Henion, 1962), stimulated progesterone synthesis in this system at the much smaller concentration of 0.2 mM (Marsh, 1969). Hall and Koritz (1965a,b) carried out a similar type of experiment using bovine corpora lutea from the estrous cycle and showed that, like LH, cyclic AMP stimulated progesterone synthesis by accelerating a step between cholesterol and pregnenolone.

The effect of exogenous cyclic AMP on progestin synthesis has been assessed in luteal tissue of other species as well. LeMaire, Askari, and Savard (1971) reported a stimulation of progesterone synthesis in terms of both mass and [1-^{14}C]acetate incorporation in incubating slices of a human corpus luteum of ectopic pregnancy. The effect of cyclic AMP exceeded that of human chorionic gonadotropin (hCG), but the pattern of distribution of ^{14}C among several of the steroids (progesterone, 17 hydroxyprogesterone, androstenedione, and estradiol) produced by the human corpus luteum was identical in both the hCG- and the cyclic AMP-stimulated tissue slices. The greater effectiveness of cyclic AMP over hCG may be related to the fact that human corpora lutea of pregnancy have a much smaller capacity to respond to hCG in terms of adenylyl cyclase, cyclic AMP accumulation, and steroid synthesis than corpora lutea of the menstrual cycle (LeMaire, Rice, and Savard, 1968; Marsh and LeMaire, 1974a). If there is a block to hCG responsiveness in human corpora lutea of pregnancy at the level of adenylyl cyclase activation, as these experiments suggest, the exogenous cyclic AMP would be able to bypass that block. Hermier, Santos, Wisnewsky, Netter, and Jutisz (1972) reported that exogenous cyclic AMP also stimulated progesterone synthesis in a corpus luteum of the menstrual cycle and that the maximal response was about the same for cyclic AMP as for hCG. It was also reported that theophylline, an inhibitor of cyclic nucleotide phosphodiesterase, potentiated the effect of hCG on progesterone synthesis and that imidazole, a stimulator of this phosphodiesterase, markedly reduced progesterone synthesis in the control and hCG incubations. These results are in complete accord with the proposal that cyclic AMP mediates the action of LH and hCG on progesterone synthesis in corpora lutea.

Dorrington and Kilpatrick (1967) reported that 5 mM cyclic AMP slightly increased the synthesis of progesterone and 20α hydroxypregn-4-en-3-one in incubating corpora lutea from pseudopregnant rabbits, although its effect on interstitial tissue was much greater. Hermier and Jutisz (1969) showed that 3 mM cyclic AMP stimulated progesterone synthesis in incubating slices of luteinized rat ovaries, prepared by treatment of immature rats with pregnant mare serum gonadotropin (PMSG) and hCG according to the procedure of Parlow (1968). They also found that the presence of Ca^{2+} in the external medium was required for this effect.

b. Endogenous cyclic AMP

The results with exogenous cyclic AMP were compatible with the hypothesis that this nucleotide was a mediator of the action of LH on steroidogenesis in the corpus luteum, but such a consideration required that the addition of LH would bring about an increase in the endogenous concentration of cyclic AMP in luteal tissue. Marsh, Butcher, Savard, and Sutherland (1966) showed that $2\mu g/ml$ of LH brought about a striking increase (as much as 100-fold) in the level of cyclic AMP, and that this response showed the same specificity of LH as had been previously demonstrated for the stimulation of progesterone synthesis (Savard et al., 1965). Furthermore, the increase in endogenous cyclic AMP preceded the increase in progesterone synthesis which would be expected if cyclic AMP were a mediator of the action of LH. At the higher concentrations of 0.2 to 2 μg LH/ml, there was a positive correlation between the increase in endogenous cyclic AMP and the stimulation of progesterone synthesis. At 0.02 μg LH/ml, however, the results were less clear. Although this concentration of LH usually stimulated progesterone synthesis, effects on cyclic AMP levels could not be regularly detected. This discrepancy was not considered to be a serious one in view of the variability of the incubating slice technique and the fact that the methodology used for the cyclic AMP measurements had sufficient experimental error to obscure small increases in the cyclic nucleotide. This interpretation may have to be reevaluated now, in light of the work of Beall and Sayers (1972), Catt and Dufau (1973), and Moyle and Ramachandran (1973), who have detected a dissociation between the stimulation of cyclic AMP and steroidogenesis in other tissues. This will be discussed in detail in a later section on the testis.

Puromycin, an antibiotic which inhibits protein synthesis, had been shown previously to be capable of inhibiting the stimulatory effects of both LH and cyclic AMP on steroidogenesis (Savard et al., 1965). This substance had no effect, however, on the stimulation of cyclic AMP accumulation by LH, which indicated that protein synthesis was not required for this effect, and that the puromycin probably inhibits the stimulation of steroidogenesis at some step after the increase in cyclic AMP. The increased amount of cyclic AMP produced in these corpora lutea slices has been found by Goldstein and Marsh (1973) to be localized predominantly in the cytosol fraction and about 25% of this cytosol cyclic AMP is bound to a macromolecule. The corpus luteum also contains a cyclic AMP-dependent protein kinase (Goldstein and Marsh, 1972, 1973; Menon, 1973), and most of this enzyme is also located in the cytosol fraction (Goldstein and Marsh, 1973). It is probable that this protein kinase makes up a major portion of the macromolecular cyclic AMP binding material.

LH and hCG also brought about a marked increase in cyclic AMP accumulation in human corpora lutea measured in terms of mass (Marsh et al., 1966) or in terms of [8-^{3}H]adenine incorporation into cyclic [8-^{3}H]AMP (Marsh and LeMaire, 1974a). The effect appeared to be specific for human gonadotropin with LH activity and showed about the same sensitivity to hCG as the effect on steroidogenesis in this tissue (Savard et al., 1965). Corpora lutea of the menstrual cycle were much more responsive to these gonadotropins in terms of cyclic AMP accumulation or steroid synthesis than corpora lutea of pregnancy (Marsh and LeMaire, 1974a). The reason for this difference, however, is not known at this time.

Lamprecht, Zor, Tsafriri, and Lindner (1973) found that isolated corpora lutea from pseudopregnant rats or luteinized rat ovaries also responded *in vitro* to 10 μg LH/ml with an increase in cyclic AMP accumulation, but the response of isolated Graafian follicles was more striking. Mason, Schaffer, and Toomey (1973) also found that LH was less effective in stimulating cyclic AMP accumulation in luteinized rat ovaries or ovaries from pregnant rats than in ovaries from immature rats. In fact, although these authors did report a stimulation of cyclic AMP accumulation by LH in incubating slices of isolated corpora lutea obtained from 6- or 18-day pregnant rats, they did not observe an increase in cyclic AMP in corpora lutea obtained from a 20-day pregnant rat or a rat which had been treated with low doses of PMSG (8 IU) and hCG (12.5 IU). No measurements of steroid synthesis were made in this study, so it is not possible to correlate a lack of response in terms of cyclic AMP with steroid synthesis. Ahren, Herlitz, Nilsson, Perklev, Rosberg, and Selstam (1974) found, in their study of corpora lutea from PMSG-treated rats, that LH markedly increased the content of cyclic AMP in 1-day-old corpora lutea. In fact, the response was much greater than that of isolated follicles. As the corpora lutea grew older, however, the response quickly declined, until at day 7 it was very small. The failure of Mason et al. (1973) to detect a response to LH in corpora lutea may have been due to the chance selection of ovaries when the corpora lutea were in this decline.

c. *Adenylyl cyclase and cyclic nucleotide phosphodiesterase*

  i. *Site of LH action on cyclic AMP accumulation.* The increase in endogenous cyclic AMP in luteal tissue could theoretically be brought about by a stimulation of adenylyl cyclase or by an inhibition of cyclic nucleotide phosphodiesterase. These alternatives were investigated in homogenates of bovine corpora lutea by assaying the effect of LH on adenylyl cyclase in the absence of phosphodiesterase activity (Marsh, 1970a). The complete inhibition of detectable phosphodiesterase activity was accomplished by using a high concentration of theophylline (0.04 M) and a short incubation period. Under these conditions, 0.1 μg LH/ml or 0.01 M NaF significantly increased adenylyl cyclase activity. Epinephrine at a concentration of 0.2 mM also produced a small but statistically significant stimulation of adenylyl cyclase activity, but LH inactivated by hydrogen peroxide, bovine serum albumin, prolactin, ACTH, and glucagon were inert in this respect. The significance of the effect of epinephrine is uncertain, since it was only assessed at one concentration and no thorough study has been carried out on its effect on cyclic AMP accumulation or progesterone synthesis in incubated slices of corpora lutea. Fontaine, Burzawa-Gerard, and Delerue-Lebelle (1970) have observed a stimulation by epinephrine of adenylyl cyclase in homogenates of whole ovaries obtained from goldfish, and this effect also required an elevated concentration of the catecholamine (0.05 mM). In testicular tissue, epinephrine has also been found to stimulate adenylyl cyclase (Murad, Strauch, and Vaughan, 1969; Pulsinelli and Eik-Nes, 1970) and cyclic AMP accumulation (Kuehl, Patanelli, Tarnoff, and Humes, 1970a). This catecholamine has even been found to stimulate testosterone secretion in perfused testes (Eik-Nes, 1971), but it is unknown at this time if it plays a physiological role.

The effect of LH on phosphodiesterase activity was also evaluated in homogenates of bovine corpora lutea; no change in activity was observed under a variety of experimental conditions (Marsh, 1970a). These data indicate, therefore, that LH brings about the increase in endogenous cyclic AMP in bovine corpora lutea by a stimulation of the adenylyl cyclase and not by an inhibition of the phosphodiesterase. Phosphodiesterase was also demonstrated in bovine, human, and rat corpora lutea by Stansfield, Horne, and Wilkinson (1971), but although the properties of this enzyme in bovine tissue were examined, no attempt was made to assess the effect of LH.

The subcellular localization of adenylyl cyclase in bovine corpora lutea has been assessed by homogenizing and fractionating the subcellular components of this tissue by isotonic and hypotonic methods (Sidhu, Camp, and Marsh, 1974). The fractions were assayed for adenylyl cyclase activity and standard marker enzymes. Only 5'-nucleotidase paralleled the distribution of adenylyl cyclase in both isotonic and hypotonic procedures, indicating that the adenylyl cyclase of bovine corpora lutea is primarily localized on the plasma membrane. Menon and Kiburz (1974) have also prepared a membrane fraction from bovine corpora lutea which contains adenylyl cyclase, [^{125}I]hCG binding activity, and the plasma membrane marker enzyme NaK-ATPase, indicating that the hormone receptor and the adenylyl cyclase are both localized in the plasma membrane.

LH has also been shown to stimulate the adenylyl cyclase activity of corpora lutea of other species, such as the rabbit (Anderson, Hubbard, and Baggett, 1970) and the pig (Anderson, Schwartz, and Ulberg, 1974). In the latter investigation, it was found that there was a decline in adenylyl cyclase activity and LH responsiveness which correlated with the onset of luteal regression. Stansfield and co-workers have carried out studies on the adenylyl cyclase in rat corpora lutea (summarized by Stansfield, Franks, Wilkinson, and Horne, 1972). Adenylyl cyclase was detectable in homogenates of rat corpora lutea, but a very active ATPase made it difficult to study. The largest amount of activity was associated with a $700 \times g$ pellet, indicating that this enzyme system in rat corpora lutea might also be associated with the plasma membrane. No studies were carried out on the effect of LH or other hormones on this enzyme system.

ii. *Possible role of prostaglandins in the action of LH on adenylyl cyclase.* Prostaglandins were implicated as possible mediators of LH action when they were shown to stimulate: (1) progesterone synthesis in incubating slices of bovine corpora lutea (Speroff and Ramwell, 1970); (2) adenylyl cyclase activity in homogenates of bovine corpora lutea (Marsh, 1970b); and (3) the incorporation of [^{14}C]adenine into cyclic AMP in incubated whole ovaries obtained from rats (Kuehl, Humes, Tarnoff, Cirillo, and Ham, 1970b). Marsh (1970b, 1971) tested the effects of six prostaglandins (PGE$_1$, PGE$_2$, PGA$_2$, PGB$_1$, PGF$_{1\alpha}$ and PGF$_{2\alpha}$) on adenylyl cyclase activity of homogenates of bovine corpora lutea and found that PGE$_1$ and PGE$_2$ were about equally potent and more effective that PGA$_2$ or PGB$_1$. PGF$_{2\alpha}$ and PGF$_{1\alpha}$ did not produce significant stimulatory effects.

An additivity experiment was carried out to test the possible mediatory role of prostaglandins in the action of LH on adenylyl cyclase. In each experiment an aliquot of a homogenate of a bovine corpus luteum was incubated with a maximally stimulatory concentration of PGE$_2$ (100 μg/ml). Another aliquot was incubated with this concentration of PGE$_2$ plus LH (10 μg/ml). A third aliquot was incubated with LH, and a fourth aliquot was incubated alone as a control. If LH transmitted its effect on luteal adenylyl cyclase solely via PGE$_2$, then it follows that a homogenate of a corpus luteum which was responding maximally to a saturating level of PGE$_2$ should show no further response when LH was added. It was found that LH produced a clear-cut additive effect above that produced by a maximally stimulatory concentration of PGE$_2$, indicating that the effect of LH was probably not transmitted via prostaglandins (Marsh, 1971).

Other investigators, using a similar experimental format, have reported that LH and prostaglandins were additive in their effects on adenylyl cyclase and cyclic AMP accumulation in a variety of tissues (Kuehl et al., 1970b; Jonsson, Shelton, and Baggett, 1972; Kolena and Channing, 1972; Lamprecht et al., 1973).

An opposite conclusion was reached, however, by another group of investigators (Kuehl et al., 1970b), who evaluated the effect of a prostaglandin antagonist, 7-oxa-13-prostynoic acid, on the stimulation by LH and prostaglandins of [^{14}C]cyclic AMP accumulation in incubated rat ovaries. They found that this antagonist not only blocked the effect of PGE$_1$ and PGE$_2$, but also competitively inhibited the stimulation of the accumulation of [^{14}C]cyclic AMP produced by

LH. They concluded, therefore, that a prostaglandin receptor functioned as a necessary intermediate in the action of LH. This conclusion should be viewed with some reservation, since the same authors indicated in a subsequent paper (Kuehl, Humes, Cirillo, and Ham, 1972) that 7-oxa-13-prostynoic acid was not a completely specific antagonist, but also inhibited cyclic AMP-dependent protein kinase. In this regard, Marsh and LeMaire (1974b) carried out experiments on the effect of 7-oxa-13-prostynoic acid on [^{3}H]adenine incorporation into cyclic AMP in incubated slices of bovine corpora lutea, and although they could demonstrate an inhibition of the stimulatory effect of LH with very large concentrations of the antagonist, they could not show that it was of the competitive type.

Channing (1972) also found that 7-oxa-13-prostynoic acid inhibited the action of PGE_2 or LH on luteinization in cultures of monkey granulosa cells. The interpretation of this effect was complicated by the fact that when this inhibitor was mixed with prostaglandin or LH it produced severe cellular necrosis.

A mediatory role for prostaglandins in the action of LH on adenylyl cyclase would suggest an LH stimulation of prostaglandin synthesis or release. Chasalow and Pharriss (1972) reported that the injection of LH antiserum into rats suppressed the synthesis of prostaglandin-like compounds from [^{3}H]arachidonic acid in incubations of ovarian homogenates, and that LH injected into the animal or added to the incubation medium overcame this suppression. Other studies also indicated that LH can stimulate prostaglandin synthesis in slices of monkey ovaries (Wilks, Forbes, and Norland, 1972) and incubated rabbit Graafian follicles (Marsh, Yang, and LeMaire, 1974). In the latter report, however, this stimulation appears to occur beyond the adenylyl cyclase step, because it could be mimicked by exogenous cyclic AMP. Dibutyryl cyclic AMP has also been reported to stimulate prostaglandin synthesis in incubated intact mouse ovaries (Kuehl, Cirillo, Ham, and Humes, 1973).

Another approach to evaluating prostaglandins as possible mediators of LH action has been the use of inhibitors of prostaglandin synthesis, such as indomethacin, aspirin, and flufenamic acid. These compounds have been reported to be ineffective in inhibiting the LH stimulation of cyclic AMP accumulation in incubating mouse or rat ovaries, although they produced a marked inhibition of prostaglandin synthesis (Kuehl et al., 1973; Zor, Bauminger, Lamprecht, Koch, Chobsieng, and Lindner, 1973). Indomethacin has also been reported to be ineffective in inhibiting the stimulation of ovarian steroidogenesis by LH when this prostaglandin synthesis inhibitor was administered *in vivo* (Grinwich, Kennedy, and Armstrong, 1972; Tsafriri, Lindner, Zor, and Lamprecht, 1972a).

In summary, four general approaches have been used to study the possibility that prostaglandins mediate the action of LH on adenylyl cyclase: additivity studies, studies of prostaglandin antagonists, studies of prostaglandin synthesis, and studies of inhibitors of prostaglandin synthesis. Although these approaches have not yielded a definitive answer to the problem, the majority of the data appear to be in opposition to an essential role for prostaglandins in the action of LH on adenylyl cyclase.

2. Graafian Follicle

LH has been shown to be the gonadotropic regulator of steroidogenesis in the Graafian follicle in both *in vivo* (Greep, Van Dyke and Chow, 1942; Lostroh and Johnson, 1966; Eshkol and Lunenfeld, 1967) and *in vitro* (Gospodarowicz, 1964; Mills, Davies, and Savard, 1971) studies. This gonadotropin has also been shown to increase the endogenous level of cyclic AMP in follicular cells. There is a difficulty, however, in demonstrating that these changes in endogenous cyclic AMP are associated only with steroidogenesis, because LH causes other changes in the Graafian follicle, such as ovulation, ovum maturation, and luteinization. The increase in cyclic AMP caused by LH may, therefore, be related to these latter effects of the gonadotropin. We will review the studies which have shown that LH can increase the endogenous concentration of cyclic AMP in follicular tissue, keeping in mind that these changes could be related to one or more of the effects on this ovarian compartment.

a. *Effect of exogenous cyclic AMP*

There is very little data available on an acute effect of exogenous cyclic AMP on steroidogenesis in Graafian follicles. Several groups have reported that cyclic AMP or dibutyryl cyclic AMP will induce granulosa cells (Cirillo, Anderson, Ham, and Gwatkin, 1969; Channing, 1970a; Channing and Seymour, 1970) or isolated follicles (Ellsworth and Armstrong, 1973, 1974) to synthesize progestins, but these effects have been observed only several days after the exposure of the tissues to the cyclic nucleotides. These effects seem, therefore, to be more related to the initiation of luteinization than to the acute stimulation of follicular steroidogenesis. Mills (1975) has found, however, that exogenous cyclic AMP can completely mimic the acute *in vitro* stimulation of steroidogenesis in the rabbit Graafian follicles, reported for LH (Mills et al., 1971). This effect of cyclic AMP is apparent after 3 hr of incubation, and the steroid products include the same proportions of estrogens, androgens, and progestins seen when LH is used as the stimulating agent.

b. *Endogenous cyclic AMP*

Kolena and Channing (1971, 1972) determined the effect of gonadotropins on cyclic AMP accumulation by [^{14}C]adenine incorporation into cyclic AMP and by the measurement of the mass amount of this cyclic nucleotide in granulosa cells in tissue culture. These cells were aspirated from medium (3 or 5 mm) Graafian follicles of young pigs (4 to 6 months of age) and are the type of granulosa cell that will not luteinize spontaneously in tissue culture (Channing, 1970b). The addition of LH to these cultures brought about a maximal elevation in the accumulation of cyclic AMP at the earliest time examined, which was 5 min, while an increase in steroidogenesis was not observed until after 3 hr. The minimum effective dose of LH was 0.02 µg/ml, which is about the same as the minimum concentration required to stimulate progestin production in culture

(Channing, 1970b). FSH also stimulated cyclic AMP accumulation, and this could not be explained on the basis of LH contamination. Thyroid-stimulating hormone (TSH) was effective, but this could be accounted for by its LH and FSH content. Prolactin and ACTH produced very small stimulatory effects, and these also could be accounted for by their LH and FSH content.

Tsafriri, Lindner, Zor, and Lamprecht (1972b) and Lamprecht et al. (1973) assessed the effect of LH on cyclic AMP accumulation, by the incorporation of [^{3}H]adenine and by the measurement of mass, in isolated rat Graafian follicles. They found that this gonadotropin caused a marked increase in cyclic AMP during a 20-min incubation. PGE_2 was also found to increase cyclic AMP to about the same extent as LH. The same investigators have reported in abstract form that FSH could also stimulate cyclic AMP accumulation, progesterone formation, and oocyte maturation in these cultured rat follicles, and this effect was not attributable to contaminating LH (Lindner, Tsafriri, Koch, Zor, Lieberman, and Pomerantz, 1973).

Marsh, Mills, and LeMaire (1972) carried out very similar experiments with isolated rabbit Graafian follicles. The addition of LH to the incubating follicles caused a 20-fold increase in the synthesis of cyclic AMP from [^{3}H]adenine. The minimum effective dose of LH was 0.05 μg/ml, and the effect was specific for this gonadotropin. FSH did have a stimulatory effect, but in contrast to the reports of Kolena and Channing (1972) and Lindner et al. (1973), this effect appeared to be accounted for by its LH contamination. In addition, this FSH effect could be abolished by treatment of the preparation with LH antiserum. Ahren et al. (1974) have also found that LH, FSH, and PGE_2 would increase the cyclic AMP production in incubating isolated rat Graafian follicles. This effect of FSH could not be eliminated by a specific anti-β LH serum, indicating that it was an intrinsic property of the FSH molecule and not due to a contamination by LH. The reasons for the reported differences between the effects of FSH on the rabbit follicle (Marsh, Mills, and LeMaire, 1973) and those on the porcine granulosa cell and the rat follicle (Kolena and Channing, 1972; Lindner et al., 1973; Ahren et al., 1974) are not clear. They could be due to species differences or differences in technique. It is not too surprising that the whole Graafian follicle responds to both FSH and LH, but from the work of Hamberger, Hamberger, and Herlitz (1971) it would be expected that the granulosa cell might respond to LH alone. These latter authors have shown that LH stimulates oxygen consumption in granulosa cells with no influence by FSH, while FSH stimulates the isolated theca capsule and there is no effect by LH.

In another paper, Marsh et al. (1973) found that if the follicles were obtained in the period between the ovulatory peak of LH and the time of the rupture of the follicle (about 10 to 12 hr), the stimulation of cyclic AMP accumulation in Graafian follicles by LH was progressively diminished. At 5 hr and 9 hr after the injection of hCG into the estrous rabbit, this *in vitro* response to LH was hardly detectable. The decrease in responsiveness to LH could also be induced by the injection of the animal with LH or by mating the animal. A similar decline in responsiveness to LH during this preovulatory period has been reported by Mills and Savard (1973) in terms of diminished follicular steroidogenesis. At this time, there does not seem to be an explanation of how this refractoriness to LH is brought about or what physiological function it might have, but Dunn and Birnbaumer (1974) have observed the same type of decline when they assayed the response of adenylyl cyclase to LH in these follicles. The refractoriness does not, however, appear to be at the level of the binding of the hormone to its receptor, because Mills and McPherson (1974) did not observe any change in the binding of [^{125}I]hCG to these rabbit Graafian follicles during the preovulatory period. Similarly, Lamprecht et al. (1973) reported a refractoriness of isolated rat follicles to stimulation by LH if the follicles had been cultured previously with LH for 18 hr. It appears that after several hours of exposure of the follicles to LH, some blocking agent is elaborated which acts between the binding of the hormone and the stimulation of adenylyl cyclase.

Possible candidates for this blocking agent could be the E and F prostaglandins, which have been shown by LeMaire, Yang, Behrman, and Marsh (1973) to increase progressively just at the time the *in vitro* response to LH is decreasing. An inhibition by PGE_1 of the hormonal stimulation of adenylyl cyclase had been implied previously for the toad bladder (Orloff, Handler, and Bergström, 1965), kidney (Orloff and Grantham, 1967), and adipose tissue (Butcher and Baird, 1968). On the other hand, prostaglandins of the E series have been shown to mimic the effect of LH and stimulate cyclic AMP accumulation in several ovarian preparations (Kuehl et al., 1970b; Marsh, 1971; Lamprecht, Zor, Tsafriri, and Lindner, 1971; Kolena and Channing, 1972; Tsafriri et al., 1972a; Zor, Lamprecht, Kaneko, Schneider, McCann, Field, Tsafriri, and Lindner, 1972a; Zor, Lamprecht, Tsafriri, Pomerantz, Koch, and Lindner, 1972b; Lamprecht et al., 1973; Ahren et al., 1974). In thyroid tissue, PGE_1 and PGE_2 have been reported to both mimic the action of TSH and diminish the effect of TSH on adenylyl cyclase (Burke, Kowalski, and Babiarz, 1971; Sato, Szabo, Kowalski, and Burke, 1972; Burke, Chang, and Szabo, 1973), and it was suggested that prostaglandins might function as feedback inhibitors modulating the responsiveness of the adenylyl cyclase system (Burke et al., 1973). In ovarian tissue, $PGF_{2\alpha}$ has been shown to inhibit steroidogenesis in the corpus luteum when it is administered *in vivo* (reviewed by Pharriss, Tillson, and Erickson, 1972) or incubated with luteal tissue for relatively long periods (O'Grady, Kohorn, Glass, Caldwell, Brock, and Speroff, 1972b; Behrman, Ng, and Orczyk, 1974). Although the role of prostaglandins in this refractory phenomenon in the Graafian follicle is uncertain at this time, studies of the effects of exogenous prostaglandins and inhibitors of prostaglandin synthesis may provide an answer.

3. *Interstitial Tissue*

In some species, such as the rabbit and the human, the interstitial cells of the ovary are also capable of steroid synthesis (Hilliard, Endroczi, and Sawyer, 1961;

Hilliard, Archibald, and Sawyer, 1963; Simmer, Hilliard, and Archibald, 1963; Dorrington and Kilpatrick, 1966a; Savard, et al., 1965). In the rabbit, this ovarian compartment synthesizes significant amounts of progestins, such as 20α hydroxypregn-4-en-3-one, which appears to function in this species as a positive feedback stimulus to maintain the continued release of LH necessary for ovulation (Hilliard, Penardi, and Sawyer, 1967). LH also stimulates the synthesis of progestins in this compartment (Hilliard, et al., 1961, 1963; Dorrington and Kilpatrick, 1966a), so it was natural to suspect that cyclic AMP might also be involved in this gonadotropic stimulation of steroidogenesis.

a. *Effect of exogenous cyclic AMP*

Dorrington and Kilpatrick (1967) were the first to investigate the role of cyclic AMP in the regulation of steroidogenesis in interstitial tissue. They added exogenous cyclic AMP to incubating slices of rabbit interstitium and found that it caused a marked increase in the *in vitro* synthesis of 20α hydroxypregn-4-en-3-one and a smaller increase in progesterone. Scoon and Major (1972) succeeded in preparing homogenates of rabbit interstitial tissue which retained a small portion of their capacity to increase steroid production in response to LH and cyclic AMP, which is in contrast to the findings of other investigators in this field (Haynes, Savard, and Dorfman, 1954; Mason and Savard, 1964b). Unlike the ovarian slice preparations, however, progesterone was the predominant steroid formed in the homogenates. In a subsequent abstract, Scoon and Major (1973) reported that when the homogenates of interstitial tissue were separated into microsomal, mitochondrial, and cytosol fractions, the effect of cyclic AMP was lost, but if a supernatant was prepared in which only the cell debris-nuclear fraction had been removed, there was a small but statistically significant stimulation of progesterone synthesis by cyclic AMP. This seems to indicate that the action of cyclic AMP on steroidogenesis is exerted through more than one subcellular component of the interstitial cells. This is in accord with the findings of Jackanicz and Armstrong (1968) that the addition of cyclic AMP directly to isolated mitochondria of rabbit interstitial tissue fails to stimulate steroidogenesis. Flint, Grinwich, and Armstrong (1973) also observed an increase in 20α hydroxypregn-4-en-3-one and progesterone in incubating slices of rabbit ovarian interstitial tissue. They presented some evidence which indicated that LH and cyclic AMP could also inhibit cholesterol ester synthetase in this tissue by a mechanism involving steroids. This inhibition of cholesterol ester synthetase could be part of the mechanism by which cyclic AMP accelerates steroidogenesis and will be discussed in more detail in a later section on the possible mechanisms of action of LH and cyclic AMP.

b. *Endogenous cyclic AMP*

Horrell, Kilpatrick, and Major (1972) indirectly implicated cyclic AMP as part of the mechanism of the control of *in vivo* ovarian steroidogenesis by assessing the effect of the injection or infusion of aminophylline or theophylline (classical inhibitors of the cyclic nucleotide phosphodiesterase) on rabbit ovarian steroidogenesis. The acute injection of aminophylline or the infusion of theophylline over 30 min increased the content of 20α hydroxypregn-4-en-3-one in interstitial tissue and the corpus luteum. The injection of aminophylline had no significant effect, however, on the secretion of progesterone or 20α hydroxypregn-4-en-3-one into the ovarian blood.

Dorrington and Baggett (1969) assessed adenylyl cyclase activity in homogenates of rabbit interstitial tissue prepared by gentle homogenization and found that these preparations responded to the addition of LH with about a twofold increase in activity. The addition of hCG or NaF also activated this enzyme system, but FSH and ACTH were completely ineffective.

4. Whole Ovarian Preparations

Since the whole ovary is made up of several steroidogenic components and several gonadotropin-responsive cell types, it would seem to be a less precise tool to use for the study of the role of cyclic AMP in steroidogenesis than the individual ovarian compartments. Nevertheless, studies with whole ovarian preparations have been used very successfully by some investigators in this field.

a. *Effect of exogenous cyclic AMP*

Dorrington and Kilpatrick (1966a) had shown that incubating slices of rabbit ovaries would respond to LH with an increase in steroid synthesis. In a subsequent paper (Dorrington and Kilpatrick, 1967), they demonstrated that this whole ovarian preparation was also responsive to cyclic AMP. A striking increase in the synthesis of 20α hydroxypregn-4-en-3-one and a smaller increase in progesterone synthesis was produced by addition of this cyclic nucleotide to incubating tissue slices. A maximum effect was brought about by 8 mM cyclic AMP, and this increase was about equal to that produced by a maximally stimulatory amount of LH (0.5 μg/ml). The ratio of the two progestins indicated that the interstitial tissue had the greatest steroidogenic capacity and was the most responsive component of this whole ovarian preparation. This was confirmed by separating the interstitium from the rest of the ovary and assessing its response to the cyclic nucleotide separately. The response of slices of whole ovary was specific for the cyclic nucleotide in that 3'-AMP, 5'-AMP, ADP, and ATP had no effect. An additivity experiment was carried out in which LH was added to a maximally stimulatory concentration of cyclic AMP (8 mM), but the LH did not cause an increase in steroidogenesis above that already produced by cyclic AMP. Another piece of evidence that cyclic AMP might be a mediator of this effect of LH was that the addition of theophylline to the medium potentiated the effects of submaximal amounts of cyclic AMP or LH.

b. *Endogenous cyclic AMP*

The addition of LH, PGE_1, or PGE_2 *in vitro* to incubating pieces of whole ovaries causes an increase in the endogenous concentration of cyclic AMP in mice

(Kuehl et al., 1970b), rabbits (Smith and Major, 1971), or rats (Lamprecht et al., 1973; Mason et al., 1973; Zor et al., 1973; Koch, Zor, Pomerantz, Chobsieng, and Lindner, 1973; Ahren, Hamberger, Herlitz, Hillensjo, Nilsson, Perklev, and Selstam, 1973; Selstam, Liljekvist, Rosberg, Gronquist, Perklev, and Ahren, 1974; Ahren et al., 1974; Mason, 1974). FSH has also been found to stimulate cyclic AMP accumulation in some of these tissue preparations (Mason et al., 1973; Koch et al., 1973), and it seems that this effect cannot be attributed to a LH contaminant. Mason et al. (1973) showed that the response to FSH occurred with amounts which contained too little LH to act by itself and that the maximal stimulation caused by FSH was less than that produced by LH. Koch et al. (1973) demonstrated that the stimulation of cyclic AMP accumulation caused by FSH could not be impaired by treatment of the FSH preparation with antiserum directed against the β-subunit of LH. The treatment of an LH preparation with this anti-β LH serum, however, completely abolished its effectiveness.

Lamprecht et al. (1973) have carried out, in addition to the measurement of cyclic AMP accumulation, a rather extensive study on the effect of LH and PGE_2 on protein kinase activity and the binding capacity for cyclic AMP in incubating whole ovaries. They found that ovaries obtained from rats 10 to 30 days of age responded to both LH and PGE_2 with a rapid increase in the endogenous concentration of cyclic AMP. Simultaneous addition of maximally stimulatory amounts of both of these substances to the incubation medium increased the cyclic AMP accumulation to a significantly greater extent than either agent alone, but the augmentation fell short of a fully additive effect. The effects of LH and PGE_2 on cyclic AMP were found to be dissociated in two experimental situations. First, ovaries obtained from rats up to 10 days of age responded well to PGE_2, but not at all to LH. Second, if the ovaries of older rats were subjected to a long incubation with LH and washed, they were found to be unresponsive to LH, but to respond fully to PGE_2. These findings suggested that LH and PGE_2 acted by separate mechanisms, and that prostaglandins did not act as mediators of the action of LH on adenylyl cyclase.

This suggestion was supported by subsequent studies from that group (Zor et al., 1973), which demonstrated that an almost complete block of prostaglandin synthesis (by the addition of aspirin, indomethacin, or flufenamic acid) did not prevent the stimulatory effect of LH on cyclic AMP accumulation. The incubation of ovaries from 28-day-old rats with LH or PGE_2 gave rise simultaneously to an increase in cyclic AMP accumulation, an increase in protein kinase activity, and a decrease in the binding capacity for exogenous [^{3}H]cyclic AMP. The activation of protein kinase was presumably due to the release of active, catalytic subunits from the native enzyme, and the reduction of binding sites for [^{3}H]cyclic AMP on the regulatory subunits due to the occupation of those sites by endogenous cyclic AMP. These results indicate that increases of ovarian cyclic AMP may produce their effects via protein kinase.

The effect of gonadotropins has been assessed on adenylyl cyclase activity in whole homogenates of ovaries obtained from the rat (Fontaine, Fontaine-Bertrand, Salmon, and Delerue-Lebelle, 1971; Danzo, 1972), the goldfish (Fontaine et al., 1970; Fontaine, Salmon, Fontaine-Bertrand, Burzawa-Gerard, and Donaldson, 1972), the trout, and the eel (Salmon, Fontaine-Bertrand, Delerue-Lebelle and Fontaine, 1974). In the rat, both LH and FSH stimulated the activity of this enzyme, and their effects were not additive when they were tested together at maximal stimulatory levels (Fontaine et al., 1971). Homogenates of goldfish ovary respond to a gonadotropin preparation purified from the pituitaries of the carp, but not a mammalian LH or FSH, indicating a specificity of the gonadotropins in stimulating adenylyl cyclase (Fontaine et al., 1970). The mechanism of action of the gonadotropins may also be quite different, as Salmon et al. (1974) have reported in an abstract that the gonadotropin preparation from the carp lowers the K_m of the enzyme system for ATP in homogenates of goldfish ovary, but does not change the V_{max}. FSH, on the other hand, is reported by them to augment the V_{max} of the adenylyl cyclase activity of homogenates of rat ovary. The adenylyl cyclase of the trout and the eel were not stimulated by the mammalian gonadotropins, LH, or FSH, or the carp gonadotropin preparation (Salmon et al., 1974).

5. Summary

There is an almost overwhelming amount of data which indicate that cyclic AMP mediates the action of LH on steroidogenesis in the ovary. LH causes a rise in the endogenous level of cyclic AMP in various ovarian preparations by activating the adenylyl cyclase enzyme system. Exogenous cyclic AMP mimics the effect of the gonadotropin on the stimulation of steroidogenesis, and the effect of LH is not additive to that produced by a maximum amount of cyclic AMP. Theophylline and aminophylline (inhibitors of the cyclic nucleotide phosphodiesterase) potentiated the effect of LH in several studies, and in one study (Hermier et al., 1972) imidazole (a stimulator of cyclic nucleotide phosphodiesterase) markedly reduced steroidogenesis in control or hCG-treated human corpora lutea.

The mechanism of action of cyclic AMP on steroidogenesis has not been elucidated, but the current concepts of its action will be taken up in a later section. Prostaglandins can mimic many of the effects of LH on the ovary, including the stimulation of steroidogenesis and the accumulation of cyclic AMP, but it seems at this time that prostaglandins do not function as a necessary component of the stimulation of adenylyl cyclase by LH. These substances may act as modulators, as suggested by Kuehl (1974).

FSH does not seem to be involved with the acute control of ovarian steroidogenesis (Greep et al., 1942; Savard et al., 1965; Dorrington and Kilpatrick, 1966a; Mills et al., 1971), but this gonadotropin does seem to cause an activation of adenylyl cyclase (Fontaine et al., 1971) and an increase in cyclic AMP accumulation in whole ovarian preparations (Mason et al., 1973; Koch et al., 1973). The finding that maximal stimulatory amounts of LH and FSH were not additive (Fontaine et al., 1971) suggests that they were acting on the same population of adenylyl cyclase, but such an interpretation is obviously premature until more thorough investigations are carried out. In this regard, it had been reported that LH and FSH were not additive in stimulating the adenylyl cyclase of the testis (Murad et al., 1969), but now it appears that LH acts on the Leydig cells and FSH acts on this enzyme system in the seminiferous tubules (Cooke, Van Beurden, Rommerts, and Van der Molen, 1972; Dorrington and Fritz, 1974; Braum and Sepsenwol, 1974).

B. Testis

It is generally recognized that the synthesis of androgens in the testis is under the control of LH (also called interstitial cell-stimulating hormone, ICSH), and it was reasonable to suspect from the work that had gone on in the adrenal cortex (Haynes et al., 1960) and the corpus luteum (Marsh and Savard, 1964b) that cyclic AMP was involved in this action of the gonadotropin. The investigation of the role of cyclic AMP in steroidogenesis in the testis followed a very similar pattern to the studies carried out in the ovary. It began with the study of the effects of exogenous cyclic AMP on steroidogenesis, progressed to the assessment of adenylyl cyclase, and finally to the measurement of endogenous cyclic AMP. This last phase of the work has recently uncovered some new problems for investigators in this field, and these problems will be discussed later in this section.

1. Effects of Exogenous Cyclic AMP

In the early studies, whole testis preparations were used, and Sandler and Hall (1966) were the first to show that exogenous cyclic AMP could stimulate testosterone production. They used teased rat testis and observed an increase in both the mass of testosterone synthesized and the conversion of [7α-^{3}H]cholesterol into [^{3}H]testosterone. The amount of cyclic AMP used (0.027 M) was far in excess of physiological levels, but the effect was specific, since ATP, ADP, and AMP had no effect at this concentration. Connell and Eik-Nes (1968) confirmed this effect of cyclic AMP on the synthesis of mass amounts of testosterone in slices of rabbit testis. Their report corrected an impression brought about by a previous study (Hall and Eik-Nes, 1962) that rabbit testicular tissue would not respond to cyclic AMP, as judged by the incorporation of [1-^{14}C]acetate into testosterone. Connell and Eik-Nes (1968) also assessed the effect of theophylline and found that it stimulated testosterone synthesis at low concentrations (10^{-7} M), indicating that there was a cyclic nucleotide phosphodiesterase and an adenylyl cyclase system in testis tissue. Other investigators (Dufau, Catt, and Tsuruhara, 1971, 1972a; Dufau, Tsuruhara, Watanabe, and Catt, 1972b; Catt, Watanabe, and Dufau, 1972; Rommerts, Cooke, Van der Kemp, and Van der Molen, 1972) have demonstrated that dibutyryl cyclic AMP is capable of accelerating testosterone production in whole testis preparations. This analogue is about 50 times more effective than cyclic AMP (Catt et al., 1972), and its time course of stimulation of steroidogenesis is almost identical to that produced by hCG (Dufau et al., 1972a).

The infusion of cyclic AMP into the dog testis via the spermatic artery *in vivo* also gave rise to an increased production and secretion of testosterone (Eik-Nes, 1967, 1969). Again large doses (150 to 300 µg/ml) of this nucleotide were required, but the effect was specific for the cyclic nucleotide. Van der Molen and Eik-Nes (1971), using the same technique, demonstrated that the effect of cyclic AMP occurred at steps in the steroidogenic pathway prior to the formation of pregnenolone. These data, taken together with the stimulation of the conversion of [7α-^{3}H]cholesterol into [^{3}H]testosterone by cyclic AMP (Sandler and Hall, 1966a), indicate that the site of cyclic AMP action in the testis is somewhere between cholesterol and pregnenolone, just as it seems to be in the corpus luteum (Hall and Koritz, 1965b) and the adrenal cortex (Karaboyas and Koritz, 1965). Mieno, Kawao, Shimizu, and Yamashita (1973) found that the injection of cyclic AMP into the systemic circulation of the dog was not effective in stimulating testicular steroidogenesis, but the injection of the cyclic nucleotide directly into the testis *in vivo* caused a considerable increase in steroid secretion.

Leydig cells are generally considered to be the main site of testosterone production (Hooker, 1970), and the effect of exogenous cyclic AMP has also been studied in isolated Leydig cell tumor preparations. Shin (1967) and Shin and Sato (1971) demonstrated that a monolayer culture of Leydig cells derived from a mouse interstitial cell tumor, which was insensitive to gonadotropins, would respond to very low levels of exogenous cyclic AMP with a significant increase in progestin synthesis. Nearly all of the studies carried out with exogenous cyclic AMP or even dibutyryl cyclic AMP have used amounts of these nucleotides above the physiological level of cyclic AMP, but these workers found that this preparation of Leydig tumor cells could be stimulated by as little as 5×10^{-6} M cyclic AMP. This is well within the range of the endogenous concentration of cyclic AMP in normal Leydig cells (Rommerts, Cooke, Van der Kemp, and Van der Molen, 1973; Dorrington and Fritz, 1974). Although this concentration of exogenous cyclic AMP was quite low, its continuous presence was required to sustain the stimulation of steroidogenesis. The same phenomenon was observed by Moyle, Moudgal, and Greep (1971), using another mouse Leydig cell tumor preparation which was sensitive to both LH and cyclic AMP. This preparation, however, required the usual high concentrations of cyclic AMP (2 mM to 0.02 M) to stimulate steroidogenesis (Moyle and Armstrong, 1970; Moyle et al., 1971). The stimulatory effect of exogenous cyclic AMP on steroidogenesis has thus been well established, but the mechanism by which it works is still uncertain.

2. Adenylyl Cyclase

Another approach to the investigation of the role of cyclic AMP in testicular steroidogenesis was the study of the adenylyl cyclase system in this tissue and its response to gonadotropins. Murad et al. (1969) found that both LH and FSH would stimulate adenylyl cyclase activity in whole homogenates of dog and rat testes. The effect of FSH appears to be due to a stimulation of an adenylyl cyclase in the seminiferous tubules and the LH effect to a stimulation of this enzyme in interstitial cells (Cooke et al., 1972; Dorrington and Fritz, 1974; Braun and Sepsenwol, 1974). NaF also produced a marked stimulation, and high concentrations of epinephrine and ACTH caused small increases in the activity of this enzyme in whole homogenates of dog testis. Glucagon had no effect. It is difficult to suggest any physiological explanation for the small stimulation produced by ACTH, since it is not known to have any effect on testicular steroidogenesis or spermatogenesis and there is disagreement on whether or not it has any effect on cyclic AMP accumulation in whole-cell testicular preparations (Kuehl et al., 1970a; Cooke et al., 1972). The epinephrine effect, however, has been observed by other groups (Pulsinelli and Eik-Nes, 1970; Kuehl et al., 1970a), and this catecholamine has also been found to cause a small increase in testosterone secretion when it is infused into the dog testis *in vivo* via the spermatic artery (Eik-Nes, 1971). Eik-Nes (1971) suggested that this effect may involve nerve fibers and the regulation of the vascular system of the testis. This might explain the stimulation of adenylyl cyclase by epinephrine, since elements of nerve and vascular tissues must be included in homogenates of whole testes. A similar effect of epinephrine has been reported in homogenates of corpora lutea (Marsh, 1970a) and in whole ovaries obtained from fish (Fontaine et al., 1970).

Hollinger (1970) and Eik-Nes (1971) reported that most of the adenylyl cyclase of the testis was found in the "nuclear" and the "mitochondrial" fractions, but these fractions were identified only by their sedimentation characteristics. No enzyme marker or electron microscopic data were presented. Pulsinelli (1972) assessed the correlation of adenylyl cyclase with marker enzymes for plasma membranes (Na,K-dependent ATPase) and mitochondria (succinic dehydrogenase), and found the cyclase to be primarily associated with plasma membranes. Sulimovici and Lunenfeld (1973) also described a hormonally responsive adenylyl cyclase in a testicular mitochondrial preparation which was relatively free of endoplasmic reticulum (Sulimovici, Bartoov, and Lunenfeld, 1973), but the degree of possible contamination with plasma membranes was not established. It seems, therefore, that most of the adenylyl cyclase is located in plasma membranes, but there may also be an association of this enzyme with mitochondria and other subcellular components of the testis.

The rat testis also contains two isozymes of cyclic nucleotide phosphodiesterase, designated c and f (Monn, Desautel, and Christiansen, 1972; Christiansen and Desautel, 1973). The f form is a low K_m type (2.5×10^{-6} M) and is found only in the seminiferous tubules. The c form is a high K_m type (6.5×10^{-5} M) and is presumably found in the other parts of the testis. The amount of total phosphodiesterase in the testis increases fivefold from day 20 to day 50 of life in the rat and then remains essentially constant to day 80 (Monn et al., 1972). This increase in enzyme production is very similar to that of adenylyl cyclase in this period of the rat's life (Hollinger, 1970). LH has no acute effect on phosphodiesterase (Kuehl et al., 1970a), and its effect on steroidogenesis is probably transmitted only through adenylyl cyclase.

3. Measurement of Endogenous Cyclic AMP

Studies carried out on the changes in endogenous cyclic AMP in the testis have, however, not completely supported the proposal that cyclic AMP mediates the effect of gonadotropins on steroidogenesis. When relatively high concentrations of gonadotropins (greater than 10 ng/ml of LH or 1 ng/ml of hCG) were used, they have repeatedly been found to cause a marked increase of endogenous cyclic AMP in whole testis *in vitro* (Kuehl et al., 1970a; Rommerts et al., 1972; Catt et al., 1972; Dufau, Watanabe, and Catt, 1973; Catt and Dufau, 1973; Braun and Sepsenwol, 1974). A similar effect is apparent in Leydig cell preparations (Cooke et al., 1972; Cooke, Rommerts, Van Beurden, Van der Kemp, and Van der Molen, 1973; Moyle and Ramachandran, 1973; Rommerts et al., 1973; Dorrington and Fritz, 1974). This stimulation appears to be quite specific in that prolactin, ACTH, and growth hormone do not increase the cyclic AMP levels in either type of testicular preparation (Cook et al., 1972). FSH was found to increase cyclic AMP in whole testis preparations, but, as mentioned before, it appears to act only on the seminiferous tubules and not on the Leydig cells (Cooke et al., 1972; Braun and Sepsenwol, 1974; Dorrington and Fritz, 1974). In regard to the time course of the hCG effect, the rise in endogenous cyclic AMP appeared after about 10 min of incubation with hCG, while the increase in testosterone production was not apparent until after 30 to 60 min of incubation (Rommerts et al., 1972, 1973). This is very similar to the kinetics of the stimulation of cyclic AMP accumulation and progesterone production by LH in incubating slices of corpora lutea (Marsh et al., 1966), and is compatible with the proposal of a mediatory role for cyclic AMP in the stimulation of steroidogenesis by gonadotropins.

When careful comparisons of the dose response of cyclic AMP accumulation and testosterone production have been carried out, however, a dissociation of these two effects has been uncovered. In brief, a hormonal stimulation of steroidogenesis can be elicited with low concentrations of LH or hCG without a detectable change in the endogenous level of cyclic AMP. This is a crucial discrepancy, since one of the criteria for a role of cyclic AMP in hormonal action is that there should be a demonstrable effect on the concentration of this cyclic nucleotide by the smallest level of the hormone capable of producing the physiological response (Robison, Butcher, and Sutherland, 1971). Two groups of investigators have carried out quite thorough studies on this problem in the testis, and I will, therefore, review their work in some detail.

Catt and Dufau (1973) compared the effects of various concentrations of hCG (specific activity 10,000 IU/mg) on the release of testosterone and cyclic AMP into the medium during a 2-hr incubation of intact decapsulated rat testes. It was found that testosterone release was significantly increased by hCG concentrations as low as 0.1 ng/ml and reached a maximum value at 0.8 to 1.0 ng/ml. On the other hand, the release of cyclic AMP began at 1 ng hCG/ml and reached a maximum at about 100 ng/ml. The cyclic AMP was assessed only at one time period (2 hr) and only in the medium, in this report, but thorough time studies from 1 to 240 min had been carried out previously (Catt et al., 1972; Dufau et al., 1973) with 1 to 500 ng hCG/ml, and no transient rises in cyclic AMP were detected. It has also been reported that the relative changes in cyclic AMP appeared to be more prominent in the incubation medium than in the testis tissue itself (Dufau et al., 1973). In addition, in both of the studies just mentioned, an assessment of cyclic AMP accumulation in the medium was carried out using the incorporation of ^{14}C-adenine into cyclic AMP. An increase in ^{14}C-adenine incorporation was found to parallel the changes in the mass amounts of cyclic AMP accumulated when 20 to 500 ng hCG/ml were added to the incubation medium. This indicated that newly synthesized cyclic AMP comprised a constant proportion of the cyclic AMP released.

For the sake of completeness, in order to rule out any possibility that there might be a transient increase in cyclic AMP in response to the low concentrations (0.1 to 1 ng/ml) of hCG, one would like to see another thorough time study carried out on the accumulation of cyclic AMP in both the medium and the tissue, when these low levels of hCG are used.

The fact that 5 mM theophylline by itself caused a significant increase in the release of testosterone into the medium by the incubated testicular tissue, and potentiated the effects of low concentrations of hCG, indicated to these authors (Catt and Dufau, 1973) that a small and as yet undetectable amount of cyclic AMP was probably involved in the steroidogenic response to low hCG levels. This stimulation of testosterone production by theophylline was also observed without a detectable increase in the amount of cyclic AMP released to the medium. It is reasonable to expect that this classical inhibitor of cyclic nucleotide phosphodiesterase acts via cyclic AMP, or cyclic GMP, since both nucleotides are substrates for this enzyme (Goldberg, O'Dea, and Haddox, 1973).

The other major study on this problem in gonadal tissue was carried out by Moyle and Ramachandran (1973). They assessed the effects of various concentrations of LH on testosterone synthesis and cyclic AMP accumulation in an isolated Leydig cell preparation from rat testis and a mouse Leydig cell tumor preparation. The testosterone was measured after 2 hr of incubation and the cyclic AMP after 10 min in both the medium and the tissue. Again there was a striking difference between the amount of LH required for an increase in testosterone production and that required for a stimulation of cyclic AMP accumulation in incubations of either tissue. The amount required for a minimal increase in cyclic AMP was about one order of magnitude (approximately 10 ng LH/ml) above that required for a minimal stimulation of steroidogenesis (approximately 1 ng LH/ml) in the isolated rat Leydig cells. A time study of the changes in cyclic AMP with these rat Leydig cells was carried out and showed that cyclic

AMP accumulated progressively from 2.5 to 60 min, but again this study was carried out only with relatively high concentrations of LH (500 ng and 5 µg LH/ml).

The results of another very interesting experiment, reported in this paper, indicated that the increase in cyclic AMP, observed with relatively high concentrations of LH, was not required for this gonadotropic stimulation of steroidogenesis. It had been shown previously (Moudgal, Moyle, and Greep, 1971; Moyle et al., 1971) that LH specifically binds to these tumor Leydig cells, stimulating testosterone production, and that a washing procedure cannot remove the bound LH nor terminate the enhanced steroidogenesis. The bound hormone could be removed, however, by treatment of the washed cells with an antiserum to LH, and this did terminate the accelerated rate of steroidogenesis. On the other hand, it had been shown that exogenous cyclic AMP mimicked the effect of the hormone on steroidogenesis, but this effect by the cyclic nucleotide could be rapidly terminated by simply washing the tumor cells. The interpretation of this data was that LH was retained by the cells even after repeated washings, and that the continued presence of the bound LH was required for maintenance of the steroidogenic response. Since the dose-response studies just described (Moyle and Ramachandran, 1973) raised some questions about the role of cyclic AMP in the hormonal stimulation of steroidogenesis, a similar washing experiment was carried out by these authors to see what effect this procedure would have on the endogenous level of cyclic AMP. It was found that although the washing procedure did not affect the steroidogenic response to a relatively high concentration of LH (100 ng/ml), it did result in a rapid cessation of cyclic AMP accumulation caused by the LH. It was proposed from this data that there might be two types of receptors for LH: one which had high affinity (since the effect could not be terminated by washing) and was involved with steroidogenesis, and another which had lower affinity (since the effect was terminated by washing) and was associated with the major increase in cyclic AMP synthesis. Further work would seem to be necessary to determine the validity of this suggestion, but it does seem that the large increase in cyclic AMP accumulation in the Leydig cells of the testis caused by relatively large concentrations of LH is not required for an increase in steroidogenesis.

Similar dissociation of cyclic AMP accumulation from the effects of ACTH on corticosteroidogenesis (Beall and Sayers, 1972; Moyle, Kong, and Ramachandran, 1973b; Richardson and Schulster, 1973) and from the effects of TSH on thyroid colloid droplet formation (Williams, 1972) have been reported.

4. Reevaluation of the Role of Cyclic AMP in Testicular Steroidogenesis

The data discussed in the previous section can be interpreted to mean that either cyclic AMP is not the second messenger for the LH effect on steroidogenesis in the testis or that this stimulation involves changes of cyclic AMP that are too small to be detected under the experimental conditions employed. It should be emphasized here that none of the investigators whose work I have just cited have concluded that cyclic AMP is absolutely not involved. They have simply suggested these two possibilities.

a. *Evidence in favor of a role for cyclic AMP*

The strongest evidence remaining which indicates that cyclic AMP is involved in steroidogenesis appears to be that exogenous cyclic AMP or dibutyryl cyclic AMP has consistently been shown to stimulate testicular steroidogenesis *in vivo* and *in vitro* (Sandler and Hall, 1966a; Shin, 1967; Eik-Nes, 1967; Connell and Eik-Nes, 1968; Eik-Nes, 1969; Moyle and Armstrong, 1970; Moyle et al., 1971; Van der Molen and Eik-Nes, 1971; Shin and Sato, 1971; Dufau et al., 1971, 1972 a; Catt et al., 1972; Rommerts et al., 1972; Mieno et al., 1973). The amount of exogenous cyclic AMP usually required for this effect is above the endogenous concentration, but in at least one Leydig cell preparation (Shin, 1967; Shin and Sato, 1971), the effective amount was within the physiological range. The time course of the stimulation with exogenous dibutyryl cyclic AMP was also found to be almost identical to that brought about by hCG (Dufau et al., 1972a).

Inhibitors of cyclic nucleotide phosphodiesterase such as theophylline have been shown to increase steroidogenesis and potentiate the effect of LH on the testis (Connell and Eik-Nes, 1968; Catt and Dufau, 1973). As mentioned earlier, this indicates that a cyclic nucleotide is probably involved, but does not distinguish between cyclic AMP and cyclic GMP.

Means, MacDougall, Soderling, and Corbin (1974) reported a stimulation of cyclic AMP-dependent protein kinase after 20 min of incubation of decapsulated rat testes with 1 μg/ml of LH or 0.15 IU/ml of hCG. This is an indirect indication that an increase in endogenous cyclic AMP occurred, since the activation of this enzyme is known to be regulated by cyclic AMP in other tissues (Soderling, Corbin, and Park, 1973). Unfortunately, this result does not help resolve the problem, because these experiments were also carried out with concentrations of the gonadotropins above that required to observe dissociation between the responses of testosterone production and cyclic AMP accumulation. This approach, however, may be useful in future experiments with the crucially low concentrations of gonadotropins to ascertain an increase in cyclic AMP production indirectly, when a change in the concentration of this nucleotide is not detectable by direct analytical methods. This approach has been explored recently by Richardson and Schulster (1973) with isolated adrenal cells, but they found that low doses of ACTH that elicited steroidogenesis failed to cause a clear stimulation of protein kinase activity. The preparation of these adrenal enzyme extracts and the assays themselves were carried out, however, in the absence of high salt (0.5 M NaCl), which has been found to be necessary to prevent the reassociation of the regulatory and catalytic subunits of protein kinase in diluted homogenates of adipose (Corbin, Soderling, and Park, 1973) and testicular tissue (Means et al., 1974). It is possible, therefore, that an activation of adrenal protein kinase at low levels of ACTH did occur, but went undetected.

b. *Possible methodologic problems in detecting changes in cyclic AMP and suggestions for further studies*

If cyclic AMP is involved in the hormonal acceleration of steroidogenesis, the inability of investigators to detect a change in the concentration of this nucleotide or its turnover rate must be due to some inadequacy of the methodology. Possibly, the changes involved are too small to be detected by present-day analytical techniques. If this is the case, progress in this area will have to await the development of more precise methodologies.

The stimulation by gonadotropin might be observed as an increased flux or rate of turnover of cyclic AMP, even though the changes in its absolute concentration are too small to be detectable. This implies that there is a rapid destructive mechanism associated with or acting just after the interaction of cyclic AMP with its receptor. Under these circumstances, critical changes in the concentration of cyclic AMP could accelerate a process, but, because of the rapid destruction of the nucleotide, only the increased turnover would be measurable. Changes in turnover such as these might be more easily detected using a radioactive precursor methodology, such as the incorporation of [^{14}C]adenine into cyclic AMP (Kuo and DeRenzo, 1969). This was used by Kuehl et al. (1970*a*), Dufau et al. (1973), and Braun and Sepsenwol (1974), but not at the low levels of LH or hCG where the dissociation between cyclic AMP and testosterone production is apparent. Beall and Sayers (1972) used this method with low levels of ACTH and found that they could still observe an increase in the synthesis of corticosterone without any change in the incorporation of [8-^{14}C]adenine into cyclic AMP.

Another possible reason for the failure to detect small increases in cyclic AMP could be the destruction of the nucleotide by the presence of cyclic nucleotide phosphodiesterase during the process of isolation and measurement. Theophylline was present at a concentration of 1 mM (Moyle and Ramachandran, 1973) or 5 mM (Catt and Dufau, 1973) in the medium when the cyclic AMP assays were carried out in these studies, but no assessment was made of the effectiveness of these inhibitor concentrations. In homogenates of corpora lutea these levels of theophylline have been found to be inadequate to completely block phosphodiesterase activity (Marsh, 1970*a*). In future experiments, larger concentrations of theophylline or more potent inhibitors such as papaverine or 1-methyl, 3-isobutylxanthine (Peytremann, Wendell, Liddle, Hardman, and Sutherland, 1973; Braun and Sepsenwol, 1974) should be tried. In this regard, it may also be necessary to stop the metabolism of the cell very rapidly by some quick-freeze technique in order to see changes in the concentration of cyclic AMP at any one time. In order to detect such a change, Namm and Mayer (1968) found it necessary to rapidly freeze rat hearts by clamping them between two metal blocks at the temperature of liquid nitrogen.

If the effective change in cyclic AMP is strictly confined to a small compartment of the cell *in vivo*, this will be difficult to resolve with our present methodology. Immunofluorescent techniques such as those described by Wedner, Hoffer, Battenberg, Steiner, Parker, and Bloom (1972) may answer this question, but at the present time this technique appears to be more useful for qualitative than for quantitative assessments. One component of the cell which might be considered as a compartment is the regulatory subunits of protein kinase. If small changes in cyclic AMP do occur when low levels of hormones are used, they may be associated with the regulatory subunit. It may be possible to detect changes in the amounts of cyclic AMP associated with this subunit when it is impossible to detect changes in the overall concentrations of this nucleotide in the cell or the incubation medium. Changes in the interaction of cyclic AMP with this subunit could be detected directly or indirectly by an increase in protein kinase activity (Means et al., 1974).

Another possible difficulty in the methods used so far is the variability in the tissue preparations. Baseline levels of cyclic AMP in tissue preparations may reflect primarily the cyclic AMP levels in cells not associated with steroid production and thus obscure the small increases in cyclic AMP levels in the gonadotropin-responsive steroidogenic cells. A good illustration of this problem is seen in the work by Butcher and Baird (1968). Prostaglandin E_1 (PGE_1), which was known to be a potent antagonist to the action of epinephrine on lipolysis, caused an increase in cyclic AMP large enough to maximally activate lipolysis in incubated fat pads. When the fat cells were isolated from the pads, however, and incubated with PGE_1, the expected decrease in cyclic AMP was observed, indicating that the rise in cyclic AMP in the whole fat pad, caused by PGE_1, was due to cell types not involved in lipolysis. This may be more of a problem for investigators using whole testes than those using Leydig cell preparations. The Leydig cell preparations, however, are not entirely homogenous. Normal Leydig cell preparations contain unspecified amounts of unidentified cell types (Moyle and Ramachandran, 1973), and the tumor tissue has only been described as containing a large percentage of Leydig cells (Moyle and Armstrong, 1970). The Leydig cell tumors have also been reported to change significantly in the type of steroid produced, although not apparently in responsiveness to LH. Recently, a paper by Varga, Dipasquale, Pawelek, McGuire, and Lerner (1974) indicated that in order to eliminate variability in a tissue, one may need to have not only a uniform population of cell types, but also a synchrony of their cell cycles. They found that melanocyte-stimulating hormone (MSH) increased the endogenous levels of cyclic AMP in mouse melanoma cells only in the G2 phase of the cell cycle. These cells would, however, respond to exogenous cyclic AMP with an increase in tyrosinase activity at any phase of the cell cycle. Ideally then, in future studies it may be necessary to use a uniform population of gonadotropic responsive cells, which have had their phases of the cell cycle synchronized with colchicine in order to detect an increase in cyclic AMP associated with steroidogenesis.

Finally, the role of cyclic AMP in this process may be determined by the use of agents which specifically reduce the concentration of cyclic AMP within the cell or block its action. Imidazole has been found to antagonize the action of hCG on steroidogenesis in human corpora lutea (Hermier et al., 1972), as well as the actions of several other hormones which are believed to act via cyclic AMP

(Robison et al., 1971). It appears that this approach has not been used with LH or hCG on testicular tissue. Another modification in procedure might be the use of inhibitors of cyclic AMP action along with the gonadotropins. If the effects of low levels of LH or hCG on steroidogenesis were blocked by a specific cyclic AMP inhibitor, this would implicate cyclic AMP in the effect of the gonadotropin. One possible candidate for such an inhibitor might be the affinity-labeling analogue $O^{2'}$-(ethyl 2-diazomalonyl) cyclic AMP which was found by Brunswick and Cooperman (1971) to bind covalently to rabbit muscle phosphofructokinase.

c. *The possibility of another second messenger*

There is no evidence for another second messenger in gonadal tissue at the present time. Cyclic guanylic acid is an obvious possibility, but the only report concerning its hypothetical role in gonadal steroidogenesis is an abstract by Mendelson, Dufau, and Catt (1974), in which no release of cyclic GMP was detected from isolated rat testes after incubation with low or high concentrations of hCG. In addition, exogenous cyclic GMP at a concentration of 25 mM was unable to stimulate testosterone production in testes which responded to 1 mM cyclic AMP.

In adrenal cortex tissue, however, there are three abstracts recently reported on the possible role of cyclic GMP in steroidogenesis, but they do not appear to be entirely consistent. Kitabchi, Nathans, James, Bower, Wilson, and Kitchell (1974) reported that 1 to 10 μU of ACTH caused a rise in cyclic GMP and corticosteroidogenesis in isolated adrenal cells without a rise in cyclic AMP. At high levels of ACTH ($>$ 50 μU), cyclic AMP was increased, but cyclic GMP formation was inhibited. Honn and Chavin (1974) also reported that ACTH or a reptilian pituitary homogenate caused an increase in cyclic GMP and a decrease in cyclic AMP in incubated adrenals from reptiles. Whittey, Stowe, Ong, Ney, and Steiner (1974), on the other hand, found that hypophysectomy in rats caused an immediate drop in cyclic AMP and an increase in cyclic GMP in adrenal tissue within an hour. Administration of ACTH (20 units, s.c.) to these hypophysectomized rats *in vivo* caused a fall of cyclic GMP to baseline and a 50-fold increase in cyclic AMP within 15 min. These data indicate that there is a reciprocal relationship between cyclic AMP and cyclic GMP, but it is still too early to say whether cyclic GMP is a mediator of the effect of ACTH on steroidogenesis.

d. *Summary*

It seems, then, that the question of the role of cyclic AMP as a mediator of steroidogenesis in testicular tissue is still unanswered. There are several reasons why small changes in cyclic AMP concentration might not be detected even if they are present, but until there is firm evidence for a change in cyclic AMP at the low level of gonadotropin its role as a mediator of LH action will remain in doubt.

At the present time the only thorough studies that have been carried out are those on the adrenal and testes, but the same situation may hold true for the ovary. Mason (1974) has presented preliminary data suggesting that endogenous levels of LH, although capable of maintaining normal ovarian function, are insufficient to cause detectable changes in endogenous cyclic AMP in whole rat ovaries. A reexamination of every gonadotropic action is probably warranted wherever it has not been demonstrated that the effect on cyclic AMP occurs at the smallest concentration of the hormone required for its physiological action.

C. Possible Modes of Action of Cyclic AMP on Steroidogenesis

If we believe that there is considerable evidence in favor of a role for cyclic AMP in the control of steroidogenesis, and that this is likely to be confirmed by future studies, we are confronted with the problem of how this effect might be carried out. As mentioned previously, the site of action of LH and cyclic AMP on the steroidogenic pathway has been shown to be between cholesterol and pregnenolone (Ichii et al., 1963; Hall and Koritz, 1964, 1965b; Koritz and Hall, 1965; Hall and Young, 1968; Armstrong et al., 1970). There are, however, several ways by which cyclic AMP might bring about the acceleration of this reaction. These are shown in Fig. 1 and are listed briefly as follows: it could bring about (1) an increase in the amount of a cofactor such as NADPH; (2) an increase in the concentration of the substrate, cholesterol; (3) an increase in the availability of this cholesterol by promoting its transport into the mitochondrion, where the cholesterol side-chain cleavage system is located; (4) an activation or an increase in the synthesis of one of the components of the cleavage system; (5) a decrease in some restraining influence on this enzyme system, perhaps by enhancing the transport of an end-product inhibitor, such as pregnenolone, out of the mitochondrion. There is evidence supporting each of these proposals, and at the present time it is not possible to choose any one of them as the correct one. In this regard, however, it should be recognized that these proposals are not mutually exclusive, and it is possible that cyclic AMP may act by one or more of these mechanisms. The evidence in favor of (or against) these hypotheses is discussed below with emphasis on the data obtained from experiments with gonadal tissue. The results of some important experiments with adrenal cortex tissue are also included, since the control of steroidogenesis in this tissue has the same general features as that in the ovary and the testis (see also review by Halkerston, *this volume*).

1. Action via Increased Cofactors

Haynes and Berthet (1957) and Haynes et al. (1960) were the first to propose that cyclic AMP might accelerate steroidogenesis in the adrenal cortex by increasing the concentration of a necessary cofactor in the steroidogenic pathway. Briefly, their proposal stated that: (1) ACTH increased the synthesis of cyclic AMP in the adrenal cortex, which in turn activated the phosphorylase system; (2) phosphorylase activation accelerated glycogen breakdown, resulting in the formation of increased amounts of glucose 6-phosphate; (3) the metabolism of

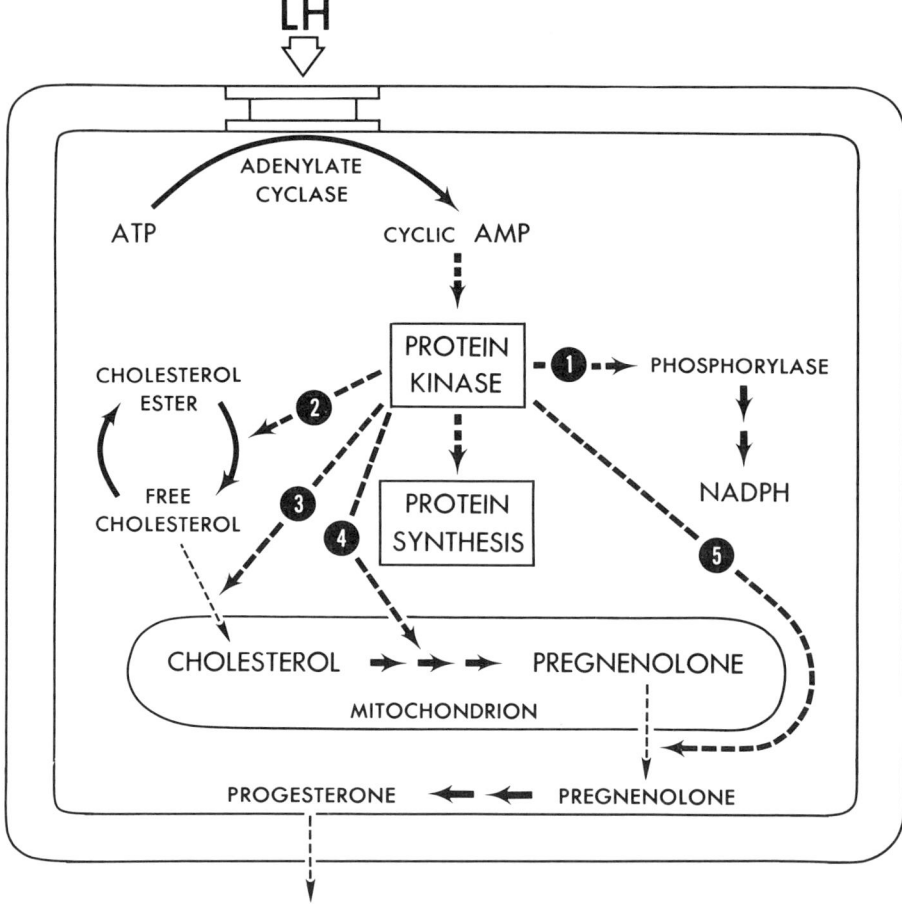

FIG. 1. Possible sites of action of cyclic AMP on steroidogenesis in a hypothetical gonadal cell. Solid arrows indicate biochemical reactions, and the large dashed arrows indicate possible effects of cyclic AMP on the steroidogenesis process. Five of the dashed arrows indicate that the possible effects of cyclic AMP might be mediated through protein kinase, but it is also possible that cyclic AMP influences these steps in steroidogenesis by a mechanism not involving protein kinase. The lighter dashed arrow indicates transport of a substance through membranes of the cell.

glucose 6-phosphate through the pentose phosphate pathway produced an increased level of NADPH, which stimulated corticosteroid production by means of its role as a required cofactor in many of the steps of steroidogenesis, including the cholesterol side-chain cleavage reaction. The enzyme system for this latter reaction has been shown to be localized in the mitochondria of the adrenal cortex cells (Halkerston, Eichhorn, and Hechter, 1959, 1961), and the NADPH produced by the pentose pathway is in the cytoplasm. Thus a shuttle system such as that described by Simpson and Estabrook (1969) is required to transport the reducing equivalents into this organelle.

There is some evidence in support of a mechanism such as that suggested by Haynes and co-workers for the action of LH in ovarian tissue. Phosphorylase activity has been detected in bovine corpora lutea (Williams, Johnson, and Field, 1961; Marsh and Savard, 1964a; Stansfield and Robinson, 1965; Yunis and Assaf, 1970), luteinizing rat ovaries (Stansfield and Robinson, 1965), and prepubertal rat ovaries (Selstam and Ahren, 1971; Ahren and Selstam, 1971; Ahren et al., 1973). Furthermore, LH has been shown to increase this activity *in vitro* in incubating slices of bovine corpora lutea (Marsh and Savard, 1964a) and whole isolated prepubertal rat ovaries (Selstam and Ahren, 1971; Ahren and Selstam, 1971; Ahren et al., 1973). It has also been found that the injection of LH into rats *in vivo* will cause an increase in the phosphorylase activity in the prepubertal or the luteinized type of ovary (Stansfield and Robinson, 1965; Selstam and Ahren, 1971; Ahren et al., 1973).

In the experiments on bovine corpora lutea, Williams et al. (1961) reported that although a crude pituitary preparation and hCG significantly increased phosphorylase activity, a preparation of luteinizing hormone was ineffective. This study appears, however, to have been carried out with only a limited number of corpora lutea, and when a large number of corpora lutea obtained from cows in the first 6 months of pregnancy were examined (Marsh and Savard, 1964a), LH was found to produce a small (35%) but significant increase in phosphorylase activity. The luteal phosphorylase activity was measured in an assay medium containing 5'-AMP, which indicated that this enzyme was similar to the liver enzyme which is only partially activated by the nucleotide, and differed from skeletal muscle phosphorylase which is completely activated by 5'-AMP. This was confirmed when the luteal enzyme was purified and characterized by Yunis and Assaf (1970). The stimulation of phosphorylase in the incubating slices of corpora lutea was specific for LH-containing hormones and correlated with an increase in progesterone synthesis in these slices. The addition of cyclic AMP at a concentration of 2 mM to the incubating slices, however, did not cause a rise in phosphorylase activity, although it did cause a small increase in progesterone synthesis. One difference in the LH and nucleotide incubations was the inclusion of 0.02 M caffeine in the cyclic AMP medium. This methylxanthine, in addition to its inhibitory effect on cyclic nucleotide phosphodiesterase, has also been reported to stimulate phosphorylase phosphatase (Sutherland, 1951), which may account for the small decrease in phosphorylase activity actually observed in the cyclic AMP incubations. The presence of caffeine does not, however, seem to account for the lack of response of this phosphorylase system to cyclic AMP, for in other experiments (Marsh, *unpublished data*) the omission of caffeine from the cyclic AMP-treated samples eliminated the slight drop in activity, but the exogenous cyclic nucleotide still did not increase the phosphorylase activity. A larger concentration of cyclic AMP (0.02 M) has also been tested on these bovine corpora lutea in the absence of caffeine, and again there was a stimulation of progesterone synthesis without an effect on phosphorylase activity (Savard et al., 1965).

This discrepancy has never been resolved, but it has been suggested that adding

LH to incubating corpora lutea slices may be a much better way of increasing intracellular cyclic AMP than adding the nucleotide exogenously. Nevertheless, this inability to activate phosphorylase by a concentration of exogenous cyclic AMP which stimulated progesterone synthesis indicated to us that phosphorylase activation might not be an essential effect of LH in stimulating progesterone synthesis. This was supported by the fact that about 10% of the corpora lutea tested showed little or no increase in terms of phosphorylase activity, while they exhibited a good response to LH in terms of an increase in progesterone synthesis (Marsh and Savard, 1964a). This is also supported by the finding that bovine corpora lutea (Stansfield and Robinson, 1965), luteinized rat ovaries (Deane, 1952; Armstrong, 1963; Stansfield and Robinson, 1965), and rabbit ovaries (Dorrington and Kilpatrick, 1967) contain little or no measurable glycogen. Stansfield and Robinson (1965) found that the injection of LH into rats *in vivo* caused about a fourfold increase in phosphorylase activity in their luteinized ovaries 4 hr later, but no significant change in glycogen concentration. Ahren and co-workers (Selstam and Ahren, 1971; Ahren and Selstam, 1971; Ahren et al., 1973) reported that the *in vivo* injection of LH or the *in vitro* incubation of prepubertal rat ovaries with LH or FSH caused a rise of ovarian phosphorylase activity. The phosphorylase of this prepubertal ovary was reported to resemble that of skeletal muscle, and the increase was due entirely to a shift from the 5'-AMP-dependent *b* form to the 5'-AMP-independent *a* form of the enzyme. The addition of dibutyryl cyclic AMP at concentrations from 1 to 25 mM to the incubating ovaries was more effective than either LH or FSH. Epinephrine at a concentration of 10 mM also increased the percentage of phosphorylase *a*.

Thus it appears that LH and perhaps FSH can stimulate phosphorylase in several ovarian preparations, but it seems that this effect may not be essential for an increase in steroidogenesis. A similar situation exists in the adrenal cortex. Haynes (1958) demonstrated that ACTH stimulated phosphorylase activity in bovine adrenal cortex, but Ferguson (1963) and Kobayshi, Yago, Morisaki, Ichii, and Matsuba (1963) found no stimulation by ACTH in rat adrenal cortex tissue even though this hormone produced its usual stimulation of corticosteroidogenesis.

It has been found that the addition of NADPH to incubating slices of bovine corpora lutea (Mason, Marsh, and Savard, 1962; Savard, Marsh, and Howell, 1963; Savard and Casey, 1964; Mason and Savard, 1964b; Armstrong, 1966) to slices of rabbit corpora lutea or interstitial tissue (Dorrington and Kilpatrick, 1966b) or to homogenates of whole rabbit ovaries (Scoon and Major, 1972), caused a marked increase in progestin biosynthesis. In addition, Savard et al. (1963) demonstrated that the bovine corpus luteum had high concentrations of the NADP dehydrogenases of the pentose pathway. It became apparent, however, that the effect of NADPH was qualitatively quite different from that of LH. The first indications of this difference came from the study of Savard et al. (1963) where it was found that the addition of LH to luteal slices, incubating in the presence of optimal amounts of NADPH, caused a further increase in progesterone synthesis. This should not have occurred if the effect of LH was mediated solely by NADPH. Dorrington and Kilpatrick (1966b) also observed additive effects of maximally effective concentrations of LH and NADPH. The effects of LH and NADPH on incubating slices of bovine corpora lutea also appeared to be different in terms of precursor utilization. LH increased the incorporation of [1-^{14}C]acetate into progesterone to a greater extent than the incorporation of [7-^{3}H]cholesterol. NADPH, on the other hand, did not increase the incorporation of [1-^{14}C]acetate, but markedly enhanced the incorporation of [7-^{3}H]cholesterol into this steroid (Savard and Casey, 1964; Mason and Savard, 1964b). Finally, it became apparent from the work of Armstrong (1966) on the bovine corpus luteum and Halkerston (1968) on the adrenal cortex that exogenous NADPH probably acted only on damaged cells while LH and ACTH acted on intact cells.

In regard to the endogenous concentration of NADP and NADPH and the effect of LH, it has been reported that this gonadotropin has no effect on the concentration of either the reduced or the oxidized form of this pyridine nucleotide in bovine corpora lutea (Marsh, 1968) or luteinized rat ovaries (Flint and Denton, 1970). Furthermore, the latter investigators also found that glucose raised the intracellular concentration of NADPH in slices of luteinized rat ovaries, but had no effect on steroidogenesis in this tissue. It was concluded (Flint and Denton, 1970), therefore, that it was unlikely that LH brought about its increase in steroidogenesis solely by increasing the production of NADPH. Hall (1971) criticized this interpretation, saying that it did not take into account the possibility that an increased rate of utilization of NADPH might balance an LH-induced increase in the rate of its production, and thereby maintain an unaltered steady-state concentration of NADPH. This alternative seems theoretically possible, but until there is more information on how NADPH is utilized by the side-chain cleavage system *in situ*, it is impossible to reach a firm conclusion on this matter.

2. Action via Increased Substrate

It is well known that the injection of LH preparations into rabbits and rats will cause a decrease in ovarian cholesterol (Claesson and Hillarp, 1947; Levin and Jailer, 1948; Bell, Mukerji, and Loraine, 1964; Herbst, 1967; Armstrong, 1968) and that nearly all of this decrease can be accounted for by a decrease in cholesterol ester (Claesson, Diczfalusy, Hillarp, and Hogberg, 1948; Claesson, Hillarp, and Hogberg, 1953; Herbst, 1967; Armstrong, 1968). Furthermore, Moyle, Jungas, and Greep (1973a) have shown that the addition of LH or cyclic AMP to incubating Leydig tumor cells causes a rapid (10 min) conversion of ester cholesterol into free cholesterol. This effect could come about either by a stimulation of the cholesterol esterase or by an inhibition of the cholesterol ester synthetase, and evidence for both sites of action has been presented.

Behrman and Armstrong (1969) demonstrated that the *in vivo* administration of LH increased cholesterol esterase activity in luteinized ovaries of rats. No direct effect of cyclic AMP (0.05 mM) or theophylline (1 mM) was observed, however, when these substances were added to the cholesterol esterase assay mixture. Flint et al. (1973) have also reported that cyclic AMP had no direct effect when added to the assay system for cholesterol esterase activity in rabbit ovarian interstitial tissue. Another piece of evidence indicating that LH activates the cholesterol esterase enzyme is that of Behrman, Moudgal, and Greep (1972a), who showed that the administration of a LH antiserum to pregnant rats *in vivo* resulted in a marked decrease in cholesterol esterase activity and a reduction in progesterone secretion.

It is possible that this stimulatory effect of LH on cholesterol esterase is secondary to the action of this hormone on steroid synthesis. An enhanced conversion of cholesterol into pregnenolone might remove a hypothetical feedback inhibition of the esterase enzyme. The evidence on this point is equivocal. Behrman, Armstrong, and Greep (1970) concluded that the LH-induced depletion of cholesterol ester was not due to a secondary effect of LH on steroidogenesis, since they could observe the action of LH on cholesterol ester depletion when steroidogenesis was blocked by aminoglutethimide. On the other hand, Flint et al. (1973), working with the same tissue, found that the administration of aminoglutethimide or cycloheximide *in vivo* blocked the effect of LH on the depletion of cholesterol ester.

Goldstein and Marsh (1973) have assessed the effect of LH and cyclic AMP on incubating slices of bovine corpora lutea and found that LH- and cyclic AMP-treated tissues had slightly higher values of cholesterol esterase activity. Moreover, homogenates of this tissue responded to cyclic AMP and ATP with a significant increase in cholesterol esterase activity (Goldstein, 1973). This effect of cyclic AMP in homogenates and the report of Moyle et al. (1973a) of a stimulation of cholesterol ester conversion into cholesterol by cyclic AMP in isolated Leydig cells suggested that the effect of LH and cyclic AMP on cholesterol esterase might be mediated through a protein kinase phosphorylation reaction, such as that which acts on the adipose tissue triglyceride lipase (Huttunen, Steinberg, and Mayer, 1970; Corbin, Reiman, Walsh, and Krebs, 1970; Steinberg and Huttunen, 1972). Cyclic AMP-dependent protein kinase activity has been detected in bovine corpora lutea (Goldstein and Marsh, 1972, 1973; Menon, 1973), rat ovaries and Graafian follicles (Tsafriri et al., 1972b; Lamprecht et al., 1973; DeAngelo, Lee, and Jungmann, 1973; DeAngelo, Skrypack, and Jungmann, 1974), and rat testes (Reddi, Ewing, and Williams-Ashman, 1971; Means et al., 1974). This enzyme has also been partially purified from bovine corpora lutea (Menon, 1973; Goldstein and Marsh, 1973) and rat testes (Reddi et al., 1971).

Cholesterol esterase was partially purified from the soluble fraction of bovine corpus luteum homogenate by ammonium sulfate fractionation (Goldstein, 1973). The addition of cyclic AMP-dependent skeletal muscle protein kinase to this partially purified luteal cholesterol esterase increased the esterase activity, but cyclic AMP or ATP did not cause any further increase in activity (Goldstein, 1973). The role of cyclic AMP and protein kinase in the activation of luteal cholesterol esterase is thus still uncertain at this time. Recently, there have been two reports of direct stimulation of cholesterol esterase activity by cyclic AMP, ATP, and protein kinase in adrenal (Naghshineh, Vahouny, and Treadwell, 1974) and adipose tissue (Pittman and Khoo, 1974).

The other possible way in which LH could control cholesterol ester concentrations is via the inhibition of the cholesterol ester synthetase system; this has been explored by Flint et al. (1973). These investigators showed that the administration of LH to rabbits *in vivo* decreased the cholesterol ester synthetase activity in the interstitial tissue, and that this effect was mimicked by cyclic AMP *in vitro*. The inhibition of this enzyme system by LH or cyclic AMP seems to be the result of an increase in steroidogenesis. Progesterone and 20α hydroxypregn-4-en-3-one are synthesized at increased rates in interstitial tissue after LH or cyclic AMP treatment, and these steroids were found to exert a direct inhibitory effect on the synthetase system. It was also found that if the effect of cyclic AMP on steroidogenesis was inhibited by cycloheximide or aminoglutethimide, then the inhibition of the synthetase was also lost. This result is in keeping with the interpretation that LH inhibits cholesterol ester synthetase by increasing the production of progestins.

It seems, then, that LH increases the level of free cholesterol at the expense of cholesterol ester in rat and rabbit tissues by both an activation of cholesterol esterase and an inhibition of the cholesterol ester synthetase. The role of cyclic AMP in the inhibition of the cholesterol ester synthetase seems to be an indirect one via increased progestin synthesis, and its role in the activation of the esterase is uncertain at this time.

The importance of this action of LH in the control of steroidogenesis in gonadal tissue is difficult to assess. It is possible that by keeping a greater proportion of the cholesterol in its free form, LH regulates a limiting step in steroidogenesis. On the other hand, this may be just a supporting effect, supplying more substrate when the steroidogenic pathway has been accelerated by an action of the gonadotropin at another and more crucial step. In the bovine corpus luteum, at least, the effect of LH on cholesterol ester storage would appear to be of relatively minor importance, since it has been reported that this tissue, which is capable of a marked steroidogenenic response to LH, has large amounts of free cholesterol and small-to-unmeasurable amounts of cholesterol ester (Zimbelman, Loy, and Casida, 1961; Hafs and Armstrong, 1968; Seifart and Hansel, 1968).

3. *Action via Increased Transport of Cholesterol*

Since the cholesterol side-chain cleavage system has been shown to be localized in the mitochondria of bovine adrenal cortex (Halkerston et al., 1959, 1961), bovine corpora lutea (Hall and Koritz, 1964; Yago, Dorfman, and Forchielli, 1967), rat testis (Toren, Menon, Forchielli, and Dorfman, 1964; Drosdowsky,

Menon, Forchielli, and Dorfman, 1965), mouse Leydig cell tumors (Moyle, Jungas, and Greep, 1973b), and human placenta (Mason and Boyd, 1971), and since cholesterol is synthesized and stored outside this organelle, there must be a migration of cholesterol into the mitochondria before it can be converted into pregnenolone. There has been a report of cholesterol side-chain cleavage activity in both the microsomes and the mitochondria of bovine corpora lutea and rat luteinized ovaries (Flint and Armstrong, 1971), but this will be discussed later in this section.

The transport of cholesterol into mitochondria might be a limiting step in steroidogenesis and therefore could be another candidate for the site of tropic hormone action. Although there has been very limited experimentation on cholesterol transport in gonadal tissues, there has been a considerable amount of work done on this possibility in the study of the mechanism of action of ACTH and cyclic AMP in the adrenal cortex. This possibility was first proposed by Hechter (1955) and then by Garren (1968). In the latter work, it was suggested that such a translocation of cholesterol might involve the synthesis of a regulatory protein, since cycloheximide, an inhibitor of protein synthesis, blocked the acceleration of steroidogenesis by ACTH, and increased the accumulation of cholesterol in extramitochondrial lipid droplets. It was proposed that ACTH, via cyclic AMP, enhanced the synthesis of a protein which facilitated cholesterol transport, and that when cycloheximide inhibited protein synthesis it blocked cholesterol transport and the increase in steroidogenesis (Garren, 1968; Garren, Gill, Masui, and Walton, 1971).

A heat-stable cholesterol-binding protein has been found in the mitochondrial and cytosol fraction of bovine adrenal cortex tissue (Kan, Ritter, Ungar, and Dempsey, 1972; Kan and Ungar, 1973; Ungar, Kan, and McCoy, 1973). It has a specific binding affinity for cholesterol and stimulates the activity of a cholesterol side-chain cleavage preparation of adrenal mitochondria. Ungar et al. (1973) have proposed that this protein represents a cholesterol carrier protein which can transport cholesterol from the cytoplasm into the mitochondrion, but no direct evidence for this function is available at this time. This cholesterol-binding protein resembles the regulatory protein proposed by Garren et al. (1971), but this hypothesis required that ACTH increase the synthesis of this protein. Ungar et al. (1973) have reported, however, that ACTH does not change the level of this cholesterol-binding protein.

Recently, Mahaffee, Reitz, and Ney (1974) have demonstrated that ACTH and dibutyryl cyclic AMP treatment of hypophysectomized rats increased the level of free cholesterol in adrenal mitochondria and the conversion of this steroid into pregnenolone. They also showed that this stimulation of cholesterol accumulation in mitochondria was probably not secondary to the stimulation of steroidogenesis at some other site, because ACTH and dibutyryl cyclic AMP promoted this accumulation even when aminoglutethimide was administered in doses known to block steroidogenesis. Cycloheximide treatment of the rats also did not prevent the mitochondrial accumulation of cholesterol brought about by ACTH, indicating that this protein-synthesis inhibitor blocks the stimulation of steroidogenesis at some later step in the process. The authors' interpretation of the data was that ACTH and cyclic AMP increased the accumulation of cholesterol in the mitochondria by increasing the ratio of free cholesterol to esterified cholesterol in the cell. It is possible, however, that ACTH, via cyclic AMP, activated a carrier protein, such as that described by Ungar et al. (1973).

A modification of this latter type of proposal was suggested by Simpson, Jefcoate, Brownie, and Boyd (1972) and Jefcoate, Simpson, and Boyd (1974). They found that ether stress increased the initial rate of pregnenolone formation from endogenous cholesterol in adrenal mitochondria isolated from these animals, but there was no effect on the initial content of cholesterol in these organelles. The ether stress also induced changes in the type II difference spectra of cytochrome P-450, which were interpreted as indicating an increased amount of the cholesterol-cytochrome P-450 complex. It was, therefore, proposed that only a fraction of the total mitochondrial cholesterol is available for side-chain cleavage, and that stress via ACTH increases that fraction which is active. This increase could be due to the transport or binding of cholesterol to the side-chain cleavage cytochrome P-450 from other intramitochondrial sites. It is also possible that these stress-induced changes may result in an increased entry of cholesterol into the mitochondria *in vivo*. This effect of stress was found to be readily inhibited by cycloheximide, indicating that a labile protein may be involved with this transport or binding.

The effect of gonadotropins or cyclic AMP on cholesterol transport into mitochondria has not been studied very extensively in gonadal tissues. Flint and Armstrong (1971), using an assay involving the production of [^{14}C]isocaproic acid from [26-^{14}C]cholesterol, reported that there was about twice as much side-chain cleavage activity in bovine corpora lutea microsomes as in mitochondria. Rat luteinized ovaries were also found to have about one-quarter as much of this activity in microsomes as in mitochondria. This suggests that there may be no requirement for a transport of cholesterol into mitochondria in luteal tissue. We have been unable, however, to confirm this distribution in bovine corpora lutea, using an assay involving the conversion of [4-^{14}C]cholesterol into labeled steroid products (Caron, 1973). In our hands, the cholesterol side-chain cleavage activity shows the same distribution in the different fractions of luteal homogenates as the marker enzyme for mitochondria, succinic dehydrogenase. There was no correlation between the cleavage activity and the microsomal marker enzyme, NADPH cytochrome C reductase. It is unknown at this time if this discrepancy between the results of Flint and Armstrong (1971), on the one hand, and the data of Caron (1973) and most other investigators, on the other hand, is due to methodological differences, but it seems that the former workers did not take into account the endogenous levels of cholesterol in these fractions when they calculated the levels of side-chain activity. We have found that the mitochondria isolated from homogenates of bovine corpora lutea contain more than four times as much endogenous cholesterol as the microsomes (Caron,

1973). If the exogenous labeled cholesterol is uniformly diluted by the endogenous pools, then the specific activity of this substrate would be about four times lower in the mitochondrial incubations than in the microsomal ones. Thus the same amount of radioactivity in the products of the side-chain cleavage reaction would represent four times as much cleavage activity in the mitochondria as in the microsomes.

Moyle et al. (1973b) determined the localization of the cholesterol side-chain cleavage enzyme system and the endogenous cholesterol in mitochondria from Leydig tumor cells and found that the enzyme activity was localized on the inner membrane of the mitochondria, while essentially all of the cholesterol was associated with the outer membrane. This indicates that a limiting step in steroidogenesis could be the availability of substrate cholesterol in this tissue as well. Robinson and Stevenson (1973) have also reported in an abstract that the treatment of superovulated immature rats with hCG causes a marked increase in the transport of cholesterol into the mitochondria obtained from the luteinized ovaries.

Thus, although an effect of LH and cyclic AMP on cholesterol transport into mitochondria remains to be examined thoroughly in gonadal tissue, the studies on the adrenal cortex implicate this process in the control of steroidogenesis.

4. Action via Increased Side-Chain Cleavage Activity

Cyclic AMP could increase the conversion of cholesterol into pregnenolone by a direct action on the side-chain cleavage enzyme system by activating, or increasing the synthesis of, one of the components of this complex. Such an effect of cyclic AMP was proposed to explain its stimulation of the conversion of labeled cholesterol into pregnenolone in various adrenal cortical mitochondrial preparations (Roberts, Creange, and Young, 1965; Roberts, McCune, Creange, and Young, 1967; Roberts and Creange, 1968). Koritz, Yun, and Ferguson (1968), however, were unable to substantiate this proposal, and suggested instead that the increased pregnenolone accumulation was due to an inhibition of the conversion of pregnenolone to progesterone rather than an increase in pregnenolone synthesis. The Δ^5-3β-hydroxysteroid dehydrogenase and the Δ^5-3-ketosteroid isomerase enzymes which are involved in this reaction appear to be present in both the mitochondria and the microsomes (Sulimovici and Boyd, 1969; McCune, Roberts, and Young, 1970). This inhibitory effect of cyclic AMP on the pregnenolone-to-progesterone reactions has been confirmed by several investigators using adrenal and ovarian preparations (Sulimovici and Boyd, 1969; McCune et al., 1970; Sulimovici and Lunenfeld, 1971; Srinivasan, Clark, and Gawienowski, 1973), and it seems that the cyclic nucleotide inhibits the Δ^5-3β-hydroxysteroid dehydrogenase, at greater than physiological concentrations, by competing with NAD for the enzyme.

Similar findings have been reported in gonadal tissue experiments. Ichii et al. (1963) suggested that LH itself had a direct stimulatory effect on the side-chain cleavage system of the bovine corpus luteum, but this could not be confirmed by some of these same authors in a later publication (Yago et al., 1967). Sulimovici and Boyd (1968) then reported that cyclic AMP both stimulated the conversion of cholesterol to pregnenolone and inhibited the conversion of pregnenolone to progesterone. When the total amount of pregnenolone and progesterone synthesized in the cyclic AMP incubations was compared with that in the controls, however, it was apparent that there was no overall stimulation of the side-chain cleavage activity, and the increased accumulation of pregnenolone was probably due to the inhibition of the Δ^5-3β-hydroxysteroid dehydrogenase (Sulimovici and Boyd, 1969; Sulimovici and Lunenfeld, 1971; Srinivasan et al., 1973).

Recently we have examined the effect of cyclic AMP, protein kinase, and ATP on a reconstituted cholesterol side-chain cleavage system from bovine corpora lutea and have observed a direct stimulation of this enzyme activity (Caron, Goldstein, Savard, and Marsh, 1974). The cleavage system was solubilized from corpora lutea mitochondria by phospholipase A treatment, and the cytochrome P-450 component was isolated from the other components of the system by column chromatography. The cholesterol side-chain cleavage activity was then reconstituted using this isolated cytochrome P-450 and the nonheme iron and flavoprotein components of adrenal cortex tissue (adrenodoxin and adrenodoxin reductase). The proportions of these three components were arranged so that the activity was proportional to the amount of cytochrome P-450 present. When cyclic AMP, ATP, and a protein kinase preparation, partially purified from corpora lutea, were added to this reconstituted cleavage system, there was a consistent stimulation of enzyme activity of the order of 20 to 70%. The stimulation was completely dependent on the presence of ATP, indicating that a phosphorylation was probably involved. This *in vitro* effect of cyclic AMP on the reconstituted system may represent a physiological action of the cyclic nucleotide, but there will have to be further confirmation of this type of activation *in situ* in whole mitochondria and whole cells.

5. Action via Increased Efflux of Pregnenolone from Mitochondria

Koritz and Hall (1964) observed that pregnenolone inhibited its own synthesis from cholesterol in an extract from adrenal mitochondria, and since it was believed that the pregnenolone must leave the mitochondria to be converted into progesterone (Beyer and Samuels, 1956), it was suggested that ACTH might control steroidogenesis by determining the egress of pregnenolone from the mitochondria (Koritz and Hall, 1964; Koritz and Kumar, 1970). It was also reported that substances which cause mitochondria of steroidogenic tissues to swell stimulated pregnenolone synthesis, while ATP, which inhibited swelling, decreased this stimulation (Hirshfield and Koritz, 1964, 1966).

There are several pieces of evidence, however, which indicate that this hypothesis is probably not correct. First, there is a Δ^5-3β-hydroxysteroid dehydrogenase present in mitochondria (Sulimovici and Boyd, 1969; McCune et al., 1970; Sulimovici and Lunenfeld, 1971; Srinivasan et al., 1973) which can convert pregnenolone into progesterone. Second, Simpson et al. (1972) reported that the amount of pregnenolone present in freshly isolated adrenal mitochondria from stressed rats was greater than that observed in the mitochondria from cycloheximide-treated rats, even though there was increased cholesterol side-chain cleavage activity in the mitochondria obtained from the stressed animals. Third, the amount of pregnenolone required for this inhibition is greater than the physiological concentrations which occur in mitochondria (Simpson et al., 1972). Fourth, when pregnenolone was made to accumulate in mitochondria, by blocking its conversion to progesterone with cyanoketone, it did not affect the stimulation of side-chain cleavage activity caused by the action of endogenous ACTH *in vivo* (Simpson et al., 1972). Similar results were obtained by Farese (1971) using incubated rat adrenal sections, ACTH, and cyclic AMP.

6. *The Role of Protein Synthesis in the Stimulation of Steroidogenesis*

Concomitant protein synthesis has been shown to be required for the actions of ACTH, LH, and cyclic AMP on steroidogenesis in their respective target tissues. I will not go into detail on this matter, since it was covered by Wicks (1974) in an earlier review in this series, but I will just mention some of the highlights in the studies on gonadal tissues.

Hall and Eik-Nes (1962) were the first to show that protein synthesis inhibitors, such as puromycin and chloramphenicol, blocked the stimulatory effect of LH on steroidogenesis in incubating slices of rabbit testes. Several groups have confirmed this effect of protein synthesis inhibitors, using predominantly puromycin and cycloheximide, in testes (Shin and Sato, 1971; Moyle et al., 1971), corpora lutea (Savard et al., 1965; Marsh, 1968; Hermier, Combarnous, and Jutisz, 1971), Graafian follicles (Tsafriri, Lieberman, Barnea, Bauminger, and Lindner, 1973), and whole ovaries (Gorski and Padnos, 1966). The site of action appears to be after the formation of cyclic AMP in that Marsh et al. (1966) showed that puromycin did not block the LH stimulation of cyclic AMP accumulation in incubating slices of bovine corpora lutea, but it did block the stimulatory effect of exogenous cyclic AMP on steroidogenesis in this same tissue (Marsh and Savard, 1966*a*).

Hermier et al. (1971) pinpointed the site of action of these inhibitors even further, when they showed that cycloheximide acted before the cholesterol side-chain cleavage reaction. They incubated slices of luteinized rat ovary in the absence of O_2 for 90 min with LH and found that no progesterone was synthesized, presumably due to the lack of O_2 needed for the side-chain cleavage reaction steps. When the slices of luteinized ovary were then incubated in the presence of O_2, there was a rapid synthesis of progesterone and this synthesis could not be inhibited by cycloheximide. This indicated that cycloheximide did not act on the side-chain cleavage step or beyond that point in steroidogenesis, since, if it did, it would have blocked progesterone synthesis in the second incubation.

The role of RNA synthesis in the action of LH is uncertain. Savard et al. (1965), using incubated slices of bovine corpora lutea, Shin and Sato (1971), using incubated mouse Leydig tumor cells, and Tsafriri et al. (1973), using isolated Graafian follicles, found that actinomycin D blocked the action of LH and cyclic AMP on steroidogenesis. On the other hand, Gorski and Padnos (1966) reported that actinomycin D had no effect on the LH stimulation of steroidogenesis in incubated slices of rabbit ovary. In regard to the studies on corpora lutea, further experiments have shown that this inhibition of the LH stimulation by actinomycin D is inconsistent (Marsh, *unpublished data*).

7. Summary

In conclusion, it is not possible to decide, at this time, which of the possible mechanisms of action of cyclic AMP on steroidogenesis is the correct one. There are experimental data in support of each. None of them is mutually exclusive, and it may be that LH and cyclic AMP have pleiotropic types of effects stimulating many aspects of the steroidogenic pathway. One part of the pathway may be the limiting step before the stimulation begins, but once steroidogenesis starts to accelerate, other parts might become limiting. A concerted action of cyclic AMP at several sites may be necessary to accelerate the whole process.

III. OTHER ROLES FOR CYCLIC AMP IN GONADAL FUNCTION

It is generally believed that LH initiates the processes of ovulation, ovum maturation, and luteinization, but it is uncertain if the initiating effect of LH on the latter two processes is simply due to a physical separation of the ovum from the follicle cells, which occurs at ovulation, or if the hormone has a more direct action on the ovum or the follicle cells. These two cell types appear to be mutually inhibitory. If the ovum is artificially removed from the Graafian follicle of the rabbit (El-Fouly, Cook, Nekola, and Nalbandov, 1970) or the pig (Nalbandov, 1973), the "ovectomized" follicle will spontaneously partially luteinize and synthesize large amounts of progesterone. The removed ovum, on the other hand, begins its maturation process, which involves the completion of the first meiotic division and other cytological changes (Pincus and Enzmann, 1935; Scheutz, 1974). It is possible that there are inhibitory substances produced by the ovum and the other cells of the follicle which hold these processes in check, but so far their existence has not been proven.

The removal of mutually inhibitory influences may play a role in these processes, but the preponderance of evidence indicates that LH does have some direct effect. Ovum maturation in the mammal, for example, usually begins prior to the rupture of the follicle (Tsafriri et al., 1972*b*; Hertig and Barton, 1973), and luteinization has also been reported to occur in some follicles before ovulation

(Nalbandov, 1973). In addition, LH can be shown to induce ovum maturation (Tsafriri et al., 1972b) and luteinization (Ellsworth and Armstrong, 1971) in unovulated follicles. The removal of the ovum from everted rabbit follicles does not cause this tissue to luteinize spontaneously when transplanted beneath the kidney capsule, but treatment with LH is effective in inducing this process (Miller and Keyes, 1974). Likewise, granulosa cells harvested from medium-sized monkey or pig follicles will not luteinize spontaneously in cell cultures, but will luteinize when treated with LH or FSH (Channing, 1970a,b). Finally, rabbit follicles luteinized by ovectomy stop functioning after about 5 days, but if LH is injected immediately into the empty follicle, the luteinization progresses until a normal-appearing corpus luteum is formed and this structure functions as long as a normal corpus luteum or pseudopregnancy (Cropper, El-Fouly, Cook, and Nalbandov, 1970). We will assume, then, that LH does play a role in these processes and continue with a discussion of the evidence that cyclic AMP is involved with these hormonal effects.

A. Ovulation

The ovulatory surge of LH is believed to bring about the rupture of the Graafian follicle and the release of the ovum in mammals. Rondell (1970) hypothesized that LH acted via an increased accumulation of endogenous cyclic AMP which, in turn, stimulated progesterone production. Progesterone was then believed to cause the induction of an ovulatory enzyme such as a collagenase, which would break down the collagen framework of the follicular wall, increase the distensibility of the follicle and, finally, result in the rupture of the follicle. The evidence in support of this hypothesis is as follows: Rondell reported that if he incubated strips of the dome of hog follicles in culture, in the presence of LH, he observed an increase in the distensibility of this tissue, which he measured by determining the amount of tension developed under a standard stretching procedure. This effect of LH was eliminated by the inclusion of cyanoketone, an inhibitor of progesterone synthesis; moreover, exogenous progesterone added to the culture medium also increased the distensibility. These results indicated that steroidogenesis was involved in the action of LH. Exogenous cyclic AMP mimicked the effect of LH by increasing the distensibility of the strips of follicle tissue, and this effect was also inhibited by cyanoketone.

While there has been quite a lot of information (reviewed by Rondell, 1974) to indicate that steroid synthesis is a part of the ovulatory process, there has not been much direct support for the role of cyclic AMP in this process. LeMaire, Mills, Ito, and Marsh (1972) injected this cyclic nucleotide and the dibutyryl derivative into the antrum of Graafian follicles of the rabbit, but they were unable to mimic the effect of intrafollicular LH in causing ovulation, reported by Jones and Nalbandov (1972). On the other hand, Kao and Nalbandov (1972) reported that the injection of antiadrenergic agents into the wall of hen follicles blocked ovulation and this inhibitory effect could be overcome by dibutyryl cyclic AMP.

These authors concluded that catecholamines were probably involved in ovulation and that cyclic AMP might be a mediator of their action.

When measurements of endogenous cyclic AMP were made on Graafian follicles, it was found that LH markedly increased the concentration of this nucleotide in estrous follicles (Tsafriri et al., 1972b; Marsh et al., 1972; Lamprecht et al., 1973). As the process of ovulation proceeds in the rabbit, however, a refractoriness develops to the effect of LH in terms of the stimulation of both cyclic AMP synthesis and steroidogenesis (Marsh et al., 1973; Mills and Savard, 1973). The process of ovulation, therefore, may be somewhat more complicated than that proposed by Rondell (1970). It may involve biphasic changes in cyclic AMP and steroidogenesis such that there is an initial increase in cyclic AMP which in turn increases steroid synthesis, followed by a decrease in cyclic AMP accumulation leading to a fall in steroid output. These biphasic changes may or may not be essential components of the ovulatory process.

Prostaglandins have been shown to play an essential role in the process of ovulation at the ovarian level, since the effect of LH on this process can be inhibited by aspirin and indomethacin, inhibitors of prostaglandin synthesis (Orczyk and Behrman, 1972; Armstrong and Grinwich, 1972; Behrman, Orczyk, and Greep, 1972b; Tsafriri et al., 1972a; Grinwich et al., 1972; O'Grady, Caldwell, Auletta, and Speroff, 1972a). Furthermore, endogenous PGE and PGF have been shown to increase markedly during this process (LeMaire et al., 1973). Recently, it has also been found that exogenous cyclic AMP will increase the amount of PGE and PGF in isolated Graafian follicles *in vitro* (Marsh et al., 1974), indicating that cyclic AMP may mediate this action of LH on prostaglandin accumulation as well.

B. Ovum Maturation

The meiotic division of the oocyte of mammals is a highly specialized process. Starting in fetal life, it proceeds to the diplotene stage before birth and then to a prolonged resting period continuing up to the time of ovulation. Shortly before ovulation occurs, the oocyte completes the first meiotic division; this, along with other cytoplasmic changes, is referred to as ovum maturation (Schuetz, 1974).

Tsafriri, Lindner, Zor, and Lamprecht (1971) and Tsafriri et al. (1972b) demonstrated that, if Graafian follicles of the rat were explanted before the LH surge, the oocytes in these follicles would remain in the resting stage throughout an 18-hr culture period. If LH was added to the culture medium, the first meiotic division was completed during this time period. The addition of hCG or FSH was also effective, but it is possible that the effectiveness of FSH might be due to the LH content of this preparation. Prolactin progesterone, 20α hydroxypregn-4-en-3-one, estradiol, and linolenic acid were completely ineffective.

Cyclic AMP was implicated in this action of LH, because there was a marked increase in the accumulation of this nucleotide in the cultured follicles when LH was added. The addition of exogenous cyclic AMP or its dibutyryl derivative to the culture medium was without effect, but the injection of dibutyryl cyclic AMP into the antrum of the follicle brought about the completion of the maturation division. This effect appeared to be specific in that the injection of 5'-AMP was essentially ineffective. It was suggested that the cyclic nucleotides might not be able to penetrate the follicle through the outside theca layer when they were added in the culture medium, but other explanations might be possible, such as the destruction of the nucleotides by enzymes which might leak out from broken cells on the surface of the follicle.

This effect of LH and cyclic AMP probably does not proceed through the mediation of steroid synthesis. Although there is a concomitant stimulation of progesterone production when LH is added to the cultured follicles (Tsafriri et al., 1973), the effect of LH on ovum maturation is not blocked by cyanoketone, an inhibitor of progesterone synthesis (Tsafriri et al., 1972b). Furthermore, progesterone synthesis in these cultured follicles can be selectively inhibited by a relatively low concentration of actinomycin D (8 μg/ml), and this treatment does not inhibit the action of LH on ovum maturation (Tsafriri et al., 1973). Finally, as mentioned previously, exogenous progesterone, estradiol, or 20α hydroxypregn-4-en-3-one will not mimic this effect of LH or cyclic AMP on ovum maturation.

Cyclic AMP may produce its effects on ovum maturation via the stimulation of a cyclic AMP-dependent protein kinase, because Tsafriri et al. (1972b) found that exogenous cyclic AMP increased protein kinase activity in a supernatant fraction of follicular homogenates. It is not possible, however, to associate this increase in enzyme activity to a particular cell type or to determine if it is related to ovum maturation. Cyclic AMP-dependent protein kinases have also been detected in the eggs of sea urchins (Lee and Iverson, 1972) and amphibians (Tenner and Wallace, 1972), but their possible association with ovum maturation is also unknown. Protein synthesis and nucleic acid synthesis are also probably required for this action of LH and cyclic AMP on ovum maturation, since Tsafriri et al. (1973) have shown that puromycin, cycloheximide, and a relatively high concentration of actinomycin D (80 μg/ml) suppressed the effect of LH. In view of the fact that the stimulation of progesterone production could be prevented by a lower concentration of actinomycin D (8 μg/ml) without an effect on ovum maturation, Tsafriri and co-workers proposed that LH controls the latter process by acting at the translational level, inducing the synthesis of a protein from preformed messenger RNA. It is possible that the inhibition of the effect of LH with high amounts of actinomycin D is due to nonspecific toxic effects, but until we have more direct information about the amounts and types of messenger RNA and proteins, this proposal will continue to be speculative.

Prostaglandin E_2 was also found to induce ovum maturation when added to the cultured follicles (Tsafriri et al., 1972a). This effect appears to be pharmacological rather than physiological since, in rats, the injection of indomethacin, a potent inhibitor of prostaglandin synthesis, blocks such prostaglandin-dependent processes as ovulation, but does not block ovum maturation (Tsafriri et al., 1971, 1972a).

C. Luteinization

During the process of ovulation the cells of the Graafian follicle, particularly the granulosa cells, become luteinized. Luteinization is generally described in both morphological and functional terms. The cells increase in size and in their cytoplasmic-to-nuclear ratio. They accumulate lipid droplets and granules in the cytoplasm, and begin to produce large amounts of progesterone (Brambell, 1956; Channing, 1969). The endpoint of this process is the formation of the mature corpus luteum.

The process of luteinization has been studied in several model systems. Channing (1969, 1970a, 1973, 1974b) studied the luteinization of isolated granulosa cells from monkey and pig follicles *in vitro* in long-term cell cultures and determined the *in vivo* and *in vitro* effects of hormones on this process. She found that if the cells were taken from large preovulatory follicles, they would luteinize spontaneously in culture, but if they were taken from smaller follicles they did not. The addition of either LH or FSH to the culture of cells obtained from medium-sized follicles would induce luteinization, but cells obtained from very small follicles required both LH and FSH for this induction to take place. The luteinization did not last beyond a few days unless LH or FSH was continually added to the culture medium (Channing and Crisp, 1972; Channing, 1974a).

The addition of cyclic AMP or dibutyryl cyclic AMP to the monkey and pig granulosa cells in culture mimicked the effects of the gonadotropins. These cyclic nucleotides induced luteinization in cells obtained from medium-sized follicles (Channing and Seymour, 1970; Channing, 1970a, 1973, 1974a), and dibutyryl cyclic AMP maintained the cells in a luteinized state if it was added every day to the culture medium (Channing, 1974a).These effects appeared to be relatively specific in that 2 mM concentrations of 5'-AMP, ADP, ATP, CTP, cyclic CMP, or cyclic GMP had no effect (Channing, 1973). Guanosine triphosphate (2 mM), however, did seem to have some stimulatory effect, but it was smaller than that produced by 2 mM cyclic AMP (Channing, 1973). The addition of aminophylline, an inhibitor of cyclic nucleotide phosphodiesterase, stimulated progestin secretion by these granulosa cells (Channing, 1973), and when added in combination with either LH or cyclic AMP it potentiated their effects on luteinization (Channing and Seymour, 1970). Imidazole, a phosphodiesterase stimulator, was found to inhibit the luteinization brought about by LH (Channing, 1970a).

Further evidence for the role of cyclic AMP in the process of luteinization is that LH or FSH increased the endogenous levels of cyclic AMP in these cells (Channing, 1970a; Kolena and Channing, 1972). The minimum effective dose of LH (0.02 μg/ml) required to cause an increase in cyclic AMP in medium-sized pig follicles (Kolena and Channing, 1972) was in the same range as that (0.01 μg/ml) required to induce luteinization in these cells (Channing, 1970b). The small difference in response may be due to the fact that ovine LH was used in the former study (Kolena and Channing, 1972) and porcine LH in the latter (Channing, 1970b). This effect of LH on cyclic AMP accumulation can be inhibited if follicular fluid obtained from small pig follicles is added to the culture medium, and this results in an inhibition of luteinization as well (Channing, 1974b). Cirillo et al. (1969) have also studied the effect of exogenous cyclic AMP on granulosa cells in culture. They prepared monolayer cultures of bovine granulosa cells and found that exogenous cyclic AMP increased progesterone production over a 2-day period. Nekola and Nalbandov (1971), on the other hand, have been able to demonstrate an effect of the ovum on luteinization of rat ovarian cells. Ovaries of immature rats were dispersed and two types of cultures were prepared. One contained many ova and the other none or only a few ova. The culture with many ova continued to maintain cell populations which resembled granulosa cells, while the cultures with few or no ova became luteinized. Furthermore, the cells close to the ova retained their appearance of granulosa cells after a prolonged culture period, while the cells farther away were transformed into lutein cells. The authors postulated, on the basis of these results, that either the ova secreted a substance which prevented luteinization or it metabolized a substance which caused it. The finding of Channing (1974b) mentioned earlier, that follicular fluid from pig follicles inhibits the effect of LH on luteinization in porcine granulosa cells in culture, indicates that the first postulate may be the correct one. The procedures of Nekola and Nalbandov (1971) and of Channing (1974b) appear to be useful as assay systems and may lead to a purification and identification of the ovum factor responsible for the suppression of luteinization.

Another approach to the study of luteinization has been to dissect follicles of rabbits (Keyes, 1969; Keyes, Canastar, and Miller, 1972; Miller and Keyes, 1974) or rats (Ellsworth and Armstrong, 1971, 1973, 1974), expose them to LH or other test substances, and then transplant them beneath the kidney capsule of suitable host animals to see if corpora lutea would develop. In both rabbit (Keyes, 1969) and rat (Ellsworth and Armstrong, 1971) studies, it was found that the incubation of the follicles with LH before transplantation brought about luteinization.

Ellsworth and Armstrong (1973) then showed that incubating the rat follicles with 7 mM dibutyryl cyclic AMP mimicked the effect of LH in inducing luteinization in the transplanted follicles. When hypophysectomized rats were used as hosts for the transplants, it was found that the luteinized follicles would secrete significant amounts of progesterone only if prolactin was administered. This was true whether the luteinization was induced with LH (Ellsworth and Armstrong, 1971) or with dibutyryl cyclic AMP (Ellsworth and Armstrong, 1973). The minimum effective dose of LH required to induce luteinization was 0.1 µg/ml, the same dose of LH required to increase cyclic AMP accumulation in rat ovaries (Mason and Toomey, 1971). In a subsequent paper, Ellsworth and Armstrong (1974) demonstrated that PGE_2 could also cause the dissected follicles to luteinize upon transplantation, and that the effects of this prostaglandin, LH, or dibutyryl cyclic AMP could be blocked by treatment of the follicles with polyphloretin phosphate. This latter substance is generally regarded as a prostaglandin antagonist (Beitch and Eakins, 1969; Eakins and Karim, 1970; Eakins and Sanner, 1972), but it has also been reported to inhibit protein kinase (Kuehl,

Humes, Mendel, Cirillo, and Ham, 1971) and protein synthesis (Ahren and Perklev, 1972), so it is impossible to pinpoint its site of action in this study.

The effect of LH on the luteinization of rabbit follicles also seems to be mediated by cyclic AMP. Keyes et al. (1972) first reported that they were unable to induce isolated rabbit Graafian follicles to luteinize by incubating them with either cyclic AMP or dibutyryl cyclic AMP, but in a later report (Miller and Keyes, 1974) they stated that dibutyryl cyclic AMP was effective. The difference in the results may be due to the fact that in the latter study they carried out their experiments with everted follicles, while everted follicles were used in only a few experiments in the former study. They also indicated that different lots of dibutyryl cyclic AMP were not uniformly effective, and that the earlier results might have been due to an inactive preparation of this cyclic nucleotide. The apparent inability of dibutyryl cyclic AMP to act when added to the outside of follicles is reminiscent of the report of Tsafriri et al. (1972b), who were able to induce ovum maturation with dibutyryl cyclic AMP only after injecting it into the antrum of the follicle.

LeMaire et al. (1972) used a third experimental approach, the intrafollicular injection of rabbit follicles with LH or other substances. They found that although they could induce luteinization with LH, they could not do so with cyclic AMP or dibutyryl cyclic AMP. In fact, they reported that the injection of cyclic AMP with LH inhibited the response to the hormone. The injection of cyclic nucleotide phosphodiesterase also induced luteinization in some follicles, indicating that endogenous cyclic AMP might be inhibiting luteinization, but the specificity of this effect was not determined. At this time it is not possible to reconcile these inhibitory effects with the stimulatory effects of Miller and Keyes (1974), Ellsworth and Armstrong (1973), or Channing (1973), but it does seem that the evidence is highly in favor of a stimulatory effect of cyclic AMP and that this nucleotide probably mediates the effect of LH on luteinization.

The mechanism of luteinization is essentially unknown. It obviously involves differentiation, and Keyes et al. (1972) have implicated RNA synthesis in the process by showing that treatment of the isolated rabbit Graafian follicles with actinomycin D completely blocked the effect of LH on luteinization. It is not difficult to imagine that cyclic AMP might be involved in gene transcription, and in a later section a recent hypothesis of cyclic AMP action on ovarian RNA synthesis is discussed.

D. Spermatogenesis

The requirement for gonadotropins in the process of spermatogenesis has been established in the hypophysectomized animal, but a clear understanding of the functions and the sites of action of LH and FSH is still lacking (Steinberger, 1971). Nevertheless, FSH seems to be required for complete quantitative restoration of spermatogenesis, and Means and co-workers (Means and Hall, 1967, 1969; Means, 1970, 1971) have demonstrated that FSH stimulated protein and

RNA synthesis in the testes of immature rats. These effects were specific for hormones with FSH activity and could be observed rather quickly after the administration of the hormone. This stimulation of protein and RNA synthesis was found to be dependent on the age of the animal (Means and Hall, 1967, 1969; Means, 1970, 1971; Means et al., 1974), in that it could be demonstrated in rats only before 21 to 24 days of age. The sensitivity of the older animals to FSH could, however, be restored by hypophysectomy.

The first indication that cyclic AMP might be involved in this action of FSH came from the work of Murad et al. (1969), who showed that FSH stimulated the adenylyl cyclase activity in homogenates of dog or rat testes. It was unlikely that this effect of FSH was due to the small amount of contaminating LH, because the effect of FSH was usually greater than that of LH. These results were confirmed by Kuehl et al. (1970a), who showed that FSH caused a marked increase in the incorporation of [8-^{14}C]adenine into cyclic AMP in slices of testes from immature or hypophysectomized rats; LH also increased this incorporation, but to a smaller degree. When slices of testes from intact mature rats were used, no effect of FSH could be detected, although the effect of LH was still present. A preparation which contained "substantial quantities" of seminiferous tubules was made by treating the rat testes with collagenase, and this preparation also responded to FSH and LH. In retrospect, it would appear that this preparation must have contained Leydig cells, because in subsequent studies where seminiferous tubules were separated from Leydig cells, LH had no effect (Dorrington, Vernon, and Fritz, 1972; Cooke et al., 1973; Dorrington and Fritz, 1974; Braun and Sepsenwol, 1974).

The localization of the site of action of FSH to the seminiferous tubules was demonstrated by Dorrington et al. (1972) and confirmed in several subsequent publications (Cooke et al., 1973; Dorrington and Fritz, 1974; Braun and Sepsenwol, 1974). In each study the seminiferous tubules were dissected away from the interstitial tissue of rat testis and then incubated under control conditions or in the presence of FSH or LH. FSH stimulated cyclic AMP accumulation in the isolated tubules, but LH had no effect. In the isolated interstitial tissue incubations, LH, but not FSH, increased cyclic AMP accumulation. The effect of FSH was demonstrable in seminiferous tubules obtained from immature or hypophysectomized rats, but was not demonstrable in tubules obtained from intact mature rats, unless phosphodiesterase inhibitors were included in the incubations (Dorrington and Fritz, 1974; Braun and Sepsenwol, 1974). These results indicated that the testes of intact mature rats had more effective phosphodiesterase activity than the testes of immature or hypophysectomized rats.

Monn et al. (1972) have reported that there is a low level of phosphodiesterase activity in the testes of the 20-day-old immature rat which gradually increases until the animal is 50 days of age. The low level of activity is associated with a phosphodiesterase isozyme, c, which has a high K_m for cyclic AMP, and the increase in activity is due to the appearance of a new testicular isozyme, f, which has a low K_m for cyclic AMP. In a subsequent abstract, Christiansen and Desautel

(1973) reported that the *f* isozyme of phosphodiesterase was located in the seminiferous tubules and on subcellular fractionation could be found in the nuclear fraction. Hypophysectomy in immature rats prevented the appearance of the *f* isozyme, and in the mature animal caused this enzyme to disappear over a period of days. The *f* isozyme reappeared again in these hypophysectomized animals when they were treated with FSH and LH, or FSH and testosterone. It seems that FSH causes a rise in cyclic AMP in the seminiferous tubules of the immature rat and then, in conjunction with testosterone, limits the level of this cyclic nucleotide during later phases of sexual maturation by inducing a specific and more effective phosphodiesterase. When measurements of the endogenous level of cyclic AMP were carried out on the testes of the rat during this time of sexual maturation, it was found to be 140 pmoles/100 mg of tissue at 25 days of age and to decrease steadily to 70 pmoles/100 mg of tissue at 220 days of age (Hollinger, 1973). It would have been of interest to know how the levels might have changed at earlier ages, but these measurements were not carried out.

The target cell in the seminiferous tubules for FSH appeared to be either the spermatogonia or the Sertoli cells, because these two cells types were the most prominent ones in the testes of immature (Means et al., 1974) or hypophysectomized rats (Dorrington and Fritz, 1974). In order to study this further, Dorrington and Fritz (1974) and Dorrington (1974) caused a marked reduction in the spermatogonia in the left testis of normal intact rats by exposing them to X-irradiation. The tubules prepared from the irradiated testis were incubated in the presence of theophylline with and without FSH and were found to demonstrate the same magnitude of response to FSH as tubules prepared from nonirradiated testis of the same animals. This indicated that the Sertoli cells were probably the principal target cell of FSH.

Means and Huckins (1974) also reported rather convincing data that the Sertoli cell is the primary target for FSH in the testis. Pregnant rats were exposed to whole-body ^{60}Co-radiation on day 19.5 of pregnancy. This caused a selective destruction of the spermatogonia in the testes of the male offspring after birth. The more radioresistant Sertoli cells continued to proliferate and finally formed "Sertoli cell only (SCO)" seminiferous tubules. These SCO tubules were reported to have the same number of FSH binding sites as normal tubules and to respond to FSH with an increase in cyclic AMP accumulation.

These data of Dorrington and Fritz (1974), Dorrington (1974), and Means and Huckins (1974) indicate that the Sertoli cell is the principal, if not the only, target cell for FSH in the testis; this conclusion is consistent with the report of Castro, Seiguer, and Mancini (1970) that FSH labeled with ferritin was predominantly localized in or on the Sertoli cells. Thus, it seems reasonable to suggest that the effect of FSH on spermatogenesis is exerted via the Sertoli cell. The role of the Sertoli cell in spermatogenesis is not clearly understood, but it is considered to serve a supportive, nutritive, and perhaps endocrine function (Courot, Hochereau-de Riviero, and Ortavant, 1970).

The mechanism of action of FSH and cyclic AMP on seminiferous tubules and the Sertoli cell has also been studied by Means and co-workers (Means et al., 1974; Means and Huckins, 1974). Since cyclic AMP is known to act in several systems via protein kinase, Means et al. (1974) assessed the effect of FSH on the cyclic AMP-dependent protein kinase activity in seminiferous tubules. They found that FSH stimulated protein kinase activity within 5 min of being added to incubating tubule preparations. FSH continued to increase this activity until a maximal threefold stimulation was reached at 20 min, but LH had no effect on the protein kinase activity of this preparation. The increased protein kinase activity produced by FSH correlated with an increased intracellular accumulation of cyclic AMP.

This enzymatic response to FSH was also dependent on the age of the animal. The increase in protein kinase activity was maximal at 6 days of age and declined steadily until there was no response at 36 days of age. This was presumably due to the low concentration of cyclic AMP in seminiferous tubules caused by the appearance of the f form of phosphodiesterase (Monn et al., 1972; Christiansen and Desautel, 1973). In this regard, Means et al. (1974) found that the response to FSH in these older animals, in terms of increased protein kinase activity, could be restored by hypophysectomy or by incubating the tubules in the presence of an inhibitor of phosphodiesterase. This increase in protein kinase activity also seems to be localized in the Sertoli cells, because Means and Huckins (1974) found that their SCO seminiferous tubules responded to FSH with an increase in protein kinase activity.

The data from these studies suggest that FSH binds to the Sertoli cell, leading to the activation of adenylyl cyclase, which in turn increases the intracellular concentration of cyclic AMP. The cyclic AMP then causes the activation of protein kinase which mediates, by phosphorylation, the action of FSH on RNA and protein synthesis. There is, however, a great deal of work to be done before the action of FSH on spermatogenesis is understood. This includes the determination of the nature of the substrate of the protein kinase, how this substrate might control RNA and protein synthesis, and how these latter metabolic changes might affect spermatogenesis. In addition, the temporal changes in phosphodiesterase isozymes and their effects on cyclic AMP, RNA synthesis, protein synthesis, and spermatogenesis remain to be clarified.

E. Effects on Nucleic Acid Synthesis

The role of cyclic AMP in the regulation of protein synthesis and nucleic acid synthesis in the gonads was recently reviewed by Wicks (1974), and I will only mention a few studies not covered in the earlier review.

It is well recognized that gonadotropins regulate growth and differentiation in the gonads as well as steroidogenesis, but there is very little known about the mechanism of these former actions, and still less about the possible involvement of cyclic AMP. Recently, however, Hiestand, Eppenberger, and Jungmann (1973), and Jungmann, Hiestand, and Schweppe (1974) have obtained some results which suggest a mechanism by which cyclic AMP might control ovarian RNA synthesis. Several groups of investigators had reported previously that gonadotropins stimulated ovarian RNA synthesis (Callantine, Humphrey, and Lee, 1965; Civen, Brown, and Hilliard, 1966; Reel and Gorski, 1968a,b; Jungmann and Schweppe, 1972b), and this increase in RNA synthesis in immature rat ovaries was accompanied by an increased phosphorylation and acetylation of histones and nuclear acidic proteins (Jungmann and Schweppe, 1972a). In view of the fact that hCG causes an increase in cyclic AMP accumulation in ovarian tissue and that cyclic AMP increases phosphorylation via protein kinase, it seemed possible that cyclic AMP might be involved in this effect.

Hiestand et al. (1973) found that there was a specific cyclic AMP-binding protein in the cytosol of calf ovaries, but Jungmann et al. (1974) found that ovarian nuclei or chromatin, from immature rat ovaries, had no binding affinity for free cyclic AMP and very low levels of protein kinase activity. When the cyclic AMP was bound to the cytosol protein, however, this complex readily attached itself to isolated ovarian nuclei and chromatin, and caused a rise in nuclear protein kinase activity. As a consequence of the binding of the cyclic AMP-protein complex to the nuclei and the rise in protein kinase, there was a general increase in the phosphorylation of nonhistone proteins and a selectively large increase in the phosphorylation of two of these proteins (Jungmann et al., 1974). At the present time, it is unknown if the cyclic AMP-protein complex induces an increase in activity of endogenous nuclear protein kinase or causes the translocation of the cytosol enzyme into the nucleus. The authors (Jungmann et al., 1974) preferred the latter interpretation, since preliminary studies indicated that there was a decrease in the cytosol protein kinase specific activity when there was an increase in the nuclear protein kinase activity.

It was suggested by these authors (Jungmann et al., 1974) that this increase in phosphorylation of nonhistone protein was involved with the increase in RNA caused by gonadotropins, and studies on other cell types would indicate that this is possible (Martelo, 1973; Langan, 1973). In gonadal tissues, however, cyclic AMP and dibutyryl cyclic AMP have been reported to cause a decrease in the synthesis of RNA (Jarlstedt, Nilsson, Hamberger, and Ahren, 1973) and DNA (Hollinger and Hwang, 1974).

There have been other studies carried out on the effects of cyclic AMP and dibutyryl cyclic AMP on amino acid transport (Ahren and Hamberger, 1969; Hamberger, Herlitz, and Sjogren, 1973) and lactic acid production (Ahren, Hamberger, and Rubinstein, 1969; Hamberger et al., 1973), but the relationship of these effects to the actions of cyclic AMP on ovarian functions remains to be clarified.

IV. CONCLUSIONS

There is a large body of evidence which indicates that cyclic AMP plays a role in the stimulation of gonadal steroidogenesis. This is particularly true in the case of the corpus luteum, but there is still some doubt that it mediates the effect of very small concentrations of gonadotropins on the steroidogenic process. One is prompted by the data to predict that the difficulty with small concentrations of the hormones is due to methodological deficiencies, but a final answer to this problem will have to await further results.

Although the evidence for its role in other gonadal processes is not as extensive, it seems that cyclic AMP mediates the effects of gonadotropins on ovum maturation, luteinization, and spermatogenesis.

The major areas for future research in this field would seem, therefore, to be the elucidation of the role of cyclic AMP or another second messenger in the effect of low concentrations of gonadotropins on steroidogenesis, and the determination of the mechanisms of actions of this cyclic nucleotide in these very different processes. This latter area of research is a major undertaking that will probably occupy the attention of investigators for many years to come.

V. ACKNOWLEDGMENTS

I would like to express my gratitude to Dr. Kenneth Savard, who kindled my interest in this field and who with other colleagues, Drs. Norman R. Mason and William J. LeMaire, were instrumental in the generation of some of the ideas expressed in this review. I would also like to thank Drs. Peter J. A. Davies, Kirpal S. Sidhu, and William J. LeMaire for their editorial suggestions, and to express my appreciation to Maria A. Rodriguez and Nieves Cerver for their care in the preparation of the manuscript.

VI. REFERENCES

Ahren, K., and Hamberger, L. (1969): Effect of cyclic 3′,5′-AMP on ovarian amino acid transport. *Acta Physiologica Scandinavica,* 77 Supplement 330:75 (Abstract 102).
Ahren, K., Hamberger, L., Herlitz, H., Hillensjo, T., Nilsson, L., Perklev, T., and Selstam, G. (1973): Aspects of the mechanism of action of gonadotrophins. In: *The Endocrine Function of the Human Testis,* edited by V. H. T. James, M. Serio, and L. Martini, Vol. I, pp. 251–272. Academic Press, New York.
Ahren, K., Hamberger, L., and Rubinstein, L. (1969): Acute *in vivo* and *in vitro* effects of gonadotropins on the metabolism of the rat ovary. In: *The Gonads,* edited by K. W. McKerns, pp. 327–354. Appleton-Century-Crofts, New York.
Ahren, K., Herlitz, H., Nilsson, L., Perklev, T., Rosberg, S., and Selstam, G. (1974): Gonadotropins and cyclic AMP in various compartments of the rat ovary. In: *Symposium on Advances in Chemistry, Biology and Immunology of Gonadotropins,* edited by N. R. Moudgal. Academic Press, New York and London *(in press).*
Ahren, K., and Perklev, I. (1972): Effects of PGE_1 and 7-oxa-13-prostynoic acid on the isolated prepubertal rat ovary. In: *Advances in Biosciences,* Vol. 9, pp. 717–721. International Conference on Prostaglandins, Vienna, Pergamon Press, Viewig.
Ahren, K., and Selstam, G. (1971): Hormonal regulation of ovarian phosphorylase activity. *Acta Physiologica Scandinavica,* 82:31A–32A.
Andersen, R. N., Hubbard, W. R., and Baggett, B. (1970): Hormonal responsiveness of adenyl cyclase of rabbit corpus luteum. *Federation Proceedings,* 29:705 (Abstract 2600).
Andersen, R. N., Schwartz, F. L., and Ulberg, L. C. (1974): Adenylate cyclase activity of porcine corpora lutea. *Biology of Reproduction,* 10:321–326.

Armstrong, D. T. (1963): Stimulation of glycolytic activity of rat corpus luteum tissue by luteinizing hormone. *Endocrinology,* 72:908–913.
Armstrong, D. T. (1966): Comparative studies of the action of luteinizing hormone upon ovarian steroidogenesis. *Journal of Reproduction and Fertility,* Supplement 1:101–112.
Armstrong, D. T. (1968): III. Hormones and reproduction. Gonadotropins, ovarian metabolism and steroid biosynthesis. *Recent Progress in Hormone Research,* 24:255–319.
Armstrong, D. T., and Grinwich, D. L. (1972): Blockade of spontaneous and LH-induced ovulation in rats by indomethacin, an inhibitor of prostaglandin biosynthesis. *Prostaglandins,* 1:21–26.
Armstrong, D. T., Lee, T. P., and Miller, L. S. (1970): Stimulation of progesterone biosynthesis in bovine corpora lutea by luteinizing hormone in the presence of an inhibitor of cholesterol synthesis. *Biology of Reproduction,* 2:29–36.
Beall, R. J., and Sayers, G. (1972): Isolated adrenal cells: Steroidogenesis and cyclic AMP accumulation in response to ACTH. *Archives of Biochemistry and Biophysics,* 148:70–76.
Behrman, H. R., and Armstrong, D. T. (1969): Cholesterol esterase stimulation by luteinizing hormone in luteinized rat ovaries. *Endocrinology,* 85:474–480.
Behrman, H. R., Armstrong, D. T., and Greep, R. O. (1970): Studies on the rapid cholesterol-depleting and steroidogenic actions of luteinizing hormone in the rat ovary: Effects of aminoglutethimide phosphate. *Canadian Journal of Biochemistry,* 48:881–884.
Behrman, H. R., Moudgal, N. R., and Greep, R. O. (1972a): Studies with antisera to luteinizing hormone *in vivo* and *in vitro* on luteal steroidogenesis and enzyme regulation of cholesteryl ester turnover in rats. *Journal of Endocrinology,* 52:419–426.
Behrman, H. R., Ng, T. S., and Orczyk, G. P. (1974): Interactions between prostaglandins and gonadotropins on corpus luteum function. In: *Symposium on Advances in Chemistry, Biology and Immunology of Gonadotropins,* edited by N. R. Moudgal. Academic Press, New York and London (in press).
Behrman, H. R., Orczyk, G. P., and Greep, R. O. (1972b): Effect of synthetic gonadotrophin-releasing hormone (GN-RH) on ovulation blockade by aspirin and indomethacin. *Prostaglandins,* 1:245–258.
Beitch, B. R., and Eakins, K. E. (1969): The effects of prostaglandins on the intraocular pressure of the rabbit. *British Journal of Pharmacology,* 37:158–167.
Bell, E. T., Mukerji, S., and Loraine, J. A. (1964): A new bioassay method for luteinizing hormone depending on the depletion of rat ovarian cholesterol. *Journal of Endocrinology,* 28:321–328.
Beyer, K. F., and Samuels, L. T. (1956): Distribution of steroid-3β-ol-dehydrogenase in cellular structures of the adrenal gland. *Journal of Biological Chemistry,* 219:69–76.
Brambell, F. W. R. (1956): Ovarian changes. In: *Marshall's Physiology of Reproduction,* 3rd ed., edited by A. S. Parkes, Part I, pp. 397–542. Longmans, Green and Co., London.
Braun, T., and Sepsenwol, S. (1974). Stimulation of ^{14}C-cyclic AMP accumulation by FSH and LH in testis from mature and immature rats. *Endocrinology,* 94:1028–1033.
Brunswick, D. J., and Cooperman, B. S. (1971): Photo-affinity labels for adenosine 3′,5′-monophosphate. *Proceedings of the National Academy of Sciences (U.S.),* 68:1801–1804.
Burke, G., Chang, L. L., and Szabo, M. (1973): Thyrotropin and cyclic nucleotide effects on prostaglandin levels in isolated thyroid cells. *Science,* 180:872–875.
Burke, G., Kowalski, K., and Babiarz, D. (1971): Effects of thyrotropin, prostaglandin E_1 and A prostaglandin antagonist on iodine trapping in isolated thyroid cells. *Life Sciences,* 10:513–521.
Butcher, R. W., and Baird, C. E. (1968): Effects of prostaglandins on adenosine 3′,5′-monophosphate levels in fat and other tissues. *Journal of Biological Chemistry,* 243:1713–1717.
Callantine, M. R., Humphrey, R. R., and Lee, S. L. (1965): Effect of follicle-stimulating hormone on ovarian nucleic acid content. *Endocrinology,* 76:332–334.
Caron, M. G. (1973): A study of the cholesterol side-chain cleavage enzyme system in the bovine corpus luteum. Doctoral Dissertation, Department of Biochemistry, University of Miami School of Medicine, Miami, Fla.
Caron, M. G., Goldstein, S., Savard, K., and Marsh, J. M. (1974): Protein kinase stimulation of a reconstituted cholesterol side-chain cleavage enzyme system in the bovine corpus luteum. *Federation Proceedings,* 33:1323 (Abstract 563).
Castro, A. E., Seiguer, A. C., and Mancini, R. E. (1970): Electron microscopic study of the localization of labeled gonadotropins in the Sertoli and Leydig cells of the rat testis. *Proceedings of the Society for Experimental Biology and Medicine,* 133:582–586.
Catt, K. J., and Dufau, M. L. (1973): Spare gonadotropin receptors in rat testis. *Nature New Biology,* 244:219–221.

Catt, K. J., Watanabe, K., and Dufau, M. L. (1972): Cyclic AMP released by rat testis during gonadotropin stimulation *in vitro. Nature,* 239:280–281.
Channing, C. P. (1969): The use of tissue culture of granulosa cells as a method of studying the mechanism of luteinization. In: *The Gonads,* edited by K. W. McKerns, pp. 245–275. Appleton-Century-Crofts, New York.
Channing, C. P. (1970*a*): Influences of the *in vivo* and *in vitro* hormonal environment upon luteinization of granulosa cells in tissue culture. *Recent Progress in Hormone Research,* 26:589–622.
Channing, C. P. (1970*b*): Effect of stage of the estrous cycle and gonadotrophins upon luteinization of porcine granulosa cells in culture. *Endocrinology,* 87:156–164.
Channing, C. P. (1972): Effects of prostaglandin inhibitors, 7-oxa-13-prostynoic acid and eicosa-5,8,11,14-tetraynoic acid, upon luteinization of rhesus monkey granulosa cells. *Prostaglandins,* 2:351–367.
Channing, C. P. (1973): Regulation of luteinization in granulosa cell cultures. In: *The Regulation of Mammalian Reproduction,* edited by S. J. Segal, R. Crozier, P. A. Corfman, and P. G. Condliffe, pp. 505–518. Charles C Thomas, Springfield, Ill.
Channing, C. P. (1974*a*): Temporal effects of LH, hCG, FSH and dibutyryl cyclic 3',5'-AMP upon luteinization of rhesus monkey granulosa cells in culture. *Endocrinology,* 94:1215–1223.
Channing, C. P. (1974*b*): The use of granulosa cell cultures and short term incubations as an assay for gonadotropins. In: *Symposium in Advances in Chemistry, Biology and Immunology of Gonadotropins,* edited by N. R. Moudgal. Bangalore Press, Bangalore *(in press).*
Channing, C. P., and Crisp, T. M. (1972): Comparative aspects of luteinization of granulosa cell cultures at the biochemical and ultraestructural levels. *General and Comparative Endocrinology,* Supplement 3, 3:617–625.
Channing, C. P., and Seymour, J. F. (1970): Effects of dibutryl cyclic-3',5'-AMP and other agents upon luteinization of porcine granulosa cells in culture. *Endocrinology,* 87:165–169.
Chasalow, F. I., and Pharriss, B. B. (1972): Luteinizing hormone stimulation of ovarian prostaglandin biosynthesis. *Prostaglandins,* 1:107–117.
Christiansen, R. O., and Desautel, M. (1973): Induction of testicular cyclic nucleotide phosphodiesterase by LH and FSH. *Abstracts of the 55th Endocrine Society Meeting,* p. A-100 (Abstract 104).
Cirillo, V. J., Andersen, O. F., Ham, E. A., and Gwatkin, R. B. L. (1969): The effect of luteinizing hormone and adenosine 3',5'-monophosphate on progesterone biosynthesis by bovine granulosa cells in culture. *Experimental Cell Research,* 57:139–142.
Civen, M., Brown, C. B., and Hilliard, J. (1966): Ribonucleic acid and protein synthesis in ovary. *Biochimica et Biophysica Acta,* 114:127–134.
Claesson, L., Diczfalusy, E., Hillarp, N. A., and Hogberg, B. (1948): Lipids of the pregnant rabbit ovary and their changes at gonadotropic stimulation. *Acta Physiologica Scandinavica,* 16:183–200.
Claesson, L., and Hillarp, N. A. (1947): The formation mechanism of oestrogenic hormones. I. The presence of an oestrogen-precursor in the rabbit ovary. *Acta Physiologica Scandinavica,* 13:115–129.
Claesson, L., Hillarp, N. A., and Hogberg, B. (1953): Lipid changes in the interstitial gland of the rabbit ovary at oestrogen formation. *Acta Physiologica Scandinavica,* 29:329–339.
Connell, G. M., and Eik-Nes, K. B. (1968): Testosterone production by rabbit testis slices. *Steroids,* 12:507–516.
Cooke, B. A., Rommerts, F. F. G., Van Beurden, W. M. O., Van der Kemp, J. W. C. M., and Van der Molen, H. J. (1973): Effect of tropic hormones *in vitro* on 3',5'-cyclic AMP and testosterone levels in rat testicular tissues. *Journal of Endocrinology,* 57:V.
Cooke, B. A., Van Beurden, W. M. O., Rommerts, F. F. G., and Van der Molen, H. J. (1972): Effect of trophic hormones on 3',5'-cyclic AMP levels in rat testis interstitial tissue and seminiferous tubules. *FEBS Letters,* 25:83–86.
Corbin, J. D., Reiman, E. M., Walsh, D. A., and Krebs, E. G. (1970): Activation of adipose tissue lipase by skeletal muscle cyclic adenosine 3',5'-monophosphate-stimulated protein kinase. *Journal of Biological Chemistry,* 245:4849–4851.
Corbin, J. D., Soderling, T. R., and Park, C. R. (1973): Regulation of adenosine 3',5'-monophosphate-dependent protein kinase. I. Preliminary characterization of the adipose tissue enzyme in crude extracts. *Journal of Biological Chemistry,* 248:1813–1821.
Courot, M., Hochereau-de Reviers, M., and Ortavant, R. (1970): Spermatogenesis. In: *The Testis,* edited by A. D. Johnson, W. R. Gomes, and N. L. Vandermark, Vol. I, pp. 339–432. Academic Press, New York and London.
Cropper, M., El-Fouly, M., Cook, B., and Nalbandov, A. V. (1970): Corpus luteum formation of

ovectomized follicles treated with LH. *Abstracts of the 52nd Meeting of the Endocrine Society,* p. A-80 (Abstract 87).
Danzo, B. (1972): Studies of the mechanism of action of hCG. *Dissertation Abstracts,* 33:2405B.
DeAngelo, A. B., Lee, P. C., and Jungmann, R. A. (1973): Ovarian cAMP-binding protein and protein kinase activities during postnatal development of the rat. *American Zoologist,* 13:1285 (Abstract 144).
DeAngelo, A. B., Skrypack, L. M., and Jungmann, R. A. (1974): Gonadotropin-stimulation of ovarian cyclic AMP-binding and protein kinase levels in neonatal and hypophysectomized rats. *Abstracts of the 56th Endocrine Society Meeting,* p. A-110 (Abstract 110).
Deane, H. W. (1952): Histochemical observations on the ovary and oviduct of the albino rat during the estrous cycle. *American Journal of Anatomy,* 91:363–393.
Dorrington, J. H. (1974): Cell types influenced by FSH in the rat testis. In: *Advances in Chemistry, Biology and Immunology of Gonadotropins,* edited by N. R. Moudgal. Bangalore Press, Bangalore (in press).
Dorrington, J. H., and Baggett, B. (1969): Adenyl cyclase activity in the rabbit ovary. *Endocrinology,* 84:989–996.
Dorrington, J. H., and Fritz, I. B. (1974): Effects of gonadotropins on cyclic AMP production by isolated seminiferous tubule and interstitial cell preparations. *Endocrinology,* 94:395–403.
Dorrington, J. H., and Kilpatrick, R. (1966a): Effects of pituitary hormones on progestational hormone production by the rabbit ovary *in vivo* and *in vitro*. *Journal of Endocrinology,* 35:53–63.
Dorrington, J. H., and Kilpatrick, R. (1966b): Effects of luteinizing hormone and nicotinamide adenine dinucleotide phosphate on synthesis of progestational steroids by rabbit ovarian tissue *in vitro*. *Journal of Endocrinology,* 35:65–73.
Dorrington, J. H., and Kilpatrick, R. (1967): Effect of adenosine 3′,5′-(cyclic)-monophosphate on the synthesis of progestational steroids by rabbit ovarian tissue *in vitro*. *Biochemical Journal,* 104: 725–730.
Dorrington, J. H., Vernon, R. G., and Fritz, I. B. (1972): The effect of gonadotropins on the 3′,5′-AMP levels of seminiferous tubules. *Biochemical and Biophysical Research Communications,* 46: 1523–1528.
Drosdowsky, M., Menon, K. M. J., Forchielli, E., and Dorfman, R. I. (1965): Requirements of the cholesterol side-chain-cleaving enzyme system of rat-testis mitochondria. *Biochimica et Biophysica Acta,* 104:229–236.
Dufau, M. L., Catt, K. J., and Tsuruhara, T. (1971): Gonadotropin stimulation of testosterone production by the rat testis *in vitro*. *Biochimica et Biophysica Acta,* 252:574–579.
Dufau, M. L., Catt, K. J., and Tsuruhara, T. (1972a): A sensitive gonadotropin responsive system: Radioimmunoassay of testosterone production by the rat testis *in vitro*. *Endocrinology,* 90:1032–1040.
Dufau, M. L., Tsuruhara, T., Watanabe, K., and Catt, K. J. (1972b): *Abstracts of the IVth International Congress of Endocrinology,* p. 199 (Abstract 501).
Dufau, M. L., Watanabe, K., and Catt, K. J. (1973): Stimulation of cyclic AMP production by the rat testis during incubation with hCG *in vitro*. *Endocrinology,* 92:6–11.
Dunn, M., and Birnbaumer, L. (1974): HCG induced loss of responsiveness to LH of an LH-sensitive adenyl cyclase (AC): A possible early effect related to ovulation and luteolysis. *Abstracts of the 56th Endocrine Society Meeting,* p. A-111 (Abstract 111).
Eakins, K. E., and Karim, S. M. M. (1970): Polyphloretin-phosphate—a selective antagonist for prostaglandins $F_{1\alpha}$ and $F_{2\alpha}$. *Life Sciences,* 9:1–5.
Eakins, K. E., and Sanner, J. H. (1972): Prostaglandin antagonist. In: *The Prostaglandins,* edited by S. M. M. Karim, pp. 263–292. Wiley-Interscience, New York.
Eik-Nes, K. B. (1967): Factors influencing the secretion of testosterone in the anaesthetized dog. *Ciba Foundation Colloquia on Endocrinology,* 16:120–139.
Eik-Nes, K. B. (1969): Patterns of steroidogenesis in the vertebrate gonads. *General and Comparative Endocrinology,* Supplement, 2:87–100.
Eik-Nes, K. B. (1971): Production and secretion of testicular steroids. *Recent Progress of Hormone Research,* 27:517–535.
El-Fouly, M. A., Cook, B., Nekola, M., and Nalbandov, A. V. (1970): Role of the ovum in follicular luteinization. *Endocrinology,* 87:288–293.
Ellsworth, L. R., and Armstrong, D. T. (1971): Effect of LH on luteinization of ovarian follicles transplanted under the kidney capsule in rats. *Endocrinology,* 88:755–762.

Ellsworth, L. R., and Armstrong, D. T. (1973): Luteinization of transplanted ovarian follicles in the rat induced by dibutyryl cyclic AMP. *Endocrinology,* 92:840–846.

Ellsworth, L. R., and Armstrong, D. T. (1974): Inhibition of luteinization of transplanted rat ovarian follicles by polyphloretin phosphate. *Endocrinology,* 94:892–896.

Eshkol, A., and Lunenfeld, B. (1967): Purification and separation of follicle stimulating hormone (FSH) and luteinizing hormone (LH) from human menopausal gonadotropin (HMG). *Acta Endocrinologica,* 54:91–95.

Farese, R.V. (1971): Stimulation of pregnenolone synthesis by ACTH in rat adrenal sections. *Endocrinology,* 89:958–962.

Ferguson, J. J., Jr. (1963): Protein synthesis and adrenocorticotropin responsiveness. *Journal of Biological Chemistry,* 238:2745–2759.

Flint, A. P. F., and Armstrong, D. T. (1971): Intracellular localization of cholesterol side-chain cleavage enzyme in corpora lutea of cow and rat. *Nature, New Biology,* 231:60–61.

Flint, A. P. F., and Denton, R. M. (1970): Mechanism of action of luteinizing hormone. *Nature,* 228:376–377.

Flint, A. P. F., Grinwich, D. L., and Armstrong, D. T. (1973): Control of ovarian cholesterol ester biosynthesis. *Biochemical Journal,* 132:313–321.

Fontaine, Y. A., Burzawa-Gerard, E., and Delerue-Lebelle, N. (1970): Stimulation hormonale de l'activite adenylcyclasique d l'ovaire d'un poisson teleosteen le cyprin (*Carassius auratus* L.). *Comptes Rendue Hebdomadaires Des Séances De l'Académie Des Sciences, D,* 271:780–783.

Fontaine, Y. A., Fontaine-Bertrand, E., Salmon, C., and Delerue-Lebelle, N. (1971): Stimulation *in vitro* par les deux hormones gonadotropes hypophysaires (LH et FSH) de l'activité adenylcyclasique de l'ovarie chez la ratte prépubere. *Comptes Rendue Hebdomadaires Des Séances De l'Académie Des Sciences, D.,* 272:1137–1140.

Fontaine, Y. A., Salmon, C., Fontaine-Bertrand, E., Burzawa-Gerard, E., and Donaldson, E. M. (1972): Comparison of the activities of two purified fish gonadotropins on adenyl cyclase in the goldfish ovary. *Canadian Journal of Zoology,* 50:1673–1676.

Garren, L. D. (1968): The mechanism of action of adrenocorticotropic hormone. *Vitamins and Hormones,* 26:119–145.

Garren, L. D., Gill, G. N., Masui, H., and Walton, G. M. (1971): On the mechanism of action of ACTH. *Recent Progress in Hormone Research,* 27:433–478.

Goldberg, N. D., O Dea, R. T., and Haddox, M. K. (1973): Cyclic GMP. *Advances in Cyclic Nucleotide Research,* 3:156–223.

Goldstein, S. (1973): A study of cyclic AMP and protein kinase in the bovine corpus luteum. Doctoral Dissertation, Department of Biochemistry, University of Miami School of Medicine, Miami, Fla.

Goldstein, S., and Marsh, J. M. (1972): Subcellular localization of endogenous cyclic AMP and protein kinase in the bovine corpus luteum. *Federation Proceedings,* 31:861 (Abstract 3661).

Goldstein, S., and Marsh, J. M. (1973): Protein kinase in the bovine corpus luteum. In: *Protein Phosphorylation in Control Mechanisms.* Miami Winter Symposium. Vol. 5, pp. 123–144, edited by Academic Press, New York and London.

Gorski, J., and Padnos, D. (1966): Translational control of protein synthesis and the control of steroidogenesis in the rabbit ovary. *Archives of Biochemistry and Biophysics,* 113:100–106.

Gospodarowicz, D. (1964): The action of follicle stimulating hormone and of human chorionic gonadotrophin upon steroid synthesis by rabbit ovarian tissues *in vitro. Acta Endocrinologica,* 47:293–305.

Greep, R. O. (1971): Regulation of luteal cell function. *Proceedings of the 3rd International Congress on Hormonal Steroids,* edited by V. H. T. James and L. Martini, pp. 670–679. Excerpta Medica Foundation, Amsterdam.

Greep, R. O., Van Dyke, H. B., and Chow, B. F. (1942): Gonadotropins of swine pituitary. I. Various biological effects of purified thylakentrin (FSH) and pure metakentrin (ICSH). *Endocrinology,* 30:635–649.

Grinwich, D. L., Kennedy, T. G., and Armstrong, D. T. (1972): Dissociation of ovulatory and steroidogenic actions of luteinizing hormone in rabbits with indomethacin, an inhibitor of prostaglandin biosynthesis. *Prostaglandins,* 1:89–96.

Hafs, H. D., and Armstrong, D. T. (1968): Corpus luteum growth and progesterone synthesis during the bovine estrous cycle. *Journal of Animal Science,* 27:134–141.

Halkerston, I. D. K. (1968): Heterogeneity of the response of adrenal cortex tissue slices to adrenocorticotropin. In: *Functions of the Adrenal Cortex,* edited by K. W. McKerns, Vol. I, pp. 399–461. Appleton-Century-Crofts, New York.

Halkerston, I. D. K., Eichhorn, J., and Hechter, O. (1959): TPNH requirement for cholesterol side chain cleavage in adrenal cortex. *Archives in Biochemistry and Biophysics,* 85:287–289.

Halkerston, I. D. K., Eichhorn, J., and Hechter, O. (1961): A requirement for reduced triphosphopyridine nucleotide for cholesterol side-chain cleavage by mitochondrial fractions of bovine adrenal cortex. *Journal of Biological Chemistry,* 236:374–380.

Hall, P. F. (1970): Endocrinology of the testis. In: *The Testis,* edited by A. D. Johnson, W. R. Gomes, and N. L. Vandemark, Vol. II, pp. 39–46. Academic Press, New York and London.

Hall, P. F., and Eik-Nes, K. B. (1962): The action of gonadotropic hormones upon rabbit testis *in vitro*. *Biochimica et Biophysica Acta,* 63:411–422.

Hall, P. F., and Koritz, S. B. (1964): The conversion of cholesterol and 20α hydroxycholesterol to steroids by acetone powder of particles from bovine corpus luteum. *Biochemistry,* 3:129–134.

Hall, P. F., and Koritz, S. B. (1965a): The influence of ICSH (LH) and 3',5'-AMP on steroidogenesis in the corpus luteum. *Federation Proceedings,* 24:320.

Hall, P. F., and Koritz, S. B. (1965b): Influence of interstitial cell-stimulating hormone on the conversion of cholesterol to progesterone by bovine corpus luteum. *Biochemistry,* 4:1037–1043.

Hall, P. F., and Young, D. G. (1968): Site of action of trophic hormones upon the biosynthetic pathways to steroid hormones. *Endocrinology,* 82:559–568.

Hamberger, L., Hamberger, A., and Herlitz, H. (1971): Methods for metabolic studies on isolated granulosa and theca cells. *Acta Endocrinologica,* Supplement 153:41–61.

Hamberger, L., Herlitz, H., and Sjogren, A. (1973): Influence of gonadotropins and cyclic AMP on carbohydrate and protein metabolism in the immature rat ovary. *Acta Endocrinologica,* 72:425–437,

Haynes, R. C., Jr. (1958): The activation of adrenal phosphorylase by the adrenocorticotropic hormone. *Journal of Biological Chemistry,* 233:1220–1222.

Haynes, R. C., Jr., and Berthet, L. (1957): Studies on the mechanism of action of the adrenocorticotropic hormone. *Journal of Biological Chemistry,* 225:115–124.

Haynes, R. C., Jr., Koritz, S. B., and Peron, F. G. (1959): Influence of adenosine 3',5'-monophosphate on corticoid production by rat adrenal glands. *Journal of Biological Chemistry,* 234:1421–1423.

Haynes, R. C., Savard, K., and Dorfman, R. I. (1954): The action of adrenocorticotropic hormone on beef adrenal slices. *Journal of Biological Chemistry,* 207:925–938.

Haynes, R. C., Jr., Sutherland, E. W., and Rall, T. W. (1960): The role of cyclic adenylic acid in hormone action. *Recent Progress in Hormone Research,* 16:121–138.

Hechter, O. (1955): Concerning possible mechanisms of hormone action. *Vitamins and Hormones,* 13:293–346.

Herbst, A. L. (1967): Response of rat ovarian cholesterol to gonadotropins and anterior pituitary hormones. *Endocrinology,* 81:54–60.

Hermier, C., Combarnous, Y., and Jutisz, M. (1971): Role of a regulating protein and molecular oxygen in the mechanism of action of luteinizing hormone. *Biochimica et Biophysica Acta,* 244:625–633.

Hermier, C., and Jutisz, M. (1969): Biosynthèse de la progesterone *in vitro* dans le corps jaune de la ratte pseudo-gestante. *Biochimica et Biophysica Acta,* 192:96–105.

Hermier, C., Santos, A. A., Wisnewsky, C., Netter, A., and Jutisz, M. (1972): Role de l'AMPc et d'une proteine regulatrice dans l'action, *in vitro* de la gonadotropine choriale humaine (HCG) sur le corps jaune humain. *Comptes Rendue Academic Sciences, Paris,* 275:1415–1418.

Hertig, A. T., and Barton, B. R. (1973): Fine structure of mammalian oocytes and ova. In: *Handbook of Physiology,* Section 7. Endocrinology, Vol. II, Female Reproductive System, Part I, edited by R. O. Greep, pp. 317–348. Williams & Wilkins, Baltimore.

Hiestand, P. C., Eppenberger, U., and Jungmann, R. A. (1973): Mechanism of action of gonadotropin. III. Binding of adenosine 3',5'-monophosphate by the 10,500 X g supernatant fraction from homogenates of calf ovaries. *Endocrinology,* 93:217–230.

Hilliard, J., Archibald, D., and Sawyer, C. (1963): Gonadotropic activation of preovulatory synthesis and release of progestin in the rabbit. *Endocrinology,* 72:59–66.

Hilliard, J., Endroczi, E., and Sawyer, C. (1961): Stimulation of progestin release from rabbit ovary in vivo. *Proceedings of the Society for Experimental Biology and Medicine,* 108:154–156.

Hilliard, J., Penardi, R., and Sawyer, C. (1967): A functional role for 20α hydroxypregn-4-en-3-one in the rabbit. *Endocrinology,* 80:901–909.

Hirshfield, I. N., and Koritz, S. B. (1964): The stimulation of pregnenolone synthesis in the large particles from rat adrenals by some agents which cause mitochondrial swelling. *Biochemistry,* 3:1994–1998.

Hirshfield, I. N., and Koritz, S. B. (1966): Pregnenolone synthesis stimulation in the large particles from bovine adrenal and bovine corpus luteum. *Endocrinology,* 78:165–168.
Hollinger, M. A. (1970): Studies on adenyl cyclase in rat testis. *Life Sciences, Physiology and Pharmacology,* 9:533–540.
Hollinger, M. A. (1973): Effect of age, cryptorchidism and hypophysectomy on cyclic AMP concentration in rat testis. *Journal of Reproduction and Fertility,* 35:169–172.
Hollinger, M. A., and Huang, F. (1974): Effect of dibutyryl cyclic AMP on *in vitro* rat testis DNA, RNA and protein labeling. *Endocrinology,* 94:444–449.
Honn, K. V., and Chavin, W. (1974): Control of cyclic nucleotide levels in the reptilian adrenals. *Abstracts of the 56th Endocrine Society Meeting,* p. A-161 (Abstract 211).
Hooker, C. W. (1970): The intertubular tissue of the testis. In: *The Testis,* edited by A. D. Johnson, W. R. Gomes, and N. L. Vandemark, Vol. I, pp. 493–596. Academic Press, New York and London.
Horrel, E., Kilpatrick, R., and Major, P. W. (1972): A comparison of the effects of pituitary hormones and aminophylline on progestational hormone production by the rabbit ovary *in vivo. Journal of Endocrinology,* 55:205–206.
Huttunen, J. K., Steinberg, D., and Mayer, S. E. (1970): ATP-dependent and cyclic AMP-dependent activation of rat adipose tissue lipase by protein kinase from rabbit skeletal muscle. *Proceedings of the National Academy of Sciences (U.S.),* 67:290–295.
Ichii, S., Forchielli, E., and Dorfman, R. I. (1963): *In vitro* effect of gonadotrophins on the soluble cholesterol side-chain cleaving enzyme system of bovine corpus luteum. *Steroids,* 2:631–656.
Jackanicz, T. M., and Armstrong, D. T. (1968): Progesterone biosynthesis in rabbit ovarian interstitial tissue mitochondria. *Endocrinology,* 83:769–776.
Jarlstedt, J., Nilsson, L., Hamberger, L., and Ahren, K. (1973): Effects of gonadotropins and cyclic 3′,5′-AMP on *in vitro* incorporation of [^{3}H] uridine into RNA of the prepubertal rat ovary. *Acta Endocrinologica,* 72:771–785.
Jefcoate, C. R., Simpson, E. R., and Boyd, G. S. (1974): Spectral properties of rat adrenal-mitochondria cytochrome P 450. *European Journal of Biochemistry,* 42:539–551.
Jones, E. E., and Nalbandov, A. V. (1972): Effects of intrafollicular injection of gonadotropins on ovulation or luteinization of ovarian follicle. *Biology of Reproduction,* 7:87–93.
Jonsson, H. T., Shelton, V. L., and Baggett, B. (1972): Stimulation of adenyl cyclase by prostaglandins in rabbit corpus luteum. *Biology of Reproduction,* 7:107.
Jungmann, R. A., Hiestand, P. C., and Schweppe, J. S. (1974): Mechanism of action of gonadotropin. IV. Cyclic adenosine monophosphate-dependent translocation of ovarian cytoplasmic cyclic adenosine monophosphate-binding protein and protein kinase to nuclear acceptor sites. *Endocrinology,* 94:168–183.
Jungmann, R. A., and Schweppe, J. S. (1972a): Mechanism of action of gonadotropin. I. Evidence for gonadotropin-induced modifications of ovarian nuclear basic and acidic protein biosynthesis, phosphorylation and acetylation. *Journal of Biological Chemistry,* 247:5535–5542.
Jungmann, R. A., and Schweppe, J. S. (1972b): Mechanism of action of gonadotropin. II. Control of ovarian nuclear ribonucleic acid polymerase activity and chromatin template capacity. *Journal of Biological Chemistry,* 247:5543–5548.
Kan, K. W., Ritter, M. C., Ungar, F., and Dempsey, M. E. (1972): The role of a carrier protein in cholesterol and steroid hormone synthesis by adrenal enzymes. *Biochemical and Biophysical Research Communications,* 48:423–429.
Kan, K. W., and Ungar, F. (1973): Characterization of an adrenal activator for cholesterol side chain cleavage. *Journal of Biological Chemistry,* 248:2868–2875.
Kao, L. W. L., and Nalbandov, A. V. (1972): The effect of antiadrenergic drugs on ovulation in hens. *Endocrinology,* 90:1343–1349.
Karaboyas, G. C., and Koritz, S. B. (1965): Identity of the site of action of 3′,5′-adenosine monophosphate and adrenocorticotropic hormone in corticosteroidogenesis in rat adrenal and beef adrenal cortex slices. *Biochemistry,* 4:462–468.
Karg, H., Hoffmann, B., and Schams, D. (1971): Luteinizing hormone, prolactin and progesterone relationships *in vivo* (data from the cow). *Proceedings of the 3rd International Congress on Hormonal Steroids,* edited by V. H. T. James and L. Martini, pp. 691–698. Excerpta Medica Foundation, Amsterdam.
Keyes, P. L. (1969): Luteinizing hormone: Action on the Graafian follicle *in vitro. Science,* 164: 846–847.
Keyes, P. L., Canastar, G. D., and Miller, J. B. (1972): Studies on luteinization of transplanted rabbit

follicles exposed to LH, actinomycin D, or adenosine 3',5'-cyclic-monophosphate *in vitro*. *Endocrinology,* 91:197–205.
Kitabchi, A. E., Nathans, A. H., James, P., Bower, F., Wilson, D. B., and Kitchell, L. C. (1974): Cyclic 3',5' GMP (cGMP) as possible mediator of steroidogenic action ACTH at physiologic concentrations. *Abstracts of the 56th Endocrine Society Meeting,* p. A-160 (Abstract 210).
Kobayashi, S., Yago, N., Morisaki, M., Ichii, S., and Matsuba, M. (1963): *In vitro* effect of corticotropin on the activity of phosphorylase of rat adrenal. *Steroids,* 2:167–174.
Koch, Y., Zor, U., Pomerantz, S., Chobsieng, P., and Lindner, H. R. (1973): Intrinsic stimulatory action of follicle-stimulating hormone on ovarian adenylate cyclase. *Journal of Endocrinology,* 58:677–678.
Kolena, J., and Channing, C. P. (1971): Stimulatory effects of gonadotropins on the formation of cyclic adenosine 3',5'-monophosphate by porcine granulosa cells. *Biochimica et Biophysica Acta,* 252:601–606.
Kolena, J., and Channing, C. P. (1972): Stimulatory effects of LH, FSH and prostaglandins upon cyclic 3',5'-AMP levels in porcine granulosa cells. *Endocrinology,* 90:1543–1550.
Koritz, S. B., and Hall, P. F. (1964): End-product inhibition of the conversion of cholesterol to pregnenolone in an adrenal extract. *Biochemistry,* 3:1298–1304.
Koritz, S. B., and Hall, P. F. (1965): Further studies on the locus of action of interstitial cell-stimulating hormone on the biosynthesis of progesterone by bovine corpus luteum. *Biochemistry,* 4:2740–2747.
Koritz, S. B., and Kumar, A. M. (1970): On the mechanism of action of the adrenocorticotrophic hormone: The stimulation of the activity of enzymes involved in pregnenolone synthesis. *Journal of Biological Chemistry,* 245:152–159.
Koritz, S. B., Yun, J., and Ferguson, J. J., Jr. (1968): Inhibition of adrenal progesterone biosynthesis by 3',5'-cyclic AMP. *Endocrinology,* 82:620–622.
Kuehl, F. A., Jr. (1974): Prostaglandins, cyclic nucleotides and cell function. *Prostaglandins,* 5: 324–340.
Kuehl, F. A., Jr., Cirillo, V. J., Ham, E. A., and Humes, J. L. (1973): The regulatory role of the prostaglandins on the cyclic AMP system. *Advances in Biosciences,* 9:155–172.
Kuehl, F. A., Jr., Humes, J. L., Cirillo, V. J., and Ham, E. A. (1972): Cyclic AMP and prostaglandins in hormone action. *Advances in Cyclic Nucleotide Research,* 1:493–502.
Kuehl, F. A., Jr., Humes, J. L., Mandel, L. R., Cirillo, V. J., and Ham, E. A. (1971): Prostaglandin antagonist: Studies on the mode of action of polyphloretin phosphate. *Biochemical and Biophysical Research Communications,* 44:1464–1470.
Kuehl, F. A., Jr., Humes, J. L., Tarnoff, J., Cirillo, V. J., and Ham, E. A. (1970b): Prostaglandin receptor site: Evidence for an essential role in the action of lutenizing hormone. *Science,* 169: 883–885.
Kuehl, F. A., Jr., Patanelli, D. J., Tarnoff, J., and Humes, J. L. (1970a): Testicular adenyl cyclase: Stimulation by the pituitary gonadotropins. *Biology of Reproduction,* 2:154–163.
Kuo, J. F., and DeRenzo, E. C. (1969): A comparison of the effects of lipolytic and antilipolytic agents on adenosine 3',5'-monophosphate levels in adipose cell as determined by prior labelling with adenine-8-^{14}C. *Journal of Biological Chemistry,* 244:2252–2260.
Lamprecht, S. A., Zor, U., Tsafriri, A., and Lindner, H. R. (1971). Action of prostaglandin E$_2$ and luteinizing hormone on cyclic adenosine 3',5'-monophosphate production and protein kinase activity in fetal, early postnatal and adult rat ovaries. *Israel Journal of Medical Sciences,* 7:704–705.
Lamprecht, S. A., Zor, U., Tsafriri, A., and Lindner, H. R. (1973): Action of prostaglandin E$_2$ and of luteinizing hormone on ovarian adenylate cyclase, protein kinase and ornithine decarboxylase activity during postnatal development and maturity in the rat. *Journal of Endocrinology,* 57: 217–233.
Langan, T. A. (1973): Histone phosphorylation and regulation of nuclear function. In: *Protein Phosphorylation in Control Mechanisms,* edited by F. Huijing and E. Y. C. Lee, pp. 287–292. Academic Press, New York and London.
Lee, M. Y. E., and Iverson, R. M. (1972): Protein kinase in sea urchin gametes and embryos. *Experimental Cell Research,* 75:300–304.
LeMaire, W. J., Askari, H., and Savard, K. (1971): Steroid hormone formation in the human ovary: VII. *Steroids,* 17:65–84.
LeMaire, W. J., Mills, T., Ito, Y., and Marsh, J. M. (1972): Inhibition by 3',5'-cyclic AMP of luteinizing hormone. *Biology of Reproduction,* 6:109–116.

LeMaire, W. J., Rice, B. F., and Savard, K. (1968): Steroide hormone formation in the human ovary: V. Synthesis of progesterone *in vitro* in corpora lutea during the reproductive cycle. *Journal of Clinical Endocrinology and Metabolism*, 28:1249–1256.

LeMaire, W. J., Yang, N. S. T., Behrman, H. R., and Marsh, J. M. (1973): Preovulatory changes in the concentration of prostaglandins in rabbit graafian follicles. *Prostaglandins*, 3:367–376.

Levin, L., and Jailer, J. W. (1948): The effect of induced secretory activity on the cholesterol content of the immature rat ovary. *Endocrinology*, 43:154–166.

Lindner, H. R., Tsafriri, A., Koch, Y., Zor, U., Lieberman, M. E., and Pomerantz, S. (1973): Intrinsic action of ovine follicle stimulating hormone on cyclic AMP production, steroidogenesis and ovum maturation in cultured rat follicles. *Abstracts of the Symposium on Advances in Chemistry, Biology and Immunology of Gonadotropins*, edited by N. R. Moudgal, p. 48. Bangalore Press, Bangalore.

Lostroh, A., and Johnson, R. E. (1966): Amounts of interstitial cell-stimulating hormone and follicle-stimulating hormone required for follicular development, uterine growth and ovulation in the hypophysectomized rat. *Endocrinology*, 79:991–996.

Mahaffee, D., Reitz, R. C., and Ney, R. L. (1974): The mechanism of action of adrenocorticotropic hormone. The role of mitochondrial cholesterol accumulation in the regulation of steroidogenesis. *Journal of Biological Chemistry*, 249:227–233.

Marsh, J. M. (1968): The mechanism of action of luteinizing hormone on steroidogenesis in the corpus luteum *in vitro*. *Advances in Experimental Medicine and Biology*, 2:213–222.

Marsh, J. M. (1969): The role of adenosine 3',5'-monophosphate in the action of luteinizing hormone on steroidogenesis. In: *Progress in Endocrinology*, edited by C. Gual and F. J. B. Ebling, pp. 83–88. Excerpta Medica Foundation, Amsterdam.

Marsh, J. M. (1970a): The stimulatory effect of luteinizing hormone on adenyl cyclase in the bovine corpus luteum. *Journal of Biological Chemistry*, 245:1596–1603.

Marsh, J. M. (1970b): The stimulatory effect of prostaglandin E_2 on adenyl cyclase in the bovine corpus luteum. *FEBS Letters*. 7:283–286.

Marsh, J. M. (1971): The effect of prostaglandins on the adenyl cyclase of the bovine corpus luteum. *Annals of the New York Academy of Sciences*, 180:416–425.

Marsh, J. M., Butcher, R. W., Savard, K., and Sutherland, E. W. (1966): The stimulatory effect of luteinizing hormone on adenosine 3',5'-monophosphate accumulation in corpus luteum slices. *Journal of Biological Chemistry*, 241:5436–5440.

Marsh, J. M., and LeMaire, W. J. (1974a): Cyclic AMP accumulation and steroidogenesis in the human corpus luteum: Effect of gonadotropins and prostaglandins. *Journal of Clinical Endocrinology and Metabolism*, 38:99–106.

Marsh, J. M., and LeMaire, W. J. (1974b): The role of cyclic AMP and prostaglandins in the actions of luteinizing hormone. In: *Symposium on Advances in Chemistry, Biology and Immunology of Gonadotropins*, edited by N. R. Moudgal. Academic Press, New York and London (in press).

Marsh, J. M., Mills, T. M., and LeMaire, W. J. (1972): Cyclic AMP synthesis in rabbit Graffian follicles and the effect of luteinizing hormone. *Biochimica et Biophysica Acta*, 273:389–394.

Marsh, J. M., Mills, T. M., and LeMaire, W. J. (1973): Preovulatory changes in the synthesis of cyclic AMP by rabbit Graafian follicles. *Biochimica et Biophysica Acta*, 304:197–202.

Marsh, J. M., and Savard, K. (1964a): The activation of luteal phosphorylase by luteinizing hormone. *Journal of Biological Chemistry*, 239:1–7.

Marsh, J. M., and Savard, K. (1964b): The effect of 3',5'-AMP on progesterone synthesis in the corpus luteum. *Federation Proceedings*, 23:462.

Marsh, J. M., and Savard, K. (1966a): Studies on the mode of action of luteinizing hormone on steroidogenesis in the corpus luteum *in vitro*. *Journal of Reproduction and Fertility Supplement*, 1:113–126.

Marsh, J. M., and Savard, K. (1966b): The stimulation of progesterone synthesis in bovine corpora lutea by adenosine 3',5'-monophosphate. *Steroids*, 8:133–148.

Marsh, J. M., Yang, N. S. T., and LeMaire, W. J. (1974): The stimulation of prostaglandin synthesis by LH in Graafian follicles *in vitro*. *Abstracts of the 56th Endocrine Society Meeting*, A-109 (Abstract 107).

Martelo, O. J. (1973): Phosphorylation of RNA-polymerase in *E. coli* and rat liver. In: *Protein Phosphorylation in Control Mechanisms*, edited by F. Huijing and E. Y. C. Lee, pp. 199–216. Academic Press, New York and London.

Mason, J. I., and Boyd, G. S. (1971): The cholesterol side-chain cleavage enzyme system in mitochondria of human term placenta. *European Journal of Biochemistry*, 21:308–321.

Mason, N. R. (1974): LH effect on cyclic AMP levels in rat ovaries *in vivo*. *Abstracts of the 56th Endocrine Society Meeting*, p. A-272 (Abstract 433).
Mason, N. R., Marsh, J. M., and Savard, K. (1962): An action of gonadotropin *in vitro*. *Journal of Biological Chemistry*, 237:1801–1806.
Mason, N. R., and Savard, K. (1964a): Specificity of gonadotropin stimulation of progesterone synthesis in bovine corpus luteum *in vitro*. *Endocrinology*, 74:664–668.
Mason, N. R., and Savard, K. (1964b): Conversion of cholesterol to progesterone by corpus luteum slices. *Endocrinology*, 75:215–221.
Mason, N. R., Schaffer, R. J., and Toomey, R. E. (1973): Stimulation of cyclic AMP accumulation in rat ovaries *in vitro*. *Endocrinology*, 93:34–41.
Mason, N. R., and Toomey, R. E. (1971): Stimulation of cyclic AMP in rat ovaries. *Abstracts of the 53rd Meeting of the Endocrine Society*, p. A-118 (Abstract 152).
McCune, R. W., Roberts, S., and Young, P. L. (1970): Competitive inhibition of adrenal Δ^5-3β-hydroxysteroid dehydrogenase and Δ^5-3-ketoesteroid isomerase activities by adenosine 3′,5′-monophosphate. *Journal of Biological Chemistry*, 245:3859–3867.
Means, A. R. (1970): Early effects of FSH upon testicular metabolism. *Advances in Experimental Medicine and Biology*, 10:301–313.
Means, A. R. (1971): Concerning the mechanism of FSH action: Rapid stimulation of testicular synthesis of nuclear RNA. *Endocrinology*, 89:981–989.
Means, A. R., and Hall, P. F. (1967): Effect of FSH on protein biosynthesis in testes of the immature rat. *Endocrinology*, 81:1151–1160.
Means, A. R., and Hall, P. F. (1969): Protein biosynthesis in the testis. V. Concerning the mechanism of stimulation by follicle-stimulating hormone. *Biochemistry*, 8:4293–4298.
Means, A. R., and Huckins, C. (1974): The Sertoli cell is the primary target for FSH in the testis. *Abstracts of the 56th Endocrine Society Meeting*, p. A-107 (Abstract 103).
Means, A. R., MacDougall, E., Soderling, T. R., and Corbin, J. D. (1974): Testicular adenosine 3′,5′-monophosphate-dependent protein kinase. *Journal of Biological Chemistry*, 249:1231–1238.
Mendelson, C., Dufau, M., and Catt, K. (1974): Gonadotropin induced steroidogenesis in dispersed rat Leydig cells: A sensitive and precise *in vitro* bioassay for LH and hCG. *Abstracts of the 56th Endocrine Society Meeting*, p. A-106 (Abstract 101).
Menon, K. M. J. (1973): Purification and properties of a protein kinase from bovine corpus luteum that is stimulated by cyclic adenosine 3′,5′-monophosphate and luteinizing hormone. *Journal of Biological Chemistry*, 248:494–501.
Menon, K. M. J., and Kiburz, J. (1974): Isolation of plasma membranes from bovine corpus luteum possessing adenylate cyclase, ^{125}I—hCG binding and Na-K-ATPase activities. *Biochemical and Biophysical Research Communications*, 56:363–371.
Mieno, M., Kawao, K., Shimizu, T., and Yamashita, K. (1973): Effects of adenosine-3′,5′-monophosphate and 5-hydroxytryptamine on the secretory activity of the canine testis. *Tohoku Journal of Experimental Medicine*, 110:113–117.
Miller, J. B., and Keyes, P. L. (1974): Initiation of luteinization in rabbit Graafian follicles by dibutyryl cyclic AMP *in vitro*. *Endocrinology*, 95:253–259.
Mills, T. M. (1975): Effect of luteinizing hormone and cyclic adenosine 3′,5′-monophosphate on steroidogenesis in the ovarian follicle of rabbit. *Endocrinology*, 96:440–445.
Mills, T. M., Davies, P. J. A., and Savard, K. (1971): Stimulation of estrogen synthesis in rabbit follicles by luteinizing hormone. *Endocrinology*, 88:857–862.
Mills, T. M., and McPherson, J. C., III (1974): The binding of ^{131}I-hCG to mature follicles isolated from the rabbit ovary. *Proceedings of the Society for Experimental Biology and Medicine*, 145:446–449.
Mills, T. M., and Savard, K. (1973): Steroidogenesis in ovarian follicles isolated from rabbits before and after mating. *Endocrinology*, 92:788–791.
Monn, E., Desautel, M., and Christiansen, R. O. (1972): Highly specific testicular adenosine 3′,5′-monophosphate phosphodiesterase associated with sexual maturation. *Endocrinology*, 91:716–720.
Moudgal, N. R., Moyle, W. R., and Greep, R. O. (1971): Specific binding of luteinizing hormone to Leydig tumor cells. *Journal of Biological Chemistry*, 246:4983–4986.
Moyle, W. R., and Armstrong, D. T. (1970): Stimulation of testosterone biosynthesis by luteinizing hormone in transplantable mouse Leydig cell tumors. *Steroids*, 15:681–693.
Moyle, W. R., Jungas, R. L., and Greep, R. O. (1973a): Influence of luteinizing hormone and adenosine 3′,5′-cyclic monophosphate on the metabolism of free and esterified cholesterol in mono-

phosphate on the metabolism of free and esterified cholesterol in mouse Leydig-cell tumors. *Biochemical Journal,* 134:407–413.

Moyle, R. W., Jungas, R. L., and Greep, R. O. (1973*b*): Metabolism of free and sterified cholesterol by Leydig-cell tumor mitochondria. *Biochemical Journal,* 134:415–424.

Moyle, W. R., Kong, Y. C., and Ramachandran, J. (1973*c*): Divergent effects of adrenocorticotropin and its o-nitrophenyl sulfenyl derivative. *Journal of Biological Chemistry,* 248: 2409–2417.

Moyle, W. R., Moudgal, N. R., and Greep, R. O. (1971): Cessation of steroidogenesis in Leydig cell tumors after removal of luteinizing hormone and adenosine cyclic 3',5'-monophosphate. *Journal of Biological Chemistry,* 246:4978–4982.

Moyle, W. R., and Ramachandran, J. (1973): Effect of LH on steroidogenesis and cyclic AMP accumulation in rat Leydig cell preparations and mouse tumor Leydig cells. *Endocrinology,* 93: 127–134.

Murad, F., Strauch, B. S., and Vaughan, M. (1969): The effect of gonadotropins on testicular adenyl cyclase. *Biochimica et Biophysica Acta,* 177:591–598.

Naghshineh, S., Vahouny, G. V., and Treadwell, C. R. (1974): cAMP activation of adrenal sterol ester hydrolase. *Federation Proceedings,* 33:689 (Abstract 2704).

Nalbandov, A. V. (1973): Control of luteal function in mammals. In: *Handbook of Physiology,* Section 7. Endocrinology, Vol. II, Female Reproductive System, Part I, edited by R. O. Greep, pp. 153–167. Williams & Wilkins, Baltimore.

Namm, D. H., and Mayer, S. E. (1968): Effects of epinephrine on cardiac cyclic AMP, phosphorylase kinase, and phosphorylase. *Molecular Pharmacology,* 4:61–69.

Nekola, M. V. and Nalbandov, A. V. (1971): Morphological changes of rat follicular cells as influenced by oocytes. *Biology of Reproduction,* 4:154–160.

O'Grady, J. P., Caldwell, B. V., Auletta, F. J., and Speroff, L. (1972*a*): The effects of an inhibitor of prostaglandin synthesis (indomethacin) on ovulation, pregnancy and pseudopregnancy in the rabbit. *Prostaglandins,* 1:97–106.

O'Grady, J. P., Kohorn, E. I., Glass, R. H., Caldwell, B. V., Brock, W. A., and Speroff, L. (1972*b*): Inhibition of progesterone synthesis *in vitro* by prostaglandin $F_{2\alpha}$ *Journal of Reproduction and Fertility,* 30:153–156.

Orczyk, G. P., and Behrman, H. R. (1972): Ovulation blockade by aspirin or indomethacin. *In vivo* evidence for a role of prostaglandin in gonadotrophin secretion. *Prostaglandins,* 1:3–20.

Orloff, J., and Grantham, J. (1967): The effect of prostaglandin (PGE_1) on the permeability response of rabbit collecting tubules to vasopressin. *Nobel Symposium No. 2, Prostaglandins,* edited by S. Bergstrom and B. Samuelsson, Almquist and Wiksell Stockholm, pp. 143–146.

Orloff, J., Handler, J. S., and Bergström, S. (1965): Effect of prostaglandin (PGE_1) on the permeability response of toad bladder to vasopressin, theophylline and adenosine 3',5'-monophosphate. *Nature,* 205:397–398.

Parlow, A. F. (1968): A rapid bioassay method for LH and factors stimulating LH secretion. *Federation Proceedings,* 17:402 (Abstract 1587).

Peytremann, A., Wendell, E. N., Liddle, G. W., Hardman, J. G., and Sutherland, E. W. (1973): Effects of methylxanthines on adenosine 3',5'-monophosphate and corticosterone in the rat adrenal. *Endocrinology,* 92:525–530.

Pharriss, B. B., Tillson, S. A., and Erickson, R. R. (1972): Prostaglandins in luteal function. *Recent Progress in Hormone Research,* 28:51–89.

Pincus, G., and Enzmann, E. V. (1935): The comparative behavior of mammalian eggs *in vivo* and *in vitro*. I. The activation of ovarian eggs. *Journal of Experimental Medicine,* 62:665–675.

Pittman, R. C., and Khoo, J. C. (1974): Hormonally regulated cholesterol esterase in rat adipocytes. *Federation Proceedings,* 33:1357 (Abstract 757).

Posternak, T., Sutherland, E. W., and Henion, W. F. (1962): Derivatives of cyclic 3',5'-adenosine monophosphate. *Biochimica et Biophysica Acta,* 65:558–560.

Pulsinelli, W. A. (1972): Adenyl cyclase activity in subcellular fractions of dog testis and the effect of hCG and cyclic AMP on testicular cholesterol desmolase. *Dissertation Abstracts,* 32:6823-B.

Pulsinelli, W. A., and Eik-Nes, K. B. (1970): Adenyl cyclase activity in subcellular fractions of dog testis. *Federation Proceedings,* 29:918 (Abstract 3828).

Reddi, A. H., Ewing, L. L., and Williams-Ashman, H. G. (1971): Protein phosphokinase reactions in mammalian testis. *Biochemical Journal,* 122:333–345.

Reel, J. R., and Gorski, J. (1968*a*): Gonadotropic regulation of precursor incorporation into ovarian RNA, protein, and acid-soluble fractions. I. Effects of pregnant mare serum gonadotrophin

(PMSG), follicle-stimulating hormone (FSH), and luteinizing hormone (LH). *Endocrinology,* 83: 1083–1091.
Reel, J. R., and Gorski, J. (1968b): Gonadotropic regulation of precursor incorporation into ovarian RNA, protein, and acid-soluble fractions. II. Changes in nucleotide labeling, nuclear RNA synthesis, and effects of RNA and protein synthesis inhibitors. *Endocrinology,* 83:1092–1100.
Richardson, M. C., and Schulster, D. (1973): The role of protein kinase activation in the control of steroidogenesis by adrenocorticotrophic hormone in the adrenal cortex. *Biochemical Journal,* 136: 993–998.
Roberts, S., and Creange, J. E. (1968): The role of 3',5'-adenosine phosphate in the subcellular localization of regulatory processes in corticosteroidogenesis. In: *Functions of the Adrenal Cortex,* edited by K. W. McKerns, Vol. I, pp. 339–397. Appleton-Century-Crofts, New York.
Roberts, S., Creange, J. E., and Young, P. L. (1965): Stimulation of steroid transformation in adrenal mitochondria by cyclic 3',5'-adenosine phosphate. *Biochemical and Biophysical Research Communications,* 20:446–451.
Roberts, S., McCune, R. W., Creange, J. E., and Young, P. L. (1967): Adenosine 3',5'-cyclic monophosphate: Stimulation of steroidogenesis in sonically disrupted adrenal mitochondria. *Science,* 158:372–374.
Robison, G. A., Butcher, R. W., and Sutherland, E. W. (1971): *Cyclic AMP.* Academic Press, New York, pp. 43–44.
Robinson, J., and Stevenson, P. M. (1973): On the mechanism of action of luteinizing hormone on steroidogenesis. *Journal of Endocrinology,* 59:XIX–XX.
Rommerts, F. F. G., Cooke, B. A., Van der Kemp, J. W. C. M., and Van der Molen, H. J. (1972): Stimulation of 3',5'-cyclic AMP and testosterone production in rat testis *in vitro. FEBS Letters,* 24:251–254.
Rommerts, F. F. G., Cooke, B. A., Van der Kemp, J. W. C. M., and Van der Molen, H. J. (1973): Effect of luteinizing hormone on 3',5'-cyclic AMP and testosterone production in isolated interstitial tissue of rat testis. *FEBS Letters,* 33:114–118.
Rondell, P. (1970): Follicular processes in ovulation. *Federation Proceedings,* 29:1875–1879.
Rondell, P. (1974): Role of steroid synthesis in the process of ovulation. *Biology of Reproduction,* 10:199–215.
Salmon, C., Fontaine-Bertrand, E., Delerue-LeBelle, N., and Fontaine, Y. A. (1974): Etude comparative de quelques propriétés de l'activité adenyl cyclasique de l'ovaire. *General and Comparative Endocrinology,* 22:350–351.
Sandler, R., and Hall, P. F. (1966): Stimulation *in vitro* by adenosine-3',5'-cyclic monophosphate of steroidogenesis in rat testis. *Endocrinology,* 79:647–649.
Sato, S., Szabo, M., Kowalski, K., and Burke, G. (1972): Role of prostaglandin in thyrotropin action on the thyroid. *Endocrinology,* 90:343–356.
Savard, K., and Casey, P. (1964): Effects of pituitary hormones and NADPH on acetate utilization in ovarian and adrenocortical tissues. *Endocrinology,* 74:599–610.
Savard, K., Marsh, J. M., and Howell, D. S. (1963): Progesterone biosynthesis in luteal tissue: Role of nicotinamide adenine dinucleotide phosphate and NADP-linked dehydrogenases. *Endocrinology,* 73:554–563.
Savard, K., Marsh, J. M., and Rice, B. F. (1965): Gonadotropins and ovarian steroidogenesis. *Recent Progress in Hormone Research,* 21:285–365.
Schams, D., Hoffmann, B., Fischer, S., Marz, E., and Karg, H. (1972): Simultaneous determination of LH and progesterone in peripheral bovine blood during pregnancy, normal and corticoid-induced parturition and the post-partum period. *Journal of Reproduction and Fertility,* 29:37–48.
Schuetz, A. W. (1974): Role of hormones in oocyte maturation. *Biology of Reproduction,* 10:150–178.
Scoon, V., and Major, P. (1972): Regulation of steroid synthesis in rabbit ovarian homogenates by luteinizing hormone, cyclic AMP and pyridine nucleotides. *Journal of Endocrinology,* 52:147–159.
Scoon, V., and Major, P. (1973): The effect of pyridine nucleotides and cyclic AMP on the synthesis of progestational steroids in subcellular fractions of rabbit ovarian interstitial tissue. *Journal of Endocrinology,* 58:i.
Seifart, K. H., and Hansel, W. (1968): Some characteristics and optimum incubation conditions of *in vitro* progesterone synthesis by bovine corpora lutea. *Endocrinology,* 82:232–240.
Selstam, G., and Ahren, K. (1971): Effects of gonadotrophins on ovarian phosphorylase activity. *Acta Endocrinologica,* Supplement 155:55.
Selstam, G., Liljekvist, J., Rosberg, S., Gronquist, L., Perklev, T., and Ahren, K. (1974): Comparison

between the effect of luteinizing hormone and prostaglandin E_1 on ovarian cyclic AMP. *Prostaglandins,* 6:303–311.

Sidhu, K. S., Camp, C. E., and Marsh, J. M. (1974): Subcellular localization of adenyl cyclase in corpora lutea. *Federation Proceedings,* 33:204 (Abstract 8).

Shin, S. (1967): Studies on interstitial cells in tissue culture: Steroid biosynthesis in monolayers of mouse testicular interstitial cells. *Endocrinology,* 81:440–448.

Shin, S., and Sato, G. H. (1971): Inhibition by actinomycin D, cycloheximide and puromycin of steroid synthesis induced by cyclic AMP in interstitial cells. *Biochemical and Biophysical Research Communications,* 45:501–507.

Simmer, H. H., Hilliard, J., and Archibald, D. (1963): Isolation and identification of progesterone and 20α-hydroxypregn-4-en-3-one in ovarian venous blood of rabbits. *Endocrinology,* 72:67–70.

Simpson, E. R., and Estabrook, R. W. (1969): Mitochondrial malic enzyme: The source of reduced nicotinamide adenine dinucleotide phosphate for steroid hydroxylation in bovine adrenal cortex mitochondria. *Archives of Biochemistry and Biophysics,* 129:384–395.

Simpson, E. R., Jefcoate, C. R., Brownie, A. C., and Boyd, G. S. (1972): The effect of ether anesthesia stress on cholesterol-side-chain cleavage and cytochrome P 450 in rat-adrenal mitochondria. *European Journal of Biochemistry,* 28:442–450.

Smith, B. M., and Major, P. W. (1971): The effect of age and maturity of the responsiveness of rabbit ovarian adenyl cyclase to luteinizing hormone. *Life Sciences,* 10:1433–1439.

Soderling, T. R., Corbin, J. D., and Park, C. R. (1973): Regulation of adenosine 3′,5′-monophosphate-dependent protein kinase. II. Hormonal regulation of the adipose tissue enzyme. *Journal of Biological Chemistry,* 248:1822–1829.

Speroff, L., and Ramwell, P. W. (1970): Prostaglandin stimulation of *in vitro* progesterone synthesis. *Journal of Clinical Endocrinology and Metabolism,* 30:345–350.

Srinivasan, C. N., Clark, J. N., and Gawienowski, A. M. (1973): Effect of adenosine-3′,5′-cyclic monophosphate on β-hydroxy steroid dehydrogenase in the bovine corpus luteum. *Journal of Steroid Biochemistry,* 4:533–535.

Stansfield, D. A., Franks, D. J., Wilkinson, G. H., and Horne, J. R. (1972): Studies in the formation and degradation of adenosine 3′,5′-cyclic monophosphate in corpus luteum. *Journal of Steroid Biochemistry,* 3:643–653.

Stansfield, D. A., Horne, J. R., and Wilkinson, G. H. (1971): Adenosine 3′,5′-cyclic phosphate phosphodiesterase of corpus luteum. *Biochimica et Biophysica Acta,* 227:413–418.

Stansfield, D. A., and Robinson, J. W. (1965): Glycogen and phosphorylase in bovine and rat corpora lutea and the effect of luteinizing hormone. *Endocrinology,* 760:390–395.

Steinberg, D., and Huttunen, J. K. (1972): The role of cyclic AMP in activation of hormone sensitive lipase of adipose tissue. *Advances in Cyclic Nucleotide Research,* 1:47–62.

Steinberger, E. (1971): Hormonal control of mammalian spermatogenesis. *Physiological Reviews,* 51:1–22.

Sulimovici, S., Bartoov, B., and Lunenfeld, B. (1973): Localization of 3β-hydroxysteroid dehydrogenase in the inner membrane subfraction of rat testis mitochondria. *Biochimica et Biophysica Acta,* 321:27–40.

Sulimovici, S., and Boyd, G. S. (1968): The cholesterol side-chain cleavage enzymes in immature rat ovary. *European Journal of Biochemistry,* 3:332–345.

Sulimovici, S., and Boyd, G. S. (1969): The Δ⁵-3β-hydroxysteroid dehydrogenase of rat ovarian tissue. *European Journal of Biochemistry,* 7:549–558.

Sulimovici, S., and Lunenfeld, B. (1971): The effect of adenosine 3′,5′-monophosphate and of N⁶-2′-O-dibutyryl-adenosine 3′,5′-monophosphate on the mouse ovarian Δ⁵-3β-hydroxysteroid dehydrogenase. *Hormones and Metabolic Research,* 3:114–119.

Sulimovici, S., and Lunenfeld, B. (1973): The effect of gonadotropins on the mitochondrial adenylate cyclase of rat testis. *Biochemical and Biophysical Research Communications,* 55:673–679.

Sutherland, E. W. (1951): The effect of the hyperglycemic factor and epinephrine on enzyme systems of liver and muscle. *Annals of the New York Academy of Sciences,* 54:693–706.

Tenner, A. J., and Wallace, R. A. (1972): A cyclic AMP-stimulated protein kinase from amphibian ovary and oocytes. *Biochimica et Biophysica Acta,* 276:416–424.

Toren, D., Menon, K. M. J., Forchielli, E., and Dorfman, R. I. (1964): *In vitro* enzymatic cleavage of the cholesterol side chain in rat testis preparations. *Steroids,* 3:381–390.

Tsafriri, A., Lieberman, M. E., Barnea, A., Bauminger, S., and Lindner, H. R. (1973): Induction by luteinizing hormone of ovum maturation and of steroidogenesis in isolated Graafian follicles of the rat: Role of RNA and protein synthesis. *Endocrinology,* 93:1378–1386.

Tsafriri, A., Lindner, H. R., Zor, U., and Lamprecht, S. A. (1971): Induction of meiotic division in follicle-enclosed ova in culture. *Israel Journal of Medical Science,* 8:170.

Tsafriri, A., Lindner, H. R., Zor, U., and Lamprecht, S. A. (1972a): Physiological role of prostaglandins in the induction of ovulation. *Prostaglandins,* 2:1–10.

Tsafriri, A., Lindner, H. R., Zor, U., and Lamprecht, S. A. (1972b): *In vitro* induction of meiotic division in follicle-enclosed rat oocytes by LH, cyclic AMP and prostaglandin E_2. *Journal of Reproduction and Fertility,* 31:39–50.

Ungar, F., Kan, K. W., and McCoy, K. E. (1973): Activator and inhibitor factors in cholesterol side-chain cleavage. *Annals of the New York Academy of Science,* 212:276–289.

Van der Molen, H. J., and Eik-Nes, K. B. (1971): Biosynthesis and secretion of steroids by the canine testis. *Biochimica et Biophysica Acta,* 248:343–362.

Varga, J. M., Dipasquale, A., Pawelek, J., McGuire, J. S., and Lerner, A. B. (1974): Regulation of melanocyte stimulating hormone action at the receptor level: Discontinuous binding of hormone to synchronized mouse melanoma cells during the cell cycle. *Proceedings of the National Academy of Sciences (U.S.),* 71:1590–1593.

Wedner, H. J., Hoffer B. J., Battenberg, E., Steiner, A. L., Parker, C. W., and Bloom, F. E. (1972): A method for detecting intracellular cyclic adenosine monophosphate by immunofluorescence. *Journal of Histochemistry and Cytochemistry,* 20:293–295.

Whittey, T. H., Stowe, N. W. Ong, S. H., Ney, R. L., and Steiner, A. L. (1974): Control of intracellular localization of adrenal cyclic GMP, comparison with cyclic AMP. *Abstracts of the 56th the Endocrine Society Meeting,* p. A-160 (Abstract 209).

Wicks, W. D. (1974): Regulation of protein synthesis by cyclic AMP. *Advances in Cyclic Nucleotide Research,* 4:335–438.

Wilks, J. W., Forbes, K. K., and Norland, J. F. (1972): Synthesis of prostaglandin $F_{2\alpha}$ by the ovary and uterus. *Journal of Reproductive Medicine,* 9:271–276.

Williams, H. E., Johnson, P. L., and Field, J. B. (1961): An effect of anterior pituitary hormones on bovine corpus luteum phosphorylase. *Biochemical and Biophysical Research Communications,* 6:129–133.

Williams, J. A. (1972): Cyclic AMP formation and thyroid secretion by incubated mouse thyroid lobes. *Endocrinology,* 91:1411–1417.

Yago, N., Dorfman, R. I., and Forchielli, E. (1967): Extramitochondrial NADPH and cholesterol side-chain cleaving enzyme system in the heavy mitochondria of bovine corpora lutea. *Journal of Biochemistry,* 62:345–352.

Yunis, A. A., and Assaf, S. A. (1970): Purification and properties of glycogen phosphorylase from bovine corpus luteum. Kinetics of salt activation. *Biochemistry,* 9:4381–4388.

Zimbelman, R. G., Loy, R. G., and Casida, L. E. (1961): Variations in some biochemical and histological characteristics of bovine corpora lutea during early pregnancy. *Journal of Animal Science,* 20:99–105.

Zor, U., Bauminger, S., Lamprecht, S. A., Koch, Y., Chobsieng, P., and Lindner, H. R. (1973): Stimulation of cyclic AMP production in the rat ovary by luteinizing hormone: Independence of prostaglandin mediation. *Prostaglandins,* 4:499–507.

Zor, U., Lamprecht, S. A., Kaneko, T., Schneider, H. P. G., McCann, S. M., Field, J. B., Tsafriri, A., and Lindner, H. R. (1972a): Functional relations between cyclic AMP, prostaglandins and luteinizing hormone in rat pituitary and ovary. *Advances in Cyclic Nucleotide Research,* 1:503–520.

Zor, U., Lamprecht, S. A., Tsafriri, A., Pomerantz, S., Koch, Y., and Lindner, H. R. (1972b): Stimulatory action of cholera *(Vibrio cholerae)* enterotoxin on cyclic adenosine-3′,5′-monophosphate formation, ovum maturation and ornithine decarboxylase activity in rat ovaries. *Israel Journal of Medical Sciences,* 8:1774.

Cyclic AMP and the Physiology of the Islets of Langerhans

W. Montague and
S. L. Howell*

*Department of Chemical Pathology and Diabetic Department, King's College Hospital Medical School, Denmark Hill, London SE5 8RX, England, and *School of Biological Sciences, University of Sussex, Brighton BNI 9QG, Sussex, England*

CONTENTS

I.	Introduction	202
II.	The Synthetic and Secretory Functions of Islets of Langerhans	204
	A. Cellular Components of Islets of Langerhans	204
	B. B Cell	204
	C. Cyclic AMP and Insulin Biosynthesis	207
	D. Cyclic AMP and Insulin Secretion	207
	E. A Cell	208
	F. Cyclic AMP and Glucagon Secretion	208
III.	Components of the Cyclic AMP System in Islets of Langerhans	208
	A. Adenylyl Cyclase	209
	B. Phosphodiesterase	212
	C. Protein Kinase	212
	D. Phosphoprotein Phosphatase	213
IV.	Interaction of Secretagogues with Components of the B Cell Cyclic AMP System	214
	A. Glucose	214
	B. Amino Acids	216
	C. Sulfonylureas	217
	D. Methylxanthines	217
	E. Prostaglandins	218
	F. Hormones and Neurohumoral Agents	218
V.	Adrenergic Receptors in Islets of Langerhans	219
	A. B Cell	219
	1. α-Receptor Activity	220
	2. β-Receptor Activity	220

           3. Adrenergic Receptors and Cyclic AMP   220
           4. Glucose and Adrenergic Receptor Activity 221
       B. A Cell   222

VI.   The Role of Cyclic AMP in the Regulation of Insulin Secretion 222
       A. Short-Term Regulation   222
       B. Long-Term Regulation   224
           1. Dietary Changes   224
           2. Pregnancy   226
           3. Hormonal Changes   226
       C. Fetal Pancreas   227

VII.   Mode of Action of Cyclic AMP in Regulating Insulin Secretion 228
       A. Cyclic AMP as a Modulator of Islet Cell Metabolism . .   229
       B. Cyclic AMP and Microtubules   230
       C. Cyclic AMP and Calcium   231
       D. Membranes   232
       E. Speculation on the Mechanism of Action of Cyclic AMP . 232

VIII. Other Cyclic Nucleotides   233

IX.   Summary   234
X.   Acknowledgments   235
XI.   References   235

I. INTRODUCTION

A widely accepted model (Fig. 1) for the regulation of hormone secretion suggests that intracellular levels of cyclic AMP in endocrine cells may play an important role in determining rates of release, and that many potential secretagogues influence cyclic AMP levels, and hence rates of secretion, by interaction with receptors which are related to adenylyl cyclase, or by inhibition of phosphodiesterase. A mechanism of this type seems to operate during activation of secretion by physiological agents in, for instance, the anterior pituitary (Labrie, Lemaire, and Courte, 1971; Howell and Montague, 1971) and thyroid (Dumont, Willems, Van Sande, and Nève, 1971). One purpose of this review is to determine how far this simple model is applicable to the regulation of hormone secretion by the mammalian islets of Langerhans. While the mode of action of a variety of agents which are known to influence rates of secretion will be discussed, our purpose is not to provide an exhaustive list of these factors, but rather to try to define the role and mode of action of cyclic AMP in the regulation of insulin secretion in the light of the evidence which is presently available.[1] The factors

---

[1] The literature survey for this review was completed in May 1974.

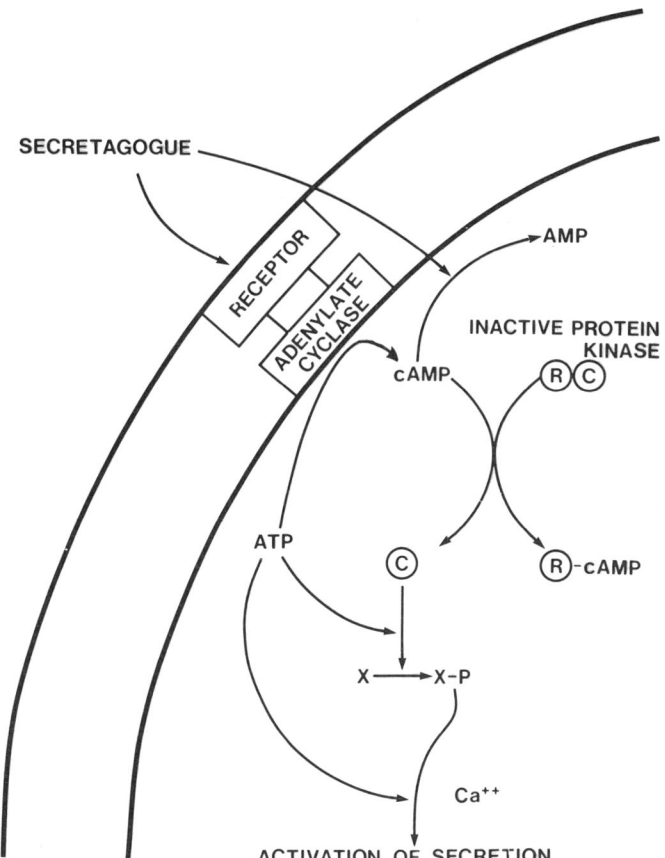

FIG. 1. A model of the regulation of hormone secretion by cyclic AMP. Secretagogues interact with either receptors for adenylyl cyclase (adenylate cyclase) or with phosphodiesterase to produce elevation of cyclic AMP levels. The resultant activation of cyclic AMP-dependent protein kinase induces phosphorylation of protein. This in turn promotes secretion, which occurs by an ATP- and calcium-dependent process.

which influence rates of insulin secretion have been extensively reviewed by Mayhew, Wright, and Ashmore (1969), Randle and Hales (1972), and Malaisse (1972), while the regulation of glucagon secretion has been discussed by Foa (1972) and Unger (1972).

There follows a brief discussion of the cellular mechanisms which may be involved in the synthesis, storage, and secretion of insulin and glucagon. Present knowledge concerning the components of the adenylyl cyclase-cyclic AMP-phosphophosphodiesterase system in the pancreatic A and B cells, and the effects on these components of agents known to alter rates of secretion, are discussed in detail. An attempt is made to define the role of cyclic AMP in the short- and longer-term regulation of rates of insulin secretion, and possible mechanisms of action of cyclic AMP in the secretory process are also described.

II. THE SYNTHETIC AND SECRETORY FUNCTIONS OF ISLETS OF LANGERHANS

A. Cellular Components of Islets of Langerhans

Studies of hormone secretion from the pancreas can be achieved by perfusion of whole pancreas *in situ* (Anderson and Long, 1948) or *in vitro* (Grodsky, Batts, Bennett, Vcella, McWilliams, and Smith, 1963), or by incubation of pieces of pancreas (Coore and Randle, 1964), despite the fact that islets themselves constitute only 1 to 2% of the total pancreatic tissue volume. In addition, estimations of enzymes and metabolites have been made on sections of microdissected frozen islet tissue (Matschinsky and Ellerman, 1968; Hellman, 1970). Investigations of the biochemistry of mammalian islets have been greatly facilitated during the last 9 years by the availability of methods for the separation of large numbers of islets from the vast excess of pancreatic exocrine tissue (Moskalewski, 1965; Kostianovsky and Lacy, 1966; Howell and Taylor, 1966). Using these methods, metabolically viable and ultrastructurally intact islets can be isolated from many small mammals with yields in a range of 0.5 to 4 mg of wet weight of tissue per animal, depending on the species.

Islets isolated in this way still represent a heterogeneous cell population, composed typically of 70% B cells responsible for insulin production, 25% A_2 or A cells producing glucagon, and 5% of a third endocrine cell, the A_1 or D cell which may contain gastrin (Hellerström and Hellman, 1960; Lomsky, Langr, and Vortel, 1969). Homogenates or extracts of isolated islets are usually assumed to reflect the characteristics of the predominating B cells, an assumption which has also been made in this review. However, the glucagon-producing cells represent a significant component of isolated mammalian islets, and as such the contribution of A cell metabolism to the overall islet metabolism should be considered. Analysis of A cell metabolism has been facilitated to some extent by the use of islets obtained from guinea pigs which have previously been treated with streptozotocin, an agent which specifically destroys the pancreatic B cells (Petersson, Hellerström, and Gunnarsson, 1970; Howell, Edwards, and Whitfield, 1971). Islets isolated from streptozotocin-treated guinea pigs contain 70% A cells and have proved useful in biochemical studies of A cell function.

B. B Cell

Biosynthesis of insulin occurs via a biosynthetic precursor, proinsulin (Steiner, Cunningham, Spigelman, and Aten, 1967). Proinsulin biosynthesis takes place in the elements of the rough-surfaced endoplasmic reticulum, and the newly synthesized proteins are then transferred rapidly across transitional areas of endoplasmic reticulum to the Golgi complex, probably via transfer vesicles and by an energy-dependent process. The conversion of proinsulin to insulin, and the concentration of the newly synthesized proteins into storage granules, is initiated within the Golgi complex (Howell, Kostianovsky, and Lacy, 1969a; Howell, 1972). The rate of transfer of material through the intracellular pathway which leads to granule formation and the rate of conversion of proinsulin to insulin appear to be relatively independent of the rates of either synthesis or secretion.

Newly formed storage granules (β granules), containing zinc insulin in a crystalline or paracrystalline form (Greider, Howell, and Lacy, 1969; Howell, Young, and Lacy, 1969b), are available for secretion within 1 hr of biosynthesis of the protein they contain (Howell and Taylor, 1967); it is the number of these granules which are secreted which determines the rate of release of the hormone. There appears to be no good evidence for a pathway of secretion in normal cells which bypasses this granule-formation stage, and all the secretory mechanisms which we shall consider relate to the process of insulin secretion by granule extrusion. This final extrusion occurs by fusion of the granule membrane with the plasma membrane of the cell, with consequent release of the granule contents into the extracellular space, and is termed exocytosis or emiocytosis (Lacy, 1961). The movement of storage granules from the cytoplasmic pool, where they are present in very large numbers ($\sim$13,000 per cell in the mouse: Dean, 1973) may involve the participation of microtubules (Lacy, Howell, Young, and Fink, 1968), since secretion has been found to be inhibited in the presence of agents such as colchicine, vinblastine, vincristine, or deuterium oxide, which interfere with microtubule function and integrity (Lacy et al., 1968; Malaisse, Malaisse-Lagae, Walker, and Lacy, 1971). Current concepts of the possible role of microtubules in the process of insulin secretion have been reviewed by Lacy and Malaisse (1973). There has recently been some interest in the role of microfilaments, and in particular of a microfilamentous cell web (Orci, Gabbay, and Malaisse, 1972; Van Obberghen, Somers, Devis, Vaughan, Malaisse-Lagae, Orci, and Malaisse, 1973), in the insulin secretory process, since cytochalasin B, an agent which disrupts microfilament organization, has been shown to stimulate insulin release; however, the possibility of an effect of cytochalasin B on glucose transport in islets (Lacy, Klein, and Fink, 1973) may raise some question of the specificity of this agent for microfilaments. A schematic diagram showing some of the stages which are now recognized to be involved in the biosynthesis, storage, and secretion of insulin from the B cell is shown in Fig. 2.

Two factors have been identified as essential for insulin secretion to occur by this mechanism: the presence of adequate intracellular concentrations of ATP (Coore and Randle, 1964), and the availability of extracellular calcium, the optimal calcium concentration being 2 mM (Grodsky and Bennett, 1966). ATP is required for the final secretory process itself, presumably for the movement of the granules through the cytoplasm and/or in the final membrane fusion involved in granule extrusion. In addition, however, a number of earlier steps in the pathway require ATP. These include the transfer of newly synthesized proinsulin from the rough-surfaced endoplasmic reticulum to the Golgi complex and the formation of the granules themselves (Howell, 1972). The requirement for calcium which, like that for ATP, is shared with many other secretory cell

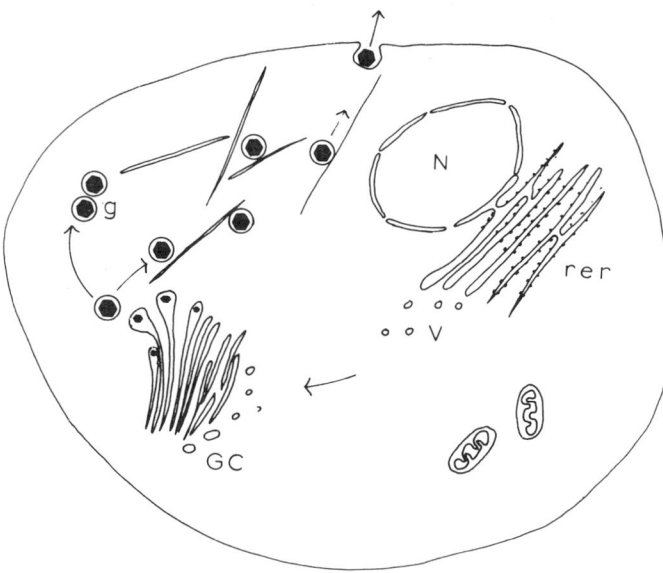

FIG. 2. Diagrammatic representation of some of the events involved in the biosynthesis, storage, and secretion of insulin. Proinsulin biosynthesis occurs in the rough-surfaced elements of the endoplasmic reticulum (rer), and the newly synthesized protein is then transported via transfer vesicles (V) to the Golgi complex (GC), where conversion of proinsulin to insulin is initiated and storage granule (g) formation occurs. Insulin is stored in the granules as zinc insulin in crystalline or paracrystalline form. Transfer of granules to the plasma membrane involves microtubules, and the final secretion occurs by fusion of granule and plasma membranes with liberation of the granule contents. Nucleus (N).

types appears to be specific for the secretory process itself, since insulin biosynthesis and the rate of conversion of proinsulin to insulin in the earlier stages of the pathway leading to granule formation are relatively unaffected by the absence of extracellular calcium (Steiner, Kemmler, Clark, Oyer, and Rubenstein, 1972).

The dynamics of insulin release have been investigated during stimulation of secretion by a variety of stimuli including glucose, theophylline, and sulfonylureas (Curry, Bennett, and Grodsky, 1968; Curry, 1971; Grodsky, 1972). With glucose as the stimulant there is an initial rapid release of insulin which subsides within approximately 2 to 5 min, followed by a second, slower release phase which is maintained over a 60-min period or until the glucose concentration is decreased. Cyclic AMP has been shown to potentiate the effect of glucose on both phases of release (Burr, Balant, Stauffacher, and Renold, 1970; Burr, Kanazawa, Marliss, and Lambert, 1971). On the other hand, sulfonylureas stimulate only the first phase of secretion under comparable conditions (Fussganger, Goberna, Hinz, Jaros, Karsten, Pfeiffer, and Raptis, 1969).

The significance of these two phases of secretion seen in response to glucose is uncertain. It has been suggested that the pattern may have a morphological basis in that granules already associated with microtubules are secreted rapidly in a first phase, while the second phase consists of exocytosis of granules which become associated with the microtubule system during prolonged stimulation (Howell and Lacy, 1971). Grodsky (1972) proposed a model showing the dynamic pattern as representing release of insulin from two separate but interrelated compartments; these two proposals do not appear incompatible.

C. Cyclic AMP and Insulin Biosynthesis

The dramatic effects of glucose in increasing rates of proinsulin or insulin biosynthesis in isolated islets are well established (Howell and Taylor, 1966; Steiner et al., 1967; Lin and Haist, 1969), but until recently relatively little attention has been devoted to the effects of cyclic nucleotides on rates of biosynthesis.

Tanese, Lazarus, Devrim, and Recant (1970) and Lin and Haist (1973) have shown that theophylline, caffeine, and dibutyryl cyclic AMP stimulate the synthesis of proinsulin and insulin in isolated rat islets of Langerhans, while Schatz, Maier, Hinz, Nierle, and Pfeiffer (1973) have reported similar effects of glucagon, cyclic AMP, and dibutyryl cyclic AMP in isolated mouse islets. The maximal stimulation of proinsulin biosynthesis by cyclic AMP was less than that produced in response to high glucose concentration (Schatz et al., 1973), so that it seems unlikely that cyclic AMP plays as dominant a role as glucose in the regulation of rates of proinsulin biosynthesis, or that the effects of glucose on biosynthesis are mediated through cyclic AMP.

The effect of cyclic AMP on insulin biosynthesis appears to be dependent on the presence of glucose, since no effect of cyclic AMP was observed in the absence of glucose (Schatz et al., 1973; Lin and Haist, 1973) or in the presence of pyruvate (Lin and Haist, 1969). It seems unlikely that the effects of cyclic AMP on insulin biosynthesis are merely reflections of increased rates of secretion, since secretion and biosynthesis are stimulated to different extents with different agents. Thus, theophylline is a more effective stimulus to secretion than glucagon, but glucagon has a more dramatic effect on biosynthesis (Schatz et al., 1973). Little information is so far available about the mechanism of action of cyclic AMP in stimulating insulin biosynthesis, and this aspect of the role of cyclic AMP in the B cell will not be considered further in this review.

D. Cyclic AMP and Insulin Secretion

Although glucose has long been recognized as the major physiological stimulus to insulin secretion, hormones which activate adenylyl cyclase in a variety of tissues have also been shown to stimulate insulin release, including glucagon (Samols, Marri, and Marks, 1965; Vecchio, Luyckx, Zahand, and Renold, 1966), corticotropin (Lebovitz and Pooler, 1967; Sussman and Vaughan, 1967), and thyrotropin (Malaisse, Malaisse-Lagae, and Mayhew, 1967a). Inhibitors of cyclic nucleotide phosphodiesterase activity such as theophylline (Turtle and Kipnis,

1967; Lebovitz and Pooler, 1967), caffeine (Lambert, Jeanrenaud, and Renold, 1967; Ashcroft, Bassett, and Randle, 1972a), and 3-isobutyl-1-methylxanthine (Montague and Cook, 1971; Cooper, Ashcroft, and Randle, 1973) increase rates of insulin release. Suggestive evidence that the effects of these agents are related to an increase in the intracellular concentration of cyclic AMP in the B cell has been provided by the observations that cyclic AMP and dibutyryl cyclic AMP themselves stimulate insulin secretion (Sussman and Vaughan, 1967; Malaisse et al., 1967a; Levine, Oyama, Kagan, and Glick, 1970).

E. A Cell

The intracellular pathway involved in the formation of storage granules in A cells (Howell, Hellerström, and Tyhurst, 1974b) is similar to that in B cells, and glucagon is apparently synthesized via a higher-molecular-weight protein which may be converted relatively slowly into glucagon (Hellerström, Howell, Edwards, and Andersson, 1973). The secretion of glucagon involves exocytosis of A cell granules (Gomez-Acebo, Parrilla, and Candela, 1968; Esterhuizen and Howell, 1970), but in contrast to the B cell this secretion is *stimulated* by colchicine or vinblastine, perhaps suggesting a rather different role for microtubules in glucagon secretion (Edwards and Howell, 1973). Furthermore, both removal of calcium and ATP depletion increase rates of glucagon release (Edwards and Taylor, 1970), in contrast to their inhibitory effects on insulin secretion.

F. Cyclic AMP and Glucagon Secretion

In contrast to insulin secretion, glucagon release is inhibited in the presence of high glucose concentration but is stimulated at low glucose levels (Vance, Buchanan, Challoner, and Williams, 1968a; Nonaka and Foa, 1969). Theophylline and epinephrine have been shown to stimulate glucagon release from islets isolated from mouse (Chesney and Schofield, 1969), rat (Leclerq-Meyer, Brisson, and Malaisse, 1971), and guinea pig (Howell, Edwards, and Montague, 1974a), and from the isolated perfused canine pancreas (Iversen, 1973). The effects of theophylline and epinephrine appear to be mediated by an increase in intracellular concentration of cyclic AMP, since dibutyryl cyclic AMP has also been shown to stimulate glucagon release (Howell et al., 1974a). There is thus increasing evidence that cyclic AMP may play an important role in the regulation of glucagon secretion. Studies of the possible involvement of cyclic AMP in the regulation of glucagon biosynthesis have not so far been reported.

III. COMPONENTS OF THE CYCLIC AMP SYSTEM IN ISLETS OF LANGERHANS

In this section the general characteristics of the components of the cyclic AMP system in islet cells will be described. The effects on these components of agents which affect insulin release will be discussed in detail in Section IV.

A. Adenylyl Cyclase

Cytochemical localization of adenylyl cyclase in islets of Langerhans using adenylyl imidodiphosphate as substrate and a lead phosphate capture procedure (Howell and Whitfield, 1972) has shown the enzyme to be distributed apparently uniformly in the plasma membrane of both A and B cells. Activity was not detected in any other organelles (Fig. 3a). In studies of granules undergoing secretion by exocytosis, it was clear that the granule membrane was devoid of cyclase activity even at the moment of its fusion with the plasma membrane. Despite the fusion of granule and plasma membranes during exocytosis, the two are clearly enzymatically distinguishable in this tissue (Fig. 3b, c).

Two completely separate assay procedures have been used in biochemical studies of islet adenylyl cyclase: (1) Broken cell systems (homogenates or subcellular fractions) are incubated in the presence of labeled ATP, either ^{14}C or ^{3}H in the purine ring or ^{32}P at the alpha phosphate position, and the labeled cyclic AMP formed is then separated by one of several methods, including thin-layer chromatography (Atkins and Matty, 1971; Davis and Lazarus, 1972), ion-exchange chromatography on Dowex (Goldfine, Roth, and Birnbaumer, 1972; Kuo, Hodgins, and Kuo, 1973), or on columns of neutral alumina (Howell and Montague, 1973). (2) In surviving cell techniques, islets are preincubated with ^{14}C-adenosine or ^{3}H-adenine to label ATP within the cell, and after further incubation of the tissue in the presence of a secretory stimulus, labeled cyclic AMP formed from this pool of ATP is separated and its radioactivity determined (Miller, Wright, and Allen, 1972; Grill and Cerasi, 1973).

Properties of adenylyl cyclase have been investigated in obese mouse islets (Atkins and Matty, 1971), normal mouse islets (Davis and Lazarus, 1972), and normal rat islets (Howell and Montague, 1973), islets isolated from rats pretreated with pilocarpine (Kuo et al., 1973), and hamster (Rosen, Hirsch, and Goren, 1971) and human islet cell tumors (Goldfine et al., 1972). In general, the characteristics of the enzyme are similar in these different species: It is predominantly (80 to 94%) particulate, being sedimented with a force of 10,000 $\times$ g for 10 min. That proportion of the enzyme which is not sedimentable even by a force of 105,000 $\times$ g for 60 min is no longer responsive to glucagon, although stimulation by fluoride remains (Howell and Montague, 1973), and this may represent enzyme solubilized from the plasma membrane rather than a completely different form. The K_m for ATP lies in the range 1 to 4 $\times$ 10^{-4}M, the pH optimum being close to 7.6. Comparison of activities from species to species, and with different tissues, is difficult because of the varying ATP concentrations and conditions used in the assays; in general, activities appear to be of the same order of magnitude as those reported in other mammalian systems.

As is the case in other tissues, nucleotides apparently play a role in the modulation of adenylyl cyclase activity in islets of Langerhans. Thus GTP or CTP increases both basal and hormone-stimulated levels of adenylyl cyclase in normal rat islets (Howell and Montague, 1973; Thompson, Johnson, and Williams, 1973) and in islets from rats pretreated with pilocarpine (Kuo et al., 1973). In view of

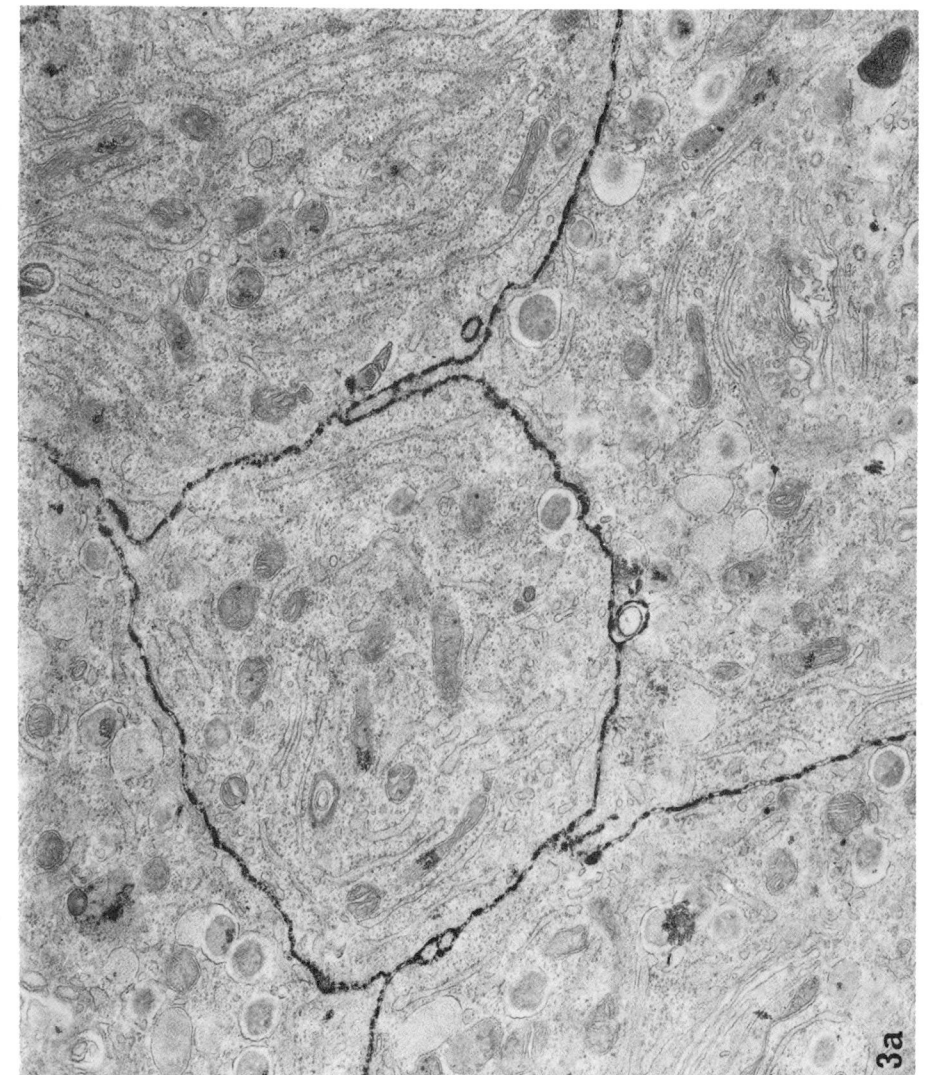

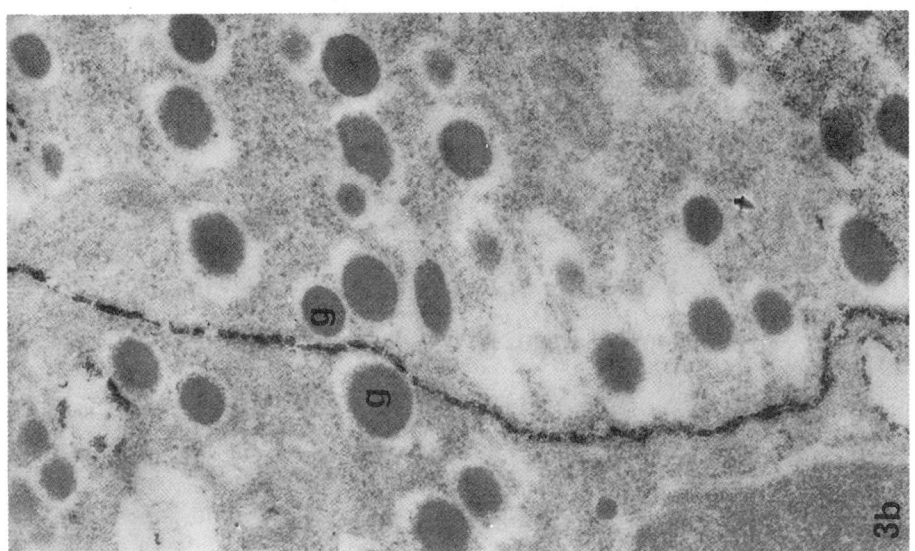

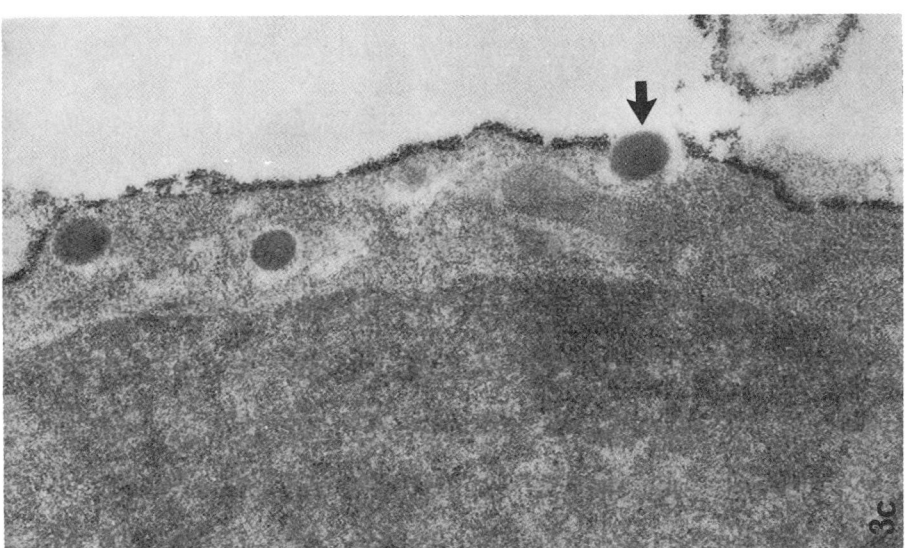

FIG. 3. Cytochemical localization of adenylyl cyclase in rat islets of Langerhans. (a) Portions of B cells incubated with 0.5 mM adenylyl imidodiphosphate (AMP-PNP) and 10 mM sodium fluoride in the presence of lead nitrate. Lead pyrophosphate is precipitated apparently evenly along the plasma membranes and in the intercellular clefts, but there is no evidence of a specific reaction product over any other organelles. Uranyl acetate, ×19,000. (b, c) Portions of a B cell showing granules (g) lying very close to the plasma membrane without fusing with it (b), while in another part of the same cell (c) plasma membrane and granule membrane have fused during exocytosis (arrow). The granule membrane recently incorporated into the plasma membrane shows no adenylyl cyclase activity. Uranyl acetate, ×36,000. Reproduced with permission from *Journal of Histochemistry and Cytochemistry*, 20, 873 (1972), © The Histochemical Society.

the possibility that the effects of some agents on adenylyl cyclase in islets might be demonstrable only in the presence of GTP, Howell and Montague (1973) have assessed the effects of most insulin secretagogues both in the presence and absence of this nucleotide. In no case did the presence of GTP unmask a previously unobserved activation of islet adenylyl cyclase. In common with all other mammalian adenylyl cyclase systems, islet cell adenylyl cyclase in broken cell preparations is markedly stimulated (usually three- to fourfold) by the addition of 10 mM sodium fluoride. The mechanism of this effect of fluoride is uncertain, but it seems unlikely to play an important role in the physiological regulation of islet cell adenylyl cyclase activity.

Adenylyl cyclase has also been identified in islets isolated from streptozotocin-treated guinea pigs which are composed of 70% A cells and has properties similar to those seen in normal islets composed of 70 to 80% B cells, the K_m for ATP being 3×10^{-4}M and the pH optimum 7.6. Adenylyl cyclase from these A cell-rich islets is stimulated 3.5-fold by 10 mM fluoride (Howell et al., 1974a).

B. Phosphodiesterase

Cyclic AMP phosphodiesterase activity has been investigated in islet cell tumors of the Syrian hamster (Goldfine, Perlman, and Roth, 1971; Rosen et al., 1971; Schubart, Udem, Baum, and Rosen, 1973) and in islets isolated from guinea pigs (Sams and Montague, 1972), normal mice (Atkins and Matty, 1971; Ashcroft, Randle, and Täljedal, 1972b; Bowen and Lazarus, 1973), and obese hyperglycemic mice (Atkins and Matty, 1971). Subcellular distribution studies have indicated that both soluble and particulate activities exist, although most (70%) of the activity appears to be soluble. There is general agreement on the pH optimum of the phosphodiesterase activity in islet cells, the recorded values being 8.2 (Bowen and Lazarus, 1973), 8.0 (Atkins and Matty, 1971), and 8.7 (Sams and Montague, 1972). Several investigators have obtained K_m values for cyclic AMP in the micromolar range: 2 µM (Goldfine et al., 1971), 3 µM (Sams and Montague, 1972), 9 µM (Bowen and Lazarus, 1973), and 10 µM (Ashcroft et al., 1972b). However, detailed electrophoretic (Schubart et al., 1973) and kinetic (Ashcroft et al., 1972b; Sams and Montague, 1972) studies of the enzyme in islet cells have given evidence of distinct forms. The K_m values of the enzyme activities for cyclic AMP were 10 µM and 500 µM in mouse islets (Ashcroft et al., 1972b), 3 µM and 30 µM in guinea pig islets (Sams and Montague, 1972), and 2, 30, and 160 µM in an islet cell tumor (Schubart et al., 1973). Since the intracellular concentration of cyclic AMP in islets of Langerhans lies in the micromolar range, studies have been largely confined to the low K_m phosphodiesterase activity.

C. Protein Kinase

Protein kinase activity has been demonstrated in islets of Langerhans isolated from rat (Montague and Howell, 1972; Dods and Burdowski, 1973), guinea pig

(Montague and Howell, 1972), mouse (Davis and Lazarus, 1973), and cod (Davis and Lazarus, 1973), and in an islet cell tumor (Schubart et al., 1973). Protein kinase activity has also been demonstrated in A cell-rich islets (Howell et al., 1974a).

The subcellular distribution of the enzyme has been investigated in rat islets of Langerhans (Montague and Howell, 1972). The major portion of the activity was found in the postmicrosomal (105,000 $\times$ g for 60 min) supernatant, although significant activity was also demonstrated in the granule and microsomal fractions. Two approaches have been made to the study of islet cell protein kinase. In the first the enzyme has been partially purified from islet cells and its characteristics investigated using histone as an exogenous phosphate acceptor (Montague and Howell, 1972), and in the second approach protein kinase activity in subcellular fractions of islet cells has been investigated using the endogenous substrates present in the same fractions (Davis and Lazarus, 1973; Dods and Burdowski, 1973).

The general characteristics of the soluble enzyme from rat islets of Langerhans, with histone as substrate, are very similar to those of protein kinases extracted from other mammalian tissues. Thus the islet enzyme is half-maximally activated by cyclic AMP at a concentration of 1.15×10^{-8} M; its apparent K_m for cyclic AMP and ATP are 1.1×10^{-8} M and 1.1×10^{-5} M, respectively (Montague and Howell, 1972). These values are similar to those obtained for protein kinase from, for instance, brain (Miyamoto, Kuo, and Greengard, 1969) and skeletal muscle (Reimann, Walsh, and Krebs, 1971).

The mechanism by which cyclic AMP stimulates the phosphorylating activity of islet cell protein kinase appears to be similar to that for the enzymes isolated from liver, adrenal cortex, and skeletal muscle (Erlichman, Hirsch, and Rosen, 1971; Gill and Garren, 1971; Reimann et al., 1971). Thus, islet cell protein kinase consists of two subunits, one of molecular weight ~90,000, which binds cyclic AMP (regulatory subunit), and the other of molecular weight ~75,000, which possesses protein kinase activity (catalytic subunit). In the presence of cyclic AMP, the regulatory subunit binds the nucleotide and becomes dissociated from the catalytic subunit. The catalytic subunit is then no longer dependent on cyclic AMP for its activity and exhibits maximal kinase activity in the dissociated state (Montague and Howell, 1972). No evidence has yet been obtained for multiple molecular forms of the enzyme in islet cells, although this may be a reflection of the small amounts of tissue available for analysis rather than the absence of such forms.

D. Phosphoprotein Phosphatase

Phosphoprotein phosphatase activity has been demonstrated in islets of Langerhans (Davis and Lazarus, 1973; Dods and Burdowski, 1973). Although systematic studies of the activity of this enzyme have not yet been performed, there appears to be a high activity of the enzyme in both soluble and particulate fractions.

IV. INTERACTION OF SECRETAGOGUES WITH COMPONENTS OF THE B CELL CYCLIC AMP SYSTEM

In this section the acute effects of a variety of insulin secretagogues on adenylyl cyclase, cyclic nucleotide phosphodiesterase, and cyclic AMP levels in islets of Langerhans are described with a view to determining the possible involvement of cyclic AMP in eliciting the effects of each agent. The major findings are summarized in Table 1.

TABLE 1. *Effects of some agents which alter rates of insulin secretion on components of the B cell cyclic AMP system*

	Effect on			
Agent	Adenylyl cyclase	Phosphodi-esterase	Cyclic AMP levels	Insulin secretion
Glucose	0	0	?	↑
Leucine	0	?	?	↑
Arginine	0	0	?	↑
Tolbutamide	↑	↓	↑	↑
Glibenclamide	0	↓	↑	↑
Diazoxide	0	0	↓	↓
Methylxanthines	0	↓	↑	↑
Prostaglandins E_1, E_2	↑			↑
Glucagon	↑	0	↑	↑
Secretin	↑	0		↑
Pancreozymin	↑	0		↑
Corticotropin	↑	0		↑
Acetylcholine	↑	0		↑
Epinephrine	↓	0	↓	↓
Norepinephrine	↓	0	↓	↓

0 indicates no effect on parameter concerned.
↑ indicates stimulation of enzyme activity or secretion, or elevation of cyclic AMP level.
↓ indicates inhibition of enzyme activity or secretion, or reduction of cyclic AMP level.
? indicates controversial result, discussed in detail in the text.
Blank indicates information not available.

A. Glucose

It is a question of crucial importance to determine whether the effects of glucose on insulin secretion are mediated via a direct short-term activation of adenylyl cyclase, as postulated by Cerasi, Efendic, and Luft (1969). Results of studies using broken cell preparations have so far been unequivocal: In none of five reported studies has a correlation been observed between an increase of glucose concentration of the incubation medium in the physiologically important range of 5 to 20 mM and a stimulation of adenylyl cyclase activity (Atkins and Matty, 1971; Davis and Lazarus, 1972; Levey, Schmidt, and Mintz, 1972; Howell and Montague,

1973; Kuo et al., 1973). It would be valuable to determine the effects of glucose on islet adenylyl cyclase over very short periods (30 sec to 1 min) if this were technically feasible, since an early spike of activity could be masked by the overall enzyme activity found during the 10- to 30-min assay periods used in these studies, but limitations of tissue availability mean that negligible quantities of cyclic AMP would be formed in such short incubations.

The results of two studies using intact cell preparations of rat islets for adenylyl cyclase estimations (see Section III,A) have been conflicting. Miller et al. (1972) found no direct effect of glucose on adenylyl cyclase activity during 15 min of incubation, although glucose was found to affect the rate of conversion of ^{14}C-adenosine to ^{14}C-ATP during the 1-hr prelabeling period. By contrast, Grill and Cerasi (1973), using a technique involving 60 min of prelabeling with ^{3}H-adenine, showed that after 3 min postlabeling incubation in the presence of 5.0 mg/ml of glucose, there was a significant increase in the label present in cyclic AMP. If incubations were performed in the presence of a phosphodiesterase inhibitor (3-isobutyl-1-methylxanthine), the effect of glucose was enhanced. It would seem worthwhile to examine the possibility that effects of glucose on the specific activity of the labeled pool of ATP might affect results obtained in this type of experiment. Longer-term effects of glucose on adenylyl cyclase, which may be observed after a period of some hours exposure to glucose, and the overall role of cyclic AMP in the regulation of glucose-stimulated insulin release are discussed further in Section VI.

Effects of glucose on phosphodiesterase activity have been investigated in enzyme extracted from mouse (Ashcroft et al., 1972b; Bowen and Lazarus, 1973) and guinea pig (Sams and Montague, 1972) islets, and from an islet cell tumor (Schubart et al., 1973). There was no evidence of a direct effect of glucose in inhibiting phosphodiesterase activity in any of these experiments.

The effect of glucose on cyclic AMP concentrations in isolated islets of Langerhans has been extensively investigated. Kipnis (1970) and Montague and Cook (1971) found that 30 or 60 min of incubation with 20 mM glucose did not affect cyclic AMP levels, a result confirmed in a perifusion system in which observations were made from 0.5 to 60 min by Cooper et al. (1973). However, significantly elevated levels of cyclic AMP were observed 2 and 20 min after exposure of islets to glucose in a perifusion system by Charles, Fanska, Schmid, Forsham, and Grodsky (1973a), while Selawry, Marcks, Fink, Lavine, Cresto, and Recant (1973) showed in a static incubation system that glucose increased cyclic AMP levels. Very recently Hellman, Idahl, Lernmark, and Täljedal (1974) have also made estimations of cyclic AMP levels after incubation of islets from obese hyperglycemic mice with 3 or 20 mM glucose, in the presence or absence of a phosphodiesterase inhibitor. In the absence of inhibitor, glucose did not affect cyclic AMP levels; but if 3-isobutyl-1-methylxanthine was present throughout, then 20 mM glucose significantly increased islet cyclic AMP concentrations in comparison to those seen with 3 mM glucose. However, the poor correlation between changes of cyclic AMP level and alteration of rates of secretion in these and other experiments led the authors to suggest that glucose does not exert its effects on secretion directly by increasing cyclic AMP concentrations in the B cells.

Effects of glucose on protein kinase have also been investigated. Montague and Howell (1972) and Steiner (1972) reported a lack of effect of glucose on cyclic AMP-dependent protein kinase activity in extracts of rat islets, while in experiments in which protein kinase was assayed immediately after incubation of intact islets in the presence of high glucose concentrations there was no change in protein kinase activity under conditions in which effects of phosphodiesterase inhibitors and stimulants of adenylyl cyclase were readily detectable (Montague and Howell, 1973).

Thus, despite disturbing anomalies in the data relating to effects of variation of glucose concentration on cyclic AMP levels, the balance of evidence at present suggests that glucose does not exert its effects on insulin secretion via activation of adenylyl cyclase or inhibition of phosphodiesterase with consequent elevation of cyclic AMP levels, nor via a direct activation of protein kinase.

B. Amino Acids

Several amino acids have been shown to stimulate insulin release *in vivo* (Floyd, Fajans, Conn, Knopf, and Rull, 1966) and *in vitro* (Malaisse and Malaisse-Lagae, 1968; Lambert, Jeanrenaud, Junod, and Renold, 1969; Milner, 1970). Leucine and arginine are the most effective in stimulating secretion and have been extensively investigated, although their mechanism of action is still largely unknown. Metabolism by the B cell would not appear to be an essential component of their action, since nonmetabolizable analogues stimulate secretion (Lambert et al., 1969) while metabolites of the amino acids do not stimulate secretion (Knopf, Fajans, Floyd, and Conn, 1963). Furthermore, some amino acids which stimulate release are not readily metabolized by the B cell (Hellman, Sehlin, and Täljedal, 1971a). Interaction of amino acids with components of the cyclic AMP system has been considered as a possible site of action (Milner, 1970; Efendic, Cerasi, and Luft, 1972). However, none of the amino acids investigated, including leucine and arginine, has any stimulatory effect on islet cell adenylyl cyclase (Rosen et al., 1971; Levey et al., 1972; Howell and Montague, 1973; Kuo et al., 1973). Phosphodiesterase activity of islet extracts does not appear in general to be affected by amino acids (Ashcroft et al., 1972b; Kuo et al., 1973), although inhibition of the enzyme by leucine was reported by Sams and Montague (1972). Cyclic AMP levels in islet cells were shown to be elevated in the presence of arginine (Charles, Lawecki, and Grodsky, 1973b) and leucine (Selawry et al., 1973), although Montague and Howell (1973) could demonstrate no change in the cyclic AMP-dependent protein kinase activity of islets incubated with either of these amino acids. Further investigation will be required before a final evaluation can be made of the role of cyclic AMP in the insulin secretory responses to amino acids.

C. Sulfonylureas

The hypoglycemic sulfonylureas exert their major effects by stimulating the release of insulin from the B cell, and are effective secretagogues both *in vivo* (Yalow, Black, Villazon, and Berson, 1960) and *in vitro* (Coore and Randle, 1964). The question of whether sulfonylurea derivatives can penetrate the B cells has been investigated by Howell and Lacy (1969) and by Hellman et al. (1971b), who showed that glibenclamide apparently does enter the cells. However, Hellman et al. (1971b) found that tolbutamide did not penetrate into islet cells, at least during the time scale of its secretory effects. Therefore, either the two sulfonylureas have separate sites of action, on the cell membrane (tolbutamide) or within the cell (glibenclamide), or alternatively penetration is not essential for eliciting their effects.

Since the insulin-releasing sufonylureas appear to be potent inhibitors of phosphodiesterase activity (see below), their use in adenylyl cyclase assays is likely to be difficult to interpret unless the assay incubation takes place in the presence of an already very potent phosphodiesterase block. Thus, assays containing 3-isobutyl-1-methylxanthine which inhibits islet phosphodiesterase by ~80% give no apparent effect of the sulfonylurea glibenclamide on rat islet adenylyl cyclase (Howell and Montague, 1973). On the other hand, tolbutamide apparently stimulates cyclase activity in rat islets (Kuo et al., 1973) and in an islet cell tumor (Levey et al., 1972) using an assay in which phosphodiesterase inhibition (50% effective) was achieved with theophylline. It seems possible that under these conditions the apparent stimulation of adenylyl cyclase might result from the increased inhibition of phosphodiesterase activity in the presence of tolbutamide.

Both tolbutamide (Sams and Montague, 1972; Bowen and Lazarus, 1973) and glibenclamide (Ashcroft et al., 1972b) have been shown to inhibit the activity of the low K_m phosphodiesterase isolated from mouse or guinea pig islets, and tolbutamide was shown to inhibit phosphodiesterase activity in an islet cell tumor from the Syrian golden hamster (Goldfine et al., 1971; Rosen et al., 1971).

Cyclic AMP levels are elevated following exposure of isolated islets to glibenclamide (Sams and Montague, 1974) or tolbutamide (Charles et al., 1973b) and cyclic AMP-dependent protein kinase activity is increased in islets which were previously incubated in the presence of glibenclamide or tolbutamide (Montague and Howell, 1973). It thus seems likely that the mechanism of action of the sulfonylureas may involve, at least in part, elevation of cyclic AMP levels as a a result of activation of B cell adenylyl cyclase, inhibition of phosphodiesterase, or both.

D. Methylxanthines

There is general agreement on the effects of the methylxanthines theophylline, caffeine, and 3-isobutyl-1-methylxanthine, which increase rates of insulin release, on the components of the cyclic nucleotide system in islets of Langerhans. All three agents have been shown to inhibit islet cell cyclic nucleotide phosphodiesterase activity from a variety of species, including guinea pig (Sams and Montague, 1972), mouse (Atkins and Matty, 1971; Ashcroft et al., 1972b; Bowen and Lazarus, 1973), and Syrian golden hamster (Goldfine et al., 1971; Rosen et al., 1971). The relative effectiveness of these agents as inhibitors of phosphodiesterase activity parallels their order of potency in increasing insulin secretion, intracellular cyclic AMP concentrations, and islet cell protein kinase activity; 3-isobutyl-1-methylxanthine has the most dramatic effect on all four parameters (Montague and Cook, 1971; Cooper et al., 1973; Montague and Howell, 1973).

These results suggest that the effects of methylxanthines on insulin release are related to an increased intracellular cyclic AMP concentration, secondary to phosphodiesterase inhibition.

E. Prostaglandins

The effects of prostaglandins (PG) on insulin release have been studied by several investigators: Spellacy, Buhi, and Holsinger (1971) found that $PGF_{2\alpha}$ and PGE_2 had no effect on plasma insulin levels in pregnancy in humans, while Vance et al. (1971) and Rossini, Lee, and Frawley (1971) were unable to obtain any effect of PGE_1 and PGA_1 on insulin release from isolated rat islets. The results of these studies contrast with those of Johnson, Fujimoto, and Williams (1973), who found that PGE_1, PGE_2, and $PGF_{2\alpha}$ potentiated insulin release in response to glucose from isolated rat pancreatic islets. Stimulatory effects of PGE_1 on insulin release in mouse (Bressler, Vargas-Cordon, and Lebovitz, 1968) and dog (Lefebvre and Luyckx, 1972) have also been reported.

The characteristics of the secretory response to prostaglandins resemble those seen with agents that act through cyclic AMP in that the stimulatory effects of PGE_1 and PGE_2 on release are dependent on the glucose concentration of the incubation medium (Johnson et al., 1973).

Studies on rat islet cell adenylyl cyclase have shown that PGE_1, E_2, A_1, and $F_{1\alpha}$ increase its activity (Howell and Montague, 1973; Johnson et al., 1973; Kuo et al., 1973; Thompson et al., 1973), the order of potency paralleling the relative effectiveness of the various prostaglandins in stimulating insulin release. No studies have been reported on the effects of prostaglandins on phosphodiesterase activity or on cyclic AMP levels, but the information available at present suggests that prostaglandins may stimulate insulin release by activation of adenylyl cyclase in the B cell, thereby increasing intracellular cyclic AMP concentrations.

F. Hormones and Neurohumoral Agents

Peptide hormones which have been shown to stimulate rates of insulin secretion and to activate islet cell adenylyl cyclase include glucagon (Rosen et al., 1971; Goldfine et al., 1972; Howell and Montague, 1973), corticotropin (Kuo et al.,

1973), secretin (Davis and Lazarus, 1972; Thompson et al., 1973), and pancreozymin (Davis and Lazarus, 1972). Furthermore, glucagon has been shown to increase islet cell cyclic AMP concentrations (Turtle and Kipnis, 1967; Montague and Cook, 1971; Selawry et al., 1973) and to increase protein kinase activity in islet cells (Montague and Howell, 1973).

Acetylcholine increases rates of insulin secretion (Malaisse, Malaisse-Lagae, Wright, and Ashmore, 1967e; Vance et al., 1971) and increases adenylyl cyclase activity in islet cells from pilocarpine-pretreated rats (Kuo et al., 1973), suggesting a possible role for cyclic AMP in the response to this agent.

Epinephrine and norepinephrine inhibit the release of insulin in response to almost all agents tested, and this effect, which will be considered in detail in Section V, appears to be mediated through the cyclic AMP system of the B cell (Turtle and Kipnis, 1967). Thus interaction of hormones and neurohumoral agents with B cell adenylyl cyclase (presumably via specific receptors) may play an important role in mediating the effects of these agents on the insulin release mechanism.

V. ADRENERGIC RECEPTORS IN ISLETS OF LANGERHANS

Extensive studies have been made of the adrenergic receptor activities in islets of Langerhans of several species and, while some differences in results have been obtained, there is sufficient agreement to allow a generalized description of the adrenergic receptor system of both A and B cells, of its relationship to the cyclic AMP system, and its role in the secretion of glucagon and insulin.

A. B Cell

Coore and Randle (1964) were the first to demonstrate that epinephrine inhibited insulin secretion induced by glucose from pieces of rabbit pancreas, and this observation has been confirmed both *in vivo* and *in vitro* in all species so far investigated. Thus, epinephrine inhibits insulin release in response to glucose in the rat (Malaisse et al., 1967e), dog (Kosaka, Ide, Kuzuya, Niki, Kuzuya, and Okinaka, 1964), pig (Hertelendy, Machlin, Gordon, Horino, and Kipnis, 1966), sheep (Hertelendy, Machlin, and Kipnis, 1969), monkey (Kris, Miller, Wherry, and Mason, 1966), and man (Porte, Graber, Kuzuya, and Williams, 1966). Inhibition of insulin release by epinephrine was also observed in the presence of tolbutamide and glucagon (Porte et al., 1966), theophylline (Turtle and Kipnis, 1967), amino acids (Milner and Hales, 1969), or dibutyryl cyclic AMP (Malaisse, Brisson, and Malaisse-Lagae, 1970). Norepinephrine has also been shown to inhibit insulin release *in vivo* (Porte and Williams, 1966) and *in vitro* (Wong, Symchowicz, Staub, and Tabachnick, 1967; Malaisse et al., 1967e).

The effect of epinephrine on insulin release is blocked by dihydroergotamine (Coore and Randle, 1964; Loubatières, Mariani, Chapal, Taylor, Houareau, and Rondot, 1965), suggesting that adrenergic receptors similar to those responsible for mediating the contractile effects of catecholamines on smooth muscle are involved. The classification proposed by Ahlquist (1948), which separates adrenergic receptors into α and β on a functional basis, can account for the effects of catecholamines in a variety of tissues including the islets of Langerhans (Porte, 1967).

1. α-Receptor Activity

The inhibitory effects of epinephrine and norepinephrine on insulin secretion appear to be related to their ability to activate α-receptors in the B cell. Thus, the order of potency of epinephrine and norepinephrine as inhibitors of insulin release parallels their effectiveness as activators of α-receptors (Malaisse et al., 1967e; Wong et al., 1967). Furthermore, phentolamine, an agent which produces α-receptor blockade, will abolish the inhibitory action of epinephrine on insulin release although β-receptor blockade by propranolol is without effect (Porte, 1967; Malaisse et al., 1967e).

2. β-Receptor Activity

There is evidence that under appropriate conditions β-receptors in the B cell may be activated by adrenergic agents, resulting in increased rates of insulin secretion. Thus isoproterenol, a potent activator of β-receptors, has a stimulatory effect on insulin secretion (Porte, 1967; Iversen, 1973). This effect of isoproterenol was prevented by simultaneous β-blockade with propranolol but not by α-blockade with phentolamine (Porte, 1967; Iversen, 1973), supporting the suggestion that the effect is mediated by β-receptor activation.

The effects of epinephrine on the B cell do not appear to be confined to α-receptor interaction, since β-receptor effects can also be demonstrated under appropriate conditions. Thus, in the presence of α-blockade the normally inhibitory effect of epinephrine on insulin release is replaced by one of stimulation (Porte, 1967; Turtle and Kipnis, 1967). Furthermore, β-blockade with propranolol increases the inhibitory effect of epinephrine (Porte, 1967; Turtle and Kipnis, 1967), suggesting a partial interaction of epinephrine with β-receptors.

3. Adrenergic Receptors and Cyclic AMP

The relationship between adrenergic receptor activity and intracellular cyclic AMP concentrations in islets of Langerhans was first investigated by Turtle and Kipnis (1967), who demonstrated that epinephrine decreased the concentration of cyclic AMP in rat islets incubated *in vitro*; this effect was increased in the presence of β-blockade with propranolol but was overcome by α-blockade with phentolamine. The effect of epinephrine in lowering islet cell cyclic AMP levels was later confirmed by Montague and Cook (1971). These observations prompted investigations into the relationship between adrenergic agents and islet cell adenylyl cyclase activity, in islets from mice (Atkins and Matty, 1971) and rat

(Howell and Montague, 1973; Thompson et al., 1973) and rats pretreated with pilocarpine (Kuo et al., 1973). Epinephrine and norepinephrine were found to inhibit adenylyl cyclase activity, an effect which was not prevented by β-blockade with propanolol but was overcome by α-blockade with phenoxybenzamine (Kuo et al., 1973; Howell and Montague, 1973). A stimulatory effect of isoproterenol on adenylyl cyclase activity which was abolished by β-blockade was reported by Kuo et al. (1973) and Thompson et al. (1973), although Howell and Montague (1973) and Atkins and Matty (1971) could demonstrate a stimulatory effect only in the presence of α-blockade. Since it is clear that the B cell contains both α- and β-receptors, these differences in results may reflect the differential stabilities of the receptors and their coupling systems in the different preparation and assay conditions used in the studies. In general, it seems clear that α-receptor activation causes a decrease, while β-receptor activation causes an increase in B cell adenylyl cyclase activity.

4. Glucose and Adrenergic Receptor Activity

The effects of adrenergic blocking agents on glucose-mediated insulin release have been extensively investigated in order to determine whether the effect of glucose on secretion is mediated by interaction with β-adrenergic receptors as postulated by Cerasi et al. (1969). α-Receptor blockade with phentolamine has no effect on the basal rate of insulin release (Malaisse et al., 1967e; Porte, 1967; Iversen, 1973), although it increases the response of the B cell to glucose (Cerasi et al., 1969; Buse, Johnson, Kuperminc, and Buse, 1970; Cryer, Herman, and Sode, 1971). Since α-receptor blockade might be expected to increase B cell intracellular cyclic AMP concentrations, these results suggest increased effectiveness of glucose in stimulating insulin release when cyclic AMP concentrations are raised. Results concerning the effect of β-blockade on glucose-induced insulin release have been equivocal, since propranolol was found to have no effect on the basal rate of insulin release and did not appear to alter the response to glucose in the studies of Feldman and Lebovitz (1970), Robertson and Porte (1973), Allison, Chamberlain, Miller, Ferguson, Gillett, Bemano, and Saunders (1969), Iversen (1973), and Malaisse et al. (1967e), although Akerblom, Martin, and Cingolani (1969), Cerasi et al. (1969), Loubatières, Mariani, Sorel, and Savi (1971), and Raptis, Dollinger, Chrissiku, Rothenbuchner, and Pfeiffer (1973) have presented evidence that propranolol lowers the insulin secretory response to glucose *in vivo,* a result which has also been confirmed *in vitro* (Laube, Fussganger, and Pfeiffer, 1972). The reason for this discrepancy may be related to the concentrations of propranolol used, since Northrop, Ryan, and Schwartz (1973) found that the qualitative effects of propranolol on glucose-induced insulin release from isolated rat islets varied dramatically with the concentration of propranolol employed. Thus, below 20 μg/ml propranolol had no effect, at 20 μg/ml it inhibited, while at 50 μg/ml it increased insulin secretory response to glucose.

The interpretation of all these studies is open to the serious objection that the adrenergic receptors of the B cell play an important role in controlling intracellular cyclic AMP concentrations and hence rates of release. Thus it is not possible to distinguish between a primary effect of the blocking agent in inhibiting possible interaction between glucose and its postulated receptor and a secondary effect consequent on the blocking agent itself interacting with an adrenergic receptor and producing a change in the concentration of cyclic AMP. There seems at present to be no clear evidence to favor the original suggestion of Cerasi et al. (1969) that the effects of glucose on insulin secretion are mediated by an interaction of glucose with β-adrenergic receptors.

B. A Cell

The results of studies on the effect of adrenergic agents on the secretion of glucagon from pancreatic tissue *in vitro* have been contradictory. The original observations of Chesney and Schofield (1969), confirmed by Vance et al. (1971), indicated that epinephrine, norepinephrine, isoproterenol, and phentolamine had no effect on the release of glucagon from mouse or rat islets. However, Leclercq-Meyer et al. (1971), Marliss, Wollheim, Blondel, Orci, Lambert, Stauffacher, Like, and Renold (1973), and Howell et al. (1974a) have suggested that epinephrine stimulates the release of glucagon *in vitro*. Iversen (1973) has found that epinephrine, norepinephrine, and isoproterenol stimulate glucagon release from the isolated perfused canine pancreas, an effect which was suppressed when the β-blocking agent propranolol was simultaneously infused, but was not affected by the simultaneous presence of α-blockade. These latter observations suggest that the effects of catecholamines on glucagon secretion which are obtained under some conditions may be mediated by β-receptor interaction. Moreover, epinephrine has been shown to increase the adenylyl cyclase activity in a preparation of islets consisting predominantly of A cells (Howell et al., 1974a), suggesting that the pancreatic A cell may contain functional β-receptors which are related to the activation of adenylyl cyclase.

VI. THE ROLE OF CYCLIC AMP IN THE REGULATION OF INSULIN SECRETION

A. Short-Term Regulation

As discussed in Section IV, the weight of evidence available at present suggests that glucose does not exert its short-term effects on insulin secretion via direct interaction with components of the cyclic AMP system. On the other hand, there is no doubt that agents which are known to raise cyclic AMP levels (e.g., phosphodiesterase inhibitors) will increase rates of both first and second phases of secretion (Burr et al., 1970, 1971), provided that a critical concentration of glucose is present. The extensive studies of Malaisse on the characteristics of a variety of secretagogues have led him to postulate the existence of two separate types of stimuli: "primary" stimuli, which will increase rates of secretion in the absence of other agents (e.g., glucose, mannose, or leucine); and "potentiators," which will stimulate secretion only in the presence of a primary stimulus (Malaisse, 1973). It is apparent that this second group of potentiators includes agents which exert their effects through the cyclic AMP system (Curry, 1970; Montague and Cook, 1971; Cooper et al.,1973).

A decrease in concentration of cyclic AMP such as occurs in the presence of epinephrine, norepinephrine, or imidazole decreases the responsiveness of the B cell to glucose and increases the threshold concentration at which glucose becomes effective (Fig. 4); catecholamines at high concentration can abolish the secretory response to any primary stimulus. On the other hand, an increase in B cell concentration of cyclic AMP such as occurs in the presence of peptide hormones, prostaglandins, or phosphodiesterase inhibitors increases the effectiveness of a given stimulatory concentration of glucose and also lowers the threshold glucose concentration at which the response is observed: the cell becomes sensitized to the effects of primary stimuli (Fig. 4). Thus the role of cyclic AMP in

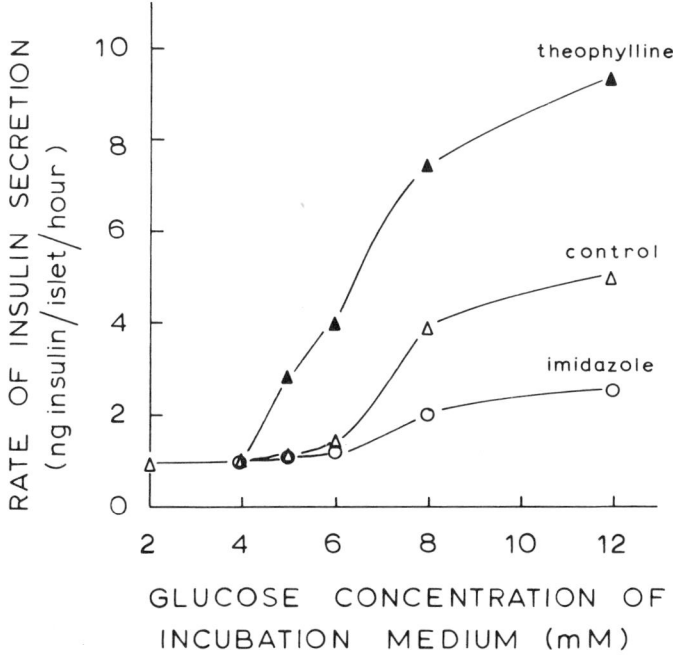

FIG. 4. Effect of theophylline (5 mM) or imidazole (13 mM) on insulin secretory responses to 2 to 12 mM glucose of islets isolated from fed rats. Theophylline, which raises islet cyclic AMP levels, lowers the threshold concentration at which glucose becomes an effective secretagogue, and increases the response to any stimulatory glucose concentration. Imidazole, which may lower islet cyclic AMP levels, slightly increases the threshold concentration at which glucose becomes effective and markedly decreases the effectiveness of any stimulatory glucose concentration.

the short-term regulation of rates of insulin release would seem to be to modify the sensitivity of the B cell to primary stimuli, in response to the short-term changes in the levels of circulating hormones or to neurohumoral agents.

B. Long-Term Regulation

There is some evidence that in addition to the acute, minute-by-minute alteration of rates of secretion, there may exist in the B cell regulatory mechanisms responsible for longer-term adaptations of the secretory response to more prolonged changes in the physiological status of the animal. These longer-term adaptations have been identified in the case of diet, hormonal imbalance, obesity, and pregnancy, and in two cases—diet and pregnancy—the possible role of cyclic AMP in achieving the observed changes has been considered in some detail.

1. *Dietary Changes*

Starvation, a dietary change of the most extreme type, has been recognized to have a profound effect on patterns of insulin secretion. Thus starvation of rats for 48 hr results in a drastic inhibition of the normal insulin secretory response to glucose (Malaisse, Malaisse-Lagae, and Wright, 1967d; Fig. 5), although the responses to tolbutamide or theophylline remain relatively unimpaired (Grey, Goldring, and Kipnis, 1970). A similar refractory response to glucose stimulation has been noted in man (Cahill, Herrera, Morgan, Soeldner, Steinke, Levy, Reichard, and Kipnis, 1966) and dogs (Vance et al., 1968b). The inhibition can be overcome within 24 hr by refeeding the animals with a carbohydrate-rich diet, or by the intraperitoneal or intravenous injection of glucose (Grey et al., 1970). An extensive study of the enzymes of glucose phosphorylation and of glycolysis in islets of starved rats has not revealed a defect which might be responsible for the impairment of the secretory response to glucose (Matschinsky, 1972; Kipnis, 1972), while the normal insulin content of the islets and their responsiveness to other stimuli exclude a general defect in the secretory mechanism as the cause of the lack of response to glucose stimulation. The fact that the two agents which most readily restored the normal response to glucose in islets from starved rats were both phosphodiesterase inhibitors (Grey et al., 1970) suggested that a defect in the cyclic AMP system might be important, and in support of this concept Howell, Green, and Montague (1973) showed that islets from fasted rats had lower adenylyl cyclase activity and reduced cyclic AMP-dependent protein kinase activity in comparison with their fed controls. Furthermore, adenylyl cyclase activity of isolated islets could be increased by prior loading of the animals with glucose (2 g/kg every hr) for 4 hr by either intraperitoneal or intravenous route (Howell et al., 1973). A direct effect of glucose on islet adenylyl cyclase activity was established by incubation of islets with 5.5 or 17 mM glucose for periods of up to 24 hr: A progressive activation was achieved over a period of 2 to 7 hr and maintained through a 24-hr period. Further investigations *in vitro* during incuba-

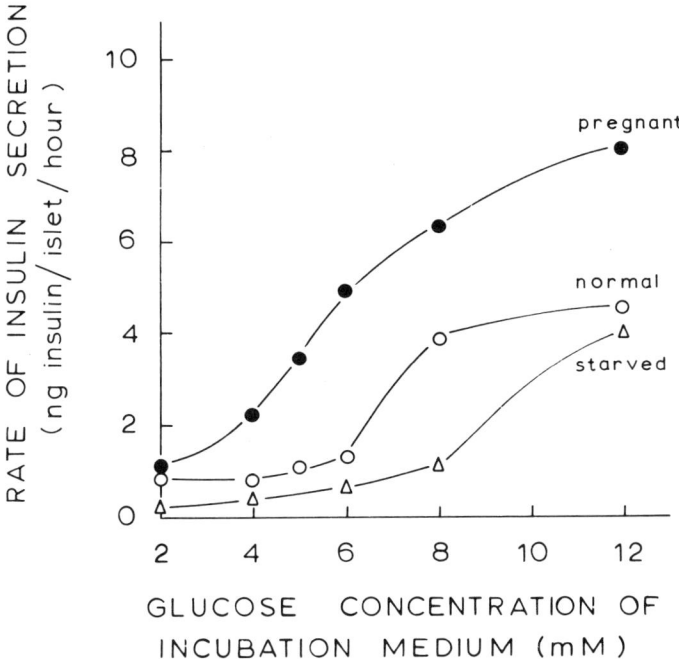

FIG. 5. Insulin secretory responses to 2 to 12 mM glucose of isolated islets of Langerhans from starved, normal fed, or 19-day pregnant rats. Islets from starved rats show a raised threshold for stimulation of secretion and a decreased response to any stimulatory glucose concentration; those from pregnant rats show a reduced threshold for glucose stimulation and an increased responsiveness to any stimulatory glucose concentration. These alterations resemble those of imidazole and theophylline (which respectively lower and raise cyclic AMP levels) on islets from normal fed rats (see Fig. 4).

tion for 4 hr with cycloheximide or actinomycin D suggested that enzyme activation rather than induction might be responsible for the increase (Howell et al., 1973). Thus the longer-term regulation of adenylyl cyclase and cyclic AMP in islets by glucose might be responsible for the altered secretory responses of islets from fasted or glucose-loaded rats. Some support for this hypothesis derives from the observation that over a range of glucose concentrations of 2 to 8 mM, the secretory response of islets from fasted rats incubated in the presence of theophylline closely resembles that of fed rats with glucose alone (Howell et al., 1973). Similarly, insulin secretion from islets isolated from fed rats and incubated with theophylline resembles the responses of islets from glucose-loaded fed rats incubated with glucose alone. Finally, incubation of islets from fed rats with imidazole, a stimulant of islet phosphodiesterase, which would be expected to lower B cell cyclic AMP levels, results in a reduced response to 2 to 8 mM glucose resembling that seen in fasting (Howell et al., 1973: Figs. 4 and 5).

Thus, with the proviso that these results refer to the physiologically important range of glucose concentrations (2 to 8 mM) and not necessarily to maximum rates of secretion, it seems that the effects of fasting and glucose loading on insulin secretion may be the result, at least in part, of alterations of B cell cyclic AMP levels, and that these changes may possibly be mediated in part by direct long-term effects of glucose in regulating B cell adenylyl cyclase activity.

2. Pregnancy

The insulin secretory responses of rat and human B cells in pregnancy are characterized by a lower threshold concentration required for glucose stimulation and an increased response to a given glucose stimulus above the threshold level (Green and Taylor, 1972), and these responses are comparable to those noted above as achieved by incubation of islets from normal rats in the presence of a phosphodiesterase inhibitor (Fig. 5). Green, Howell, Montague, and Taylor (1973) have compared some components of the cyclic AMP system in islets from pregnant rats with their nonpregnant littermates and observed elevated adenylyl cyclase activity and increased cyclic AMP levels and cyclic AMP-dependent protein kinase activity in islets from pregnant rats. Pair feeding experiments suggested that the increased cyclase activity might be a result of the higher food and particularly carbohydrate intake of the pregnant rats (Green and Taylor, 1974), although it is probable that hormonal influences, particularly of placental lactogen, progesterone, and estriol, may also be involved. These hormones have been shown to exert effects on the insulin secretory pattern of islets of animals to which they were administered (Costrini and Kalkhoff, 1971), but it is not clear whether the effects were direct ones on B cell adenylyl cyclase or indeed directly on the islets at all.

3. Hormonal Changes

A number of hormones in addition to those mentioned above have been shown to affect the insulin secretory responses following prolonged administration to intact animals, although they are without effect on isolated islets in short-term experiments. Thus, adrenalectomy diminishes the insulin secretory response to glucose, while glucocorticoid administration will increase it (Malaisse, Malaisse-Lagae, McGraw, and Wright, 1967c). A similar situation exists in the case of growth hormone which, when present at high concentration over long periods, increases rates of insulin synthesis and secretion (Martin, Akerblom, and Garay, 1968). Thyroidectomy depresses insulin secretion, but the injection of thyroxine does not enhance it (Malaisse, Malaisse-Lagae, and McGraw, 1967b). Again it seems likely that long-term regulation of islet function by these hormones may be important and could conceivably be mediated through changes in the B cell cyclic AMP system.

In addition to the altered secretory pattern of the B cells, many of the factors known to affect insulin secretion in the long term also affect other parameters of islet function, particularly their size. For instance, pregnancy (Green and

Taylor, 1972) or elevation of plasma growth-hormone levels (Martin et al., 1968) results in an increase of islet size, the increase being reflected, at least in pregnancy, in a higher number of cells in each islet as a result of an increase in mitosis. It is not known whether higher cyclic AMP levels in the B cells in these conditions might directly or indirectly increase their mitotic index to produce this hypertrophy.

C. Fetal Pancreas

While glucose is the major physiological regulator of insulin secretion from adult pancreas, islets of fetal or perinatal animals of several species including man have been shown to be unresponsive to glucose stimulation of insulin release, although they respond to stimulation by phosphodiesterase inhibitors or glucagon, or to glucose provided that phosphodiesterase inhibitors are also present. This led Lambert, Kanazawa, Burr, Orci, and Renold (1971) and Chez, Mintz, and Hutchinson (1971) to postulate that adenylyl cyclase activity was reduced or phosphodiesterase activity was increased in fetal islets so that B cell cyclic AMP levels are low. Cyclic AMP concentrations in islets of pregnant rats and their offspring during the perinatal period were determined by Mintz, Levey, and Schenk (1973), who showed that the levels in fetal islets increased twofold during the period from 3 to 24 hr after birth, the rise being significantly greater in fed than in nonfed neonatal animals. This elevated cyclic AMP level was maintained for at least 72 hr, although by 13 days the cyclic AMP levels had fallen again to those seen in islets from newborn rats 3 hr post partum, and were similar to those found in the maternal islets. Phosphodiesterase activity closely paralleled the changes in cyclic AMP level. Although adenylyl cyclase activity was not measured in this study, it seems likely that it was dramatically increased in the postnatal period in order to allow such an increase of cyclic AMP and that the higher cyclic AMP level in turn induced increases in phosphodiesterase activity. Thus in early life cyclic AMP would seem to be of major importance both in regulating acute insulin secretory responses and in establishing the mechanisms responsible for the regulation of secretion in the adult. The mechanism for this proposed activation of adenylyl cyclase in the B cell after birth might conceivably involve a direct effect of glucose on the enzyme, as postulated for adult islets (see above). Thus the effect of starvation or variation of food intake in preventing the occurrence of glucose-mediated insulin release in the newborn, and the effect of intraperitoneal glucose injection in promoting the appearance of glucose-induced secretion, has been well documented (Asplund, 1973). If glucose or gastrointestinal hormones were to activate adenylyl cyclase in fetal islets, then the observed patterns of secretion and of changes of cyclic AMP level could readily be explained. Direct investigations of glucose metabolism in the islets of newborn rats by Asplund and Hellerström (1972) suggest that uptake, phosphorylation, and oxidation of glucose were not deficient, although a relative inability of fetal islets to oxidize glucose labeled in the C_1 position has been reported and attributed to a deficiency in the pentose phosphate pathway in these islets (Heinze and Steinke, 1971). Since an adult-type pattern of insulin release in response to glucose is readily observed in neonatal islets provided that a phosphodiesterase inhibitor is present, it seems unlikely that a basic defect in, for instance, the pentose phosphate pathway is the reason for the poor secretory response to glucose alone.

VII. MODE OF ACTION OF CYCLIC AMP IN REGULATING INSULIN SECRETION

The effects of cyclic AMP in many mammalian tissues appear to be related to the ability of the nucleotide to activate cyclic AMP-dependent protein kinases (Miyamoto et al., 1969; Walsh and Ashby, 1973) and, as noted above (Section III), comparable enzyme activity has been demonstrated in both the A and B cells of the islets of Langerhans. Incubation of intact islets with glucagon, theophylline, caffeine, or 3-isobutyl-1-methylxanthine, agents which raise cyclic AMP concentrations in islets of Langerhans and stimulate hormone release, have been shown to increase the protein kinase activity of islet homogenates where this is determined immediately after the incubation period, whereas diazoxide and epinephrine, which lower cyclic AMP and inhibit insulin release, decreased protein kinase activity (Montague and Howell, 1973). Thus, cyclic AMP may exert its effect on insulin secretion by increasing the activity of a cyclic AMP-dependent protein kinase in islets, thereby promoting the phosphorylation and altering the activity of one or more rate-determining components of the secretory mechanism. The interplay between the activities of protein kinase and phosphoprotein phosphatase in regulating the state of phosphorylation of the components of the secretory process may provide a mechanism whereby the rate of secretion could be regulated on a minute-to-minute basis.

Some evidence has been obtained for the presence in islet cells of protein kinase substrates whose phosphorylation can be stimulated by cyclic AMP. Thus proteins present in $105,000 \times g$ for 60 min supernatants, as well as proteins in the microsomal, granule, and membrane fractions, have all been shown to be phosphorylated *in vitro* in the presence of cyclic AMP (Montague and Howell, 1972; Dods and Burdowski, 1973; Davis and Lazarus, 1973). While it is tempting to attribute physiological significance to these findings, it should be noted that islet cell protein kinase will phosphorylate serine and threonine residues in many proteins *in vitro* (Montague and Howell, 1973). The characteristics of protein kinase substrates which may be important in the regulation of insulin secretion demand a rapid phosphorylation followed by dephosphorylation to terminate secretion after stimulation, and this will make the identification of proteins which are important in this respect even more difficult. Identification and characterization of these substrates and of their role in secretion must await further research. Cellular components which appear to play an essential role in secretion and which might be considered as possible substrates for the protein kinase are regulatory enzymes of glucose metabolism, protein components of the granule and plasma membranes, components of the microtubular/microfilamentous system, and calcium-binding phosphoproteins.

A. Cyclic AMP as a Modulator of Islet Cell Metabolism

Since cyclic AMP potentiates glucose-induced insulin release (Malaisse et al., 1967a) and the effect of glucose on the release process may be related to glucose metabolism (Randle, Ashcroft, and Gill, 1968), the possibility exists that cyclic AMP might promote glucose metabolism by the B cell (Samols, Marri, and Marks, 1966). Evidence in favor of this hypothesis was provided by the observation that agents thought to increase intracellular cyclic AMP levels were able to stimulate insulin release only in the presence of glucose (Malaisse et al., 1967a), while the enhancing effect of glucagon or theophylline upon glucose-induced insulin release was suppressed by agents such as mannoheptulose and 2-deoxyglucose, which inhibit glucose metabolism by the islet cells. It was proposed that the mode of action of cyclic AMP might be to promote glycogenolysis (Samols et al., 1966) or alternatively some rate-limiting steps in the glycolytic pathway or tricarboxylic acid cycle in B cells (Malaisse et al., 1967a). Furthermore, intracellular glucose produced from the breakdown of glycogen in islets from rats previously infused for 8 to 10 hr with glucose to increase islet cell glycogen can substitute for extracellular glucose in maintaining a secretory response to cyclic AMP (Malaisse et al., 1967a). Direct evidence that cyclic AMP might play a role in regulating glycogen levels in the B cell was subsequently provided by the studies of Idahl and Hellman (1971), who demonstrated that glucagon and theophylline reduced the content of glycogen in islets from obese-hyperglycemic mice whereas epinephrine increased the level of islet glycogen. However, the available biochemical data do not provide evidence to support the suggestion that cyclic AMP has any major effect on glucose metabolism in the B cell, apart from this role in the regulation of glycogenolysis, which is more likely to be important in the glycogen-rich B cell of the obese-hyperglycemic mouse than in normal animals. Thus, glucagon has been shown to have no significant effect on glucose oxidation as measured by $^{14}CO_2$ production from ^{14}C-glucose by mouse islets (Ashcroft, Hedeskov, and Randle, 1970) or goosefish islets (Hostetler and Williams, 1970), or on the level of glucose-6-phosphate in mouse islets (Ashcroft et al., 1970). Dibutyryl cyclic AMP was without effect on glucose oxidation or the levels of citrate, ATP, and 6-phosphogluconate in islets from the obese-hyperglycemic mouse (Idahl, 1971; Idahl, Hurme, Wahlquist and Hellman, 1971; Hellman et al., 1971a; Hellman and Idahl, 1972). Furthermore, neither glucagon nor dibutyryl cyclic AMP stimulated the respiration of isolated islets in the presence of 11 mM glucose (Hellerström and Gunnarsson, 1970). These observations do not rule out the possibility of more subtle changes in glucose metabolism by the B cell. However, since theophylline and glucagon will stimulate insulin release in the presence of certain amino acids such as leucine or glycine in the total absence of glucose, it seems unlikely that the effects of cyclic

AMP on insulin release can be accounted for solely on the basis of alterations in the metabolism of glucose by the B cell. Cyclic AMP must therefore affect the release process in such a way as to make no great extra demands on the energy supply of the cell, perhaps by increasing the efficiency of coupling between the energy-producing and the secretory mechanisms, or of the secretory mechanism itself.

The reason why cyclic AMP is unable to stimulate release in the absence of a primary stimulus such as glucose is uncertain. The effects of phosphodiesterase inhibitors on intracellular cyclic AMP concentrations are not altered by glucose concentration: 3-Isobutyl-1-methylxanthine increases cyclic AMP concentrations to the same extent whether glucose is present at 2 mM or 20 mM (Montague and Cook, 1971), while similar results have been obtained with caffeine by Cooper et al. (1973), and with theophylline by Charles et al. (1973a). Thus glucose does not appear to be essential for maintenance of B cell cyclic AMP levels. It has been suggested (Burr et al., 1970) that glucose might simply be providing a source of energy for the secretory response, since citrate or pyruvate, metabolites which by themselves did not alter secretion, were found to stimulate release from perfused minced pancreas pieces provided that theophylline was also present. It has not been possible to confirm these results using isolated rat islets of Langerhans (Green and Montague, *unpublished observations*), and it seems possible that external factors (such as potassium ions released in response to theophylline from the vast excess of exocrine tissue) may have contributed to the secretory response observed in experiments using minced pancreas. Thus glucose and other primary stimuli evidently have some specific role in the stimulation of secretion by cyclic AMP which is unrelated merely to maintenance of cyclic AMP levels or of energy supplies. The simultaneous presence of a signal for secretion of some type appears to be necessary before a cyclic AMP-induced potentiation effect can be observed.

B. Cyclic AMP and Microtubules

A role for microtubules in transporting granules from the cytoplasmic storage pool to the plasma membrane of the B cell was proposed by Lacy et al. (1968) on the basis of morphological studies and of evidence of the inhibition of insulin secretion by colchicine, a specific inhibitor of microtubule function. These results were extended by Malaisse et al. (1971) using other inhibitors of microtubule function, and microtubules have subsequently been suggested to play a role in secretion in a variety of other endocrine glands including the thyroid (Williams and Wolff, 1970), adrenal medulla (Poisner and Bernstein, 1971), and adenohypophysis (Kraicer and Milligan, 1971). It is uncertain whether microtubules act to provide a passive directional guide to granules during secretion, or whether the microtubules themselves have some contractile function in transporting the granules to the plasma membrane (see review by Lacy and Malaisse, 1973).

Rasmussen (1970) first suggested that microtubular protein (tubulin) might provide a substrate for cyclic AMP-dependent protein kinase in nervous tissue and that this phosphorylation might provide the mechanism for the control of intracellular movements by cyclic AMP via the microtubular network. The experimental basis for the hypothesis of the direct phosphorylation of tubulin by protein kinase has subsequently been confirmed (Reddington and Lagnado, 1973; Leterrier, Rappaport, and Nunez, 1974), although sufficient tubulin has not yet been obtained from isolated islets to make possible a direct estimation of the phosphorylation of this protein by protein kinase extracted from islets. A role of cyclic AMP in maintaining a pool of phosphorylated (activated) microtubule subunits or microtubules in a state which would promote secretion would certainly provide a conceivable mechanism of action of this nucleotide in the regulation of insulin secretion.

C. Cyclic AMP and Calcium

The presence of adequate concentrations of extracellular calcium is an absolute requirement for insulin secretory responses to most stimuli (Grodsky and Bennett, 1966). Since cyclic AMP also plays a role in regulating insulin secretion, it seems possible that there might be a direct relationship between calcium and cyclic AMP and this has also been proposed as a general mechanism of action of cyclic AMP by Rasmussen and Tenenhouse (1968). Malaisse and collaborators have undertaken a series of studies designed to examine the role of calcium in the insulin secretory process, by a study of the efflux and uptake of $^{45}Ca^{2+}$ during secretion from islets of Langerhans. On the basis of experiments on glucose and $^{45}Ca^{2+}$ fluxes, Malaisse (1973) has proposed that glucose inhibits the normal outward flow of calcium from the B cell, causing an increase in the intracellular calcium concentration which in turn triggers secretion, possibly by interaction with the microtubular/microfilamentous system. Dibutyryl cyclic AMP or theophylline did not modify the glucose-induced accumulation of calcium by islets under conditions where they were able to potentiate glucose stimulation of insulin release, suggesting that the effects were not directly exerted on rates of calcium uptake or efflux. However, dibutyryl cyclic AMP or theophylline was able, to a small extent, to restore the secretory response to glucose in the absence of extracellular calcium, indicating that cyclic AMP may modify the B cell in such a way that the normal supply of extracellular calcium is no longer needed for glucose to initate release. A possible effect of cyclic AMP on the intracellular distribution of calcium in islet cells was further indicated by the observation that theophylline caused a dramatic increase in calcium efflux from perifused islets in the absence of glucose when secretion was not stimulated. These observations have been interpreted as showing that cyclic AMP may promote the intracellular translocation of calcium within the B cell from an organelle-bound pool to a cytoplasmic pool of free ionized calcium, the increased concentration of which serves to activate the secretory process and becomes readily available for transport across the cell membrane (Malaisse, 1973). A model of this type requires the demonstration of calcium-binding sites within the B cell. Calcium has been shown to bind to phosphoproteins in adrenal medulla (Brooks and Siegel, 1973a), brain (Wolff and Siegel, 1972), and testes (Brooks and Siegel, 1973b), and it is possible that these proteins might be substrates for a cyclic AMP-dependent protein kinase and that alterations in their state of phosphorylation might induce changes in calcium binding to provide an increase in free ionized calcium.

It would seem to be useful to confirm the interpretation of the ^{45}Ca flux experiments in islets by direct estimation of calcium concentrations in islets at different stages of secretion. We have recently attempted a direct localization of calcium within the B cell at the ultrastructural level using unfixed, ultrathin, frozen sections cut on a dry knife, in which calcium is estimated by X-ray microanalysis. Preliminary results suggest that a proportion of the storage granules do contain calcium. Thus at least part of the calcium efflux occurring during secretion may be a result of simultaneous release of calcium and insulin in the granules. Further studies of this type, together with investigations of pools of calcium in other organelles within the B cells, will be needed before a proper understanding can be achieved of the role of cyclic AMP in B cell calcium metabolism (see also review by Berridge in this volume).

D. Membranes

Interaction of insulin storage granules with the B cell plasma membrane is an essential component of the secretory process, and it is possible that changes in the membrane proteins induced by phosphorylation could play a role in this interaction. Protein kinase activity has been demonstrated in membrane fractions from cod and mouse islets (Davis and Lazarus, 1974); the membrane-bound enzyme was unable to use exogenous protein (histone) as a substrate, although specific proteins in the membrane were readily phosphorylated and dephosphorylated, their phosphorylation being enhanced in the presence of cyclic AMP. Although formidable difficulties remain in the isolation and characterization of purified granule and plasma membrane fractions from mammalian islets, the results of these studies indicate that phosphorylation of membrane proteins in the B cell may form an essential component in the regulation of the secretory response.

E. Speculation on the Mechanism of Action of Cyclic AMP

Cyclic AMP and a primary stimulus such as glucose both contribute to the regulation of the secretory process in the B cell, the effect of each one being dependent on the presence of an adequate concentration of the other. Primary stimuli will activate the secretory mechanism provided that adequate concentrations of ATP, calcium, and cyclic AMP are available. Cyclic AMP might perhaps play its permissive role in the process by maintaining a hypothetical pool of phosphorylated intermediates at a level adequate to allow a satisfactory input into the secretory mechanism. If this pool falls below a critical level, then the secretory response to a primary stimulus cannot be maintained and the threshold concentration required for secretion to be initiated is higher. Conversely, when the size of the pool is increased, then the secretory response to a primary stimulus is enhanced and the threshold is lowered. The existence of this pool and the nature of the intermediates which might comprise it remain a matter for further investigation.

VIII. OTHER CYCLIC NUCLEOTIDES

In contrast to the extensive studies on cyclic AMP, little attention has been given to the possible involvement of other cyclic nucleotides in regulating the synthetic and secretory activities of the islets of Langerhans.

Dibutyryl cyclic GMP has recently been shown to stimulate insulin release (Voyles, Gutman, Selawry, Fink, Penhos, and Recant, 1973) and insulin synthesis (Howell and Montague, 1974) in islets of Langerhans, suggesting a possible involvement of cyclic GMP in regulating these processes.

Guanylyl cyclase activity has been demonstrated in islets of Langerhans and its characteristics investigated (Howell and Montague, 1974). The enzyme was present predominantly in particulate fractions and could be solubilized with 0.1% Triton-X100 with a resultant 2.3-fold increase in activity. The pH optimum of the activity was 7.3, and it required the presence of divalent cations, manganese being the most effective. In contrast to adenylyl cyclase, guanylyl cyclase was not stimulated by 10 mM fluoride; it was inhibited by ATP, whereas adenylyl cyclase was stimulated by addition of GTP. A comparison of the effects of insulin secretagogues on adenylyl and guanylyl cyclase activities is shown in Table 2.

TABLE 2. *Effects of some agents which alter rates of insulin secretion on islet cell adenylyl and guanylyl cyclase*

Agent	Effect on		
	Adenylyl cyclase	Guanylyl cyclase	Insulin secretion
Glucose	0	0	↑
Glibenclamide	0	0	↑
Arginine	0	0	↑
Leucine	0	0	↑
Glucagon	↑	0	↑
PGE$_1$, E$_2$	↑	0	↑
Acetylcholine	↑	↑	↑
Secretin	↑	↑	↑
Pancreozymin	↑	↑	↑
Diazoxide	0	0	↓
Epinephrine	↓	↓	↓
Norepinephrine	↓	↓	↓

0 indicates no effect.
↑ indicates stimulation of enzyme activity or of secretion.
↓ indicates inhibition of enzyme activity or of secretion.

No agent was found which uniquely stimulated guanylyl cyclase in the absence of a parallel alteration of adenylyl cyclase activity although glucagon and prostaglandins E_1 and E_2 activated adenylyl cyclase without affecting guanylyl cyclase. Secretin, pancreozymin, and acetylcholine stimulated, while epinephrine and norepinephrine inhibited, both enzymes in islet homogenates. Glucose was without effect on either activity.

Phosphodiesterase activity capable of the hydrolysis of cyclic GMP has been demonstrated in islet cells (Schubart et al., 1973), as has cyclic GMP-dependent protein kinase activity (Montague, *unpublished observations*).

Basal levels of cyclic GMP in islets were found to be 2 pmoles/mg of islet protein, which is approximately 25% of the levels of cyclic AMP. Increased levels of cyclic GMP were found in islets incubated with acetylcholine or 3-isobutyl-1-methylxanthine, although the levels were unaltered after incubation in the presence of high concentrations of glucose (Howell and Montague, 1974). Basal cyclic GMP levels in isolated islets were of the same order as those required for the activation of islet cell protein kinase activity by cyclic GMP.

These results suggest that cyclic GMP may play a role in mediating the effects of certain hormonal and neurohumoral agents on the synthetic and secretory activities of the islets of Langerhans, possibly by activation of cyclic GMP-dependent protein kinase.

IX. SUMMARY

In the light of evidence reviewed above, the following general conclusions can be made concerning the role of cyclic AMP in the physiology of the islets of Langerhans.

1. Islets of Langerhans possess all the enzymes involved in the generation, hydrolysis, and mode of action of cyclic AMP which have been observed in other mammalian cell types, the activities being present in both the A and B cells.

2. Changes in the intracellular concentrations of cyclic AMP play an important role in regulating the rate of insulin release from the B cells and of glucagon release from the A cells. Cyclic AMP may also play a part in regulating rates of insulin biosynthesis, although this function has not yet been defined in detail.

3. Primary stimuli of insulin release, such as glucose, mannose, and leucine, have no effect either on adenylyl cyclase or on phosphodiesterase activities in broken cell preparations, so that the weight of available evidence suggests that they do not exert their short-term effects on secretion directly through elevation of cyclic AMP levels.

4. The effects of certain modifiers of secretion, which require the presence of a critical concentration of primary stimulus before they are effective, are directly attributable to their effects on adenylyl cyclase or phosphodiesterase activity and on cyclic AMP levels. These agents include glucagon, corticotropin, secretin, pancreozymin, epinephrine, norepinephrine, prostaglandins, and acetylcholine, which affect adenylyl cyclase; and sulfonylureas and methylxanthines, which affect phosphodiesterase activity.

5. Changes of cyclic AMP levels in adult islets may be produced in at least two ways: (a) transient changes produced by short-term alterations in B cell adenylyl cyclase or phosphodiesterase activity; (b) longer-term adaptive changes which occur over a period of hours or days in response to maintained changes in adenylyl cyclase activity, as have been detected in starvation and in pregnancy. Blood glucose concentrations, as well as hormonal influences, may have an important role in the long-term modulation of adenylyl cyclase activity under these conditions.

6. The effects of alteration of cyclic AMP levels, whether produced by short-term regulatory or long-term adaptive changes, are the same. An increase in cyclic AMP will: (a) potentiate the effect of a primary stimulus of secretion; and (b) lower the threshold concentration at which glucose becomes an effective secretagogue. A decrease in cyclic AMP concentration will: (a) diminish and eventually abolish the secretory effect of a primary stimulus; and (b) increase the minimal concentration of glucose which is required to obtain an effect on secretion.

7. The mode of action of cyclic AMP in regulating rates of secretion appears to be related to the ability of the nucleotide to increase cyclic AMP-dependent protein kinase activity. This enzyme promotes the phosphorylation and possibly activity of rate-determining components of the secretory mechanism. Possible substrates for protein kinase in islet cells include granule and plasma membranes, microtubule protein (tubulin), and calcium-binding phosphoproteins.

8. The simple scheme proposed in Fig. 1 may reflect the situation found in fetal islets, where cyclic AMP appears to play a major role in regulating rates of insulin secretion. In the mature B cell there are superimposed additional regulatory factors, so that the rate of insulin release can be controlled by agents such as glucose or amino acids which do not exert their effects directly via cyclic AMP. Regulation of secretion by these agents is, however, still subject to modification by the prevailing cyclic AMP concentration in the B cell, and this in turn is regulated by short- and long-term changes in the hormonal and nutritional balance and physiological status of the animal.

X. ACKNOWLEDGMENTS

We thank Dr. K. W. Taylor for advice and encouragement, our colleagues for critically reviewing the manuscript, and Mrs. M. Tyhurst for patient help in its preparation. Financial assistance from the Medical Research Council, British Diabetic Association, and Hoechst Pharmaceuticals is gratefully acknowledged.

XI. REFERENCES

Ahlquist, R. P. (1948): A study of the adrenotropic receptors. *American Journal of Physiology,* 153:586–600.

Akerblom, H. K., Martin, J. M., and Cingolani, H. E. (1969): Circulating glucose, insulin, free fatty acids, and acetone bodies in rats given propranolol. *American Journal of Physiology,* 217:1690–1693.

Allison, S. P., Chamberlain, M. J., Miller, J. E., Ferguson, R., Gillett, A. P., Bemand, B. V., and Saunders, R. A. (1969): Effects of propranolol on blood sugar, insulin and free fatty acids. *Diabetologia,* 5:339–342.

Anderson, E., and Long, J. A. (1948): The hormonal influences on the secretion of insulin. *Recent Progress in Hormone Research,* 2:209–227.

Ashcroft, S. J. H., Bassett, J. M., and Randle, P. J. (1972a): Insulin secretion mechanisms and glucose metabolism in isolated islets. *Diabetes,* 21, Suppl. 2:538–545.

Ashcroft, S. J. H., Hedeskov, C. J., and Randle, P. J. (1970): Glucose metabolism in mouse pancreatic islets. *Biochemical Journal,* 118:143–154.

Ashcroft, S. J. H., Randle, P. J., and Täljedal, I. B. (1972b): Cyclic nucleotide phosphodiesterase activity in normal mouse pancreatic islets. *FEBS Letters,* 20:263–266.

Asplund, K. (1973): Effects of intermittent glucose infusions in pregnant rats on the functional development of the foetal pancreatic B cells. *Journal of Endocrinology,* 59:285–293.

Asplund, K., and Hellerström, C. (1972): Glucose metabolism of pancreatic islets isolated from neonatal rats. *Hormone and Metabolic Research,* 4:159–163.

Atkins, T., and Matty, A. J. (1971): Adenyl cyclase and phosphodiesterase activity in the isolated islets of Langerhans of obese mice and their lean litter mates: The effect of glucose, adrenaline and drugs on adenyl cyclase activity. *Journal of Endocrinology,* 51:67–78.

Bowen, V., and Lazarus, N. R. (1973): Glucose mediated insulin release: 3'5' cAMP phosphodiesterase. *Diabetes,* 22:738–743.

Bressler, R., Vargas-Cordon, M., and Lebovitz, H. E. (1968): Tranylcypromine: A potent insulin secretagogue and hypoglycemic agent. *Diabetes,* 17:617–624.

Brooks, J. C., and Siegel, F. L. (1973a): Purification of a calcium-binding phosphoprotein from beef adrenal medulla. *Journal of Biological Chemistry,* 248:4189–4193.

Brooks, J. C., and Siegel, F. L. (1973b): Calcium binding phosphoprotein: The principal acidic protein of mammalian sperm. *Biochemical and Biophysical Research Communications,* 55:710–716.

Burr, I. M., Balant, L., Stauffacher, W., and Renold, A. E. (1970): Perifusion of rat pancreatic tissue in vitro: Substrate modification of theophylline-induced biphasic insulin release. *Journal of Clinical Investigation,* 49:2097–2105.

Burr, I. M., Kanazawa, Y., Marliss, E. B., and Lambert, A. E. (1971): Biphasic insulin release from perifused cultured fetal rat pancreas. *Diabetes,* 20:592–597.

Buse, M. G., Johnson, A. H., Kuperminc, D., and Buse, J. (1970): Effect of α-adrenergic blockade on insulin secretion in man. *Metabolism,* 19:219–225.

Cahill, G. F., Jr., Herrera, M. G., Morgan, A. P., Soeldner, J. S., Steinke, J., Levy, P. L., Reichard, G. A., Jr., and Kipnis, D. M. (1966): Hormone-fuel interrelationships during fasting. *Journal of Clinical Investigation,* 45:1751–1769.

Cerasi, E., Efendic, S., and Luft, R. (1969): Role of adrenergic receptors in glucose-induced insulin secretion in man. *Lancet,* 2:301–302.

Charles, M. A., Fanska, R., Schmid, F. G., Forsham, P. H., and Grodsky, G. M. (1973a): Adenosine 3',5'-monophosphate in pancreatic islets: Glucose-induced insulin release. *Science,* 179:569–571.

Charles, M. A., Lawecki, J., and Grodsky, G. M. (1973b): Cyclic AMP in pancreatic islets. *Diabetes,* 22, Suppl. 1:297.

Chesney, T. McC., and Schofield, J. G. (1969): Studies on the secretion of pancreatic glucagon. *Diabetes,* 18:627–632.

Chez, R. A., Mintz, D. H., and Hutchinson, D. L. (1971): Effect of theophylline on glucagon and glucose-mediated plasma insulin responses in subhuman primate fetus and neonate. *Metabolism,* 20:805–815.

Cooper, R. H., Ashcroft, S. J. H., and Randle, P. J. (1973): Concentrations of adenosine 3',5'-cyclic monophosphate in mouse pancreatic islets measured by a protein-binding radioassay. *Biochemical Journal,* 134:599–605.

Coore, H. G., and Randle, P. J. (1964): Regulation of insulin secretion studied with pieces of pancreas incubated in vitro. *Biochemical Journal,* 93:66–78.

Costrini, N. V., and Kalkhoff, R. K. (1971): Relative effects of pregnancy, estradiol and progesterone on plasma insulin and pancreatic islet insulin secretion. *Journal of Clinical Investigation,* 50:992–999.

Cryer, P. E., Herman, C. M., and Sode, J. (1971): Insulin release during α-adrenergic receptor blockade: Primacy of the glycaemic stimulus. *Endocrinology,* 89:918–920.

Curry, D. L. (1970): Glucagon potentiation of insulin secretion by the perfused rat pancreas. *Diabetes,* 19:420–428.
Curry, D. L. (1971): Insulin secretory dynamics in response to slow-rise and square-wave stimuli. *American Journal of Physiology,* 221:324–328.
Curry, D. L., Bennett, L. L., and Grodsky, G. M. (1968): Dynamics of insulin secretion by the perfused rat pancreas. *Endocrinology,* 83:572–584.
Davis, B., and Lazarus, N. R. (1972): Insulin release from mouse islets. Effect of glucose and hormones on adenylate cyclase. *Biochemical Journal,* 129:373–379.
Davis, B., and Lazarus, N. R. (1973): Insulin release from pancreatic islets: Properties of a membrane bound phosphokinase from cod and mouse islets. Preceedings of 8th Congress of the International Diabetes Federation. *Excerpta Medica ICS,* 280, p. 7.
Davis, B., and Lazarus, N. R. (1974): The presence of two protein phosphokinase enzymes in cod *(Gadus callarius)* islet membranes. *Biochemical Society Transactions,* 2:31–33.
Dean, P. M. (1973): Ultrastructural morphometry of the pancreatic B cell. *Diabetologia,* 9:115–119.
Dods, R. F., and Burdowski, A. (1973): Adenosine 3',5'-cyclic monophosphate dependent protein kinase and phosphoprotein phosphatase activities in rat islets of Langerhans. *Biochemical and Biophysical Research Communications,* 51:421–427.
Dumont, J. E., Willems, C., Van Sande, J., and Nève, P. (1971): Regulation of the release of thyroid hormones: Role of cAMP. *Annals of the New York Academy of Science,* 185:291–316.
Edwards, J. C., and Howell, S. L. (1973): Effects of vinblastine and colchicine on the secretion of glucagon from isolated guinea pig islets of Langerhans. *FEBS Letters,* 30:89–92.
Edwards, J. C., and Taylor, K. W. (1970): Fatty acids and the release of glucagon from isolated guinea pig islets of Langerhans incubated in vitro. *Biochimica et Biophysica Acta,* 215:310–315.
Efendic, S., Cerasi, E., and Luft, R. (1972): Arginine-induced insulin release in relation to the cyclic AMP system in man. *Journal of Clinical Endocrinology,* 34:67–72.
Erlichman, J., Hirsch, A. H., and Rosen, O. M. (1971): Interconversion of cyclic nucleotide-activated and cyclic nucleotide-independent forms of a protein kinase from beef heart. *Proceedings of the National Academy of Sciences (U.S.),* 68:731–735.
Esterhuizen, A. C., and Howell, S. L. (1970): Ultrastructure of the A cells of cat islets of Langerhans following sympathetic stimulation of glucagon secretion. *Journal of Cell Biology,* 46:593–601.
Feldman, J. M., and Lebovitz, H. E. (1970): Mechanism of epinephrine and serotonin inhibition of insulin release in the golden hamster in vitro. *Diabetes,* 19:480–486.
Floyd, J. C., Jr., Fajans, S. S., Conn, J. W., Knopf, R. F., and Rull, J. (1966): Stimulation of insulin secretion by amino acids. *Journal of Clinical Investigation,* 45:1487–1502.
Foa, P. P. (1972): The secretion of glucagon. In: *Handbook of Physiology,* edited by D. F. Steiner and N. Freinkel, Section 7, Vol. 1, pp. 261–278. American Physiological Society.
Fussganger, R. D., Goberna, R., Hinz, M., Jaros, P., Karsten, C., Pfeiffer, E. F., and Raptis, S. (1964): Comparative studies on the dynamics of insulin secretion following HB419 and tolbutamide of the perifused isolated rat pancreas and the perifused isolated pieces and islets of rat pancreas. *Hormone and Metabolic Research,* 1, Suppl. 1:34–40.
Gill, G. N., and Garren, L. D. (1971): Role of the receptor in the mechanism of action of adenosine 3'5'-cyclic monophosphate. *Proceedings of the National Academy of Sciences (U.S.),* 68:786–790.
Goldfine, I. D., Perlman, R., and Roth, J. (1971): Inhibition of cyclic 3'5'-AMP phosphodiesterase in islet cells and other tissues by tolbutamide. *Nature,* 234:295–297.
Goldfine, I. D., Roth, J., and Birnbaumer, L. (1972): Glucagon receptors in B cells. Binding of ^{125}I glucagon and activation of adenylate cyclase. *Journal of Biological Chemistry,* 247:1211–1218.
Gomez-Acebo, J., Parrilla, R., and Candela, J. L. R. (1968): Fine structure of the A and D cells of the rabbit endocrine pancreas in vivo and incubated in vitro. *Journal of Cell Biology,* 36:33–44.
Green, I. C., Howell, S. L., Montague, W., and Taylor, K. W. (1973): Regulation of insulin release from isolated islets of Langerhans of the rat in pregnancy. *Biochemical Journal,* 134:481–487.
Green, I. C., and Taylor, K. W. (1972): Effects of pregnancy in the rat on the size and insulin secretory response of th ̀ islets of Langerhans. *Journal of Endocrinology,* 54:317–325.
Green, I. C., and Taylor, K. W. (1974): Insulin secretion, response of isolated islets of Langerhans in pregnant rats. Effects of dietary restriction. *Journal of Endocrinology,* 62:137–143.
Greider, M. G., Howell, S. L., and Lacy, P. E. (1969): Isolation and properties of secretory granules from rat islets of Langerhans. II. Ultrastructure of the beta granule. *Journal of Cell Biology,* 41:162–166.

Grey, N. J., Goldring, S., and Kipnis, D. M. (1970): The effect of fasting, diet and actinomycin D on insulin secretion in the rat. *Journal of Clinical Investigation,* 49:881–889.

Grill, V., and Cerasi, E. (1973): Activation by glucose of adenyl cyclase in pancreatic islets of the rat. *FEBS Letters,* 33:311–314.

Grodsky, G. M. (1972): A threshold distribution hypothesis for packet storage of insulin. II. Effect of calcium. *Diabetes,* 21, Suppl. 2:584–593.

Grodsky, G. M., Batts, A. A., Bennett, L. L., Vcella, C., McWilliams, N. B., and Smith, D. F. (1963): Effects of carbohydrates on secretion of insulin from isolated rat pancreas. *American Journal of Physiology,* 205:638–644.

Grodsky, G. M., and Bennett, L. L. (1966): Cation requirements for insulin secretion in the isolated perfused pancreas. *Diabetes,* 15:910–913.

Heinze, E., and Steinke, J. (1971): Glucose metabolism of isolated pancreatic islets; difference between fetal, newborn and adult rats. *Endocrinology,* 88:1259–1263.

Hellerström, C., and Gunnarsson, R. (1970): Bioenergetics of islet function: Oxygen utilization and oxidative metabolism in the B cell. *Acta Diabetologica Latina,* 7, Suppl. 1:127–151.

Hellerström, C., and Hellman, B. (1960): Some aspect of silver impregnation of the islets of Langerhans in the rat. *Acta Endocrinologica,* 35:518–532.

Hellerström, C., Howell, S. L., Edwards, J. C., and Andersson, A. (1972): An investigation of glucagon biosynthesis in isolated pancreatic islets of guinea pigs. *FEBS Letters,* 27:97–101.

Hellman, B. (1970): Methodological approaches to studies on the pancreatic islets. *Diabetologia,* 6:110–120.

Hellman, B., and Idahl, L. A. (1972): Pancreatic islet levels of citrate under conditions of stimulated and inhibited insulin release. *Diabetes,* 21:999–1002.

Hellman, B., Idahl, L. A., Lernmark, A., and Täljedal, I. B. (1974): The pancreatic B cell recognition of insulin secretagogues: Does cyclic AMP mediate the effect of glucose? *Proceedings of the National Academy of Sciences (U.S.),* 71:3405–3409.

Hellman, B., Sehlin, J., and Täljedal, I. B. (1971a): Effects of glucose and other modifiers of insulin release on the oxidative metabolism of amino acids in microdissected pancreatic islets. *Biochemical Journal,* 123:513–521.

Hellman, B., Sehlin, J., and Täljedal, I. B. (1971b): The pancreatic B cell recognition of insulin secretagogues. II. Site of action of tolbutamide. *Biochemical and Biophysical Research Communications,* 45:1384–1388.

Hertelendy, F., Machlin, L. J., Gordon, R. S., Horino, M., and Kipnis, D. M. (1966): Lipolytic activity and inhibition of insulin release by epinephrine in the pig. *Proceedings of the Society for Experimental Biology and Medicine,* 121:675–677.

Hertelendy, F., Machlin, L. J., and Kipnis, D. M. (1969): Further studies on the regulation of insulin and growth hormone secretion in the sheep. *Endocrinology,* 84:192–199.

Hostetler, K. Y., and Williams, H. R. (1970): Effects of hypoglycemia, tolbutamide and glucagon on the pathways of glucose oxidation in the goosefish islets in vitro. *Diabetes,* 19:554–558.

Howell, S. L. (1972): Role of ATP in the intracellular translocation of proinsulin and insulin in the rat pancreatic B cell. *Nature (London),* 235:85–87.

Howell, S. L., Edwards, J. C., and Montague, W. (1974a): Regulation of adenylate cyclase and cyclic AMP dependent protein kinase activities in A_2-cell rich guinea pig islets of Langerhans. *Hormone and Metabolic Research,* 6:49–52.

Howell, S. L., Edwards, J. C., and Whitfield, M. (1971): Preparation of B cell deficient guinea pig islets of Langerhans. *Hormone and Metabolic Research,* 3:37–43.

Howell, S. L., Green, I. C., and Montague, W. (1973): A possible role of adenylate cyclase in the long-term dietary regulation of insulin secretion from rat islets of Langerhans. *Biochemical Journal,* 136:343–349.

Howell, S. L., Hellerström, C., and Tyhurst, M. (1974b): Intracellular transport and storage of newly synthesized proteins in the guinea pig pancreatic A cell. *Hormone and Metabolic Research,* 6:267–271.

Howell, S. L., Kostianovsky, M. K., and Lacy, P. E. (1969a): B granule formation in isolated islets of Langerhans. A study by electron microscopic radioautography. *Journal of Cell Biology,* 42:695–705.

Howell, S. L., and Lacy, P. E. (1969): Studies of the effect of HB419 (glibenclamide) on isolated islets and granules. *Hormone and Metabolic Research,* 1, Suppl. 1:45–47.

Howell, S. L., and Lacy, P. E. (1971): Biochemical and ultrastructural studies of insulin storage granules and their secretion. *Memoirs of the Society for Endocrinology,* 19:469–480.

Howell, S. L., and Montague, W. (1971): Mode of action of cyclic AMP in rat anterior pituitary. *FEBS Letters,* 18:293–296.

Howell, S. L., and Montague, W. (1973): Adenylate cyclase activity in isolated rat islets of Langerhans. Effects of agents which alter rates of insulin secretion. *Biochimica et Biophysica Acta,* 320: 44–52.

Howell, S. L., and Montague, W. (1974): Regulation of guanylate cyclase in guinea pig islets of Langerhans. *Biochemical Journal,* 142:379–384.

Howell, S. L., and Taylor, K. W. (1966): Effect of glucose concentrations on the incorporation of (^{3}H)leucine into insulin in isolated rabbit islets of Langerhans. *Biochimica et Biophysica Acta,* 130:519–521.

Howell, S. L., and Taylor, K. W. (1967): Secretion of newly synthesised insulin in vitro. *Biochemical Journal,* 102:922–927.

Howell, S. L., and Whitfield, M. (1972): Cytochemical localization of adenyl cyclase activity in rat islets of Langerhans. *Journal of Histochemistry and Cytochemistry,* 20:873–879.

Howell, S. L., Young, D. A., and Lacy, P. E. (1969*b*): Isolation and properties of secretory granules from rat islets of Langerhans III. Studies of the stability of isolated B granules. *Journal of Cell Biology,* 41:167–176.

Idahl, L. A. (1971): Glucose-6-phosphate content in mammalian pancreatic B cells. Effects of various stimulators and inhibitors of insulin release. *Hormones,* 2:371–377.

Idahl, L. A., and Hellman, B. (1971): Regulation of pancreatic B cell glycogen through cyclic 3'5' AMP. *Diabetologia,* 7:139–142.

Idahl, L. A., Hurme, P., Wahlquist, Y., and Hellman, B. (1971): Pancreatic B cell function and content of 6-phosphogluconate. *Hormone and Metabolic Research,* 3:141–144.

Iversen, J. (1973): Adrenergic receptors and the secretion of glucagon and insulin from the isolated, perfused canine pancreas. *Journal of Clinical Investigation,* 52:2102–2116.

Johnson, D. G., Fujimoto, W. Y., and Williams, R. H. (1973): Enhanced release of insulin by prostaglandins in isolated pancreatic islets. *Diabetes,* 22:658–663.

Kipnis, D. M. (1970): Studies of insulin secretion: Radioimmunoassay of cyclic nucleotides and the role of cyclic AMP. *Acta Diabetologica Latina* 7, Suppl. 1:314–337.

Kipnis, D. M. (1972): Nutrient regulation of insulin secretion in human subjects. *Diabetes,* 21, Suppl. 2:606–616.

Knopf, R. F., Fajans, S. S., Floyd, J. C., Jr., and Conn, J. W. (1963): Comparison of experimentally induced and naturally occurring sensitivity to leucine hypoglycemia. *Journal of Pediatrics,* 57: 346–362.

Kosaka, K., Ide, T., Kuzuya, T., Niki, E., Kuzuya, N., and Okinaka, S. (1964): Insulin-like activity in pancreatic vein blood after glucose loading and epinephrine. *Endocrinology,* 75:9–14.

Kostianovsky, M., and Lacy, P. E. (1966): A method for the isolation of intact islets of Langerhans from the mammalian pancreas. *Federation Proceedings,* 25:377.

Kraicer, J., and Milligan, J. V. (1971): Effect of colchicine on in vitro ACTH release induced by high K$^+$ and hypothalamus-stalk-median eminence extract. *Endocrinology,* 89:408–412.

Kris, A. O., Miller, R. E., Wherry, F. E., and Mason, J. W. (1966): Inhibition of insulin secretion by infused epinephrine in rhesus monkeys. *Endocrinology,* 78:87–97.

Kuo, W., Hodgins, D. S., and Kuo, J. F. (1973): Adenylate cyclase in islets of Langerhans, isolation of islets and regulation of adenylate cyclase activity by various hormones and agents. *Journal of Biological Chemistry,* 248:2705–2711.

Labrie, F., Lemaire, S., and Courte, C. (1971): Adenosine 3'5' monophosphate-dependent protein kinase from bovine anterior pituitary gland. I. Properties. *Journal of Biological Chemistry,* 246: 7293–7302.

Lacy, P. E. (1961): Electron microscopy of the beta cell of the pancreas. *American Journal of Medicine,* 31:851–859.

Lacy, P. E., Howell, S. L., Young, D. A., and Fink, C. J. (1968): New hypothesis of insulin secretion. *Nature,* 219:1177–1179.

Lacy, P. E., Klein, N. J., and Fink, C. J. (1973): Effect of cytochalasin B on the biphasic release of insulin in perifused rat islets. *Endocrinology,* 92:1458–1468.

Lacy, P. E., and Malaisse, W. J. (1973): Microtubules and beta cell secretion. *Recent Progress in Hormone Research,* 29:199–228.

Lambert, A. E., Jeanrenaud, B., Junod, A., and Renold, A. E. (1969): Organ culture of foetal rat pancreas. II. Insulin release induced by amino acids and organic acids, by hormonal peptides, by cationic alterations of the medium and by other agents. *Biochimica et Biophysica Acta,* 174:540–553.

Lambert, A. E., Jeanrenaud, B., and Renold, A. E. (1967): Enhancement by caffeine of glucagon-induced and tolbutamide-induced insulin release from isolated foetal pancreatic tissue. *Lancet,* i:819–820.

Lambert, A. E., Kanazawa, Y., Burr, I. M., Orci, L., and Renold, A. E. (1971): On the role of cyclic AMP in insulin release. I. Overall effects in cultured fetal rat pancreas. *Annals of the New York Academy of Science,* 185:232–244.

Laube, H., Fussganger, R. D., and Pfeiffer, E. F. (1972): The effect of a B-adrenergic blocker on insulin release from the isolated perfused pancreas of obese mice. *Journal of Endocrinology,* 55:209–210.

Lebovitz, H. E., and Pooler, K. (1967): Puromycin potentiation of corticotropin-induced insulin release. *Endocrinology,* 80:656–662.

Leclercq-Meyer, V., Brisson, G. R., and Malaisse, W. J. (1971): Effect of adrenaline and glucose on release of glucagon and insulin in vitro. *Nature New Biology,* 231:248–249.

Lefebvre, P. J., and Luyckx, A. S. (1972): Effect of prostaglandin PGE_1 on blood flow and insulin output of dog pancreas in situ. *Diabetes,* Suppl. 1, 21:369.

Leterrier, J. F., Rappaport, L., and Nunez, J. (1974): Phosphorylation and aggregation of neurotubulin and "associated" protein kinase. *Molecular and Cellular Endocrinology,* 1:65–75.

Levey, G. S., Schmidt, W. M. T., and Mintz, D. H. (1972): Activation of adenyl cyclase in a pancreatic islet cell adenoma by glucagon and tolbutamide. *Metabolism,* 21:93–98.

Levine, R. A., Oyama, S., Kagan, A., and Glick, S. M. (1970): Stimulation of insulin and growth hormone secretion by adenine nucleotides in primates. *Journal of Laboratory and Clinical Medicine,* 75:30–36.

Lin, B. J., and Haist, R. E. (1969): Insulin biosynthesis: Effects of carbohydrates and related compounds. *Canadian Journal of Physiology and Pharmacology,* 47:791–801.

Lin, B. J., and Haist, R. E. (1973): Effect of some modifiers of insulin secretion on insulin biosynthesis. *Endocrinology,* 92:735–742.

Lomsky, R., Langr, F., and Vortel, V. (1969): Immunohistochemical demonstration of gastrin in mammalian islets of Langerhans. *Nature,* 223:618–619.

Loubatières, A., Mariani, M. M., Chapal, J., Taylor, J., Houareau, M. H., and Rondot, A. M. (1965): Action nocive de l'adrenaline pour la structure histologique des ilôts de Langerhans du pancreas. *Diabetologia,* 1:13–21.

Loubatières, A., Mariani, M. M., Sorel, G., and Savi, L. (1971): The action of B-adrenergic blocking and stimulating agents on insulin secretion. *Diabetologia,* 7:127–132.

Malaisse, W. J. (1972): Hormonal and environmental modification of islet activity. In: *Handbook of Physiology,* edited by D. F. Steiner and N. Freinkel, Section 7, Vol. 1, pp. 237–260. American Physiological Society.

Malaisse, W. J. (1973): Insulin secretion: Multifactorial regulation for a single release process. *Diabetologia,* 9:167–173.

Malaisse, W. J., Brisson, G., and Malaisse-Lagae, F. (1970): The stimulus-secretion coupling of glucose induced insulin release. I. Interaction of epinephrine and alkaline earth cations. *Journal of Laboratory and Clinical Medicine,* 76:895–902.

Malaisse, W. J., and Malaisse-Lagae, F. (1968): Chronic effects of insulin and glucagon upon islet function. *Diabetologia,* 5:349–352.

Malaisse, W. J., Malaisse-Lagae, F., and Mayhew, D. (1967a): A possible role for the adenylcyclase system in insulin secretion. *Journal of Clinical Investigation,* 46:1724–1734.

Malaisse, W. J., Malaisse-Lagae, F., and McGraw, E. F. (1967b): Effects of thyroid function upon insulin secretion. *Diabetes,* 16:643–646.

Malaisse, W. J., Malaisse-Lagae, F., McGraw, E. F., and Wright, P. H. (1967c): Insulin secretion in vitro by pancreatic tissue from normal adrenalectomized and cortisol-treated rats. *Proceedings of the Society of Experimental Biology and Medicine,* 124:924–928.

Malaisse, W. J., Malaisse-Lagae, F., Walker, M. O., and Lacy, P. E. (1971): The stimulus secretion coupling of glucose-induced insulin release. V. The participation of a microtubular-microfilamentous system. *Diabetes,* 20:257–265.

Malaisse, W. J., Malaisse-Lagae, F., and Wright, P. H. (1967d): Effect of fasting upon insulin secretion in the rat. *American Journal of Physiology,* 213:843–848.

Malaisse, W. J., Malaisse-Lagae, F., Wright, P. H., and Ashmore, J. (1967e): Effects of adrenergic and cholinergic agents upon insulin secretion in vitro. *Endocrinology,* 80:975–978.
Marliss, E. B., Wollheim, C. B., Blondel, B., Orci, L., Lambert, A. E., Stauffacher, W., Like, A. A., and Renold, A. E. (1973): Insulin and glucagon release from monolayer cell cultures of pancreas from newborn rats. *European Journal of Clinical Investigation,* 3:16–26.
Martin, J. M., Akerblom, H. K., and Garay, G. (1968): Insulin secretion in rats with elevated levels of circulating growth hormone due to MtT-W15 tumour. *Diabetes,* 17:661–667.
Matschinsky, F. M. (1972): Enzymes, metabolites and cofactors involved in intermediary metabolism of islets of Langerhans. In: *Handbook of Physiology,* edited by D. F. Steiner and N. Freinkel, Section 7, Vol. 1, pp. 199–214. American Physiological Society.
Matschinsky, F. M., and Ellerman, J. E. (1968): Metabolism of glucose in the islets of Langerhans. *Journal of Biological Chemistry,* 243:2730–2736.
Mayhew, D. A., Wright, P. H., and Ashmore, J. (1969): Regulation of insulin secretion. *Pharmacological Reviews,* 21:183–212.
Miller, E. A., Wright, P. H., and Allen, D. O. (1972): Effects of hormones on accumulation of cyclic AMP-^{14}C in isolated pancreatic islets of rats. *Endocrinology,* 91:1117–1119.
Milner, R. D. G. (1970): The stimulation of insulin release by essential amino acids from rabbit pancreas in vitro. *Biochimica et Biophysica Acta,* 192:154–156.
Milner, R. D. G., and Hales, C. N. (1969): The interaction of various inhibitors and stimuli of insulin release studied with rabbit pancreas in vitro. *Biochemical Journal,* 113:473–479.
Mintz, D. H., Levey, G. S., and Schenk, A. (1973): Adenosine 3'5'-cyclic monophosphate and phosphodiesterase activity in isolated fetal and neonatal rat pancreatic islets. *Endocrinology,* 92:614–617.
Miyamoto, E., Kuo, J. F., and Greengard, P. (1969): Cyclic nucleotide-dependent protein kinases. III Purification and properties of adenosine 3'5'-monophosphate-dependent protein kinase from bovine brain. *Journal of Biological Chemistry,* 244: 6395–6402.
Montague, W., and Cook, J. R. (1971): The role of adenosine 3'5' cyclic monophosphate in the regulation of insulin release by isolated rat islets of Langerhans. *Biochemical Journal,* 122:115–120.
Montague, W., and Howell, S. L. (1972): The mode of action of 3'5'-cyclic monophosphate in mammalian islets of Langerhans. Preparation and properties of islet cell protein phosphokinase. *Biochemical Journal,* 129:551–560.
Montague, W., and Howell, S. L. (1973): The mode of action of adenosine 3'5'-cyclic monophosphate in mammalian islets of Langerhans. Effects of insulin secretagogues on islet-cell protein kinase activity. *Biochemical Journal,* 134:321–327.
Moskalewski, S. (1965): Isolation and culture of the islets of Langerhans of the guinea pig. *General and Comparative Endocrinology,* 5:342–353.
Nonaka, K., and Foa, P. P. (1969): A simplified glucagon immunoassay and its use in a study of incubated pancreatic islets. *Proceedings of the Society for Experimental Biology and Medicine,* 130:330–336.
Northrop, G., Ryan, W. G., and Schwartz, T. B. (1973): Propranolol-induced insulin release in isolated rat islets of Langerhans. *Diabetes,* Vol. 22, Suppl. 2:91–93.
Orci, L., Gabbay, K. H., and Malaisse, W. J. (1972): Pancreatic beta cell web: Its possible role in insulin secretion. *Science,* 175:1128–1130.
Petersson, B., Hellerström, C., and Gunnarsson, R. (1970): Structure and metabolism of the pancreatic islets in streptozotocin treated guinea pigs. *Hormone and Metabolic Research,* 2:313–317.
Poisner, A. M., and Bernstein, J. (1971): A possible role of microtubules in catecholamine release from the adrenal medulla. *Journal of Pharmacology and Experimental Therapeutics,* 177:102–108.
Porte, D., Jr. (1967): A receptor mechanism for the inhibition of insulin release by epinephrine in man. *Journal of Clinical Investigation,* 46:86–94.
Porte, D., Jr., Graber, A. L., Kuzuya, T., and Williams, R. H. (1966): The effects of epinephrine on immunoreactive insulin levels in man. *Journal of Clinical Investigation,* 45:228–236.
Porte, D., Jr., and Williams, R. H. (1966): Inhibition of insulin release by norepinephrine. *Science,* 152:1248–1250.
Randle, P. J., Ashcroft, S. J. H., and Gill, J. R. (1968): Carbohydrate metabolism and release of hormones. In: *Carbohydrate Metabolism and Its Disorders,* edited by F. Dickens, P. J. Randle, and W. J. Whelan, Vol. 1, pp. 427–447. Academic Press, London.
Randle, P. J., and Hales, C. N. (1972): Insulin release mechanisms. In: *Handbook of Physiology,* edited

by D. F. Steiner and N. Freinkel, Section 7, Vol. 1, pp. 219–235. American Physiological Society.

Raptis, S., Dollinger, H., Chrissiku, M., Rothenbuchner, G., and Pfeiffer, E. F. (1973): The effect of the β-receptor blockade (propranolol) on the endocrine and exocrine pancreas function in man after the administration of intestinal hormones. *European Journal of Clinical Investigation,* 3: 163–168.

Rasmussen, H. (1970): Cell communication, calcium ion, and cyclic adenosine monophosphate. *Science,* 170:404–412.

Rasmussen, H., and Tenenhouse, A. (1968): Cyclic adenosine monophosphate Ca^{++} and membranes. *Proceedings of the National Academy of Sciences (U.S.),* 59:1364–1370.

Reddington, M., and Lagnado, J. R. (1973): The phosphorylation of colchicine-binding ("microtubular") protein in respiring slices of guinea-pig cerebral cortex. *FEBS Letters,* 30:188–194.

Reimann, E. M., Walsh, D. A., and Krebs, E. G. (1971): Purification and properties of rabbit skeletal muscle adenosine 3'5'-monophosphate-dependent protein kinases. *Journal of Biological Chemistry,* 246:1986–1995.

Robertson, R. P., and Porte, D. (1973): The glucose receptor. A defective mechanism in diabetes mellitus distinct from the beta adrenergic receptor. *Journal of Clinical Investigation,* 52:870–876.

Rosen, O. M., Hirsch, A. H., and Goren, E. N. (1971): Factors which influence cyclic AMP formation and degradation in an islet cell tumour of the Syrian hamster. *Archives of Biochemistry and Biophysics,* 146:660–663.

Rossini, A. A., Lee, J. B., and Frawley, T. F. (1971): An unpredictable lack of effect of prostaglandins on insulin release in isolated rat islets. *Diabetes,* 20, Suppl. 1:374.

Samols, E., Marri, G., and Marks, V. (1965): Promotion of insulin secretion by glucagon. *Lancet,* 2:415–416.

Samols, E., Marri, G., and Marks, V. (1966): Interrelationship of glucagon insulin and glucose. The insulinogenic effect of glucagon. *Diabetes,* 15:855–866.

Sams, D. J., and Montague, W. (1972): The role of adenosine 3'5' cyclic monophosphate in the regulation of insulin release. *Biochemical Journal,* 129:945–952.

Sams, D. J., and Montague, W. (1974): Possible involvement of adenosine 3'5' cyclic monophosphate in the mechanism of action of sulphonylureas on insulin secretion from islets of Langerhans. *Biochemical Society Transactions,* 2:411–412.

Schatz, H., Maier, V., Hinz, M., Nierle, C., and Pfeiffer, E. F. (1973): Stimulation of H-3 leucine incorporation into the proinsulin and insulin fraction of isolated pancreatic mouse islets in the presence of glucagon, theophylline and cyclic AMP. *Diabetes,* 22:433–441.

Schubart, U., Udem, L., Baum, S., and Rosen, O. M. (1973): Regulation of cyclic nucleotide phosphodiesterase activity in an islet cell tumour of the Syrian hamster. *Diabetes,* 22, Suppl. 1:306.

Selawry, H., Marcks, C., Fink, G., Lavine, R., Cresto, J., and Recant, L. (1973): A mechanism for glucose-induced insulin release. *Diabetes,* 22, Suppl. 1:295.

Spellacy, W. N., Buhi, W. C., and Holsinger, K. K. (1971): The effect of prostaglandin $F_{2\alpha}$ and E_2 on blood glucose and plasma insulin levels during pregnancy. *American Journal of Obstetrics and Gynecology,* 111:239–243.

Steiner, D. F. (1972): Summary of discussion. *Diabetes,* 21, Suppl. 2, 571.

Steiner, D. F., Cunningham, D. D., Spigelman, L., and Aten, B. (1967): Insulin biosynthesis; evidence for a biosynthetic precursor. *Science,* 157:697–700.

Steiner, D. F., Kemmler, W., Clark, J. L., Oyer, P. E., and Rubenstein, A. H. (1972): The biosynthesis of insulin. In: *Handbook of Physiology,* edited by D. F. Steiner and N. Freinkel, Section 7, Vol. 1, pp. 175–198. American Physiological Society.

Sussman, K. E., and Vaughan, G. D. (1967): Insulin release after ACTH glucagon and adenosine 3'5' phosphate (cyclic AMP) in the perfused isolated rat pancreas. *Diabetes,* 16:449–454.

Tanese, T., Lazarus, N. R., Devrim, S., and Recant, L. (1970): Synthesis and release of proinsulin and insulin by isolated rat islets of Langerhans. *Journal of Clinical Investigation,* 49:1394–1404.

Thompson, W. J., Johnson, D. G., and Williams, R. H. (1973): Modulation of hormonal stimulation of adenyl cyclase from isolated pancreatic islets by guanosine triphosphate. *Diabetes,* 23, Suppl. 1:297.

Turtle, J. R., and Kipnis, D. M. (1967): An adrenergic receptor mechanism for the control of cyclic 3'5' adenosine monophosphate synthesis in tissues. *Biochemical and Biophysical Research Communications,* 28:797–802.

Unger, R. H. (1972): Circulating pancreatic glucagon and extra-pancreatic glucagon-like materials.

In: *Handbook of Physiology*, edited by D. F. Steiner and N. Freinkel, Section 7, Vol. 1, pp. 529–544. American Physiological Society.

Vance, J. E., Buchanan, K. D., Challoner, D. R., and Williams, R. H. (1968a): Effect of glucose concentration on insulin and glucagon release from isolated islets of Langerhans of the rat. *Diabetes,* 17:187–193.

Vance, J. E., Buchanan, K. D., and Williams, R. H. (1968b): Effect of starvation and refeeding on serum immunoreactive glucagon and insulin levels. *Journal of Laboratory and Clinical Medicine,* 72:290–297.

Vance, J. E., Buchanan, K. D., and Williams, R. H. (1971): Glucagon and insulin release, influence of drugs affecting the autonomic nervous system. *Diabetes,* 20:78–82.

Van Obberghen, E., Somers, G., Devis, G., Vaughan, G. D., Malaisse-Lagae, F., Orci, L., and Malaisse, W. J. (1973): Dynamics of insulin release and microtubular-microfilamentous system. I. Effects of cytochalasin B. *Journal of Clinical Investigation,* 52:1041–1051.

Vecchio, D., Luyckx, A., Zahand, G. R., and Renold, A. E. (1966): Insulin release induced by glucagon in organ cultures of fetal rat pancreas. *Metabolism,* 15:577–581.

Voyles, N., Gutman, R. A., Selawry, A., Fink, G., Penhos, J. C., and Recant, L. (1973): Interaction of various stimulators and inhibitors on insulin secretion in vitro. *Hormone Research,* 4:65–73.

Walsh, D. A., and Ashby, C. D. (1973): Protein kinases: Aspects of their regulation and diversity. *Recent Progress in Hormone Research,* 29:329–359.

Williams, J. A., and Wolff, J. (1970): Possible role of microtubules in thyroid secretion. *Proceedings of the National Academy of Sciences (U.S.),* 67:1901–1908.

Wolff, D. J., and Siegel, F. L. (1972): Purification of a calcium-binding phosphoprotein from pig brain. *Journal of Biological Chemistry,* 247:4180–4185.

Wong, K. K., Symchowicz, S., Staub, M. S., and Tabachnick, I. I. A. (1967): The in vitro effect of catecholamines, diazoxide and theophylline on insulin release. *Life Sciences,* 6:2285–2291.

Yalow, R. S., Black, H., Villazon, M., and Berson, S. A. (1960): Comparison of the plasma insulin levels following administration of tolbutamide and glucose. *Diabetes,* 9:356–362.

Cyclic Nucleotides in Cultured Cells

Francis J. Chlapowski, Lewis A. Kelly, and Reginald W. Butcher

University of Massachusetts Medical School, Department of Biochemistry, 55 Lake Avenue North, Worcester, Massachusetts 01605

CONTENTS

I. Introduction 246
II. Cell Culture Systems and Cyclic Nucleotide Metabolism . . . 247
   A. Early Studies with Intact cell systems 247
      1. Cells of Neural Origin 247
      2. Fibroblasts 251
      3. Generalizations About Intact Cell Culture Studies . . . 251
   B. Personal Experiences with Cell Culture Systems 254
   C. Complications of the Model. 263
      1. Culture Conditions. 263
      2. Escape of Cyclic AMP 267
      3. Agonist Specificities and Refractoriness 268
   D. The Enzymes of Cyclic AMP Metabolism in Cultured Cells. 273
      1. Adenylyl Cyclase 273
         a. Occurrence and Distribution 273
         b. Characteristics 274
      2. Cyclic-3′,5′-nucleotide Phosphodiesterase 278
         a. Occurrence, Distribution, and Properties 278
         b. Regulation 279
         c. Inhibitors 281
      3. Protein Kinase Activity 283
      4. Cell Functions Altered by Cyclic AMP. 285
   E. Cyclic GMP 285
   F. Mammalian Cell Genetics and Cyclic AMP Metabolism . . 290
      1. Viral Transformation 290
      2. Clonal Variants and Mutants 291
      3. Somatic Cell Hybridization 293
   G. Comments 294

III. Cyclic Nucleotides and Cell Growth 295
   A. Basal Cyclic AMP Levels and Cell Growth 295
      1. Fluctuations of Basal Levels of Cyclic AMP During Growth and Confluency of Fibroblasts 296
      2. Experimental Manipulation of Basal Levels of Cyclic AMP in Fibroblasts 297
      3. Basal Levels in Transformed Fibroblasts 299
      4. Basal Levels in Nonfibroblastic Cells. 300
      5. Is Cyclic AMP the Ultimate Regulator of Cell Growth? . 301
   B. Effects of Experimentally Increased Intracellular Cyclic AMP Levels on Cells 305
      1. Metabolism of Exogenous Cyclic AMP and Dibutyryl Cyclic AMP 305
      2. Inhibition of Cell Division 307
         a. The Cell Cycle 307
         b. Macromolecular Synthesis 308
         c. Contact Inhibition and Agglutinability 308
      3. Restoration of "Normal Morphology" to Some Transformed Cells 309
      4. Lack of Growth Inhibition 310
      5. Summary 312
   C. The Enzymes of Cyclic AMP Metabolism in Malignant Cultured Cells and Tissues 312
      1. Adenylyl Cyclase 314
         a. Aberrations in the Catalytic Site 314
         b. Loss of or Defect in Hormonal Receptors 315
         c. Translocation of Adenylyl Cyclase 316
         d. Loss of Functional Enzyme 316
      2. Phosphodiesterase 316
      3. Protein Phosphokinase 318
      4. Comments 319
   D. Enzyme Changes and Cyclic AMP Levels 320
   E. Cyclic GMP Levels and Cell Growth 324

IV. Acknowledgments 326

V. References 326

I. INTRODUCTION

When we were invited to prepare a review on the use of cell culture systems to study cyclic nucleotide metabolism, we had no inkling of how colossal an undertaking it would be. The number of reports which have appeared during the last 18 months or so is staggering. In addition, the difficulty is compounded by the diversity of interests, approaches, cell lines, technical details, and even terminology which we have encountered.

Our hope was to make a critical and comprehensive synthesis of the data available to answer a relatively simple question: to wit, are cell culture systems valid and useful models for studying cyclic nucleotide metabolism? The answer, alas, is far more complex than the question. Yes, cell culture systems are valid and useful models, but only with great reservations and with a firm grasp of the many factors which may complicate experimental results.

The information which, we feel, justifies this answer is contained in Section II of this review. The structure of Section II is somewhat unusual, and hence warrants a few words of explanation. We have attempted to make use of some of the elegant intact cell experiments from 1971–1972 which provided an entry into the whole area as evidence for the affirmative part of the answer. This is supported by a cryptic but (hopefully) comprehensive tabulation of those cell systems which have been studied and reported upon. Next, we have used our own intact cell experiments as a framework for the discussion of the potential complications which follow. Finally, we have pulled together the salient data on the enzymes of cyclic AMP metabolism and action and on cyclic GMP, with special emphasis on some pertinent aspects of cell culture systems.

Section III of this review deals with our prejudices about the hypothesized relationship of cyclic AMP levels to the control of growth. In surveying the literature, it became obvious that well over one-half of the reports dealing with cell culture and cyclic nucleotides also dealt with growth. Thus, despite the fact that several reviews on the subject are available, it seemed a a reasonable idea to update the story. Further, as we read more, the more pronounced became certain of our reservations about the validity of the various hypotheses relating cyclic AMP levels to control of cell growth, particularly as a generalization, and especially if one attempted to correlate defective cyclic AMP metabolism to malignancies. Our reservations, then, coupled with a few experiments in our own laboratories, have encouraged us to present views which may be somewhat at variance with those of others.

II. CELL CULTURE SYSTEMS AND CYCLIC NUCLEOTIDE METABOLISM

A. Early Studies with Intact Cell Systems

1. *Cells of Neural Origin*

The first reports dealing primarily with cyclic AMP metabolism in cultured cells appeared in 1971 (Gilman and Nirenberg, 1971a,b; Clark and Perkins, 1971). These reports suggested that cell culture systems were feasible as models for studying cyclic nucleotide metabolism. There were in fact several earlier studies in which cyclic AMP was considered or even measured in cultured cells (e.g., Burk, 1968; Taunton, Roth, and Pastan, 1969; Makman, 1970; Peery, Johnson, and Pastan, 1971; Heidrick and Ryan, 1971a,b). However, these studies were primarily either of adenylyl cyclase activities or dealt with the possible role of cyclic AMP in the control of proliferation rather than with the control of cyclic AMP metabolism *per se*. The latter studies are considered in detail in Section III.

The experimental design for monolayer cultures, as described by Gilman and Nirenberg (1971a), has been so widely adapted that it might be well to summarize it here. Cells, derived from stock cultures and freed by treatment with trypsin, are plated on appropriate growth surfaces and grown for whatever period of time necessary (this may vary from 2 to several days, depending on the cell type involved and the cell population desired). The growth medium, which also varies with the cell type, usually consists of an enriched physiological salt solution supplemented with serum, most often fetal calf serum. The cells are prepared for the experimental incubation by washing the cell sheets (which adhere to the growth surface) in serum-free medium and allowing the cells to equilibrate for a short time in the fresh, serum-free medium. This protocol is not always used, and for this reason it is often difficult to extrapolate from one set of experimental conditions to another. For example, in several cell lines the presence of serum has been reported to lower basal cyclic AMP levels (see Section III.A.2.) and to reduce the magnitude of the cyclic AMP responses to catecholamines or prostaglandins (Makman, Dvorkin, and Keehn, 1974). Further, the addition of serum has been shown to raise cyclic GMP levels (Section III.E.).

Experimental incubations are begun with the addition of agonists of the adenylyl cyclase system, phosphodiesterase inhibitors, or other agents as specified. Incubations are generally rather short (1 to 20 min), although longer incubations are not uncommon. The experiments are terminated by aspirating the medium [which may be fixed by addition of an appropriate acid such as trichloroacetic acid (TCA) or HCl] and by pipetting 5% TCA onto the cell sheet. The cells appear to be fixed almost instantaneously by this method, and the recovery of cellular cyclic AMP (as verified by the use of radioactive cyclic AMP and also by exhaustive extraction) appears to be essentially quantitative. The TCA extract is then fractionated by a number of protocols, usually dependent upon the prejudice of the laboratory involved, and the cyclic AMP determined by any of the assays currently in vogue. The cellular protein, which remains precipitated on the incubation surface, can then be dissolved in an appropriate base (e.g., 0.2 N NaOH) and analyzed by the method of Lowry, Rosebrough, Farr, and Randall (1951), thus providing for an expression of cyclic AMP per milligram of cellular protein.

In their first report, Gilman and Nirenberg (1971a) described the effects of catecholamines on cyclic AMP levels in three clonal lines of glial tumor cells. Clone C6, which had earlier been derived from a glial tumor induced in a Wistar rat with N-nitrosomethylurea, proved to be a Napoleonic responder, and even

TABLE 1. *Effects of catecholamines on cyclic AMP levels in rat glioma C6 cells*

Additions	pmoles cyclic AMP/mg protein
H_2O	9
Norepinephrine (0.01 mM)	21
Norepinephrine (0.10 mM)	2,390
Isoproterenol (0.01 mM)	850
Isoproterenol (0.10 mM)	2,530

C6 cells were subcultured on 25 mm glass coverslips at a density of 1×10^4 cells/cm^2 and grown 5 to 6 days. The growth medium was removed and cells were placed in serum-free medium containing 1 mM theophylline for a period of 60 min. Catecholamines were then added as indicated and incubations continued for 15 min. Values were obtained from TCA extracts of the cells. Further details are described in Gilman and Nirenberg (1971a). Reprinted with permission from Gilman and Nirenberg (1971a).

after 3 years remains one of the champion cell lines. For example, cyclic AMP levels were increased by more than 200-fold after 5 min of incubation with 100 μM norepinephrine or isoproterenol (Table 1). Further, C6 cells were very sensitive to catecholamines, in that cyclic AMP levels were increased 10-fold by 0.01 μM isoproterenol. In addition, the time course of the response to isoproterenol was extraordinarily fast—cyclic AMP levels were increased 40-fold after 30 sec with isoproterenol. The response of the C6 cells appeared to be quite specific for catecholamines; several prostaglandins, histamine, 5-hydroxytryptamine, carbamylcholine, and γ-aminobutyric acid were without effect. The catecholamine response appeared to be subserved by β-adrenergic receptors, as judged by relative potency studies with the catecholamines and by the effects of α- and β-blocking agents on cyclic nucleotide responses to catecholamines. A second rat glioma clone, C-2, also gave a pronounced if somewhat less dramatic response to catecholamines, whereas the human astrocytoma CHB responded but with modest (10- to 15-fold) increases. The two latter lines also appeared to be quite specific for catecholamines. Two months after the appearance of the Gilman and Nirenberg paper, Clark and Perkins (1971) described the regulation of cyclic AMP levels in a clonal human astrocytoma cell line, 1181N1. Clark and Perkins found responses by this cell line which were quantitatively quite similar to those seen by Gilman and Nirenberg with the human CHB clone. That is, while 1181N1 responded to the catecholamines, the magnitude of the response was small compared to C6. However, they pointed out that these experiments were done at a relatively high cell population density. When they examined the cells at lower population densities, they found that the response to norepinephrine was

TABLE 2. *Cyclic AMP response to norepinephrine and histamine as a function of cell density*

Days of cell growth	Number of cells	Total mg of protein	pmoles of cyclic AMP/mg protein[a]		
			Controls	Norepinephrine	Histamine
5[b]	1.6×10^6	0.31 ± 0.01[c]	18.2 ± 0.79	$2,682 \pm 269$	64.9 ± 5.5
8	2.9×10^6	0.82 ± 0.024	14.4 ± 1.3	609 ± 49	65.0 ± 6.9
11	4.8×10^6	1.47 ± 0.032	24.4 ± 3.2	240 ± 20	83.2 ± 7.6
11[d]	3.7×10^6	1.24 ± 0.030	12.0 ± 0.95	129 ± 21	83.5 ± 10.0

[a] Norepinephrine and histamine were 30 μM. All incubations were 5 min.
[b] Cells were inoculated at day 1 with 9×10^5 cells/plate; medium was changed at days 4, 7, and 10, and 4 hr before the experiment.
[c] Values in the table were determined in cell extracts from 3 to 10 determinations, and are expressed ±SE.
[d] Cells of this group were identical to the other 11-day group, except that the medium was not changed after day 7 or before the experiment on day 11.
Reprinted with permission from Clark and Perkins (1971).

very flamboyant (Table 2). Interestingly, histamine, which also caused increased cyclic AMP levels (albeit much less dramatically) was essentially unaffected by changes in cell population density. The 1181N1 cells, like C6, appeared to be subserved by β-adrenergic receptors. In addition, the effects of theophylline were small in C6 and insignificant in 1181N1.

Gilman and Nirenberg closed out 1971 with a report on the regulation of cyclic AMP levels in cultured neuroblastoma cells (1971b). Four distinct clones of mouse neuroblastoma C-1300 were responsive only to prostaglandin E_1 (PGE_1). The responses of one clone, N10, were studied in detail and found to be very rapid in onset although much less dramatic than the effects of catecholamines on the astrocytoma cells. Further, Gilman and Nirenberg made the interesting observation that PGE_1 inhibited the rate of cell multiplication and that this effect of PGE_1 was much more dramatic in the presence of theophylline. These studies were of particular interest because many of the differentiated functions of neuroblastoma cells were known to be best expressed when cell division was slowed either by growth to confluence or by other means. Thus, the ability of agents which increased cyclic AMP levels to slow growth as well made the interpretation of the role of cyclic AMP on the development of differentiated functions in the neuroblastoma cells somewhat difficult.

These three reports did much to establish cell culture systems as feasible and desirable for studying cyclic nucleotide metabolism. In several of the lines studied, the responses to agonists of adenylyl cyclase were very dramatic, so that it appeared that they could serve as relatively simple and well-controlled models for such studies. Further, the old caveat about confusion arising from studies of heterogeneous cell populations was largely eliminated by use of clonal cell lines. Finally, there seemed to be a nice bit of specificity in the responses of the cells to agonists. That is, C6 and C2 responded very nicely to catecholamines but not to prostaglandins, while the clones of the neuroblastoma 1300 responded only to prostaglandin E_1.

2. Fibroblasts

Among the earlier studies with fibroblasts, Manganiello and Vaughan (1972a) observed that PGE_1 and PGE_2 stimulated increases in cyclic AMP levels in L-929 and L-2071 mouse fibroblast lines, while epinephrine had no stimulatory effect. The responses to PGE_1 by both cells were relatively small (maximum 100 pmoles/mg of protein) compared with the responses usually seen in other cell types (500 to 5,000 pmoles/mg of protein, depending on the cell and the agonist used), and the effect was not potentiated by theophylline. Even so, stimulated cyclic AMP levels were 10-fold above basal levels. At about the same time, Otten, Johnson, and Pastan (1972b) reported a similar effect of PGE_1 in L-929 cells and also noted a partial antagonism of PGE_1 action by insulin.

In other studies, using cultured foreskin fibroblasts, Manganiello, Breslow, and Vaughan (1972) reported 10- to 20-fold cyclic AMP increases with epinephrine and isoproterenol and several hundred-fold increases with PGE_1. The isoproterenol effect was blocked by propranolol. Theophylline potentiated the cellular responses to both catecholamines and PGE_1, as did 1 μM dexamethasone, either in the presence or absence of theophylline.

These early experiments with astrocytoma cells and fibroblasts triggered a flurry of activity in many laboratories. Indeed, it seems likely that the sales of incubators, sterile-ware for growth, media, etc., were enhanced greatly. Further, if our own experiences were representative, the numbers of petri dishes accidentally infected with yeast, molds, and other sorts of unwanted guests probably reached an all-time high!

3. Generalizations About Intact Cell Culture Studies

It would have been a far easier task to write this review if cell culture systems behaved in predictable fashion, that is, if they would respond to appropriate hormones in a relatively straightforward way and would yield reproducible and constant results over a wide range of conditions. Sadly, such is not the case. The complications arising from variables such as culture conditions, cell growth, cell population densities, and hormone refractoriness, to say nothing of the differences in the qualitative and quantitative aspects of the responses of different cell strains or of different clones of individual cell strains, are such that generalizations are for the most part impossible. However, since generalizations are useful if for no other reason than to set a standard against which exceptions can be made, we offer a few. One generalization is that most fibroblasts and glioma cells are responsive to prostaglandins. Although there are relatively large differences in the degree of responsiveness, increased cyclic AMP levels are virtually a concomitant of prostaglandin addition. Likewise, many cell lines are responsive

TABLE 3. Changes in cyclic AMP levels of cultured cells in response to various effector agents

Cell	Agents used	Changes in cellular cyclic AMP	Reference
Rat glioma, C-6; C6 (TK⁻)	NE, INE	↑200X; ↑1,000X	Gilman and Nirenberg, 1971a; Schultz et al., 1972; Schwartz and Passonneau, 1974; Schwartz et al., 1973; Browning et al., 1974b
Rat glioma, RGC₆	NE	↑200–250X	deVellis and Brooker, 1974
Rat glioma, C-2₁	NE	↑100X	Gilman and Nirenberg, 1971a; Schultz et al., 1972
Human astrocytoma, CHB	NE; INE	↑10–15X	Gilman and Nirenberg, 1971a
Human astrocytoma, 1181N1	NE; histamine; Ado	↑5–100Xa; ↑3–10X; ↑30–50X	Clark and Perkins, 1971; Clark et al., 1974
Mouse neuroblastoma, C1300	PGE₁; Ado; 2-Cl-Ado	↑10–200X; ↑5–35X; ↑5X	Gilman and Nirenberg, 1971b; Schultz and Hamprecht, 1973; Hamprecht and Schultz, 1973a,b; Gilman, 1974; Blume et al., 1973
Fetal rat brain	INE; PGE₁; Ado	↑50X; ↑30X; ↑12X	Gilman and Schrier, 1972
Human lung fibroblast	INE; PGE₁	↑5X; ↑20X	Franklin and Foster, 1973a
Somatic cell hybrids, mixed PGE₁⁺ × PGE₁⁻	PGE₁	↑ ≥ PGE₁⁺ parent	Minna and Gilman, 1973;
Somatic cell hybrids, mixed NE⁺ × NE⁻	NE	↑ ≤ NE⁺ parent	Hamprecht and Schultz, 1973b
Hamster kidney, fibroblast, BHK/21	Insulin; adenovirus 12	↓30%; ↓80%b	deAsua et al., 1973; Raska, 1973
Mouse fibroblast, 3T3	Insulin; PGE₁; trypsin	↓80%; ↑8X; ↓75%	Seifert and Paul, 1972, Otten et al., 1972b; Rudland et al., 1974c
Mouse fibroblast, 3T3	Serum, proteases	↓30%; ↓40–60%	Kram et al., 1973; Burger et al., 1972a; Seifert and Rudland, 1974
Mouse fibroblast, BALB, 3T3	Simian virus 40	↓20–50%b	Rein et al., 1973
Secondary mouse embryo fibroblast	PGE₁; serum; insulin	↑50X; ↓40%; ↓40%	Rozengurt and deAsua, 1973

Cell type	Agents	Effects	References
Human lung fibroblast, WI-38	EPI; PGE$_1$	↑0–12X[a]; ↑30–50X	Kurtz et al., 1974; Kelly et al., 1974;
Ibid, SV40-transformed	EPI; PGE$_1$; Ado	↑100–200X; ↑200–500X; ↑100X	Makman et al., 1974; Haslam and Goldstein, 1974; Gilman, 1974
Human foreskin fibroblasts	INE; PGE$_1$	↑30X; ↑100X	Manganiello and Breslow, 1974; Manganiello et al., 1972
Mouse fibroblasts, L-929	PGE$_1$	↑12[c]	Manganiello and Vaughan, 1972a; Otten et al., 1972b
Mouse fibroblasts, 3T3	PGE$_1$	↑6–9X	Minna and Gilman, 1973
Mouse embryonic vertebral fibroblasts	bradykinin; estradiol-17β; PGE$_1$	↑7X; ↑3X; ↑5X	Schonhofer et al., 1974
Rat liver, BRL30E	PGE$_1$; INE	↑70X; ↑130X	Minna and Gilman, 1973; Gilman and Minna, 1973
Human liver, Chang	EPI	↑12X	Makman et al., 1974
Mouse adrenal cortex tumor Y-1	ACTH; CT	↑10–20X; ↑6X	Schimmer, 1972; Temple and Wolff, 1973; Kwan and Wishnow, 1974
Mouse adrenal cortex tumor	ACTH; CT	↑280X	Kowal et al., 1974
Mouse adrenal cortex	ACTH	↑3–15X	Kowal, 1973; Kowal and Harano, 1974
Mouse lymphoma, S49	INE; PGE$_1$	↑250X; ↑160X	Daniel et al., 1973a; Bourne et al., 1973b
Mouse melanoma Cloudman S91	MSH	↑12X	Pawelek et al., 1973
Chinese hamster ovary, CHO	CT	↑2–100X	Brunton and Gurrant, 1974
Rabbit lens epithelium	EPI	↑120X	Makman et al., 1974
HeLa (AT)	EPI	↑9X	Makman et al., 1974

Agents listed as increasing cyclic AMP levels were generally added with a phosphodiesterase inhibitor present. Since conditions varied considerably, approximate magnitudes of responses are given only for conditions yielding the highest cyclic AMP levels. In studies where more than one clone was investigated, results of only the most responsive clone are given. ↑ and ↓ indicate increases or decreases, respectively, in cyclic AMP levels. Abbreviations are: EPI, epinephrine; NE, norepinephrine; INE, isopropylnorepinephrine (isoproterenol); CT, cholera toxin or E. coli enterotoxin; Ado, adenosine; and MSH, melanocyte-stimulating hormone.

[a] Varied dramatically with cell population density.
[b] Viral infection of cells; effects observed within 3 to 8 hr.
[c] Antagonized by insulin.

to catecholamines, and, where studied, β-adrenergic antagonists such as propranolol are effective inhibitors. Finally, phosphodiesterase inhibitors have been studied in a number of cell lines, and often have potentiated the actions of stimulatory hormones.

References to the data supporting these generalizations are presented in Table 3. Now we will proceed to some of the complications underlying studies with cell culture systems. We will begin with a brief discussion of our own experiences, not so much because we view them as particularly significant, but rather because we know them better than the studies of other investigators. Then, using our experiments as a kind of table of contents, we will try to bring together what is known about each of the complications involved in the intact cell culture model.

B. Personal Experiences with Cell Culture Systems

Our own experiences with cell culture systems began in 1970. We were searching for a good model system in which to study cyclic AMP metabolism. Past experiences had made the danger of working with heterogeneous cell populations manifest (e.g., Butcher and Baird, 1968). On the other hand, the isolated fat cell system with which we had been working had lost much of its charm because it was such a complex system from an endocrinological point of view, was difficult to work with because of its inherent lability and the problems associated with high triglyceride content, and because of the concern that the collagenase treatment might be causing subtle but real damage to the cell membrane.

After many months of abortive attempts to use Chinese hamster lung fibroblasts (V79–753) as models (they were only slightly responsive to catecholamines and prostaglandins) and a brief but disenchanting series of experiments with CH-33 chimpanzee liver cells (they responded to catecholamines poorly and to glucagon not at all), we happened upon the human diploid lung fibroblast WI-38.

WI-38 is an elegant-looking fibroblast (Fig. 1A) which was derived by Hayflick (1965) and used by him in studies of aging. WI-38 cells, under usual culture conditions, have a finite lifetime (40 to 60 population doublings) after which they become senescent, and they maintain their diploid karyotype throughout. Hence, they are thought to be somewhat more highly differentiated than many cell lines which are rather thoroughly adapted to culture and which have taken on some characteristics of transformed cells, including virtual immortality.

WI-38 cells responded very nicely to catecholamines and prostaglandins (Fig. 2). Further, the effects of these compounds were potentiated by methylxanthines. Thus, we were encouraged and began to view WI-38 as a good model system. However, complications set in very quickly. The response of WI-38 to catecholamines proved to be most unpredictable in magnitude. In fact, in some experiments it was absent. However, as shown in Fig. 3, the reason for the variability was much the same as had been found by Clark and Perkins (1971). That is, there was an inverse relationship between the magnitude of the response of WI-38 cells to catecholamines and the cell population density. On the other hand, the cellular cyclic AMP response to PGE_1 was essentially unchanged.

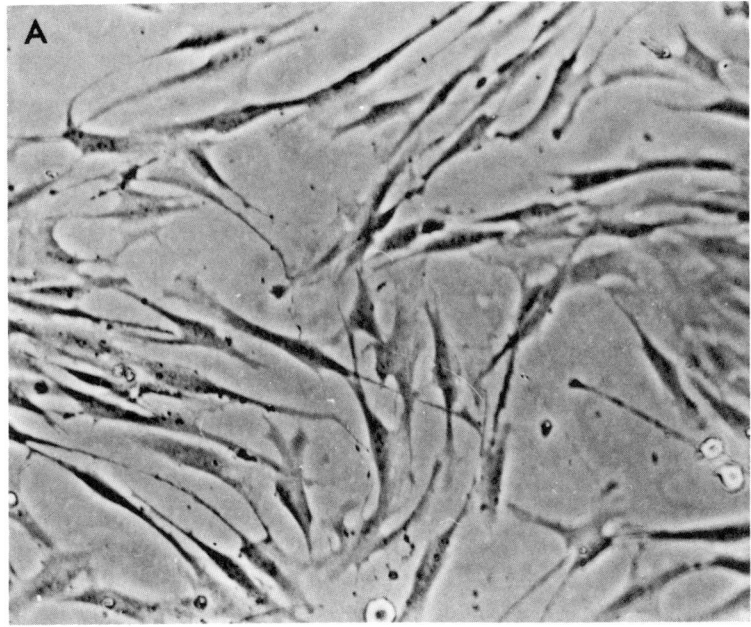

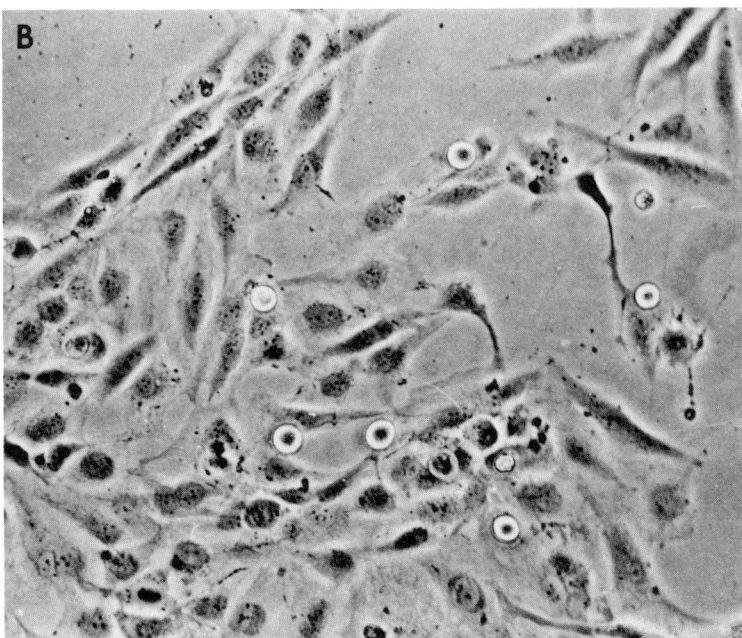

FIG. 1. Phase-contrast micrographs of cultured WI-38 human fibroblasts (A) and SV40 transformed WI-38 fibroblasts (B). X150.

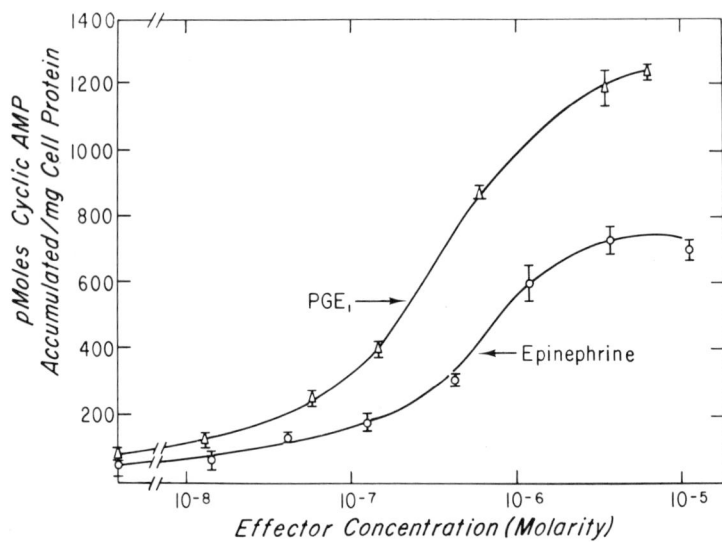

FIG. 2. The effects of various concentrations of epinephrine and PGE_1 on cyclic AMP levels in WI-38 cells. Monolayers of WI-38 cells were grown to confluence in 25-mm glass vials. The cell sheets were rinsed twice with serum-free medium, 2 ml of serum-free medium containing 2 mM theophylline were added, and the cells were equilibrated for 60 min. Final incubations with epinephrine or PGE_1 were 10 min. TCA extracts of the cells were purified on Dowex-50 and were analyzed for cyclic AMP by the protein-binding assay described by Gilman (1970). Values given are the averages of duplicate incubations. Reprinted with permission from Kelly and Butcher, 1974.

A second complication became manifest hard on the heels of the population density problem. In this case, we found that stimulated WI-38 cells released large amounts of cyclic AMP to the medium, and that the process was very rapid as shown in Fig. 4. This figure is somewhat deceiving because both medium and cellular cyclic AMP levels are shown as totals rather than as concentrations. In fact, cyclic AMP concentrations were much lower in the media than in the cells (by a factor of roughly 1:20).

Thus, WI-38 had turned out not to be a simple model system. On the one hand, the problem of the dependence of the catecholamine response on cell population density dictated that strict attention be paid to culture and growth conditions. On the other hand, the magnitude of the escape of cyclic AMP from the cells to the media necessitated measurements of all medium samples. In fact, the experiment illustrated in Fig. 4 raises at least one very perplexing question: to wit, why are intracellular levels of cyclic AMP decreasing between 5 and 30 min while the rate of release is essentially unchanged between 0 and 30 min? The time course of cellular cyclic AMP changes would suggest that WI-38 was behaving much like isolated fat cells, in that the rate of intracellular cyclic AMP production, or at least accumulation, was severely diminished as the incubation progressed (Manganiello, Murad, and Vaughan, 1971). However, at least as judged by the rate of escape of cyclic AMP to the medium, it would appear that cyclic

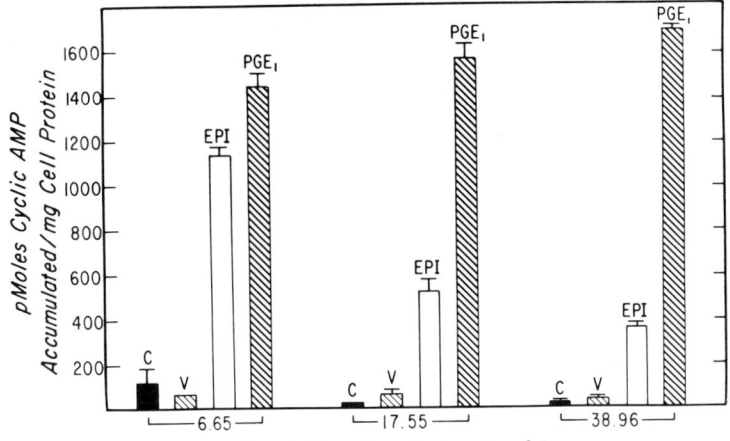

FIG. 3. The effects of epinephrine and PGE$_1$ on cyclic AMP levels in populations of WI-38 cells grown to varied densities. A concentrated suspension of WI-38 cells was seeded directly and at two dilutions into separate sets of 90-mm glass petri dishes. After 2 days, the cells were rinsed twice with serum-free medium, 10 ml of serum-free medium containing 2 mM theophylline were added to each dish, final additions were made as indicated, and incubations were carried out for 10 min. Separate controls were run with (V) and without (C) prostaglandin solubilization vehicle; vehicle was not added to vessels treated with epinephrine. The concentration of epinephrine (EPI) was 10 μM and the PGE$_1$ concentration was 5.7 μM. Values given are the duplicate averages for cyclic AMP in purified cell extracts. Reprinted with permission from Kelly and Butcher, 1974.

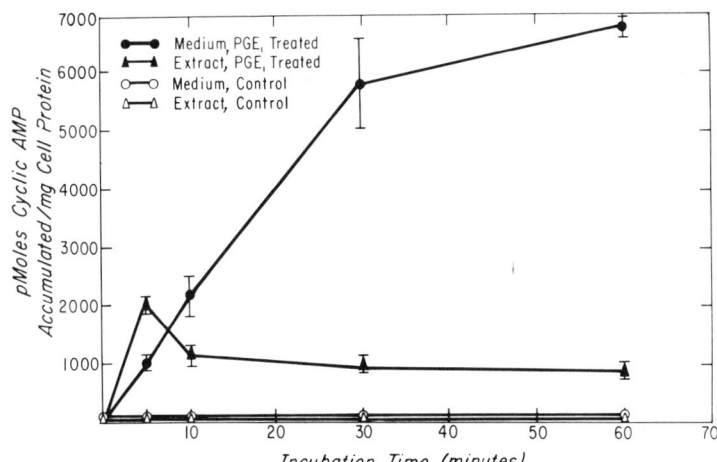

FIG. 4. Cyclic AMP levels in WI-38 cells and in the incubation medium after various periods of exposure of the cells to PGE$_1$. WI-38 cells were grown to confluence in 60-mm glass petri dishes. The cell sheets were rinsed twice and the growth media replaced with 3 ml of serum-free medium buffered with 20 mM HEPES (N-2-hydroxy-ethylpiperazine-N'-2-ethanesulfonic acid), pH 7.4. PGE$_1$ (5.7 μM) or vehicle (V) were then added and vessels were incubated for the times indicated in the absence of a phosphodiesterase inhibitor. Incubation media and cell sheets were fixed and analyzed separately for cyclic AMP. The values shown are duplicate averages. Reprinted with permission from Kelly and Butcher, 1974.

AMP production was not decreased. In any event, the phenomenon of escape may provide, as was originally suggested by Exton, Lewis, Ho, Robison, and Park (1971), working with the perfused liver, a very useful and less complex way of studying cyclic AMP metabolism.

Unfortunately, the situation with cyclic AMP escape became even more complex when studied in the context of cell population density. It seemed possible that the diminished catecholamine response at confluence might in fact reflect a redistribution of cyclic AMP rather than a real decrease in production. However, as shown in Table 4, such was not the case. The total cyclic AMP accumulation by WI-38 cells in response to 10 μM epinephrine and 2 mM theophylline decreased in going from low to high population density. Further, the total cyclic AMP production in cells exposed to PGE_1 and theophylline increased with increasing density, and the proportion of cyclic AMP in the medium increased. This latter point does in fact make the whole system much more complex. Not only is the phenomenon one which must be kept in mind, but it has also to be noted that the magnitude of cyclic AMP escape is clearly dependent upon population density, and hence is subject to change from experiment to experiment if there are changes in growth conditions. Such variations can of course be minimized by strict adherence to growth protocols, but the problem remains a real one capable of causing artifactual differences between experiments or laboratories. The problems introduced by differences in culture conditions will be discussed in more detail in Section II.C.1., and the escape phenomenon in Section II.C.2.

It also seemed reasonable to study one or more of the SV40 transformed WI-38

TABLE 4. *A comparison of cyclic AMP levels in WI-38 cells and in the incubation media after exposure of high- and low-density populations to epinephrine or PGE_1*

Cell density	Incubation conditions	pmoles cyclic AMP accumulated/mg cell protein		
		Cells	Media	Cells plus media
7.94 μg/cm²	Control	59 ± 15	22 ± 2	81
	Epinephrine (10 μM)	657 ± 48	404 ± 192	1,061
	PGE_1 (5.7 μM)	3,028 ± 207	967 ± 54	3,995
12.1 μg/cm²	Control	44 ± 9	23 ± 3	67
	Epinephrine (10 μM)	142 ± 20	150 ± 5	292
	PGE_1 (5.7 μM)	3,568 ± 521	1,474 ± 110	5,042
13.5 μg/cm²	Control	53 ± 16	17 ± 7	70
	Epinephrine (10 μM)	36 ± 17	33 ± 3	69
	PGE_1 (5.7 μM)	3,689 ± 291	1,761 ± 63	5,450

WI-38 cells were prepared and incubated exactly as described in Fig. 3, except for the inclusion of prostaglandin solubilization vehicle in the control and epinephrine-treated vessels. Values reported for "Cells plus media" (right-hand column) are the sums of the averaged values of the "Cells" and "Media" determinations. Reprinted with permission from Kelly and Butcher (1974).

lines to determine what if any differences in cyclic AMP metabolism existed in these cells. The first such strain we obtained was WI-38-VA13-2RA (hereafter referred to as VA13). It should be made clear at the outset that we did *not* view this cell pair as models for "normal" and malignant cells, although characteristics associated with malignancy are present in VA13 and it was derived, via viral transformation, from WI-38. The effects of such factors as selection pressure and mutation are so intricate that it is very difficult at best to ascribe the differences to transformation *per se*. Rather, we were looking for different patterns in cyclic AMP responses which could be related to specific defects in the enzymatic systems controlling cyclic AMP metabolism.

These we found in abundance. Not surprisingly in view of its microscopic appearance (Fig. 1*B*), VA13 is not subject to density-dependent inhibition of growth as is WI-38 (see Section III.). Likewise, as might be expected in rapidly growing cell populations, basal cyclic AMP levels were slightly lower in confluent growing VA13 than in confluent stationary WI-38 (Kelly, Hall, and Butcher, 1974).

More dramatic differences were apparent when the cells were stimulated (Table 5). First, PGE_1 produced a greater response in VA13 than in WI-38, and the effects of theophylline were much less pronounced. Second, the effects of catecholamines were even more divergent in the two cell lines. As shown in Table 6, 10 μM epinephrine caused far greater cyclic AMP accumulations in VA13, which were not influenced by cell population density, than it caused in WI-38.

Examination of the temporal aspects of the responses of the two cell lines (Fig. 5) illustrates further differences. In the absence of theophylline (Fig. 5*A*), PGE_1 caused a rapid increase in cellular cyclic AMP levels, followed by a fall and plateau after 5 min of incubation. The appearance of cyclic AMP in the media

TABLE 5. *The responses of WI-38 and VA13 cells to PGE_1 in presence and absence of theophylline*

	pmoles cyclic AMP accumulated/mg cell protein	
Incubation conditions	WI-38	VA13
Control	<50[a]	<50[a]
Theophylline	<50[a]	<50[a]
PGE_1	817 ± 21	2,129 ± 115
PGE_1 + theophylline	2,229 ± 167	3,202 ± 363

Confluent WI-38 and VA13 cells were prepared in 60 mm plastic (Falcon) dishes. The cells were rinsed and buffered medium was added as in the experiment in Fig. 4. PGE_1 (5.7 μM) was added with or without theophylline (2 mM) as indicated and incubations were stopped after 10 min. Cyclic AMP was assayed in purified cell extracts. Values shown are the mean ± SE from triplicate incubations. Reprinted with permission from Kelly et al. (1974).

[a] Calculated on the basis of the limit of sensitivity of the assay.

TABLE 6. *The effects of epinephrine and PGE_1 on cyclic AMP levels in populations of WI-38 and VA13 cells grown to varied densities*

Incubation conditions	pmoles cyclic AMP/mg cell protein after growth at		
	10.4 µg protein/cm²	24.7 µg protein/cm²	36.5 µg protein/cm²
WI-38			
Control	<75[a]	<32[a]	<21[a]
Epinephrine	784 ± 80	203 ± 3	128 ± 16
PGE_1	1,461 ± 80	1,441 ± 80	1,433 ± 86
	3.74 µg protein/cm²	23.5 µg protein/cm²	82.5 µg protein/cm²
VA13			
Control	<210[a]	38 ± 4	13 ± 4
Epinephrine	2,920 ± 20	3,520 ± 380	3,210 ± 350
PGE_1	2,940 ± 86	4,590 ± 190	3,978 ± 474

Details of the experiment were essentially identical to those in Fig. 3. Cyclic AMP was assayed in purified cell extracts. Values shown are the means ± SE from triplicate incubations.
[a] Calculated on the basis of the limit of sensitivity of the assay.
Reprinted with permission from Kelly et al. (1974).

was rapid and continued briskly throughout the experiment, as discussed previously. By contrast, the cellular response of VA13, although equally rapid, reached a maximum much later and showed little tendency toward the marked decrease seen in WI-38 (Fig. 5C). Further, the escape of cyclic AMP to the medium was distinctly lower in VA13, and appeared only after a short but significant lag period. Theophylline (2 mM) potentiated the effect of PGE_1 on intracellular cyclic AMP in WI-38 cells (Fig. 5B), and somewhat prolonged the elevation. However, the appearance of cyclic AMP in the medium was significantly reduced by theophylline. As shown in Table 7, this property appears to be shared by all compounds tested thus far which are capable of inhibiting phosphodiesterase activity, the most potent being 1-methyl-3-isobutylxanthine (MIX). Theophylline had only minor effects on VA13 (Fig. 5D).

VA13 and WI-38 had qualitatively similar specificities toward agonists (Table 8). However, VA13 responded to 1×10^{-5} M isopropylnorepinephrine and 1×10^{-4} M adenosine much more dramatically than did WI-38.

As might be expected from the foregoing, phosphodiesterase activity in VA13 was much lower than in WI-38 (from 5 to 14%) at substrate concentrations between 0.1 and 100 µM. More surprising, however, were the adenylyl cyclase activities measured in homogenates of the two lines (Table 9). While the accumulations of cyclic AMP were significantly higher in homogenates of WI-38 incubated with optimal concentrations of PGE_1 and NaF, much less was accumulated in the presence of epinephrine. In fact, adenylyl cyclase activity was higher in VA13 homogenates incubated with the catecholamine than with NaF.

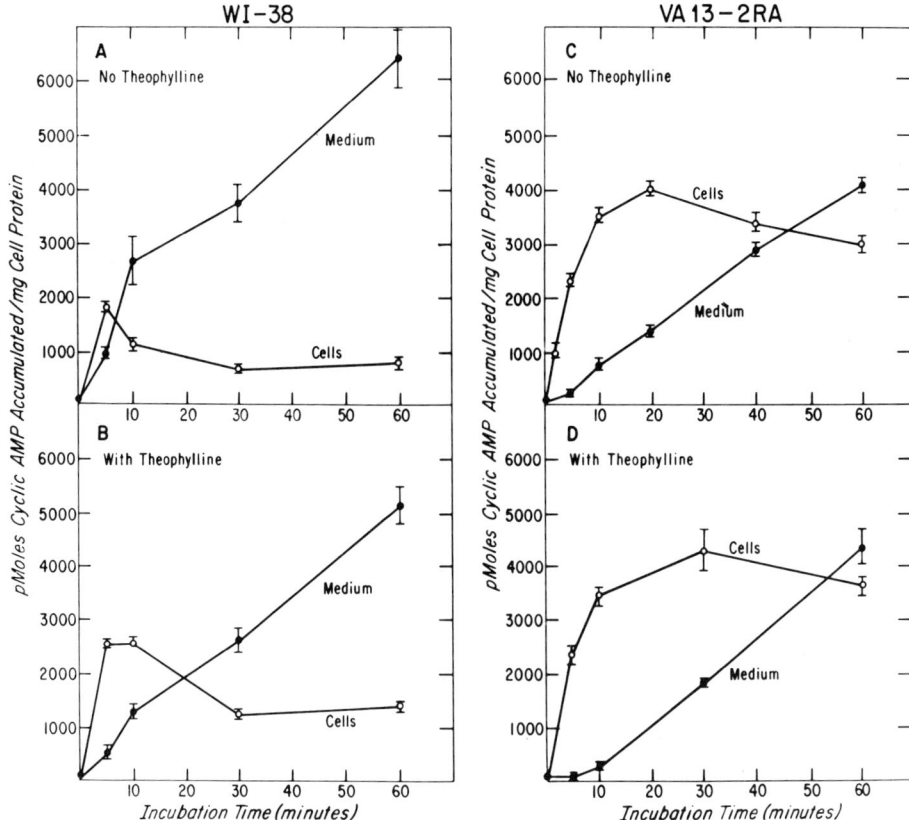

FIG. 5. Dynamics of the responses of WI-38 and VA13 cells to PGE$_1$ in the absence and presence of theophylline. Confluent WI-38 and VA13 cells were prepared in 60-mm plastic (Falcon) dishes. Other details were essentially identical to those in Fig. 4 except that theophylline (2 mM) was added where indicated. Reprinted with permission from Kelly et al., 1974.

Briefly summarized, then, there were a number of significant differences between WI-38 and VA13. These included the following:

1. VA13 was much more responsive to hormones than WI-38, especially to the catecholamines and adenosine.

2. Phosphodiesterase activities were much lower in VA13 than in WI-38, and this was reflected by the relatively insignificant effects of phosphodiesterase inhibitors on cyclic AMP levels in VA13 as compared to WI-38.

3. The escape of cyclic AMP from VA13 cells to the media was much reduced as compared to WI-38.

4. The inverse relationship between the ability of WI-38 cells to respond to catecholamines and cell population density did not obtain in VA13.

5. The activity of adenylyl cyclase, at least as measured in homogenates, was

TABLE 7. *The effects of various inhibitors of cyclic nucleotide-3',5'-phosphodiesterase on cyclic AMP accumulation in WI-38 cells and in the incubation medium in response to PGE_1*

Inhibitor	pmoles cyclic AMP/mg cell protein			
	Cells	Media	Total	Cells/medium
None	637 ± 147	839 ± 96	1,476	0.76
Theophylline (2 mM)	1,475 ± 83	475 ± 17	1,950	3.11
1-methyl,3-isobutyl-xathine (2 mM)	3,482 ± 702	355 ± 15	3,837	9.81
RO7-2956 (0.424 mM)	1,143 ± 216	831 ± 81	1,974	1.38
Papaverine (0.03 mM)	2,092 ± 32	271 ± 34	2,363	7.7

Confluent WI-38 cells in 60 mm plastic (Falcon) dishes were incubated 30 min in serum-free medium containing the indicated phosphodiesterase inhibitors. PGE_1 (5.7 µM) was then added to each dish and incubations were continued for 10 min. Cell extracts and samples of incubation media in 0.1 N HCl were purified and assayed as described previously (Kelly et al. 1974). Values in the table are averages ± one-half the difference between duplicate determinations. RO7-2956 is 4-(3,4-dimethoxybenzyl)-2-imidazolidinone.

TABLE 8. *Actions of several potential effectors of cyclic AMP metabolism on cyclic AMP levels in WI-38 and VA13 cells*

Incubation conditions	pmoles cyclic AMP/mg cell protein	
	WI-38	VA13
Control	<50[a]	<50[a]
Isopropylnorepinephrine (10 µM)	232 ± 27	5,260 ± 356
PGE_2 (5.7 µM)	781 ± 21	5,530 ± 163
Adenosine (100 µM)	144 ± 21	2,589 ± 87
Histamine (100 µM)	61 ± 6	<50[a]
Glucagon (0.6 µM)	58 ± 8	64 ± 9
ACTH (1.25 milliunits/ml)	<50[a]	77 ± 11
Insulin (2.4 milliunits/ml)	<50[a]	<50[a]

Confluent VA13 cells and WI-38 cells in log growth phase were each grown 2 days in 60 mm plastic (Falcon) dishes and assayed in separate experiments. To avoid possible complications, prostaglandin solubilization vehicle was present only in incubations containing PGE_2 and no phosphodiesterase inhibitor was used. Cyclic AMP was assayed in purified cell extracts. Values shown are the means ± SE from triplicate incubations.

[a] Calculated on the basis of the limit of sensitivity of the assay system.
Reprinted with permission from Kelly et al. (1974).

markedly different in VA13. While the activity in the presence of PGE_1 and NaF was much reduced, the activity in the presence of epinephrine was about threefold higher than in WI-38 in log growth phase.

The point of all these references to our own work has been to emphasize what we consider some of the major problems which may confront investigators working with cell culture systems.

TABLE 9. *Adenylyl cyclase activities in homogenates of WI-38 and VA13 cells*

	pmoles cyclic AMP/mg protein/min	
Incubation conditions	WI-38	VA13
Control	8.0 ± 1.3	6.1 ± 0.9
Epinephrine (10 μM)	21.8 ± 2.3	74.7 ± 4.3
PGE$_1$ (10 μM)	67.5 ± 2.7	36.6 ± 3.7
NaF (1 mM)	11.4 ± 2.2	5.8 ± 0.8
NaF (3.3 mM)	78.0 ± 3.2	24.7 ± 2.0
NaF (10 mM)	157.7 ± 7.3	55.8 ± 8.5
NaF (30 mM)	102.9 ± 8.6	59.4 ± 10.3

Homogenates of WI-38 and VA13 cells were prepared simultaneously and incubations and analyses were carried out in the presence of 6.7 mM caffein as indicated previously (Kelly et al., 1974). WI-38 and VA13 incubations contained 348 and 483 μg of protein, respectively. The incubation time was 15 min. Preliminary studies showed that under these conditions the protein concentrations were rate-limiting and the enzyme activities were constant with time. Values shown are averages ± one-half the difference between duplicate incubations. Reprinted with permission from Kelly et al. (1974).

C. Complications of the Model

1. *Culture Conditions*

Although the culture conditions under which cells are grown and maintained can be reasonably well controlled, there are a sufficient number of variables, including seeding density, methods used to free cells for seeding, media, the type and amount of sera used, growth vessels, and methods of sterilization, that consistency is not always possible. One useful adjunct to biochemical studies is to monitor the cells by phase-contrast microscopy, keeping a photographic record. This provides a chance to judge population density before an experiment is carried out and also, of course, permits the detection of any problem which leads to a morphological change in the cells.

The first clear evidence of an effect of culture conditions on the hormone responses of cultured cells was reported by Clark and Perkins (1971). They found that in the astrocytoma line 1181N1, epinephrine and norepinephrine were maximally effective at low cell population density and only poorly effective at 11 days after subculture when the cells were confluent. On the other hand, histamine maintained a modest effect throughout the period tested. Our experiences with WI-38 were quite similar, except that in the case of the fibroblast the persistent stimulator was prostaglandin E_1. Essentially identical effects of population density on the responses of low-passage human skin fibroblasts to epinephrine and PGE$_1$ have been reported (Haslam and Goldstein, 1974).

The data of Makman et al. (1974) tend to confirm a decrease in the responsiveness of WI-38 cells to catecholamines with increasing population density, despite the fact that these authors proposed the contrary. Their interpretation was based on very slight stimulation by epinephrine at the lowest population density studied (8.5 $\mu g/cm^2$). The times of incubation they used, 1 or 150 min, were very different from those in our own studies (10 min), and some of their comparisons involved cultures of very different passages. In addition, it is not clear how the different population densities were achieved in these studies.

Here a problem inherent to cell culture becomes apparent: the propagation of cell strains grown in monolayer requires, at least with present technology, removal of cells at high population density from their substratum, dilution in growth medium, and transfer to fresh growth vessels of fixed dimensions, only to be grown again from a low to a high population density at an increased passage level. Certainly the removal of cells by any means disrupts the membranes to some degree, and perforce, adenylyl cyclase. For example, the use of trypsin affects various receptors in other cell preparations (Kono, 1970). Thus, it seems likely that trypsin may in some cells largely destroy and in others simply unmask receptor sites. Damaged sites obviously require time for repair after subculture, and exposed sites might gradually become inaccessible. Further, the seeding density and the resultant rapidity with which cells can recondition the medium for growth and cell-cell interaction play an important role in establishing the responsiveness of the cells to exogenous effectors. Regarding this point, Kurtz, Polgar, Taylor, and Rutenburg (1974) reported dramatically increased responses of WI-38 cells to PGE_1 with time after subculture, indicating the probable involvement of recovery from trypsin treatment as well as increases in population density.

Further, Manganiello and Breslow (1974) have shown that human foreskin fibroblasts responded more dramatically to PGE_1 at 3 days after seeding than at 1 day. The response to 2 μM isoproterenol in populations seeded at low density increased to a maximum at 3 days and fell until the eighth day, when the response was essentially the same as at 1 day. This was essentially identical to the pattern of response to epinephrine and norepinephrine seen in 1181N1 astrocytes (Clark and Perkins, 1971). Trypsinization of confluent cells, which were very responsive to PGE_1 and relatively unresponsive to isoproterenol, followed by subculture for 3 days, resulted in a marked decrease in the response to PGE_1 and a moderate increase in the response to isoproterenol.

As part of a study dealing with effects of 5-bromodeoxyuridine on the rat glioma line C6, Schwartz, Morris, and Breckenridge (1973) found that the responses of untreated cells to norepinephrine gradually increased with time after subculture to a maximum at confluence. These studies parallel the studies of Makman (1971a) on 3T3 cells, where adenylyl cyclase assayed in cell-free preparations of cells at short intervals after subculture was stimulated by epinephrine to a much lesser extent than was the enzyme from cells grown for longer times after subculture. Similarly, Zacchello, Benson, Giannelli, and McGuire (1972) have reported an increase in adenylyl cyclase activity of human testicular and skin fibroblasts between 1 and 4 days after subculture.

A further complication in cell culture studies is the tendency of some cells to become senescent. The term senescence, as applied to cell culture systems, refers to that condition in which cells no longer divide but continue to perform vital metabolic functions. This phenomenon is fairly common in highly differentiated cells such as WI-38 or other diploid fibroblasts.

Some suggested effects of senescence on cyclic AMP metabolism have emerged from cell culture studies. Makman et al. (1974) reported increased cellular cyclic AMP responses to epinephrine in middle (29 generation) or high (40 generation) passage WI-38 cells compared with the response of low (18 generation) passage cells. However, the cells were assayed at widely varied population densities and appeared to have slightly different time courses of response, making interpretation of the results difficult. Haslam and Goldstein (1974) have similarly compared the cyclic AMP responses of low- and high-passaged human diploid skin fibroblasts to epinephrine and PGE_1. At higher passages approaching senescence, a marked increase in cellular response to epinephrine occurred with increasing population density, whereas little change occurred in the response to PGE_1. The PGE_1/epinephrine ratios (comparing peak intracellular cyclic AMP levels) increased with time after subculture to a maximum at 6 days in young cells, but changed little with time in senescent cells up to 16 days after subculture. Measurements of basal cyclic AMP levels showed slight increases with senescence, as expressed per milligram of protein, but no real increase as expressed per cell since cell volume per milligram of protein appeared to be higher in senescent cells.

Manganiello and Breslow (1974) reported slightly higher cyclic AMP responses to PGE_1 in senescent (60 generation) human foreskin fibroblasts than in "young" (30 generation) cells. The responses of both the low- and high-passaged skin fibroblasts to PGE_1 tended to increase with the increasing population density. However, as appeared to be the case with epinephrine in the skin fibroblasts studied by Haslam and Goldstein (1974), the response of cells grown for 60 generations to isoproterenol appeared to be less affected by population density (i.e., less diminished at high density) than was the response of cells grown for 30 generations. However, one obvious problem in such studies is the difficulty in distinguishing effects of senescence from the random effects of serial passage, including possible adaptation or selection.

As mentioned previously, the inclusion of serum in incubation media has been reported to lower basal cyclic AMP levels in a number of cell culture systems (see also Section III.A.2.). As shown in Table 10, the addition of 10% fetal calf serum to VA13 cells decreased the accumulation of intracellular cyclic AMP in response to PGE_2. Serum had similar effects on the response to PGE_1 and epinephrine. Levels of cyclic AMP found in the media were likewise reduced, probably in part because of phosphodiesterase activity in the serum. Also, Makman et al. (1974) have shown lowered responses of C6 rat astrocytoma cells to epinephrine and decreased responses of normal and SV40-transformed WI-38 fibroblasts to epinephrine or PGE_1 in the presence of 5 to 10% calf or fetal calf serum. The suppressive effect of serum was rapid, occurring within 5 min of

TABLE 10. *The effect of serum on the cyclic AMP response of VA13 cells to PGE_2*

Incubation time	Serum	pmoles cyclic AMP/mg protein	
		Cells	Medium
10 min	−	2,399 ± 182	438 ± 79
	+	1,601 ± 69	409 ± 7
24 hr	−	449 ± 45	11,451 ± 432
	+	253 ± 9	3,603 ± 354

VA13 (WI-38-VA13-2RA) cells were grown to confluence in 60 mm plastic (Falcon) dishes. To begin the experiment, cells were rinsed twice with either fresh serum-free medium or growth medium (containing 5% fetal calf serum), PGE_2 (5.8 μM) was added, and cells were further incubated 10 min or 24 hr as indicated. Control levels of cyclic AMP were less than 50 pmoles/mg protein. Values shown are the means ± SE from triplicate incubations.

addition to the cells, and was roughly proportional to concentrations of serum between 2 and 30%. Both cellular and media cyclic AMP levels were reduced under these conditions. The relationship between the effects of serum on cyclic AMP levels in stimulated cells and the rapid decreases in basal cyclic AMP levels observed in numerous cells is not known.

In summary, a great many factors in the overall category of culture conditions can and often do have significant effects on the cyclic AMP responses of cultured cells. In fact, almost any step along the way can cause a change in the response patterns of cells, and continued close attention to all technical and experimental procedures is absolutely mandatory if reproducible results are to be obtained.

Perhaps the most discouraging aspect of working with cell culture systems is that occasionally, and for no apparent reason, the cells respond in a different way. The causes of these unexpected changes have thus far been completely mysterious, and have been classified as spontaneous changes (for want of a better term). An example of this occurred when we were in the midst of attempts to delineate the control properties of VA13. The cells had been responding regularly to maximal concentrations of PGE_1 or isoproterenol with intracellular cyclic AMP levels of approximately 5,000 pmoles/mg protein. Suddenly, and with no obvious morphological or growth change, the response dropped to around 1,000 pmoles/mg protein. After eliminating all of the more obvious things that might have gone wrong (such as assays, incubation temperatures, etc.), it became clear that the problem was with the cells themselves. However, despite the problems with the intact cell experiments, the activity of adenylyl cyclase in these same cells was essentially unimpaired (R. B. Clark, *personal communication*). The cause of the problem with this particular batch of cells remains a mystery, since all of the stock was discarded inadvertently. Using new stock established from frozen samples of the original cultures resulted in the usual responses.

Another example of spontaneous change was reported by Clark, Gross, Su, and Perkins (1974), where over a period of 18 months the responses to adenosine of human astrocytoma cells 1181N1 and 132N1 declined to approximately 25% of initial levels. A similar loss of BRL cell responsiveness to catecholamines with serial subculture has been reported by Gilman (1974), and Peery et al. (1971) have reported diminished NaF responsive adenylyl cyclase activity in transformed 3T3 cells.

2. Escape of Cyclic AMP

The process of cyclic AMP escape from cells was first described by Davoren and Sutherland (1963), who were studying suspensions of avian erythrocytes. That cyclic AMP was released from cells was not particularly surprising, since considerable amounts had been found in human urine (Butcher and Sutherland, 1962). However, the very large amount of cyclic AMP released was surprising, as was the fact that the process was blocked by methlxanthines or probenecid, and also that the release occurred even in the presence of reasonably high extracellular cyclic AMP concentrations. Shortly after this, Makman and Sutherland (1964) described the release of cyclic AMP from *E. coli*. However, this was a very different situation, at least superficially, in that the release became prominent only after the adenylyl cyclase system was inactivated by the addition of glucose to starved cultures.

The quantitative significance of cyclic AMP escape in cultured cells is clear. As mentioned earlier, in our studies with WI-38 (Kelly and Butcher, 1974) the amount of cyclic AMP in the media was sevenfold greater than that in the cells after 60 min of incubation with PGE_1. However, medium concentrations of cyclic AMP did not exceed intracellular concentrations. Similar results were reported by Makman et al. (1974).

Franklin and Foster (1973*a,b*) also reported that substantial increases in cyclic AMP occurred in the incubation medium of an unspecified strain of human diploid fibroblasts exposed to either isoproterenol or PGE_1. PGE_1 was the greater stimulant, and it appeared to give a peak intracellular level of cyclic AMP later in time than isoproterenol. The phosphodiesterase inhibitor 2-amino,6-methyl,5-oxy,4-*n*-propyl-4,5-dihydro-S-triazolo (1,5-a) pyrimidine (I.C.I. 63, 197) was included in these studies, as was the media buffer supplement N-2-hydroxyethylpiperazine-N'-2-ethanesulfonic acid (HEPES). Previous studies (D'Armiento, Johnson, and Pastan, 1973) had suggested that HEPES increased the release of cyclic AMP to the medium of unstimulated fibroblasts over several days of culture, but the data of Franklin and Foster showed no apparent effect of HEPES on cyclic AMP appearance in the medium in their cell system. Similarly, our own studies with WI-38 cells (Kelly and Butcher, 1974) disclosed no effects of HEPES on cyclic AMP metabolism. It should be pointed out that HEPES has gained widespread use as a supplement to bicarbonate medium in cultured cell systems, both as a pH-stabilizing component of growth media (Eagle, 1971) and as a means of maintaining pH in acute experiments.

Clark et al. (1974) have demonstrated an increase in media levels of cyclic AMP in astrocytoma 1321N1 cells treated with adenosine. Both the protein-binding assay and the ^{3}H-adenine prelabeling technique for cyclic AMP analysis were employed with essentially identical results; at 120 min total medium cyclic AMP exceeded the intracellular total by a factor of seven- to eightfold and even equalled the intracellular maximum seen at 10 min. The release of ^{3}H-cyclic AMP against a gradient of 1 mM cyclic AMP was observed in 1321N1 astrocytoma cells by Clark, Su, Ortmann, Cubeddu, Johnson, and Perkins (1975). In addition, these authors reported antagonism of release by 10 μM dipyridamole, an agent which blocks nucleoside uptake.

The inhibition of the escape of cyclic AMP from cells by the compounds with phosphodiesterase inhibitory activity is fairly widespread. In addition to WI-38 (Table 8) and avian erythrocytes (Davoren and Sutherland, 1963), deVellis, Inglish, and Brooker (1974) have reported that the norepinephrine-stimulated release of cyclic AMP into the medium of rat glioma cells (RGC6A) was decreased by addition of 1 mM MIX, although MIX potentiated the norepinephrine effect on intracellular cyclic AMP almost threefold. Penit, Jard, and Benda (1974) reported an inhibition of cyclic AMP release by probenecid in C6 glial cells, similar to that found previously in pigeon erythrocytes (Davoren and Sutherland, 1963). These results with probenecid were confirmed by Mawe, Doore, McCaman, Feucht, and Saier (1974), who further reported a strong inhibition of release from C6 cells by valinomycin and dinitrophenol.

Another potentially interesting aspect of escape was provided by Kowal, Srinivasan, and Saito (1974), who reported that cyclic AMP release from mouse adrenal cortex tumor cells increased with increasing calcium concentrations in response to a given level of ACTH. The amount of steroid produced was better correlated with cyclic AMP released than with the relative binding of ACTH (which was unaffected by Ca^{2+}). Similarly, Ca^{2+} enhanced cholera toxin stimulation of cyclic AMP release and steroidogenesis, but had no effect on the irreversible binding of the toxin to the cell surface.

In summary, cyclic AMP release not only represents a potentially large pool of cyclic nucleotide, but it occurs rapidly in a number of different cell types in the presence of appropriate effectors. Furthermore, it is very sensitive to perturbations by drugs. Although its complete significance is not known, escape at least appears to represent an additional site for cyclic nucleotide regulation or abberation.

3. Agonist Specificities and Refractoriness

The straightforward effects of the catecholamines and prostaglandins on cyclic AMP levels in cell culture systems have already been summarized (Table 3).

At least four reports (Franklin and Foster, 1973b; Makman et al., 1974; Manganiello and Breslow, 1974; Haslam and Goldstein, 1974) have suggested that maximal cyclic AMP levels in human diploid fibroblasts were attained more rapidly when stimulated with catecholamines than with PGE. Similar results, including a marked evanescence of the responses to low levels of catecholamines, have been observed in our laboratory (*unpublished observations*). However, we feel that this is largely artifactual. It would appear that at least part of this difference may be due to destruction of the catecholamine. We have found, for example, that inclusion of thiourea and ascorbate in incubations of VA13 cells substantially increased the sensitivity of these cells to isoproterenol stimulation and prolonged the response. Thus, destruction of the catecholamine would partially explain both the generally lower cyclic AMP levels obtained with catecholamines than with PGE_1 and the observation that the catecholamine response is more transient.

Most cell culture systems have shown some degree of refractoriness after repeated stimulation with an agonist. For example, Schultz, Hamprecht, and Daly (1972) investigated cyclic AMP accumulation in the rat glial tumor clones C6 and C2 in response to norepinephrine using a ^{14}C-adenine prelabeling method (Shimizu, Daly, and Creveling, 1969). In this method, intracellular adenine nucleotides, including ATP, were labeled during a 40-min preincubation period and the labeled cyclic AMP formed during subsequent incubations could then be extracted, purified, and analyzed directly. Comparison with the protein-binding assay (Gilman, 1970) showed a close correlation between the two methods. Using both methods of analysis, Schultz et al. demonstrated marked increases in cellular cyclic AMP with 0.1 mM norepinephrine. Histamine, veratridine, and adenosine had no effect on the C6 or C2 glial cells, in contrast to the effects of these agents in brain. Repeated stimulations of the glial cells with norepinephrine gave progressively lower responses to the catecholamine. However, the responsiveness was substantially increased, was maintained for longer periods, and could be elicited repeatedly to near maximal levels, if incubation media were supplemented with either papaverine or isobutylmethylxanthine. Both of the latter drugs are potent inhibitors of cyclic AMP phosphodiesterase activity. Unfortunately, medium cyclic AMP contents were not measured. In any event, the latter authors concluded that increased cyclic AMP phosphodiesterase could largely account for both the decrease of cyclic AMP from maximal levels over time and the decreased ability of the cells to respond to the catecholamine upon repeated stimulation. Induction of cyclic AMP phosphodiesterase activity under conditions of increased cyclic AMP levels has in fact been demonstrated in a number of cell lines. For example, Manganiello and Vaughan (1972*a*) demonstrated a direct correlation between the refractoriness of L cells to PGE_1 stimulation and increased phosphodiesterase activity. Thus, desensitization of receptors, increased destruction of cyclic AMP, or both, may be responsible for refractoriness in a given cell type (see Section II.D.2.b.).

One of the earliest reports of a refractory phenomenon in cultured cells was the report by Makman (1971*b*) demonstrating that exposure of 3T3 cells to low levels of epinephrine for 3 hr led to the selective loss of adenylyl cyclase sensitivity to the hormone in cell-free preparations, with no appreciable diminution of the fluoride effect. In this case, unlike the glioma cells, desensitization or refractoriness to stimulation would probably not have involved phosphodiesterase induction. Franklin and Foster (1973b) also reported a selective loss of sensitivity to either isoproterenol or PGE_1 in intact human diploid fibroblasts under conditions where phosphodiesterase activity did not appear to be changed. Unfortunately, they did not indicate the substrate concentrations used in the phosphodiesterase analyses.

More recently, deVellis and Brooker (1974) have shown that cloned RGC6 rat glioma cells previously exposed to norepinephrine were refractory to further stimulation under conditions where phosphodiesterase activity was not increased. Although phosphodiesterase induction did appear to occur in the parent line in response to norepinephrine stimulation, there was no apparent increase in activity in the clone (2B) utilized in these studies. An absence of the refractory period was apparent if the cells of clone 2B received either cycloheximide or actinomycin D simultaneously with norepinephrine. Thus, a catecholamine-inducible protein other than phosphodiesterase appeared to play a role in the refractoriness of these cells. Similarly, a protein other than phosphodiesterase may modulate the response to PGE_1 in VA13 (Makman et al., 1974). A 2-hr pretreatment with 25 μM cycloheximide gave approximately 60% greater cyclic AMP accumulation with PGE_1 than was observed without cycloheximide.

Su and Perkins (1974) have reported a norepinephrine-selective refractory phenomenon in human astrocytoma cells following exposure to the catecholamine for 10 to 90 min. Under such conditions, cyclic AMP levels reached a maximum at 10 min and declined to near basal values by 120 min. Readdition of fresh norepinephrine after 10 min had no effect on the decline of cyclic AMP, while additions of adenosine or PGE_1 (at 2 hr) gave normal responses. However, almost one-half of the original response to norepinephrine could be restored by interposing a 2-hr incubation without the catecholamine prior to restimulation.

Schwartz et al. (1973), working with the glial tumor line C6, have reported a concentration-dependent reduction of the norepinephrine effect on cyclic AMP levels in cells pretreated with 5-bromodeoxyuridine. Interference with the norepinephrine effect developed gradually over a period of days and required actively dividing cells. The authors suggested that it may have been necessary for the 5-bromodeoxyuridine to be incorporated into DNA in order to suppress the norepinephrine response. The effect was partially reversible within 1 to 2 days following exposure to 5-bromodeoxyuridine. The mechanism of 5-bromodeoxyuridine action appeared to involve approximately a doubling in phosphodiesterase activity, as assayed at both low and high substrate concentrations. Basal adenylyl cyclase activity was unaffected after treatment, and norepinephrine stimulation of adenylyl cyclase was unaltered in cell-free preparations.

The elevation of cyclic AMP by adenosine has been noted in several cells of neural origin, as well as in certain cultured liver and fibroblast cell lines. Clark and Perkins (1972) reported that astrocytoma cells increased cyclic AMP levels 30- to 50-fold during a 5-min incubation period with 100 μM adenosine. Half-maximal effects were seen with 10 μM adenosine. The magnitude of the response was greater in cells at low population density than at confluence. On the other hand, adenine had no effect at concentrations up to 100 μM. At about the same time, Gilman and Schrier (1972) reported a 14-fold increase in cyclic AMP in cultures of fetal rat brain with 100 μM adenosine. PGE_1 was somewhat more effective, and isoproterenol was the most effective agent tested. The responsiveness to isoproterenol increased with time after initiation of the cultures, and reached maximum 14 to 20 days later, when the cultures reached confluence. These results may be of interest in terms of the development of adrenergic sensitivity with maturation in fetal brain tissue. Blume, Dalton, and Sheppard (1973) also have reported an elevation of cyclic AMP by adenosine in cloned mouse neuroblastoma C-1300 cells. This effect was potentiated by the phosphodiesterase inhibitor RO20-1724, but not by MIX or papaverine. Theophylline not only did not potentiate the adenosine effect, it caused an inhibition, as has been the case in other adenosine-sensitive systems (Sattin and Rall, 1970). Adenine, guanosine, guanine, isoproterenol, histamine, and acetylcholine had no effect on cyclic AMP levels.

More recently, Clark et al. (1974) reported a rapid but transient cyclic AMP increase in human astrocytoma cells in response to 100 μM adenosine. Theophylline competitively inhibited the adenosine effect in both 1321N1 and 1181N1 astrocytoma clones, whereas papaverine potentiated the effect of adenosine and the effect of norepinephrine, which also increased cyclic AMP. Propranolol completely blocked the norepinephrine but not the adenosine effect. 5'-AMP produced a response essentially identical to that seen with adenosine; that is, the 5'-AMP effect was maximal at concentrations and exposure times similar to those for adenosine. Thus it appeared that prior metabolism of the nucleotide to adenosine was not necessary. Dipyridamole (10 μM) completely inhibited adenosine uptake in 1321N1 cells without any diminution in response, supporting the authors' idea that adenosine acts at the cell exterior to increase cyclic AMP levels. Gilman (1974) has reported that an analogue of adenosine, 2-Cl-adenosine, was very potent, increasing cyclic AMP in rat liver cells and human VA2 fibroblasts as well as in neuroblastoma C1300 cells. Various methylxanthines gave various degrees of inhibition or potentiation of the 2-Cl-adenosine effect depending on the cell type and the concentrations of 2-Cl-adenosine and the methylxanthine used.

Using the mutant neuroblastoma cell line N4TG3, which is a 6-thioguanine-resistant cell, Schultz and Hamprecht (1973) demonstrated an increase in cyclic AMP levels with prostaglandin E_1, but not with dopamine, 5-hydroxytryptamine, norepinephrine, epinephrine, histamine, adenosine, or prostaglandin $F_{1\alpha}$. Interestingly, in the presence of either MIX or papaverine, adenosine clearly elevated cyclic AMP levels. This finding is similar to the observation made by Gilman (1974) that MIX partially restored the otherwise undetectable catecholamine and PGE_1 responses of high-passage rat liver (BRL) cells. It is also similar to our own observations which indicated that human epidermoid carcinoma (HEp-2) cells respond to epinephrine in the presence of MIX but not in its absence (Kelly and Butcher, 1975). The inhibitory effect of MIX on phosphodiesterase activity (Beavo, Rogers, Crofford, Hardman, Sutherland, and Newman, 1970) no doubt at least partially accounts for these observations.

Several studies with cultured cells, in addition to those of Kowal et al. (1974) mentioned previously, have demonstrated effects of various enterotoxins on cyclic AMP metabolism. Generally, toxin effects are substantially delayed compared with the effects of other agents. Kantor, Tao, and Wisdom (1974) reported that the adenylyl cyclase activity of cultured human embryonic intestinal epithelial cells was increased 5- to 15-fold after extended (2 to 18 hr) incubations of intact cells with either cholera toxin or toxin from enteropathogenic *E. coli*. No direct effect of the toxin on adenylyl cyclase could be demonstrated in broken-cell preparations. A rapid, irreversible step appeared to be involved, since the toxin effect could not be prevented by repeated washings of intact cells after exposure for a period as short as 1 min. Maximally effective concentrations of *E. coli* enterotoxin (18-hr incubations) were on the order of 10^{-8}M (1.2 µg/ml). Shorter incubations required higher concentrations of toxin in order to give the same degree of stimulation. Elimination of fetal calf serum during the intact cell incubations increased the sensitivity of cells to the toxin approximately 10-fold. *E. coli* toxin (15 g/ml for 18 hr) also increased the adenylyl cyclase activities of KB (8-fold) and HEp-2 (4-fold) cells.

Kwan and Wishnow (1974) reported a delayed (60 min) increase in both cyclic AMP accumulation and steroidogenesis in cultured (Y-1) mouse adrenal tumor cells by 0.05 — 4 µg/ml *E. coli* enterotoxin. Morphological changes from irregular, flattened cells to spherical configurations accompanied the biochemical changes. Steroid production and cyclic AMP levels remained elevated at least through 24 hr. Steroidogenesis during 3-hr incubations was the same whether toxin was continuously present or present only during an initial 10-min period. Steroidogenic effects of cholera toxin (2 ng/ml) as well as the *E. coli* toxin were inhibited by gangliosides (50 ng/ml) as well as by horse serum anticholeragenoid.

Brunton and Guerrant (1974) also have reported a delayed (10 to 30 min) elevation of cyclic AMP by cholera toxin in 18 different clones of cultured cells. Effects varied from 2- to 100-fold. Half-maximal accumulation in two clones occurred at 30 pM (approximately 2.5 ng/ml). Preincubation of cells with toxin at 3°C did not reduce the lag in cyclic AMP accumulation at 37°C. The time course was unaltered by either cycloheximide or phosphodiesterase inhibitors. In addition, toxin from treated cells produced the same delay in response as did fresh toxin. Several clones of cells showed enhanced sensitivities to stimulatory effects of PGE_1 and isoproterenol after treatment with cholera toxin.

Pawelek, Wong, Sansone, and Morowitz (1973) reported moderate increases in cyclic AMP in mouse melanoma (Cloudman, NCTC 3960) cells in response to MSH. Levels were maximal (12-fold over basal) within 30 min and declined to near basal values by 150 min. Strangely, these apparently maximal cyclic AMP levels of mouse melanoma cells (35 pmoles/mg protein at 30 min) were not much higher than the basal levels observed in many cell lines. The selective resistance of mouse melanoma adenylyl cyclase to MSH or cholera toxin following exposures of intact cells to the respective peptides also has been reported (O'Keefe and Cuatrecasas, 1974).

D. The Enzymes of Cyclic AMP Metabolism in Cultured Cells

1. *Adenylyl Cyclase*

a. *Occurrence and distribution.*

Relatively few investigations to determine the localization and distribution of adenylyl cyclase have been carried out in cultured cell systems. In most experiments a single, low-speed sedimentation of cell homogenates has been utilized. This technique demonstrates that most of the enzyme activity is associated with particulate fractions, an observation consistent with the findings obtained in most mammalian tissues (Robison, Butcher, and Sutherland, 1971).

One of the earliest studies (Burk, 1968) compared the adenylyl cyclase activity of BHK 21/13 cells with those of two transformed counterparts, Py Y (BHK 21/13 transformed by polyoma virus) and BR7 (BHK 21/13 transformed by Bryan strain Rous sarcoma virus). Particulate fractions recovered from 2,000 × g centrifugations contained the highest amount of basal activity in BR7 preparations, an intermediate amount of activity in BHK 21/13 pellets, and the lowest activity in the fraction from Py Y cells. Little enzyme activity was associated with the 2,000 × g supernatant fractions isolated from any of the cells. Interestingly, the combination of supernatants derived from either Py Y cells or BHK 21/13 cells with the particulate fraction from BHK 21/13 cells enhanced the apparent adenylyl cyclase activity.

Plasma membrane fractions from myoblast cultures (L6) have been reported to contain adenylyl cyclase activity enriched only about 10% over that of the homogenate (Wahrman, Luzzati, and Winand, 1973*a;* Wahrman, Winand, and Luzzati, 1973*b*). In the myoblast membrane preparations, the unstimulated adenylyl cyclase activity appeared surprisingly high (approximately 540 pmoles cyclic AMP/mg protein/min), even compared to the range of activities observed with NaF stimulation in most other cultured cell lines (25 to 300 pmoles cyclic AMP/mg protein/min).

In at least two cell lines, there have been reports to indicate that little or no adenylyl cyclase was present. Both Granner, Chase, Aurbach, and Tomkins (1968) and Makman (1971*a*) reported that rat hepatoma HTC cells grown in suspension culture were devoid of adenylyl cyclase activity. However, Makman also showed that when the cells were grown under monolayer culture conditions, an extremely low but significant activity appeared. Manganiello and Vaughan

(1972b) provided support for this when they demonstrated that low levels of cyclic AMP would accumulate in monolayer cultures of HTC cells treated with epinephrine in the presence of a phosphodiesterase inhibitor. A second adenylyl cyclase-deficient cell, AC−, has only recently been described by Bourne, Coffino, and Tomkins (1974). This cell is one of several clones of S49 mouse lymphosarcoma cells isolated by cytolytic selection in soft agar, and may prove to be an invaluable tool in the investigation of cyclic nucleotide metabolism (see also Section III).

b. *Characteristics.*

(1) *Effects of Hormones.* The properties of adenylyl cyclases from cells of adrenal tumor origin have probably been the most thoroughly characterized of cell culture systems. Steroid-producing, ACTH-sensitive mouse adrenal tumor cells studied by Taunton et al. (1969) showed general characteristics similar to most normal mammalian tissues. The adenylyl cyclase was localized in low-speed particulate fractions of the cells, and the specificity for activation was restricted to ACTH and sodium fluoride. Optimal stimulation of the enzyme by ACTH was obtained in a pH range from 7.4 to 7.7 with a Mg^{2+}:ATP ratio of 1:1.5. The activity with ACTH was reduced in the presence of calcium, sodium, or potassium ions. In a subsequent report, Schimmer (1972) compared a similar adrenal tumor cell line, Y-1, with two mutant cell lines, Y-6 and OS3. Both the Y-6 and OS3 lines showed no steroidogenic response to ACTH but demonstrated a reasonable steroidogenic response to cyclic AMP. The general properties of adenylyl cyclases in all three cell types were essentially identical to one another and to those described by Taunton et al. (1969), except that in the two mutant lines ACTH no longer stimulated adenylyl cyclase activity. These data indicated that the lack of steroidogenic response of intact cells to ACTH was due to a lesion at the level of ACTH receptors.

Brush, Sutliff, and Sharma (1974) reported that adenylyl cyclase from monolayer cultures of the transplantable adrenal carcinoma 494 was stimulated about threefold by 460 mU/ml ACTH and about fourfold by 10 μM epinephrine, but not by 20 mM NaF. The activity appeared to be most closely associated with the $700 \times g$ and $17,000 \times g$ particulate subcellular fractions. Unfortunately, the results are somewhat difficult to interpret in terms of relative enzyme specific activities since they are expressed as "dpm ^{3}H-ATP incorporated into cAMP." The control properties of adenylyl cyclase from the intact carcinoma 494 were studied in detail by Ney, Hochella, Grahame-Smith, Dexter, and Butcher (1969), and formed the basis of the ectopic hormone receptor concept, which is reviewed in Section III.C.1.

Schimmer (1971) showed that the adenylyl cyclase activity of rat glial tumor C6 cell homogenates responded to catecholamines in the order isoproterenol > epinephrine > norepinephrine. Monovalent cations, in the order $Li^+ > Na^+ > K^+$, increased basal enzyme activity and potentiated the effects of epinephrine and NaF. At optimal Li^+ concentrations enzyme activity in the presence of epinephrine exceeded that observed in the presence of NaF by almost twofold. Jard, Premont, and Benda (1972) reported similar effects of catecholamines on the activity in 5,000 × g particulate fractions of C6 glial cells. In these cells isoproterenol did not change the K_m for ATP (approximately 1×10^{-4} M). Dopamine (10^{-4} M) also was found to stimulate the particulate adenylyl cyclase activity. Upon examining the effect of temperature on adenylyl cyclase activity, the authors observed that isoproterenol stimulation was approximately twofold greater at 40°C than at 20°C, and exceeded the NaF stimulation at all temperatures.

Acetylcholine (1 to 100 μM) has been reported to activate adenylyl cyclase in homogenates of mouse neuroblastoma cells (Prasad, Gilmer, and Sahu, 1974). Less dramatic effects of dopamine (100 μM) and norepinephrine (100 μM) also could be observed. The combination of acetylcholine, dopamine, and norepinephrine produced additive effects on the adenylyl cyclase activity. Dopamine and norepinephrine stimulation were inhibited by 10 μM haloperidol and propranolol, respectively, while atropine (1 to 100 μM) reduced the responses to norepinephrine, acetylcholine, and dopamine. In addition, nicotine and hexamethonium antagonized the effect of acetylcholine.

A series of cultured human skin fibroblasts studied by Rao, Del Monte, and Nadler (1971) were reported to have an adenylyl cyclase stimulated by ACTH, catecholamines, and NaF, but not by glucagon or vasopressin. Comparisons of activities from low- and high-passage cultures disclosed no apparent differences in activity which would correlate with senescence of the cells. However, studies in our laboratory with WI-38 human fibroblasts revealed no stimulation of adenylyl cyclase by ACTH (Kelly and Butcher, 1974).

Transformed fibroblasts often demonstrate changes in both basal and hormone-stimulated adenylyl cyclase activities. Peery et al. (1971) reported that Py-transformed BHK cells have lower adenylyl cyclase activity than the parent line, in both the NaF- or PGE_1-stimulated and the basal state. Polyoma virus transformation of 3T3 Swiss fibroblasts decreased basal and stimulated activities in a similar manner. However, in one transformed line (PY-11, 3T3), essentially no response to PGE_1 remained. Two cell lines derived from 3T3 parent cells by SV40 and MSU/MuLU transformation each had higher basal and stimulated activities, while yet another SV40-transformed line had less basal activity and was insensitive to PGE_1. Further transformation of the PGE_1-insensitive PY-11, 3T3 cells with SV40 reportedly gave a renewed responsiveness to PGE_1, indicating that the loss of sensitivity to PGE_1 produced by the original transformation was not due to a genetic deletion. A more detailed discussion of enzyme changes brought about by transformation is given in Section III.

(2) *Effects of Cholera Toxin.* Effects of cholera toxin on adenylyl cyclase activity have been studied in Y-1 adrenal cortex carcinoma cells by Wolff, Temple, and Cook (1973). These authors reported that incubations of intact cells with 5 nM toxin for 2 hr dramatically increased adenylyl cyclase activity in 8,000 × g particulate fractions that were subsequently recovered from the treated cells.

Unlike ACTH, which stimulated adenylyl cyclase in cell-free preparations, the toxin was ineffective if added after homogenization of the cells. Maximal stimulation by choleragen (a purified preparation of cholera toxin) did not exceed the effects of 10 mM NaF alone; and although pretreatment of intact cells with choleragen, followed by inclusion of NaF in adenylyl cyclase incubations, did produce a slightly greater activity than either agent alone, the effects were not additive. Half-maximal steroidogenesis occurred with 15 pM cholera toxin, and the effect was irreversible after 1 min, except by the addition of horse anticholeragen antiserum or mixed gangliosides.

O'Keefe and Cuatrecasas (1974) studied the effects of cholera toxin and α-MSH (melanocyte-stimulating hormone) on the adenylyl cyclase from cultured mouse melanoma cells. In these functional tumor cells, cholera toxin, like MSH, induced increases in tyrosinase activity. Such increases also could be correlated with increases in adenylyl cyclase activity. Adenylyl cyclase activities ranged from 4- to 30-fold higher in 40,000 $\times$ g particulate fractions isolated from cells that had been exposed to 10^{-10} M toxin for from 1 to 5 hr, when compared to similar fractions from control cells. A lag of 20 to 30 min occurred before the enzyme's activity increased. As with adrenal cortex carcinoma cells, the toxin was ineffective when incubated only with particulate fractions. From 1 to 13 days after a 4-hr exposure of the cells to toxin, the chronically stimulated adenylyl cyclase activity slowly and progressively decreased. As late as 7 days after the initial exposure to toxin, readdition of toxin had no effect; however, at 13 days, stimulation was the same as seen in control cells. Unlike cholera toxin, MSH had to be continuously present in the culture medium to maintain elevated adenylyl cyclase activities, and removal of MSH for a period of 1 hr allowed a return to basal activity. However, the latter preparations showed an impaired response to additional MSH. NaF produced a much smaller increase in adenylyl cyclase than cholera toxin or MSH, and neither NaF nor MSH produced any significant effects in cells which had been treated with toxin for longer than 2 hr. In order to determine whether MSH and cholera toxin had different binding sites, the displacement of the binding of ^{125}I-labeled choleragen to cells was examined by O'Keefe and Cuatrecasas (1974). These authors reported that 5×10^{-7} M MSH did not affect the binding of 10^{-10} M labeled choleragen, whereas 2×10^{-9} M native cholera toxin produced half-maximal displacement.

(3) *Effects of Culture Conditions.* In an early study, Makman (1970) showed differential effects of suspension culture and monolayer culture upon the epinephrine- and NaF-stimulated activities of Chang's liver cells. In neither monolayer nor suspension culture were the cells responsive to glucagon, unlike normal human liver from which the cells were reported to be derived. Adenylyl cyclase activities in cells from monolayer cultures were stimulated less in the presence of epinephrine, but more with NaF than cells in suspension. However, the basal activities in both suspension and monolayer cells were greater than the stimulated activity of rat liver homogenates.

A similar investigation with HeLa S3 cells (Makman, 1971a) gave results that may be contrasted to those obtained with Chang's liver cells. HeLa cells grown in monolayer culture had both higher basal and stimulated adenylyl cyclase activities (i.e., in the presence of epinephrine, glucagon, PGE_1, or NaF) than cells grown in suspension. A fourfold increase in population density of the cells in suspension had no effect on adenylyl cyclase activity. However, the effect of population density on the responses of monolayers of 3T3 cells was similar to that seen in Chang's cells, in that both basal and epinephrine-stimulated activities were increased at higher densities.

Anderson, Russell, Carchman, and Pastan (1973b) compared adenylyl cyclase, phosphodiesterase, and basal cyclic AMP levels in normal rat kidney fibroblasts (NRK), which are capable of density-dependent growth inhibition, and in chick embryo fibroblasts (CEF), which show no density-dependent inhibition of growth. As population density increased, unstimulated or NaF-stimulated adenylyl cyclase activities, as well as phosphodiesterase activities, increased. In NRK cells, phosphodiesterase activities reached a peak with confluency and cessation of growth, and declined slightly thereafter. However, in CEF cells, the activities of both enzymes increased, more or less in parallel, to reach a maximum at high density. The enzyme data were offered as an explanation for the fact that basal cyclic AMP levels increased with growth in density-inhibited NRK, but not in CEF cells. Obviously, many possible unknown changes in other factors with growth, such as changes in the K_m of enzymes, in subcellular distribution, in access to substrate, or in the levels of endogenous substrates may be pertinent to the interpretation of these data. Furthermore, the status of cellular components responsive to cyclic AMP, be it protein kinase, specific nuclear material, or other factors, would certainly be important in drawing conclusions, if indeed cyclic AMP was involved in growth regulation. More discussion on this subject is given in Section III.

Another approach used to investigate the effects of growth conditions on adenylyl cyclase activity was carried out by Makman and Klein (1972) using synchronized Chang's liver cells. These workers reported that both basal and NaF-stimulated adenylyl cyclase activities were sharply decreased in thymidine-blocked Chang's liver cells. Enzyme activity decreased even more following removal of the thymidine block, to reach a low point 6 hr later. Thereafter, a gradual increase occurred with the maximum activity being achieved at mitosis. NaF-stimulated activity seemed to recover somewhat earlier in the cell cycle than either the unstimulated or epinephrine-stimulated activities.

(4) *Comments.* Although there are a number of reports on adenylyl cyclase in cell culture extant, there has been a dearth of systematic, in-depth studies which might contribute to our understanding of the role of the enzyme system in the control of cyclic AMP levels. This is understandable, for the amounts of tissue required for such studies is large. Hence, such preparations are both time-consuming and difficult.

On the positive side, it would seem that a number of guidelines have emerged for studies of adenylyl cyclase to complement the intact cell studies described earlier. It is manifest from the studies of Makman and Pastan and their colleagues and others that culture conditions, viral transformation and many other factors may influence apparent adenylyl cyclase activities. Further, if one single message stands out to us, it would seem to be that these factors as well as the cyclic AMP responses of intact cell preparations must be taken into account if a more lucid understanding of the control of adenylyl cyclase is to be achieved.

2. Cyclic-3′,5′-nucleotide Phosphodiesterases

a. Occurrence, distribution, and properties.

All cultured cells studied to date have shown some level of cyclic-3′,5′-phosphodiesterase activity, and the activity appears to be largely intracellular (Heidrick and Ryan, 1971a,b; Penit et al., 1974). In addition, the phosphodiesterase activity of cultured cells, as of most normal mammalian systems, is associated with both particulate and "soluble" fractions. Perkins, Macintyre, Riley, and Clark (1971b) reported a distribution in 1181N1 astrocytes of approximately 23% particulate (100,000 × g) and 77% soluble. In these studies, there was evidence for only a single apparent K_m for the soluble activity (15 μM cyclic AMP), unlike comparable rat brain preparations which express two apparent K_m activities (Brooker, Thomas, and Appleman, 1968). Two apparent phosphodiesterase activities were similarly distributed among subcellular fractions of Novikoff rat hepatoma, HeLa, and L cells, with approximately 60 to 70% of the total activity being present in the cell sap (Schroder and Plagemann, 1972). Other cell strains that have shown phosphodiesterase activities with more than one apparent K_m include 3T3-4 cells, 3T3-SV40 cells (D'Armiento, Johnson, and Pastan, 1972), L-929 cells (Manganiello and Vaughan, 1972a), HTC hepatoma cells (Manganiello and Vaughan, 1972b), chick embryo fibroblasts (Russell and Pastan, 1973), Novikoff rat hepatoma cells, HeLa cells, two L-cell sublines (Schroder and Plagemann, 1972), and C6 rat glial cells (Jard et al., 1972).

The phosphodiesterase activities of C6 glial cells appeared to be largely particulate in that roughly 90% of the total activity measured at both high and low cyclic AMP concentrations sedimented at 5,000 × g (Jard et al., 1972). Russell and Pastan (1973) found a lesser but nonetheless substantial proportion of phosphodiesterase activity localized in plasma membrane fractions of chick embryo fibroblasts. This membrane-associated activity was specific for hydrolysis of cyclic AMP, whereas soluble fractions hydrolyzed both cyclic AMP and cyclic GMP. Resolution of soluble activity on DEAE-cellulose gave two peaks of activity, one specific for cyclic AMP and the other active with either cyclic nucleotide. The cyclic AMP-specific peak from the soluble fraction showed dual kinetic properties similar to those of the plasma membrane fraction. On the other hand, the peak active with both nucleotides gave classical Michaelis-Menton kinetics with either substrate. Preexposure of intact cells to trypsin for 10 min increased the K_m, and after 60 min decreased the V_{max}, of the soluble cyclic GMP degradation activity.

b. *Regulation.*

Several studies have examined phosphodiesterase as a function of growth. As would seem likely *a priori,* Schroder and Plagemann (1972) found wide differences and very little correlation of activity with growth rate among a variety of cell strains having doubling times between 12 and 32 hr. The fastest growing cell (Novikoff rat hepatoma, N1S1-67) did have a slightly higher specific activity at low substrate and a much lower specific activity at high substrate levels than the other lines studied. However, no direct correlation could be made among the slower growing cells. It is of interest that the results of Anderson et al. (1973b) also indicate that rapidly growing, contact-independent CEF cells had a higher specific activity at low substrate and a lower specific activity at high substrate levels than did density-dependent, growth-inhibited NRK cells. However, it seems likely that more direct correlations might occur with growth differences within a single cell type if, indeed, variations do occur with growth. Schroder and Plagemann (1972) found no significant difference in either low or high substrate activity in growing versus density-dependent growth-inhibited suspension cultures of L cells. However, monolayer cultures of L (Heidrick and Ryan, 1971a), 3T3 (D'Armiento et al., 1972), CEF, and NRK cells (Anderson et al., 1973b) showed significant increases in phosphodiesterase activity with increases in population density. In 3T3 cells, there was no apparent change in K_m at either low or high substrate concentrations. In the density-inhibited NRK cells, phosphodiesterase (at both low and high substrate) increased to a peak level with increasing density and declined slightly to a plateau at confluence while the activity in density-independent CEF cells increased progressively through confluence.

The apparent anomaly between cells in suspension and monolayer cultures, particularly in the case of L cells, is reminiscent of the results of Makman and Klein with adenylyl cyclase in Chang's liver cells, as previously discussed, where the enzyme activity also increased with increasing population density in monolayer but not in suspension cultures. It seems possible that a stable cell-cell interaction is involved.

Bacalao and Rieber (1972) have reported the separation of multiple bands of L-cell phosphodiesterase activity by polyacrylamide gel electrophoresis. Assays with cyclic AMP and cyclic GMP yielded similar banding patterns. A markedly different banding pattern was found to occur after growth of cells for 72 hr with PGE_1, which increases cyclic AMP in the cells. In addition, PGE_1 treatment appeared to cause a higher degree of association of the activity with particulate cell fractions. In a similar study with clone C2A rat astrocytoma cells, Uzunov and Weiss (1972) and Uzunov, Shein, and Weiss (1973) obtained two peaks of phosphodiesterase activity from $100,000 \times g$ supernatant fractions by polyacrylamide gel electrophoresis. These two activities appeared to coincide with two of six activities separated from similar rat brain preparations. One of the activities was selectively increased by growth of the cells for 6 hr with 0.3 mM norepinephrine. This effect could be blocked by the adrenergic antagonist propranolol or by cycloheximide, suggesting an induction of phosphodiesterase synthesis by the increased cyclic AMP levels, as has been indicated in a number of other cell lines.

The regulation of phosphodiesterase by intracellular cyclic AMP in some cells has been established by a number of studies using cell culture systems. Manganiello and Vaughan (1972a) demonstrated a gradual increase in phosphodiesterase in L-929 cells during a 24-hr exposure to PGE_1. In the same period, cyclic AMP levels increased rapidly to a maximum, then declined gradually to near basal levels, even though transfer of the 24-hr incubation media to fresh cells showed that the PGE_1 was still fully active. Also, the addition of fresh PGE_1 to previously exposed cells at 24 hr gave no significant increase in cyclic AMP. Thus, the large increase in phosphodiesterase activity appeared to be the result of increased cyclic AMP levels. This increase in phosphodiesterase could at least partially explain both the decrease in cyclic AMP from maximum levels and the insensitivity of the cells to PGE_1 after prolonged stimulation. The increased activity was more apparent at low substrate concentrations. Phosphodiesterase activity in these cells was not appreciably affected by theophylline.

Almost simultaneously with the publication by Manganiello and Vaughan, D'Armiento et al. (1972) reported that treatment of 3T3- and SV40-transformed 3T3 cells with dibutyryl cyclic AMP and theophylline for 40 hr caused approximately a twofold increase in phosphodiesterase activity. The effect of these agents was blocked by the addition of either cycloheximide or actinomycin D, suggesting *de novo* synthesis of the enzyme. Mixtures of extracts from treated and untreated cells gave additive activity indicating that the presence of an activator or the absence of an inhibitor was probably not involved. These authors further confirmed the results of Manganiello and Vaughan with L-929 cells, in that treatment with PGE_1 for 24 hr caused a 30% increase in phosphodiesterase activity.

Prasad and Kumar (1973) reported that dibutyryl cyclic AMP (0.5 mM), PGE_1 (28.5 μM), and the phosphodiesterase inhibitor R020-1724 (0.7 mM) increased phosphodiesterase activity in neuroblastoma NBA2 cells two- to threefold in 72 hr while sodium butyrate, 5'-AMP, cyclic AMP and theophylline (all at 0.5 mM) had no effect. The addition of cycloheximide (5 μg/ml) for 6 hr after an initial 24 hr with dibutyryl cyclic AMP reduced activity to control levels. However, cycloheximide alone reduced the activity in untreated cells to 25% of the control. Actinomycin D had no effect under these conditions.

Schwartz and Passonneau (1974) demonstrated induction of both high- and low-K_m phosphodiesterase activities in C6 rat glioma cells by 1 mM dibutyryl cyclic AMP plus 1 mM theophylline or by 50 to 100 μM norepinephrine. Simultaneous administration of cycloheximide (25 μg/ml) or actinomycin D (5 μg/ml) with the inducing agents completely blocked the increases in activity, as did dichloroisoproterenol or propranolol. Four-hr incubations with norepinephrine, which approximately doubled both phosphodiesterase activities, dramatically reduced the response to subsequent additions of norepinephrine. An acute threefold increase in low-K_m phosphodiesterase activity within the first 5 min appeared to be correlated with the attainment of maximal cellular cyclic AMP levels.

Bourne, Tomkins, and Dion (1973b) reported that the S49 line of lymphoma cells contained increased phosphodiesterase activity following stimulation by either isoproterenol or PGE_1. Interestingly, these cells are killed by high levels of cyclic AMP. However, a cloned strain of these cells resistant to the lethal effect of cyclic AMP contained only a very low level of phosphodiesterase, which could not be increased upon prolonged stimulation of the cells with either isoproterenol or PGE_1. In addition, the cyclic AMP responses to isoproterenol and PGE_1 were much higher and more prolonged in these cyclic AMP resistant cells. The resistant strain of lymphoma cells was also found to be deficient in protein kinase, which is thought to mediate many of the actions of cyclic AMP. For this reason, the authors suggested that cyclic AMP-sensitive protein kinase activity may be necessary for the adaptive response of phosphodiesterase to increased levels of cyclic AMP.

A 25 to 40% decrease of phosphodiesterase activity in monolayer HTC cells after 72 hr growth in 10 μM dexamethasone has been reported by Manganiello and Vaughan (1972b). Following dexamethasone treatment, intracellular cyclic AMP, measured in the presence of either theophylline or theophylline plus epinephrine, increased approximately fourfold. This action of the corticoid analogue was considered as a possible explanation of the permissive effects of the steroids on the actions of other hormones. A similar result was reported for human foreskin fibroblasts (Manganiello et al. 1972). The effectiveness both of catecholamines and of PGE_1 was increased 50 to 100% after 2- or 3-day incubations with 1 μM dexamethasone.

c. *Inhibitors.*

Several phosphodiesterase inhibitors have been shown to potentiate increases in cyclic AMP in cultured cells by a variety of agents. Here we will attempt only to recapitulate a few reports. Theophylline, though perhaps most commonly used, is generally only moderately effective. Schroder and Plagemann (1972) reported differential effects of theophylline on the phosphodiesterase activities of cultured Novikoff rat hepatoma, HeLa, and L cells at high and low substrate concentrations. At high cyclic AMP concentrations theophylline gave a simple competitive inhibition, but at low cyclic AMP the inhibition was noncompetitive.

One of the more potent inhibitors is MIX. Schultz et al. (1972) found that the addition of either MIX or papaverine to rat glial cells prevented a refractoriness to catecholamines which developed with repeated stimulations, and which appeared to be due to induced increases in phosphodiesterase activity. In five cloned neuroblastoma cell lines, Hamprecht and Schultz (1973 a,b) found that while both papaverine and MIX acted synergistically with PGE_1 to increase intracellular levels of cyclic AMP, MIX was the more potent agent. MIX also has been shown to sensitize a variety of cells to agents which otherwise did not increase cyclic AMP levels (Schultz and Hamprecht, 1973; Gilman, 1974; Kelly and Butcher, 1975; see also Section II, C.3.)

Papaverine, which has been used frequently as a phosphodiesterase inhibitor, has been shown to cause effects in C6 astrocytoma cells which may be unrelated to any action on phosphodiesterase (Browning, Groppi, and Kon, 1974a). While papaverine did potentiate the norepinephrine elevation of cyclic AMP in C6 cells, alone it increased glycogenolysis to a greater extent than did norepinephrine alone. However, unlike norepinephrine, papaverine alone produced no apparent increase in cyclic AMP levels. Similarly, Kowal and Harano (1974) have reported multiple effects of papaverine in cultured adrenal cells. When used at concentrations between 0.05 and 0.2 mM, papaverine increased cyclic AMP in both the presence and absence of ACTH. However, the compound also markedly inhibited ACTH-stimulated steroidogenesis, increased ^{14}C-lactate production from ^{14}C-glucose, reduced the incorporation of ^{32}P into macromolecules, and reduced incorporation of ^{3}H-thymidine into DNA. It is not known whether the effect on incorporation of ^{3}H-thymidine into DNA was direct, or due to reduced uptake into the cells.

Two other phosphodiesterase inhibitors used in cell culture studies are RO20-1724 [4-(3-butoxy-4-methoxybenzyl)-2-imidazolidinone], and ICI 63, 197 [2-amino-6-methyl-5-oxy-4-n-propyl-4,5-dihydro-S-triazolo (1,5-a) pyrimidine]. Unfortunately, there is little useful information regarding the latter as a phosphodiesterase inhibitor in cell culture systems. Franklin and Foster (1973a,b) reported use of ICI 63,197, together with PGE_1 or isoproterenol, to elevate cyclic AMP levels in fibroblasts; but no information was given to indicate the degree of potentiation obtained or any other properties of the inhibitor. Blume et al. (1973) reported an increased response of neuroblastoma cells to adenosine in the presence of 10^{-6} to 10^{-3} M RO20-1724. Maximum effects occurred within 5 min after addition of the inhibitor to cells. Basal cyclic AMP levels were unaffected by RO20-1724 in one neuroblastoma clone (NS20), but appeared to be slightly elevated in another (N1E).

Despite the successful use of these disparate compounds to potentiate the cyclic AMP responses to a number of agonists in a number of cell lines, the time-honored caveat about drug specificity remains. Indeed, as previously discussed, the inhibitors have a variety of effects which often are distinctly different than a specific inhibition of phosphodiesterase.

It would seem that studies with cell culture systems have already provided several important insights into the importance of phosphodiesterase activity in the control of cyclic AMP levels. Far from being the simple "kinetic brake" originally suggested, in great naivete (Butcher and Sutherland, 1962), a number of discoveries in recent years have pointed to the complexity of the system (Appleman, Thompson, and Russell, 1973). One major contribution from cell culture studies was the demonstration that phosphodiesterase activity could be adaptive, being specifically increased when intracellular cyclic AMP levels are elevated. Further, differences in cellular phosphodiesterase activity which occur

3. Protein Kinase Activity

Among the earliest studies of protein kinase activity in cultured cells was that of Greengard and Kuo (1970), who demonstrated that the activities in two neuroblastoma lines, *Neuro*-2a and C1300, were stimulated twofold by cyclic AMP but not by cyclic GMP. C1300 cells had approximately 300-fold higher activity than the *Neuro*-2a cells. Another study by Perkins et al. (1971*b*) showed that protein kinase activity was increased approximately twofold by cyclic AMP using casein as substrate and fourfold with calf thymus histone as substrate in cell-free extracts of the astrocytoma clone 1181N1. Half-maximum stimulation occurred between 3 and 5×10^{-8} M cyclic AMP using either protein substrate. The activity was about equally distributed between particulate and supernatant fractions, with the particulate activity being largely concentrated in the 1,000 × *g* pellet. In contrast, fractionation of protein kinase from rat glial C6 cells showed that 5,000 × *g* particulate fractions contained more enzyme activity and were stimulated to a greater extent by cyclic AMP than corresponding supernatant fractions (Jard et al., 1972). However, the 5,000 × *g* supernatant fractions contained protein kinase with higher specific activity.

Klein and Makman (1971) reported the presence of cyclic AMP-sensitive protein kinase activity in Chang's liver, HeLa, and 3T6 cells. The partially purified enzymes were all stimulated two- to threefold by 1 μM cyclic AMP. Preparations of 3T6 cells contained the highest control activity (26.3 pmoles P transferred to histone per 100 μg enzyme protein per 5 min) and were stimulated to the greatest extent by cyclic AMP. Cyclic GMP did not affect the kinase activity from HeLa or 3T6 cells. Makman and Klein (1972) subsequently reported no change in the specific activity of protein kinase, assayed in either the absence or presence of cyclic AMP, throughout the cell cycle of thymidine-synchronized Chang's liver cells. That is, as total cell protein increased, protein kinase activity increased in the same proportion. A brief report by Troy, Vijay, and Kawakami (1973) also indicated that there was no difference between the protein kinase (48,000 × *g* supernatant) activities of normal and feline sarcoma virus-transformed bovine thymic cells in monolayer culture.

It has been indicated in studies of many different types of cells that cyclic AMP or its dibutyryl derivative induce morphological flattening and extension of processes, concomitant with the assembly of microtubular proteins. However, Casola, DiMatteo, and Augusti-Tocco (1974) were unable to demonstrate any relationship between cyclic AMP-dependent phosphorylation of microtubular proteins and neurite extension of neuroblastoma (NB41A3) cells. Cells grown in suspension culture, where neurite outgrowth does not occur, had essentially the same protein kinase activty in the presence or absence of cyclic AMP as did monolayer cultures, where extended processes are the characteristic morphology. In addition, ^{32}P labeling of endogenous proteins showed no differences in the abilities of the two populations to incorporate the label over a 6-hr period of incubation.

Tan and Sokol (1974) have reported the isolation of a protein kinase fraction from African green monkey kidney cells using DEAE-cellulose chromatography of the 100,000 $\times$ g supernatant. This protein kinase preparation was stimulated approximately twofold by 10^{-5} M cyclic GMP, and, at best, only slightly by 10^{-8} M cyclic AMP. In addition, it was indicated that 2 to 50 mM dithiothreitol caused a twofold stimulation of the enzyme. Unfortunately, it is difficult to understand the significance of their observations.

Cyclic CMP and cyclic GMP, as well as cyclic AMP, were found to increase the activity of partially purified protein kinase from mouse adrenal cortex tumor cells (Kowal, 1973). Effective concentrations of cyclic CMP and cyclic GMP were approximately 100-fold higher than those of cyclic AMP; but the maximum activity with cyclic CMP or cyclic GMP was approximately 50% greater than that with cyclic AMP. All three cyclic nucleotides competitively displaced ^{3}H-cyclic AMP from the associated binding protein at concentrations similar to those required for protein kinase activation. Thus, the author felt that the protein kinase appeared to be a single species with multiple specificity and differential sensitivity to the various cyclic nucleotides. Conversely, Pawelek et al. (1973) observed only cyclic AMP-binding activity in cell extracts of mouse melanoma cells. Whereas nonradioactive cyclic AMP competitively displaced bound ^{3}H-cyclic AMP, 5'-AMP, cyclic UMP, cyclic CMP, and cyclic GMP had no effects at concentrations 100 times those of ^{3}H-cyclic AMP. Incubation with ^{3}H cyclic AMP plus 4 mM theophylline for 60 min at 0°C followed by chromatography of the mixture on Sephadex G-25 showed that substantial amounts of label were associated with the high-molecular-weight material of the void volume. Binding activity was reduced to less than 10% of initial values by treatment with pronase plus trypsin but not by pancreatic nucleases.

A potentially very useful study was that of Daniel, Litwack, and Tomkins (1973b), who reported isolation of a clone of mouse lymphoma cells deficient in cyclic AMP-dependent protein kinase. As mentioned previously, "normal" S49 lymphoma cell populations were killed by increased intracellular cyclic AMP levels. Presumably, this is the usual ultimate fate of lymphocytes. These studies showed that the strain resistant to lethal doses of dibutyryl cyclic AMP (0.1 mM) contained much lower cyclic AMP-dependent protein kinase activity and lower cyclic AMP-binding capacity than cells that were not resistant, which may relate to the mutants being incapable of carrying out the purported differentiated function of dying. Granner (1972, 1974) has reported that the cyclic AMP binding subunit but not the catalytic subunit of protein kinase was deficient in several cultured hepatoma strains. The studies on mouse lymphoma cells and hepatoma cells are discussed further in Section III.

Prolonged hydrocortisone treatment has been reported to increase both total and dibutyryl cyclic AMP-stimulated protein kinase activities in Reuber H-4-II-E hepatoma cells (Sahib, Jost, and Jost, 1971). A corresponding increase in phosphorylation of nuclear histones occurred, and by 6 hr about a 50% rise in tyrosine aminotransferase activity was observed. No increase in ^{3}H-cyclic AMP binding activity was detected after hydrocortisone treatment.

The paucity of data relating to the activities of protein kinases in cultured cells is particularly lamentable in view of the importance of the system to an understanding of the interactions of extracellular signals and intracellular responses. A few further comments on this subject appear in the next section, but these do little to illuminate the darkness surrounding it. Perhaps one generalization can be made: to wit, that the difficulties inherent in working with protein kinase in a truly analytical fashion have thus far precluded more than the simple knowledge of its existence in almost all cell lines which have been examined.

4. Cell Functions Altered by Cyclic AMP

Numerous cellular functions have been found to be affected by cyclic AMP, dibutyryl cyclic AMP, or agents which increase cellular cyclic AMP levels in cultured cells. We have attempted to catalog some representative cultured cell systems and the functions affected in Table 11. Several important areas have been covered in detail in other sections and are omitted here for simplicity. The effects of cyclic AMP on cell growth, agglutination, and morphology are covered separately in Section III. The induction of phosphodiesterase by cyclic AMP was covered in substantial detail in the general discussion of phosphodiesterase, and the effects of cyclic AMP on protein kinases were just summarized. On the other hand, a detailed discussion of the many phenomena cited in Table 11 would be beyond the scope of the present survey.

E. Cyclic GMP

Cyclic GMP has emerged as the antithesis of cyclic AMP in a "dualism" hypothesis of cyclic nucleotide action. This concept has been extensively treated by Goldberg, O'Dea, and Haddox (1973) and will not be reviewed here. Perhaps the greatest attention to cyclic GMP in cell culture systems involves its possible role in cell division, which is discussed in Section III.E. However, some lines of experimentation are related to both cell division as well as cyclic GMP metabolism in cultured cells and are discussed below.

Hadden, Hadden, Haddox, and Goldberg (1972) first suggested a role for cyclic GMP in peripheral lymphocytes when they reported that the purified mitogens, concanavalin A and phytohemagglutinin, increased lymphocyte proliferation and intracellular levels of cyclic GMP, but did not affect cyclic AMP. Conversely, crude preparations of phytohemagglutinin, which caused agglutination, also increased cyclic AMP. More recently, Seifert and Rudland (1974) reported that quiescent 3T3 cells stimulated to grow with 20% serum in fresh medium showed a concurrent rise in cyclic GMP levels and fall in cyclic AMP levels within

TABLE 11. *Biochemical effects of exogenous cyclic AMP, derivatives of cyclic AMP, or other agents which increase intracellular cyclic AMP levels in various cultured cells*

Cell line	Agent added	Effect produced	Comments	Reference
C-6 rat astrocytoma	db-Cyclic AMP (10^{-3} M), papaverine (10^{-4} M), NE (10^{-4} M), or histamine (10^{-4} M)	Glycogen depletion	10^{-4} M propranolol blocked NE action	Opler and Makman, 1972; Browning et al., 1974a
Fetal rat brain cultures	mb-Cyclic AMP (1 mM)	Increased glutamic acid decarboxylase	6–12 days exposure	Schrier and Shapiro, 1973
Mouse neuroblastoma N1E-115	db-Cyclic AMP (1 mM), PGE_1 (10 μM), or Na butyrate	Increased tyrosine hydroxylase	Twofold increase in 2 days; Na butyrate control completely blocked growth	Richelson, 1973
Fetal rat brain mass cultures	mb-Cyclic AMP (1 mM) for 12 days	Increased acetylcholinesterase; decreased choline acetyltransferase		Shapiro, 1973
Mouse neuroblastoma Cl300 clone NB60	db-Cyclic AMP (1 mM) for 48–96 hr	Increased acetylcholinesterase	Effect mimicked by deprivation of serum	Furmanski et al., 1971
Mouse neuroblastoma Cl300 and clone NBP	db-Cyclic AMP (015 mM) or Na butyrate (0.5 mM)	Increased tyrosine hydroxylase	Effect blocked by cycloheximide	Waymire et al., 1972
Rat astrocytoma RGC_6	db-Cyclic AMP (0.5 mM) or catecholamines (10^{-9}–10^{-5} M)	Increased lactate dehydrogenase	Effect blocked by acetoxycycloheximide or actinomycin D	deVellis and Brooker, 1972
Chick embryo brain primary culture	L-Thyroxine (10^{-7} M), db-cyclic AMP (1 mM), plus theophylline (1 mM)	Increased choline acetyltransferase and acetylcholinesterase	6–8 days treatment	Werner et al., 1971
Reuber hepatoma (H35)	db-Cyclic AMP (0.5 mM)	Increased phosphoenolpyruvate carboxykinase and tyrosine aminotransferase	Effect blocked by cycloheximide or actinomycin D	Wicks and McKibbin, 1972; Barnett and Wicks, 1971; Butcher et al., 1972; Grossman, et al., 1971

Cell type	Treatment	Effect	Comments	Reference
Reuber hepatoma (H-4-II-E)	db-Cyclic AMP (0.1–0.5 mM)	Increased tyrosine aminotransferase	Effect blocked by cycloheximide or actinomycin D; enhanced by 10^{-8} M hydrocortisone	Butcher et al., 1971; Sahib et al., 1971
Rat hepatoma HTC	db-Cyclic AMP (2 mM)	Increased tyrosine aminotransferase	Effect blocked by cycloheximide or actinomycin D	Stellwagon, 1972
Quail embryo liver	db-Cyclic AMP (0.02–1 mM) or cyclic AMP (0.03–2 mM)	Decreased cellular aggregation	Only cyclic AMP effect was reversible	Kuroda, 1974
Primary rat liver cells	Cyclic AMP (0.15 mM)	Increased L-leucine incorporation		Armato et al., 1974
Mouse embryo fibroblasts	db-Cyclic AMP (3 mM), PGE_1 (14 μM), 17β-estradiol (0.5 μg/ml), or bradykinin (1.0 μg/ml)	Increased glycosaminoglycan (GAG) synthesis	Antagonized by benzydamine (10^{-4} M) and by prednisolone (10 μg/ml). PGE_1 increased cyclic AMP prior to GAG synthesis	Schonhofer et al., 1974; Peters et al., 1974
Mouse L cells or rat embryo cells	db-Cyclic AMP (2 mM) or epinephrine (5 mM) plus theophylline (5 mM)	Decreased interferon production upon viral challenge	Effect blocked by cycloheximide	Dianzani and Zucca, 1972
Mouse L cells	db-Cyclic AMP (1 M) or PGE_1 (10 μM)	Reduced cell interaction with fibrin		Niewiarowski and Goldstein, 1973
Mouse 3T3 fibroblasts	db-Cyclic AMP (0.2 mM) plus theophylline (1 mM) or PGE_1 (15 μM)	Reduced uridine, leucine, 2-deoxyglucose, and thymidine uptake	Reversed by cyclic GMP	Kram et al., 1973
Chick embryo fibroblasts	Cyclic AMP (4 mM)	Increased uridine incorporation into RNA	Effect blocked by puromycin or cycloheximide	Koblet et al., 1973
Mouse 3T3-4 and SV40-transformed fibroblasts	db-Cyclic AMP (1.2 mM) plus theophylline (1 mM)	Increased ^{35}S incorporation into acid mucopolysaccharides		Goggins et al., 1972
Chinese hamster ovary (CHO) cells	db-Cyclic AMP (1 mM) or db-Cyclic AMP (0.2 mM) plus testosterone (15 μM)	Reduced transport of thymidine, uridine, aminoisobutyrate, and glutamine; increased collagen synthesis	Half-maximal effects observed at 6 μM db-cyclic AMP	Hauschka et al., 1972; Hsie, et al., 1971; Rozengurt and Pardee, 1972
CHO-KI	Cyclic AMP (2 mM)	Increased uridine incor-	Effect was reversible	Tihon and Green, 1973

TABLE 11. (Continued)

Cell line	Agent added	Effect produced	Comments	Reference
Somatic cell hybrid B-6	plus testosterone propionate (0.015 mM)	poration into virus-like RNA		
	Cyclic AMP (3 mM) or db-cyclic AMP (0.25–5 mM) with theophylline (1 mM)	Increased alkaline, but not acid, phosphatase	Effect blocked by cycloheximide or actinomycin D	Koyama et al., 1972
Chick embryo retina	db-Cyclic AMP (0.4 mM), cyclic AMP (1.2 mM), or cortisol (0.03 μM)	Increased glutamine synthetase	Effect blocked by cycloheximide or actinomycin D	Chader, 1971
Monkey granulosa cells	db-Cyclic AMP (0.2 mM) or LH (0.1 μg/ml)	Increased progestin secretion and luteinization	Variable effects over time or with maturation of cells	Channing, 1974
Mouse adrenal tumor Y-6	Cyclic AMP (1 mM) or ACTH (5 mU/ml)	Increased steroid secretion	Clonal variants differ in response	Schimmer, 1969
Mouse adrenal cortex tumor	Cyclic AMP (1–10 mM), db-cyclic AMP (1–3 mM), or ACTH (10 mU/ml)	Increased steroid output	Effect mimicked by cyclic GMP, ATP, ADP, AMP, or adenosine	Kowal and Fiedler, 1969; Kowal, 1973
Mouse Cloudman CC153 melanoma (NCTC3960)	Cyclic AMP (0.1 mM), db-cyclic AMP (1–100 μM), α-MSH (0.5 μM), MSH-Sepharose (0.2 μM), or choleratoxin (3 × 10⁻¹⁰M)	Increased pigmentation and tyrosinase activity	Effect of db-cyclic AMP was irreversible; MSH effect blocked by cycloheximide or actinomycin D	Johnson and Pastan, 1972b; Varga et al., 1974; Wong and Pawelek, 1973; Pawelek et al., 1973; O'Keefe and Cuatrecasas, 1974
Melanoma B16	db-Cyclic AMP (1 mM) plus theophylline (1 mM)	Increased pigmentation; decreased growth	Cyclic AMP (3 mM) decreased pigmentation and growth	Kreider et al, 1973
Newborn rat lens epithelial cells	db-Cyclic AMP (0.01 mM) or caffeine (0.001–1 mM)	Increased crystalline formation	Both agents stimulated growth	Creighton and Trevithick, 1974
L-929 fibroblasts	Cyclic AMP (10 mM) and various substituted analogues	Enhanced antiviral interferon activity	The 6-methyl thio-derivative was 16- to 32-fold more active than cyclic AMP	Allen et al., 1974

Mouse spleen cells	Cyclic AMP (0.05–0.3 mM)	Increased antibody formation vs. sheep red blood cells	Cells were suspended in soft agar	Shimamura et al., 1973
Mouse adrenal tumor (Y-1)	Cyclic AMP (1 mM)	Increased steroid production	Effect inhibited by D_2O; colchicine acted similarly to cyclic AMP	Temple and Wolff, 1973
HeLa 65	db-Cyclic AMP (0.5 mM), sodium butyrate (0.5 mM), or cortisol (3 µM)	Increased alkaline phosphatase	Pentanoic and 2-methylbutyric acids had similar effects	Griffin et al., 1974
HeLa 53	db-Cyclic AMP (1 mM)	Decreased cellular glycogen	Cyclic AMP had an opposite and antagonistic effect	Kaukel et al., 1972a
Rat heart	db-Cyclic AMP (0.01–0.5 mM), or norepinephrine (1 µM)	Increased myosin ATPase activity	DCI blocked catecholamine but not db-cyclic AMP effect	Harary et al., 1973
Rat pituitary tumor	db-Cyclic AMP (0.5 mM)	Increased prolactin production		Tashjian and Hoyt, 1972
Mouse PY transformed 3T3 fibroblasts	db-Cyclic AMP (0.2 mM) plus theophylline (2 mM)	Decreased transport of 2-deoxyglucose		Grimes and Shroeder, 1973
Chick embryo pectoral muscle	db-Cyclic AMP (1–10 µM) or epinephrine (1 µM) plus theophylline (10–100 µM)	Increased myoglobin synthesis		Kagen and Freedman, 1974
Rat myoblasts (L_6D)	Cyclic AMP (0.4 mM) or db-cyclic AMP (0.5 mM) plus theophylline (0.5 mM)	Inhibition of myoblast fusion and growth		Wahrmann et al., 1973b

Abbreviations used are: ACTH, adrenocorticotropic hormone; DCI, dichloroisoproterenol; db-cyclic AMP, $N^6,O^{2'}$-dibutyryl cyclic AMP; LH, luteinizing hormone; mb-cyclic AMP, N^6-monobutyryl cyclic AMP; MSH, melanocyte-stimulating hormone; NE, norepinephrine; PGE_1, prostaglandin E_1.

minutes after treatment. Both cyclic GMP and its mono- or dibutyryl derivatives increased DNA synthesis in the cells somewhat. Rudland, Gospodarowicz, and Seifert (1974b) reported that fibroblast growth factor (FGF) isolated from bovine pituitary increased cyclic GMP levels without affecting cyclic AMP levels in 3T3 cultures. However, insulin increased DNA synthesis and cyclic GMP levels, but decreased cyclic AMP levels. Guanylyl cyclase activity in particulate, but not soluble, 3T3 cell fractions was increased three- to sixfold by FGF but not by insulin. Conversely, adenylyl cyclase activity in membrane fractions was decreased by insulin but not by FGF. These data were interpreted to support the "dualism" concept of cyclic nucleotide regulation.

Goridis, Massarelli, Sensenbrenner, and Mandel (1974) have reported that chick embryo brain cell cultures which contained differing proportions of neuronal and glial elements, depending on the ages of the embryo sources, contained varying degrees of guanylyl cyclase activity. Cultures derived from 8- to 12-day-old embryos were enriched in neuronal cells and contained high guanylyl cyclase activity. Little activity was observed in cultures of older embryos containing mainly astrocytes, or in cultures of dissociated meningeal cells.

Two additional situations in which cyclic GMP appeared to be antagonistic to cyclic AMP have been reported. Exogenous cyclic GMP was reported to reverse an antagonism by dibutyryl cyclic AMP on serum-induced leucine and 2-deoxyglucose uptake in 3T3 cells (Kram and Tomkins, 1973). In addition, in the presence of PGE_1, which presumably elevated intracellular cyclic AMP in 3T3 cells, transport of these substances as well as the synthesis of protein and RNA were depressed. Cyclic GMP (1 mM) reversed the effects of PGE_1 on transport and protein synthesis. Colcemid and vinblastine mimicked cyclic GMP action and made the cells insensitive to dibutyryl cyclic AMP. It was suggested that these results indicated a role for microtubular proteins in precursor transport and a role for cyclic nucleotides in regulating the function of the microtubular apparatus. Watson, Epstein, and Cohn (1973) reported that 10^{-4} M cyclic GMP reversed the antagonism of the immune response of cultured mouse spleen cells caused by 10^{-3} M dibutyryl cyclic AMP. Cyclic GMP alone had no effect on the immune response of spleen cultures from normal mice, but enhanced the response of cultures from T-cell deficient mice.

F. Mammalian Cell Genetics and Cyclic AMP Metabolism

The adaptation of microbiological techniques to cultured mammalian cells has recently proved useful in the area of cyclic nucleotide research. Three tools which have been used to advantage are viral transformation, isolation of mutants or variants, and somatic cell hybridization. We feel it important to summarize a portion of this work because of its potential utility and importance.

1. Viral Transformation

Numerous studies have been carried out comparing cyclic AMP levels to adenylyl cyclase activities in normal and transformed cells; some of these are covered in other sections of this review. However, it is of interest to consider here a series of studies by Pastan and his colleagues on cells transformed by temperature-sensitive mutant viruses. Such cells express the characteristics of transformed cells only within a restricted or so-called permissive temperature range while at other temperatures they behave as normal cells. A report by Otten, Bader, Johnson, and Pastan (1972a) indicated that chick embryo fibroblasts infected with temperature-sensitive Rous sarcoma virus showed normal morphology and basal cyclic AMP levels when grown at a nonpermissive temperature (40.5°C) but a morphology and basal cyclic AMP levels characteristic of transformed cells when grown at 36°C. Cyclic AMP levels dropped from normal to transformed cell values within 20 min of a change of infected cells from 40.5 to 36°C; the corresponding morphology change could be prevented by addition of exogenous dibutyryl cyclic AMP in combination with phosphodiesterase inhibitors.

Anderson, Johnson, and Pastan (1973a) reported a corresponding pattern of adenylyl cyclase changes with temperature shifts in the temperature-sensitive, RSV-infected chick embryo cells. In addition, the K_m for ATP in infected cells at 37°C appeared to be the same (0.67–1.0 mM) as that in cells infected with "wild-type" virus, but at 42°C the K_m (0.28 mM) resembled more closely that of uninfected cells (0.23 mM). The decrease in adenylyl cyclase activity which accompanied the transfer of infected cells from 42 to 37°C occurred within 10 min. Carchman, Johnson, Pastan, and Scolnick (1974) subsequently studied density-dependent growth inhibition in normal rat kidney (NRK) cells infected with wild-type or temperature-sensitive Kirsten sarcoma virus. Cells infected with temperature-sensitive virus and grown at nonpermissive temperature (39°C) showed increased cyclic AMP levels with increasing population density, as did uninfected NRK cells. However, when grown at a permissive temperature (32°C), the mutant virus-infected cells showed no change in cyclic AMP levels with changes in population densities, similar to the "wild-type" virus-infected cells.

2. Clonal Variants and Mutants

In addition to their studies with viral transformation, Willingham, Carchman, and Pastan (1973) reported the selection of temperature-sensitive clones of 3T3 fibroblasts with low substratum adhesiveness from populations treated with the acridine dye mutagens, ICR-170 and ICR-191. Temperature shifts of from 2.5 to 14°C in either direction between 23 and 39.5°C caused transient rounding and detachment of cells. Decreases in cyclic AMP levels appeared to occur with temperature changes in the temperature-sensitive but not in the insensitive 3T3 cells. However, since corresponding amounts of cyclic AMP accumulated in the incubation medium containing sensitive cells, but not in the medium containing insensitive cells, there actually was no change in the total medium plus cellular cyclic AMP levels. Expressed as picomoles per milligram of nucleic acid, total levels were 332 in the insensitive cells at 39.5°C and 347 at 10 min after a shift to 23°C. In the temperature-sensitive cells plus medium, the total amount of cyclic AMP was 354 at 39.5°C and 350 at 10 min after a shift to 23°C. These data are somewhat difficult to interpret in terms of the suggested role of cyclic AMP in growth regulation, and may be compared to the results of Burstin, Renger and Basilico (1974), who reported that fluctuations in cyclic AMP could be dissociated from morphological and growth changes in temperature-sensitive mutants of SV40-transformed 3T3 cells (see Section III). However, these data do indicate the existence of an extremely interesting effect of temperature on the release of cyclic AMP from cells into the extracellular medium.

An early study of cyclic AMP metabolism in a variant culture was that of Schimmer (1969), using clones of a murine adrenal cortex tumor established by Buonassisi, Sato, and Cohen (1962). Of two clones (Y-1 and Y-6) isolated from the tumor directly, only one (Y-1) increased steroid production in response to ACTH (Sato and Yasumura, 1966). The other (Y-6) was unresponsive to ACTH but increased steroid production in the presence of cyclic AMP (1 mM). In addition, three subclones of the ACTH-responsive Y-1 cells were found to have individual defects in the response machinery. Subclone OS1 failed to give a steroidogenic response to either ACTH or cyclic AMP, but did convert exogenous pregnenolone to Δ^4-3-keto steroids. Subclone OS3 resembled Y-6 cells in the lack of response to ACTH as compared to a normal response to cyclic AMP. Theophylline failed to potentiate a response to ACTH in OS3 cells. Another subclone, OS4, was essentially identical to OS1 under culture conditions. However, when OS4 cells were injected into isogenic LAF mice, tumors responsive to ACTH were produced, while injections of OS1 cells resulted in tumors unresponsive to ACTH, but responsive to cyclic AMP.

The investigation of the mouse lymphosarcoma cell line S49, which grows more slowly and apparently dies in response to cyclic AMP, appears to be a major contribution demonstrating the use of selected mutants for cyclic nucleotide research. Basically, the technique involved exposure of the S49 parent cell line to lethal concentrations of dibutyryl cyclic AMP, or to agents which increase endogenous cyclic AMP concentrations. Resistant subpopulations, which still grow, are then isolated and analyzed for defects in cyclic nucleotide metabolism. Several cyclic AMP-resistant mutants that appear to be genetically stable have thus been isolated and studied. A more comprehensive treatment of this work has recently been published (Bourne, Coffino, Melmon, Tomkins, and Weinstein, 1975).

One of the earliest cyclic AMP-resistant clones isolated was markedly deficient in both cytosolic protein kinase as well as cyclic AMP-binding protein (Daniel et al., 1973b). These deficiencies appeared in cells selected either by gradually increased dibutyryl cyclic AMP concentrations or by single-step selection in soft agar (Coffino, Bourne, and Tomkins, 1975). The cyclic AMP-resistant populations showed much greater increases in cyclic AMP in response to PGE_1 or isoproterenol than did the sensitive "wild-type" cells (Daniel, Bourne, and Tomkins, 1973a), and the exaggerated cyclic AMP levels were maintained longer in resistant cells. These results were correlated with the observation that phosphodiesterase activity of resistant cells was one-fourth to one-half that of parental cells. The cells resistant to cyclic AMP were found to be deficient in protein kinase, cyclic-AMP-binding activity, and phosphodiesterase. Additional experiments by Bourne et al. (1973b) demonstrated that while isoproterenol increased both cyclic AMP and phosphodiesterase in cyclic AMP-sensitive cells, it did not affect the phosphodiesterase of resistant cells, even though intracellular cyclic AMP levels increased more dramatically than in sensitive cells. These results were viewed as evidence indicating that cyclic AMP-dependent protein kinase activities may well be responsible for the induction and maintenance of phosphodiesterase.

A second mutant population was isolated by cytolytic selection in the presence of isoproterenol (Bourne et al., 1974). In this case, sensitive cells were killed by the increased endogenous cyclic AMP produced in response to isoproterenol, and resistant cells were shown to have a deficiency in adenylyl cyclase (designated AC^-). Essentially no adenylyl cyclase activity could be detected in broken-cell preparations using either NaF or agonists of the "wild-type" enzyme. Similarly, intact AC^- cells failed to respond to isoproterenol, cholera toxin, or PGE_1 in terms of either increased cyclic AMP levels or increased phosphodiesterase activities. As expected from the observation that phosphodiesterase could be increased in these cells by exposure to exogenous dibutyryl cyclic AMP, cyclic AMP-binding activity and cyclic AMP-dependent protein kinase proved to be normal in AC^- cells.

3. Somatic Cell Hybridization

A number of somatic cell hybridization studies have been utilized in cyclic nucleotide research. Gilman and Minna (1973) compared the effects of norepinephrine and isoproterenol on a number of somatic cell hybrids derived from parental lines with widely differing catecholamine responses. The results showed consistently that when hybrids were formed between highly responsive and unresponsive parental strains, the resultant populations were unresponsive. The loss of parental responsiveness did not appear to be due either to loss of specific marker chromosomes after hybrid formation or to the selection of specific hybrid clones. Although it was possible that chromosomes other than marker chromosomes may have been lost, the authors suggested that catecholamine responsiveness may be regulated by a negative control mechanism inherent to the unresponsive parental cells.

A similar study was carried out with cells having widely varied cyclic AMP responses to PGE_1 (Minna and Gilman, 1973). Hybridization of PGE_1-responsive with PGE_1-unresponsive cells, or hybridization of two PGE_1-responsive types of cells, yielded in most cases clonal populations with responses as good as or better than the most responsive parental cell lines. However, several hybrid clones from fusions of two PGE_1-responsive types of cells produced responses lower than those in the parental cell with the lowest response. Perhaps an important consideration is that 5 out of 7 of the parental PGE_1-responsive cell lines tested had less than 10-fold increases in cyclic AMP when stimulated by PGE_1. Nonetheless, it would seem that the regulation of PGE_1 responsiveness is very different from that of the catecholamines, and it may well be that multiple phenomena are involved. These could include the availability of receptors, differences in the reaction or relation between specific receptors and the catalytic site of adenylyl cyclase, or interactions with phosphodiesterase. Minna and Gilman (1973) also noted that the tendency of theophylline to potentiate the effects of isoproterenol or PGE_1 varied considerably among various cell lines. The theophylline potentiation in clones derived from fusion of a parental cell line demonstrating good potentiation with another having poor potentiation produced hybrid cell progeny with theophylline potentiation equal to or greater than that in the most highly potentiated parental cell type.

Adenylyl cyclase activities of several cell lines and various hybrids were recently studied by Maguire, Sturgill, Anderson, Minna, and Gilman (1974). Fusion of a catecholamine-unresponsive cell (mouse B82 cell) with a responsive cell (rat C6TG1A glioma) showed a preferential retention of catecholamine-unresponsive, EGTA-sensitive adenylyl cyclase in the hybrid. In addition, these authors fused a B82 cell with high adenylyl cyclase activity to a RAG cell with very low adenylyl cyclase activity. The hybrid clones showed activity equal to or greater than the high-activity parental line.

Hamprecht and Schultz (1973a,b) obtained results with catecholamine- and PGE_1-responsive and -unresponsive somatic cell hybrids which were essentially identical to those reported by Minna and Gilman. Parental catecholamine-unresponsive or PGE_1-responsive characteristics were preferentially retained in hybrids of opposites, while fusion of two highly responsive PGE_1 cell lines gave hybrid clones with greater cyclic AMP responses to PGE_1 than the responses of either parental cell line. The chromosomal complement of hybrids varied from much lower than the sum of parental chromosomes in mouse neuroblastoma plus mouse fibroblast clones to much greater than expected from 1:1 fusions in rat glioma plus mouse neuroblastoma clones.

Obviously, a critical element in somatic cell hybridization studies is chromosomal analysis. An intimate knowledge of chromosomal complements, chromosome loss, and gene-gene interactions will be essential to the further understanding of the regulation of cell function and the genetic control of cyclic nucleotide metabolism.

G. Comments

It seems fair to summarize Section II by reiterating our initial contention, that cell culture systems represent a singularly useful tool for studying the control of cyclic nucleotide metabolism. However, it is also fair to say that advances in our understanding of the problem will be limited by our ability to comprehend and make use of the many variables inherent in cultured cells.

Tremendous progress has been made in the years since 1971. Barring catastrophies like the threatened plunge into a new economic Dark Age for science, it seems most likely that the rate of progress will increase. If such is the case, within a few years we may be in a position to understand clearly cyclic nucleotide metabolism, and much of this knowledge will be gained through the use of cell culture systems.

III. CYCLIC NUCLEOTIDES AND CELL GROWTH

A. Basal Cyclic AMP Levels and Cell Growth

After the initial discoveries by Sutherland and Rall (1960) demonstrating that hormonal responses in liver cells were mediated through the production of cyclic AMP by adenylyl cyclase, it was shown by a number of investigators that this enzyme was associated with plasma membranes, or plasma membrane-related systems such as the T system of cardiac and skeletal muscles and synaptic membranes of nerves (Robison et al., 1971). These findings, taken together with the known physiological effects of hormones and cyclic AMP, allowed the development of the second messenger concept. In this concept adenylyl cyclase and its regulatory elements (hormonal receptors) are assigned a role within the plasma membrane, whereby adenylyl cyclase translates appropriate external stimuli into intracellular cyclic AMP production. Based upon the functional characteristics of adenylyl cyclase, schematic models generally depict its hormonal receptors on the outer surface of the plasma membrane and its catalytic site on the cytoplasmic surface. While this model may or may not be structurally correct, it adequately describes the functional state of adenylyl cyclase.

Under "basal" physiological conditions, the catalytic site of adenylyl cyclase presumably generates a certain amount of cyclic AMP continuously, perhaps due to some tonic hormonal or neurohumoral stimulation of the system. That this continuously generated cyclic AMP remains at a reasonably constant level within the cell is thought to be due to a concomitant and continuous breakdown by phosphodiesterase and, at least in some cases, cell excretion.

It would seem to us that cell culture systems are not good models of the physiological "basal" state referred to above. While it is well known that the sera used in growth media contain humoral effectors of cyclic nucleotide metabolism, the tonic aspect is missing. Thus, one might think of cultured cells as living in a kind of hormonal limbo. By contrast, the viable cells of a living multicellular organism are never in such a situation but, rather, are continuously bombarded by signals. Therefore, on physiological grounds the cell culture model seems to us to be seriously flawed in general, and specifically flawed where a hormone-sensitive control system is the object of study.

1. Fluctuations of Basal Levels of Cyclic AMP During Growth and Confluency of Fibroblasts

Fluctuations of basal intracellular levels of cyclic AMP have been suggested to be involved in the regulation of cell growth and its normal cessation with the attainment of appropriate conditions such as optimal cell density and/or contact inhibition. This hypothesis is based upon reports that the growth rates of cultured cells are inversely proportional to the intracellular concentrations of cyclic AMP, and that so-called contact-inhibited cells have higher intracellular concentrations of cyclic AMP than similar but growing cells. This observation was first reported by Otten, Johnson, and Pastan (1971, 1972b), who compared growth rates and growth states to cyclic AMP levels in 3T3 mouse embryo fibroblasts and MA-308 human diploid fibroblasts. Elevated cyclic AMP levels also have been reported in confluent WI-38 human diploid fibroblasts (Heidrick and Ryan, 1971b), human skin fibroblasts (Froehlich and Rachmeler, 1972), normal rat kidney fibroblasts (Anderson et al., 1973b), mouse fibroblasts, monkey fibroblasts, and hamster fibroblasts (Rudland, Seeley, and Seifert, 1974c). However, work by other authors indicates that it was the depletion of serum factors by the cells rather than "contact inhibition" or density restriction which signaled the rises in cyclic AMP levels concomitant with the cessation of cell growth observed in cultured fibroblasts (Sheppard, 1972; Seifert and Paul, 1972; Kram, Mamont, and Tomkins, 1973; Oey, Vogel, and Pollack, 1974).

Sheppard (1972) reported that unsynchronized cultures of 3T3, SV40 virus (SV40) transformed 3T3, 3T6 and polyoma virus (PY) transformed fibroblasts showed no increase in intracellular cyclic AMP levels in progressing from a state of growth to one of density restriction or, in the case of transformed cells, confluency—if the medium was changed often enough. Furthermore, Bonnai and Sheppard (1974) have reported that while density-dependent cessation of growth in confluent monolayers of 3T3 cells is not correlated with a temporally synchronous rise in intracellular cyclic AMP, an increase in cyclic AMP levels does occur earlier in growth when the growing cells contact one another. In contradiction to this, Haslam and Goldstein (1974) have reported that cyclic AMP levels in two different strains of human skin fibroblasts decreased from the time of subculturing at very sparse densities until density-dependent cessation of growth occurred. These data indicated that neither cell contact nor density-dependent inhibition of growth were associated with increases in cyclic AMP levels.

Similar results have been obtained in other laboratories also using 3T3 and SV40 transformed 3T3 cells, as well as WI-38 cells and temperature-sensitive strains of SV40 transformed 3T3 cells, which have the growth characteristics of transformed cells at 32°C and the density-dependent growth characteristics of normal 3T3 cells at 38.5°C (Otten et al., 1972b; Oey et al., 1974; Burstin et al., 1974). In fact, Oey et al. (1974) and Burstin et al. (1974) have quite convincingly demonstrated that, if anything, cyclic AMP levels actually *decreased* when 3T3 fibroblasts became density-restricted or when various strains of SV40 transformed

3T3 cells became confluent—even though the transformed cells continued to grow after reaching confluency. Therefore, much of the available evidence suggests that most, if not all, untransformed fibroblasts, when cultured in the presence of appropriate nutritional factors, will stop growing upon reaching a restrictive population density without any concomitant increase in intracellular cyclic AMP levels.

2. Experimental Manipulation of Basal Levels of Cyclic AMP in Fibroblasts

Depletion of serum factors leads to cessation of growth in cultured fibroblasts, and this is reflected in a rise of intracellular cyclic AMP levels (Heidrick and Ryan, 1971b; Otten et al., 1972b; Seifert and Paul, 1972; Kram et al., 1973; Oey et al., 1974). Alternatively, the administration of fresh serum to either confluent or nonconfluent serum-deprived cultures of fibroblasts results in a rapid decrease in cyclic AMP levels and eventual renewed DNA synthesis. A similar drop in cyclic AMP levels has been reported following insulin treatment of 3T3 and BHK fibroblasts (Sheppard, 1972; DeAsua, Surian, Flavia, and Torres, 1973). In BHK cells insulin has been reported to inhibit adenylyl cyclase activity in membrane preparations (DeAsua et al., 1973). Insulin or serum proteins with insulin-like activity (somatomedin) (Daughaday, Hall, Raben, Salmon, Van Den Brande, and Van Wyk, 1972) from human or calf serum have been reported to stimulate DNA synthesis in serum-exhausted cultures of BHK and 3T3 fibroblasts and replace serum in chick embryo fibroblasts (Temin, 1967; Pierson and Temin, 1972; Bombik and Burger, 1973). However, Scher, Stathakos, and Antoniades (1974) have observed that insulin and human serum factors with insulin-like activities did not stimulate DNA synthesis in confluent cultures of Balb/c 3T3 fibroblasts, while another factor (or factors) in serum with no insulin-like activity did. These authors have suggested that simply lowering cyclic AMP levels may not be sufficient to accelerate DNA synthesis. Gospodarowicz and Moran (1974) have reported that fibroblast growth factor (FGF) in combination with a glucocorticoid may completely replace serum for the growth of Balb/c 3T3 cells. In fact, FGF and hydrocortisone will stimulate growth in quiescent fibroblasts with little or no alterations in cyclic AMP levels (Rudland et al., 1974b). FGF is a small polypeptide isolated from bovine brain serum or pituitary glands (Gospodarowicz, 1974) which can specifically stimulate membrane-bound guanylyl cyclase (Rudland et al., 1974b).

Effective doses of proteases, such as trypsin, chymotrypsin, ficin, papain, and subtilisin caused a rapid, but temporary decrease in cyclic AMP levels of confluent 3T3 cells, followed by a round of cell division and overgrowth of the monolayer (Burger, Bombik, Breckenridge, and Sheppard, 1972a; Sheppard, 1972; Bombik and Burger, 1973). Reapplication of the protease after cells again became quiescent caused yet another round of cell division (Burger, Bombik, and Noonan, 1972b). Following proteolytic treatment, 3T3 fibroblasts were capable of agglutination in the presence of low concentrations of plant lectins such as wheat germ agglutinin, concanavalin A, and soybean agglutinin (Burger, 1969; Inbar and Sachs, 1969a,b). Normally, untransformed fibroblasts will not agglutinate in the presence of low concentrations of lectins, whereas transformed fibroblasts do (Burger and Goldberg, 1967; Inbar and Sachs, 1969a,b). However, during mitosis, 3T3 fibroblasts were capable of binding lectins (Fox, Sheppard, and Burger, 1971), and cyclic AMP levels in mitotic cells were lowered to about 40% of the value observed during other portions of the cell cycle (Burger et al., 1972a). Studies of Chinese hamster ovary cells have also showed that cyclic AMP levels are low during mitosis (Sheppard and Prescott, 1972). The picture that emerges, then, is that in 3T3 fibroblasts, a transient surface change and a lowering of cyclic AMP levels normally occurs during mitosis as well as following protease treatment (Burger et al., 1972a,b). Furthermore, protease treatment (Burger et al., 1972b), serum treatment (Sheppard, 1972), or, in some cases, insulin treatment (Sheppard, 1972; Bombik and Burger, 1973; Scher et al., 1974) leads to a transient lowering of cyclic AMP levels, increased DNA synthesis, and cell division.

The inclusion of a sufficient concentration of $N^6,2'O$-dibutyryl cyclic AMP during the 10-min protease treatment period of confluent 3T3 cells was sufficient to prevent renewed DNA synthesis (Burger et al., 1972a) without affecting agglutination changes brought about by the protease (Bombik and Burger, 1973). Incubation of cells in dibutyryl cyclic AMP prior to or following the brief pronase treatment did not prevent renewed DNA synthesis. Incubation of confluent WI-38 cells in 8-methylthioadenosine $3':5'$-cyclic monophosphate or prostaglandin E_1 in the presence of aminophylline has also been reported to block serum stimulation of renewed growth (Kurtz et al., 1974). These results have prompted the suggestion that the transitory lowering of cyclic AMP levels during mitosis of normal cells or via manipulation of quiescent cells with proteases, serum, or insulin will trigger a cell cycle (Burger, 1969; Burger et al., 1972b). Presumably, surface changes are responsible for the transitory decrease in cyclic AMP levels.

However, the latter suggestion is not without inconsistencies. Sheppard (1972) has shown that serum, like protease treatment, caused a rapid but transitory decrease in cyclic AMP levels of 3T3 fibroblasts. By contrast to the case in which cells were treated with proteases, dibutyryl cyclic AMP had to be present in the medium containing serum *continuously* to block a renewed round of division of confluent cells (Bombik and Burger, 1973). Assuming that inclusion of dibutyryl cyclic AMP in fresh serum blocked or compensated for the initial drop in cyclic AMP levels in the fibroblasts, one must question whether there is yet another dip in intracellular cyclic AMP levels below "normal" levels following the removal of dibutyryl cyclic AMP from the medium of cells exposed to fresh serum. According to the "trigger theory" such a decrease would seem to be necessary.

Also, as mentioned earlier, the stimulation of renewed DNA synthesis in confluent 3T3 cells by protease treatment was blocked by dibutyryl cyclic AMP, when it was included only during the brief protease treatment period (Burger et al., 1972a). However, the protease-induced agglutinability of these cells to lectins was not affected by the brief dibutyryl cyclic AMP treatment, and in fact persisted after the dibutyryl cyclic AMP had been removed from the medium (Bombik and Burger, 1973). Is there, then, any direct relationship among surface changes as measured by agglutinability, cyclic AMP levels, and DNA synthesis?

3. *Basal Levels in Transformed Fibroblasts*

Just as high intracellular levels of cyclic AMP have been associated with density-dependent inhibition of growth of nonmalignant cells, decreased cellular levels of cyclic AMP have been suggested to be important in the lack of growth regulation of neoplastic cells (Sheppard, 1972). This idea was based primarily upon the observation that 3T3 fibroblasts had basal cyclic AMP levels approximately twice those of their transformed counterparts, such as spontaneously transformed 3T6 cells, SV40 transformed 3T3 cells and PY transformed 3T3 cells.

However, despite the nearly twofold difference in average cyclic AMP levels of normal and transformed 3T3 fibroblasts, fluctuations in intracellular cyclic AMP levels of both types of cells during the cell cycle and in response to manipulations of the culture environment were remarkably similar! The addition of fresh serum, insulin, or trypsin to nongrowing, confluent 3T3 fibroblasts or growing, confluent 3T6 or PY 3T3 fibroblasts caused transient decreases in intracellular cyclic AMP levels (Sheppard, 1972), which were about the same on a *relative basis*. That is, in both the normal and the transformed 3T3 cells, cyclic AMP levels dropped to about one-half their original steady-state values, even though the initial steady-state level was lower in the transformed cells. Similarly, cyclic AMP levels in synchronized, PY transformed 3T3 cells decreased during mitosis, although the drop was not as great as that observed in normal 3T3 cells (Burger et al., 1972a). One is forced to ask why there was any decrease in cyclic AMP levels in transformed cells at all if the contention that the cells are constantly in an altered surface-low cyclic AMP level state were true? Serum deprivation also has *relatively* similar effects on both normal and transformed fibroblasts. For example, when sparse 3T3 cells were placed in medium containing 1% serum, the cells continued to grow exponentially for about 36 hr and then became quiescent (Oey et al., 1974). Intracellular cyclic AMP concentrations measured 48 hr after shifting these cells to 1% serum were approximately threefold higher, having increased from 25 to 83 pmoles/mg protein. If SV40 transformed 3T3 cells were placed in medium containing 1% serum, the cells continued to grow exponentially beyond the 48-hr period. Nonetheless, an approximate threefold increase in cyclic AMP levels was also observed in the transformed cells at 48 hr (from 12 to 31 pmoles/mg protein). It is interesting that after the 48-hr period in 1% serum the intracellular cyclic AMP levels (~31 pmoles/mg protein) in growing SV40 transformed 3T3 cells were greater than those found in normal growing or quiescent 3T3 cells in the presence of 10% serum—even though the transformed type did not stop growing (Oey et al., 1974). Finally, it should be recalled that cyclic AMP levels have been reported to decrease in both 3T3 cells and SV40 transformed strains of 3T3 cells when the cells became confluent in medium containing 10% serum, even though the transformed cells continued to grow while the 3T3 cells became density-restricted (Oey et al., 1974; Burstin et al., 1974).

Thus, intracellular cyclic AMP levels of transformed 3T3 fibroblasts fluctuated during the cell cycle or with manipulations of the growth medium in a manner that was *relatively* similar to normal 3T3 cells. It follows, then, that any hypothesis attempting to link basal cyclic AMP levels to lack of growth regulation of transformed fibroblasts must be based upon absolute differences in levels. That such a contention is valid seems highly unlikely to us.

4. Basal Levels in Nonfibroblastic Cells

Our discussion of basal cyclic AMP levels and the regulation of cell growth has dwelt largely upon 3T3 cells or other cells of fibroblastic origin up to this point because most of the work published has concerned measurements of basal cyclic AMP levels in cultured fibroblasts. How do the results obtained with cultured fibroblasts compare with other cell types?

As mentioned previously, the major drop in cyclic AMP levels has been reported to occur during mitosis in 3T3 fibroblasts (Burger et al., 1972a) and Chinese hamster ovary cells (Sheppard and Prescott, 1972). Somewhat similar to these observations are the reports of Millis, Forrest, and Pious (1972, 1974), whose data indicated that cyclic AMP levels were high in human lymphoid cells (RPMI 8866) during G1 and G2, but decreased during DNA synthesis and mitosis. Measurements of the diurnal oscillation of cyclic AMP levels in mouse epidermis also indicated that low cyclic AMP levels coincided with periods of high mitotic activity (Marks and Grimm, 1972).

In sharp contrast to reports that cyclic AMP levels were low in transformed, cultured fibroblasts, Thomas, Murad, Looney, and Morris (1973) reported that cyclic AMP levels were greatly elevated in a variety of Morris hepatoma explants *in vivo* when compared to normal rat liver. Furthermore, no correlation between growth rates of the tumors and cyclic AMP levels was apparent. Similarly, the transplanted rat adrenocortical carcinoma 494 was found to have about twice the basal level of cyclic AMP of normal rat adrenal glands (Ney et al., 1969).

On the opposite end of the spectrum, Bourne et al. (1975) have succeeded in isolating several strains of the S49 mouse lymphosarcoma cell, one of which has no apparent functional adenylyl cyclase and hence no endogenous cyclic AMP, and another which has little or no cyclic AMP-binding protein and cyclic AMP-stimulated protein kinase and is therefore unaffected even by high levels of cyclic AMP (Daniel et al., 1973a,b; Bourne et al., 1975). In both of these strains the cell cycle is similar to the parental strains.

In cultured rat myoblasts, which differentiate into myotubes by fusion after reaching confluency, cyclic AMP levels drop as the cells begin to fuse, concomitant with a decrease in adenylyl cyclase activity (Reporter, 1972; Wahrmann et al., 1973a,b). Furthermore, the addition of insulin to the growth medium potentiates the process of fusion, and the presence of insulin in serum-free medium allows fusion, which would otherwise not occur (Mandel and Pearson, 1974). Thus, in myoblasts, higher cyclic AMP levels were associated with growth, while lower cyclic AMP levels were associated with cessation of growth and the differentiation process.

5. Is Cyclic AMP the Ultimate Regulator of Cell Growth?

Using the information we have reviewed, let us pose some questions concerning cyclic AMP levels and the regulation of cell growth.

a. Is density-dependent cessation of growth (thought to be contact inhibition by some) signaled by an increase in intracellular cyclic AMP levels?

The reports are contradictory, with some authors stating that an increase in levels did occur (Heidrick and Ryan, 1971a; Otten et al., 1971, 1972b; Froehlich and Rachmeler, 1972; Anderson et al., 1973b), while others reported that steady-state levels of cyclic AMP were maintained or decreased in growth arrest (Seifert and Paul, 1972; Sheppard, 1972; Kram et al., 1973; Burstin et al., 1974; Haslam and Goldstein, 1974; Oey et al., 1974). The best evidence, in our opinion, suggests that if serum depletion is prevented, then density-dependent inhibition of cell division, at least in 3T3 and WI-38 fibroblasts, will occur without a rise in intracellular cyclic AMP levels. This leads one to ask if increased cyclic AMP levels are necessary for the cessation of cell growth. The answer would seem to be no. The fact that density-dependent inhibition can occur in the absence of changes in cyclic AMP levels would seem to obviate such a necessity, and rather, suggest only that increased basal cyclic AMP levels may occur with cessation of cell growth under certain conditions.

b. Are the low basal cyclic AMP levels observed in some transformed or malignant cells, as compared to their nonmalignant counterparts, responsible for the loss of growth regulation?

The observation that transformed 3T3 cells which lack density-dependent inhibition of growth have lower cyclic AMP levels than normal 3T3 fibroblasts has been used as evidence that cyclic AMP levels play a part in regulation of cell growth (Otten et al., 1971; Sheppard, 1972).

However, the work of Burstin et al. (1974) would seem to indicate that changes in cyclic AMP levels can be dissociated from the growth properties characteristic of either the transformed or the nonmalignant state. These authors demonstrated that no differences in cyclic AMP levels were apparent in temperature-sensitive strains of SV40 transformed 3T3 cells grown at either the permissive or the nonpermissive temperature. That is, in sparse cultures both the permissive-temperature (32°C) "transformed" cells and the nonpermissive-temperature (38.5°C) "untransformed" cells grew with cyclic AMP levels of about 6 pmoles/10^6 cells. In confluent cultures both the growing cells at the permissive temperature, and the density-restricted cells at the nonpermissive temperature, had cyclic AMP levels of about 2.5 pmoles/10^6 cells. Burstin et al. (1974) concluded that the data "indicate that cAMP is not the major determinant, at the phenotypic level, of the different growth properties exhibited by these cells. . . ."

Further, the experiments of Peytremann and Engel (1973) also indicate that basal levels of cyclic AMP *per se* are not concomitants of the malignant or nonmalignant state. These authors studied the relationship between growth and cyclic AMP levels in the mouse fibroblastlike cell line B_{82}, the mouse adrenal cancer line CCl_{79}, and hybrids derived from inactivated Sendai virus-mediated fusion of the latter cultured cells. The B_{82} cells had cyclic AMP levels of about 18.9 pmoles/mg protein, demonstrated density-dependent inhibition of growth, and were incapable of producing tumors when injected into mice. In contrast, the CCl_{79} cells had cyclic AMP levels about three times greater than normal adrenal tissue, that is, about 29.3 pmoles/mg protein, were not subject to density-dependent inhibition of growth in culture, and readily produced tumors when injected *in vivo*. The hybrid cells had cyclic AMP contents similar to the non-malignant B_{82} parent cells and showed density-dependent inhibition of growth in culture. Nonetheless, a small percentage of mice inoculated with the hybrid cells developed tumors, and the levels of cyclic AMP in the tumorous cells were virtually identical to the cultured hybrids and the nonmalignant parental line. Finally, as we mentioned earlier, Morris hepatomas (Thomas et al., 1973) and rat adrenocortical carcinoma 494 (Ney et al., 1969) had elevated basal cyclic AMP levels when compared to their normal counterparts.

Thus, if the patterns of cyclic AMP levels reviewed here are real and not due to artifacts, differences of techniques, or the inherent differences between culture systems and *in vivo* systems, then it must be concluded that cyclic AMP levels may remain the same, or may be lower or higher in neoplastic cells (as compared to their normal counterparts) depending upon the type of cell and (perhaps) the type of transformation effected.

c. Is there a correlation between the growth rate of cells and basal cyclic AMP levels?

Heidrick and Ryan (1971*b*) and Otten et al. (1972*b*) have concluded that their data indicated that there was a parallel between the time required for population doubling and basal cyclic AMP levels of cells. That is, cell growth rate is inversely proportional to cyclic AMP levels. Even neglecting the reports that a variety of neoplastic cells do not obey this rule (Ney et al., 1969; Thomas et al., 1973; Burstin et al., 1974), one must question this conclusion based on the authors' own data. For example, in fibroblasts the relationship seems to hold only if intracellular levels of cyclic AMP are expressed as micromolar concentrations or picomoles per 10^6 cells (Heidrick and Ryan, 1971*b*), but not if expressed as picomoles per milligram nucleic acid or picomoles per milligram protein. However, even if picomolar concentrations are used, the purported relationship does not always hold. For example, Heidrick and Ryan (1971*b*) reported that cultured HEp-2 laryngeal carcinoma-derived cells, fibro-5 normal human foreskin fibroblasts, and WI-38 diploid human lung fibroblasts had generation times of 26, 35, and 55 hr and cyclic AMP levels of 23.9, 28.5, and 30.8 pmoles/10^6 cells, respectively. Thus, even though WI-38 cells have a population doubling time of some 20 hr longer duration than fibro-5 cells, their cyclic AMP content is only around 7% greater. If one looks at the data in terms of cyclic AMP levels expressed as picomoles per milligram protein, the picture is even more confusing. HEp-2, fibro-5, and WI-38 then had cyclic AMP levels of 106.7, 101.4, and 80.1 pmoles/mg protein, respectively. That is in the wrong direction. Further, Sheppard (1972) has reported a strain of PY transformed 3T3 cells which grow at exactly the same rate as normal 3T3 cells yet have less than one-half the basal cyclic AMP level of normal 3T3 cells.

Pointing out these apparent inconsistencies may seem to be nitpicking on our part. However, it must be kept in mind that such minute differences have been used as support for pronouncements concerning the relationship between cyclic AMP levels and cell growth. In our opinion the data reported in the literature suggest that there is no consistent correlation between growth rates of cells and their intracellular cyclic AMP levels. In other words, there are too many exceptions to the rule.

d. Is cyclic AMP necessary for cell growth?

Certainly not in all cells. In cyclic AMP-resistant strains of S49 mouse lymphosarcoma cells, which do not have the ability to respond to cyclic AMP, cell growth and the cell cycle are similar to parental strain cells (Daniel, et al., 1973a,b; Bourne et al., 1975). Similar lymphosarcoma cell strains that have no apparent adenylyl cyclase also grow.

However, this question, like the question—is cyclic AMP necessary for the regulation of cell growth?—is in our opinion a problem of semantics. In those cells whose differentiated function it is to start growing in response to a certain environmental signal, cyclic AMP or some other cyclic nucleotide may well be the second messenger involved in that response. For example, cyclic AMP may stimulate proliferation of thymocytes under appropriate conditions (Whitfield, Rixon, MacManus, and Balk, 1973). On the other hand, in many cells cyclic AMP is the second messenger involved in eliciting a certain differentiated function other than growth. That certain differentiated functions do not occur during mitosis of cells is not a new idea among biologists. In fact, it may well be that the lowering of cyclic AMP levels observed during mitosis of fibroblasts (Burger et al., 1972a) is only indicative of a cell's inability to produce cyclic AMP during mitosis. This might represent a normal mechanism for preventing the juxtaposition of two incompatible cellular events—mitosis and the performance of a cyclic AMP-mediated differentiated function. There is little question that cyclic AMP, its analogues, or agonists which raise cyclic AMP levels may be used by men, and perhaps even cells, to *stop* cellular proliferation in the appropriate phase of the cell cycle in those cells capable of responding to it, as we shall see in the following section. However, this does not prove that cyclic AMP *regulates* cell growth, since the addition or deletion of many compounds can be used to inhibit or promote growth.

In the original second messenger concept (cf. Robison et al., 1971), cyclic AMP is viewed as an intracellular mediator of an extrinsic environmental signal—that is, of one or more hormones, neurohumoral agents, or prostaglandins. In this supposition the basal or unstimulated level of cyclic AMP in a cell was viewed as a neutral, or in a few cases, such as the case of insulin on fat cells, as a negative factor with respect to the differentiated function being considered (e.g., glycogenolysis or lipolysis). Within this hypothesis, which has at least obtained in a number of cases, the idea of intrinsic regulation seemed unlikely. In fact, the concept really attempted to describe only the superimposition of extrinsic controls upon ongoing cellular processes.

Ryan and Heidrick (1974) have likened the elevations of cyclic AMP levels observed in cultured cells under adverse growth conditions, such as serum deprivation, to *in vivo* situations that result in increased intracellular cyclic AMP levels. Among these were fasting (Selawry, Gutman, Fink, and Recant, 1973), circulatory arrest (Wollenberger, Krause, and Heier, 1969), anoxia, and electrical convulsive shock (Goldberg, Lust, O'Dea, Wei, and O'Toole, 1970). However, if one views the latter *in vivo* phenomena in terms of the second messenger concept, the analogy seems, at best, unrealistic. It is certain that most of the *in vivo* conditions described above involve hormone-mediated rises in cyclic AMP, and thus might best be described as situations of intense positive control. Further, these *in vivo* situations are not related in any apparent manner to growth. Cardiac muscle or brain cells do not stop growing with circulatory arrest or electrical convulsive shock, since they were not proliferating in the first place.

Finally, we must question the wisdom of generalizing observations on a certain type of cultured cell (e.g., rat and mouse fibroblasts) to other types of cultured cells, and especially to *in vivo* systems. Fibroblasts have long been used as a model system, since they are easy to grow in culture. The cessation of growth of fibroblasts when confluency is reached—"contact inhibition" (Abercrombe and Ambrose, 1962) or "density-dependent inhibition" (Stoker and Rubin, 1967)— is not compatible with *in vivo* observations. *In vivo* fibroblasts stop growing when the appropriate amount of collagen has been synthesized and, unlike most other cell types, do not maintain extensive areas of contact among themselves (Bloom and Fawcett, 1968).

Perhaps at this junction we should turn the pulpit over to Dr. Leonard Hayflick who, in a recent book review, stated the following:

"As is common among cell biologists generally, most of the authors of papers purportedly dealing with fibroblasts either are not dealing with fibroblasts or are not, as they seem to imply, dealing with normal fibroblasts. Studies done *in vitro* on such "fibroblast" populations as 3T3, BHK21, and mouse L cells suffer from the fact that the populations are clearly abnormal in at least one property yet are regarded either as consisting of typical normal fibroblasts or as providing normal controls for virus-transformed cells. The use of these cells is usually predicated on their common availability and the ease with which they may be cultured; but this is hardly a convincing rationale in view of the likelihood that any similarities between them and normal fibroblasts may be purely coincidental. Since such cell populations resemble no known *in vivo* cell type, their use can be likened to a kind of extraterrestrial biology. How disconcerting it is to realize that so much good science, carefully reasoned and technically sophisticated, rests on the use of cell populations twice removed from reality. Normal mouse, hamster, rabbit, and human fibroblasts are as easily available and as simply cultivated as are the continuously propagable abnormal cell populations now in wide use. Why do cell biologists resist using the genuine articles, which would make data extrapolation, with its inherent dangers, unnecessary?" (Hayflick, 1975)

B. Effects of Experimentally Increased Intracellular Cyclic AMP Levels on Cells

In 1968, Burk reported that the addition of methylxanthines to cultures of nonmalignant and transformed BHK fibroblasts decreased their growth rate. Since that time, many studies examining the effects of exogenous cyclic nucleotides on the growth and morphology of untransformed, virally transformed, chemically transformed, and spontaneously transformed cultured cells have accumulated. Unfortunately, in too few of these studies were adequate controls used, and in fewer yet was rigorous consideration given to the fates of the exogenously added compounds. Before we proceed to a discussion of such studies, it seems appropriate to consider what is known about the metabolism of exogenously added cyclic AMP and its analogues.

1. *Metabolism of Exogenous Cyclic AMP and Dibutyryl Cyclic AMP*

$N^6,2'0$-dibutyryl-$3',5'$-cyclic AMP (dibutyryl cyclic AMP) rather than cyclic AMP itself is frequently used in studies on regulation of cell growth and morphology. One rationale for this is the assumption that the lipid-soluble derivative is transported into cells more readily than cyclic AMP itself (Posternak, Sutherland, and Henion, 1962; Robison et al., 1971). However, in one published report comparing the transport of cyclic AMP and dibutyryl cyclic AMP into L cells and WI-38 cells, cyclic AMP was supposedly transported more effectively (Ryan and Durick, 1972).

In a series of experiments designed to explain the apparently diametrically opposite effects of cyclic AMP and dibutyryl cyclic AMP on glycogen metabolism in HeLa-S3 cells, Hilz, Kaukel, and their co-workers (Hilz and Tarnowski, 1970; Kaukel, Fuhrmann, and Hilz, 1972a, Kaukel and Hilz, 1972; Kaukel, Mundhenk, and Hilz, 1972b) provided some valuable insights concerning the metabolism of these compounds. These workers reported that, in cultures of HeLa cells, exogenous dibutyryl cyclic AMP or theophylline decreased glycogen content and incorporation of ^3H-thymidine into DNA, whereas cyclic AMP caused an increase in glycogen content as well as an increase in incorporation of ^3H-thymidine and ^3H-uridine into nucleic acids. Both compounds inhibited cell proliferation.

Dibutyryl cyclic AMP was found to be extremely resistant to extracellular degradation in medium containing serum and was taken up by the cells. Within the cells dibutyryl cyclic AMP was rendered biologically active by being converted to N^6-monobutyryl-3',5'-cyclic AMP (monobutyryl cyclic AMP), presumably by the actions of intracellular esterases. Monobutyryl cyclic AMP was capable of binding to the Gilman (1970) binding protein and was the only cyclic nucleotide that accumulated inside HeLa cells incubated in the presence of dibutyryl cyclic AMP. The monobutyryl derivative remained biologically active within the cells because of its immunity to phosphodiesterase digestion. In contrast, in medium containing serum that was not heat-inactivated, cyclic AMP was almost completely degraded extracellularly to adenosine and, to a lesser extent, inosine and corresponding bases such as hypoxanthine. These results were most likely due to the presence of degradative enzymes in the calf serum. The adenosine thus derived from cyclic AMP was taken up by HeLa cells and apparently caused the aforementioned paradoxical effects by altering nucleic acid precursor pools. Briefly, then, in HeLa cells under these conditions, exogenously added dibutyryl cyclic AMP ultimately mimicked endogenous cyclic AMP while exogenously added cyclic AMP did not.

While the findings of these workers (Hilz and Tarnowski, 1970; Kaukel et al., 1972a,b) should not be generalized, their data can be used as a guide for proposing logical and testable hypotheses that would explain the observed effects of exogenous nucleotides in other systems. For example, if dibutyryl cyclic AMP has no effect on cells that would normally respond to the endogenous synthesis of cyclic AMP, then it is possible either that the derivative does not penetrate the cell or that it is not converted to a biologically active form. Furthermore, as previously reviewed in some detail (Ryan and Heidrick, 1974), butyrate controls often produce results similar to the butyryl derivatives of cyclic AMP. However, no *a priori* assumptions can be made as to the mode of action of exogenous nucleotides on cells or the effects that may be observed following their administration unless a detailed study of the metabolism of the nucleotide used is carried out. The work of Eker (1974) may be illustrative on this point. This author observed that 0.1 to 1 mM cyclic AMP inhibited DNA synthesis and the growth of monolayer cultures of Chang's human liver cells incubated in medium containing heat-inactivated serum. The extracellular degradation of cyclic AMP was negligible. However, using (8-^{3}H) adenosine-3',5'-cyclic monophosphate, Eker (1974) could not demonstrate any acid-soluble intracellular cyclic AMP even when 10 mM theophylline was added to the medium. Acid-soluble radioactivity was reportedly present in ATP (71%), ADP (20%), AMP (7%), and adenosine (2%), but not cyclic AMP, and the author suggested that cyclic AMP might be exerting its effect on the external cell surface. However, given our previous discussion of HeLa cells and cyclic AMP, and the data indicating that radioactive adenosine is incorporated into the adenine nucleotide pool of the liver cells, it is also possible that this growth effect is in some way similar to that observed with HeLa cells.

2. Inhibition of Cell Division

a. *The cell cycle*

The incubation of cultured cells in medium containing cyclic AMP, dibutyryl cyclic AMP, or agonists which increase intracellular cyclic AMP levels may completely inhibit or partially suppress growth in some cell types. The reader is referred to a previous review where this phenomenon is extensively catalogued (Ryan and Heidrick, 1974). Where such inhibition occurs, cells are usually arrested in the G_1 phase of the cell cycle. For example, the growth of HeLa cells and a variety of normal and transformed fibroblasts is inhibited in G_1, in the presence of exogenous dibutyryl cyclic AMP (Frank, 1971; Froelich and Rachmeler, 1972, 1974; Kaukel et al., 1972a; Rozengurt and Pardee, 1972; Willingham, Johnson, and Pastan, 1972: Zimmermann and Raska, 1972; Bombik and Burger, 1973). In contrast, Taylor-Papadimitrious (1974) reported that L-929 fibroblasts were inhibited in the S to G_2 phase of the cell cycle when cultured in the presence of exogenous cyclic AMP. However, this author also observed that despite the fact that the cells were not proliferating, apparent incorporation of ^{3}H-thymidine into DNA was stimulated. In addition, much of the exogenous cyclic AMP had been converted to adenosine in the medium after 48 hr. Based upon our previous discussion of the divergent actions and metabolism of cyclic AMP and dibutyryl cyclic AMP in HeLa cells, it seems reasonable to suggest that these data might be indicative of a nucleotide pool effect caused by the products of degradation of the exogenously added cyclic AMP. This latter interpretation would explain the increased ^{3}H-thymidine incorporation in nongrowing cells and the arresting of cells in S instead of G_1. In contrast to cyclic AMP, dibutyryl cyclic AMP is known to inhibit the growth of L-929 fibroblasts concomitant with an inhibition of incorporation of ^{3}H-thymidine into DNA (Curtis, Elliot, Wilson, and Ryan, 1973).

Recent studies by Zeilig, Johnson, Sutherland, and Friedman, working with synchronized HeLa cells, have provided several new insights into cyclic AMP levels and their relationship to the growth of these cells. They have found that cyclic AMP levels declined as the cells proceeded into G_2 and reached a low point in mitosis. Cyclic AMP levels were at their highest at or near the beginning of the S phase. The intracellular levels of cyclic AMP were mildly elevated by β-adrenergic stimulation, and were very responsive to MIX. When cells in G_2 were exposed to MIX, to a combination of isoproterenol and MIX, or to the analogue 8-bromo-cyclic AMP, the entrance into mitosis was significantly delayed. On the other hand, if mitotic cells were shaken out, the addition of MIX caused a significant decrease in the time required for the cells to reach G_1. Increasing intracellular cyclic AMP levels had no effect on the duration of the S phase; and interestingly, no effects on the length of time in G_1 could be attributed to elevated cyclic AMP levels (D. L. Friedman, *personal communication*). This should be contrasted to the results with other cells described previously, in which elevated cyclic AMP levels were purported to arrest the cycle in G_1.

b. *Macromolecular synthesis.*

In transformed cells whose growth is inhibited by dibutyryl cyclic AMP or agonists which maintain high intracellular cyclic AMP levels, there was, naturally enough, a decrease or a cessation of DNA synthesis. More importantly, an increase in RNA and protein synthesis was usually observed (Lim and Mitsunobu, 1972; van Wijk, Wicks, and Clay, 1972; Curtis et al., 1973; Korinek, Spelsberg, and Mitchell, 1973; Kram et al., 1973). In fact, it has been reported by Korinek et al. (1973) that an activation of nuclear DNA-directed RNA polymerase occurred in rat XC sarcoma fibroblasts treated with a combination of exogenous theophylline and cyclic AMP.

c. *Contact inhibition and agglutinability*

As we mentioned earlier, transformed fibroblasts agglutinate in the presence of low concentrations of lectins and show no "contact inhibition" of growth (Burger and Goldberg, 1967; Inbar and Sacks, 1969*a,b*). In 1971, Sheppard reported that when PY transformed 3T3 and spontaneously transformed 3T6 fibroblasts were grown in the presence of a combination of dibutyryl cyclic AMP and theophylline, the transformed cells regained their normal morphology, lost their abnormal agglutinability, and were restored to "contact inhibited" growth. A month later Hsie, Jones, and Puck (1971) reported similar results with Chinese hamster ovary cells using a combination of dibutyryl cyclic AMP and testosterone. These and similar findings led Burger et al. (1972*b*) to suggest that the facts "may point to a direct relationship between membrane architecture and cyclic AMP concentration."

However, Paul (1972) subsequently demonstrated that SV40 transformed 3T3 cells did not become quiescent in the presence of a combination of dibutyryl cyclic AMP and theophylline, and did not regain "contact inhibition." What actually happened was that the transformed fibroblasts died in high concentrations of the drugs. In the presence of nonlethal concentrations of the drugs the cells simply detached from the dish, which produced the appearance of monolayer growth and contact inhibition. Bombik and Burger (1973), using PY transformed 3T3 fibroblasts, have since repeated and confirmed Paul's results. In addition, they have shown that in concentration ranges where dibutyryl cyclic AMP was nontoxic, it had no effect upon agglutinability. Thus, restoration of "contact inhibition," concomitant with a return to normal surface architecture, does not seem to occur in virally transformed fibroblasts in the presence of exogenous dibutyryl cyclic AMP. In complete contrast to the results of Hsie et al. (1971), Tihon and Green (1973) have reported that Chinese hamster ovary cells grown in the presence of dibutyryl cyclic AMP and testosterone actually demonstrated a large increase in agglutinability by concanavalin A, as well as an increase in the production of RNA tumor virus-like particles.

3. Restoration of "Normal Morphology" to Some Transformed Cells

In addition to a lack of density-dependent growth and increased agglutinability by plant lectins, many transformed or malignant cells are also identifiable by a loss of normal microscopic appearance in culture. For example, transformed fibroblasts no longer grow in whorls made up of thin, elongate cells in parallel alignment, but instead are characterized by rounded or polygonal cells growing in a random fashion. A number of workers have reported that some transformed cells reacquired their "normal morphology" and growth patterns when cyclic AMP, dibutyryl cyclic AMP, and/or agonists which raise intracellular cyclic AMP levels are added to the culture medium. Chinese hamster ovary cells treated with dibutyryl cyclic AMP plus testosterone (Hsie and Puck, 1971; Hsie et al., 1971), sarcoma and strain L fibroblasts treated with monobutyryl cyclic AMP, dibutyryl cyclic AMP, or cyclic AMP plus theophylline (Sheppard, 1971), were all reported to acquire a fibroblastic appearance and growth pattern. Mouse and human neuroblastoma cells treated by serum starvation (Seeds, Gilman, Amano, and Nirenberg, 1970; Schubert, Humphreys, deVitry, and Jacob, 1971), PGE_1 (Prasad, 1972a,b), dibutyryl cyclic AMP (Furmanski, Silverman, and Lubin, 1971; Prasad and Hsie, 1971a,b; Macintyre, Wintersgill, Perkins, and Vatter, 1972; Prasad and Vernadakis, 1972), or phosphodiesterase inhibitors (Prasad and Sheppard, 1972), grew long processes thought to be axons. These effects upon neuroblastoma cells were irreversible (Macintyre et al., 1972; Prasad and Vernadakis, 1972), unlike effects upon fibroblasts and glioma cells. Human glioma cell lines treated with dibutyryl cyclic AMP acquired a stellate configuration (Prasad and Vernadakis, 1972; Prasad and Sheppard, 1972). However, Macintyre et al. (1972) noted that although the glioma cells grew more slowly and were changed in appearance, the cells also showed a generalized ablation of membrane systems and still grew in overlapping patterns. The authors suggested that the observed effects were not consistent with a return to a differentiated state.

Neoplastic cells that are morphologically affected by manipulations resulting in increased intracellular levels of cyclic AMP or its analogues have at least one thing in common. The alterations result in a polarity in cell shape from an otherwise rounded cell. Microtubules are thought to be related to cell shape, and agents that specifically disrupt them (cold, high pressure, colcemid, and vinblastine sulfate) lead to a loss in cell shape (cf. Tilney, 1971). In non-neoplastic, embryonic chick dorsal root ganglion cultures, the rate of development of axons was reported to be accelerated by nerve growth factor which supposedly could be replaced by cyclic AMP or dibutyryl cyclic AMP (Roisen and Murphy, 1973). The extension of axons was thought to be dependent on a cyclic AMP stimulated assembly of preexisting microtubule subunits, since colchichine could antagonize or, in large enough concentrations, prevent the process. RNA or protein synthesis was not required for acute elongation periods, but protein synthesis was necessary for chronic, complete, and irreversible differentiation of axons (Seeds et al., 1970; Roisen, Murphy, and Braden, 1972a; Roisen, Murphy, Pichichero, and Braden,

1972b; Roisen and Murphy, 1973). The mechanism of the cyclic AMP effect on microtubule assembly is quite obscure, although it has been reported that isolated neural microtubular protein can serve as a substrate that is phosphorylated by cyclic AMP-dependent protein kinase (Goodman, Rasmussen, DiBella, and Guthrow, 1970).

In neoplastic cells similar evidence indicates that morphological changes mediated by cyclic AMP are a result, at least in part, of microtubule formation. Dibutyryl cyclic AMP and colchichine have antagonistic effects on cell shape in Chinese hamster ovary cells (Hsie and Puck, 1971; Puck, Waldren, and Hsie, 1972; Patterson and Waldren, 1973). Enucleated Chinese hamster ovary cells are still capable of "morphological transformation" by dibutyryl cyclic AMP or a combination of dibutyryl cyclic AMP and PGE_1 (Schroder and Hsie, 1973). Compounds which inhibit microtubule assembly, such as vinblastine and colcemid, also inhibited the effect of dibutyryl cyclic AMP in enucleates. Essentially the same findings have been observed with neuroblastoma axon elongation (Prasad and Hsie, 1971a,b; Kirkland and Burton, 1972; Prasad, 1972a,b), human glioma process formation (Macintyre et al., 1972; Edström, Kanje, and Walum, 1974), and morphological changes of a clone of Schwannoma cells (Sheppard, Hudson, Larson, and Cunningham, 1974). However, similar to non-neoplastic axon development, continued dibutyryl cyclic AMP-induced process formation in neuroblastoma and glioma cells required protein synthesis (Seeds et al., 1970; Edström et al., 1974).

4. Lack of Growth Inhibition

In some cells, cyclic nucleotides do not inhibit cell proliferation and, in fact, may enhance it. In serum-free media, calcium or cyclic AMP, in the presence of free calcium ion concentrations similar to those found in plasma (0.5 mM), will stimulate thymocyte proliferation (Whitfield et al., 1973). With higher levels of calcium, cyclic AMP will actually inhibit cell proliferation. Low levels of dibutyryl cyclic AMP seem to stimulate the proliferation of cultures of embryonic liver and kidney cells (Medoff and Parker, 1972). In primary and secondary cultures of chick embryo fibroblasts cultured in medium containing 2% serum, cyclic AMP, monobutyryl cyclic AMP, dibutyryl cyclic AMP, adenosine, ATP, cyclic GMP, and dibutyryl cyclic GMP all enhanced the growth rate (Hovi and Vaheri, 1973). In no instance did these nucleotides inhibit proliferation of chick embryo fibroblasts even though the exogenously added cyclic compounds did accumulate within the cells.

Treatment of cultured lymphocytes derived from blood or thymus with prostaglandin E_1 or mitogens leads to increased intracellular cyclic AMP levels, initiation of DNA synthesis and mitosis (Franks, MacManus, and Whitfield, 1971; Webb, Stites, and Perlman, 1973). Increases in cyclic AMP levels of lymphocytes caused by the mitogen concanavalin A appear to be mediated by specific carbohydrate receptors on the cell surface (Lyle and Parker, 1974). The effect on mitosis was mimicked in thymic lymphocytes *in vivo* by the injection of 5 mg cyclic AMP/kg rat body weight (Rixon, Whitfield, and MacManus, 1970). Similarly, isoproterenol increased cyclic AMP concentrations in rat parotid gland and induced DNA synthesis (Guidotti, Weiss, and Costa, 1972). In partially hepatectomized or hormone-infused rats a biphasic accumulation of cyclic AMP occurs in the liver, followed by an increased synthesis of DNA (MacManus, Franks, Youdale, and Braceland, 1972). The injection of adrenergic blocking agents (*d,l*-propranolol or phenoxybenzamine) following partial hepatectomy, at times appropriate to blocking the second wave of the biphasic increase in cyclic AMP, delayed and reduced the cyclic AMP response, and delayed and inhibited subsequent DNA synthesis in the regenerating liver (MacManus, Braceland, Youdale, and Whitfield, 1973). A review of the latter results and others has led MacManus and Whitfield (1974) to conclude "that an *increase* in cyclic AMP concentration does occur as cells enter upon a cell-cycle, and that this increase is related to initiation of DNA synthesis."

Cultured rat or chick myoblasts will differentiate into myotubules by the fusion of cells and synthesis of muscle proteins after growing to confluency (Konigsberg, 1961; Yaffe, 1968; Shainberg, Yagil, and Yaffe, 1971; Loomis, Wahrmann, and Luzzati, 1973). The addition of sufficient concentrations of cyclic AMP, dibutyryl cyclic AMP, or phosphodiesterase inhibitors slowed the growth rate of the myoblasts and prevented or delayed the fusion and formation of muscle elements (Wahrmann et al., 1973*a*; Zalin, 1973). If the concentrations of either of the compounds were lowered such that only a minor slowdown in myoblast proliferation was noted, the cells grew to confluency but did not fuse and differentiate (Wahrmann et al., 1973*a*). Removal of the compounds allowed differentiation to occur. These findings are in accordance with the observation that endogenous cyclic AMP levels normally decrease during myoblast fusion (Reporter, 1972; Wahrmann et al., 1973*b*). Thus, in myoblasts, it seems as if high levels of cyclic AMP are antagonistic to normal differentiation and cessation of growth, although application of abnormally high concentrations of endogenous nucleotides will suppress growth.

There are instances where cyclic AMP appears to have neither a positive nor a negative effect upon cell division. While cultured rat lens epithelial cells showed a transient mitotic inhibition within the first 2 hr of incubation with high, but nontoxic, concentrations of isoproterenol, monobutyryl cyclic AMP, or dibutyryl cyclic AMP, mitosis proceeded normally thereafter (von Sallmann and Grimes, 1974). Sea urchin eggs (*Stronglylocentrotus purpuratus* and *Lytechinus pictus*) possess the enzymes of cyclic AMP metabolism (Castenada and Tyler, 1968; Nath and Rebhun, 1973*a,b*), and culturing these cells in the presence of cyclic AMP, dibutyryl cyclic AMP, or monobutyryl cyclic AMP allowed the intracellular accumulation of levels of cyclic AMP or monobutyryl cyclic AMP at least two orders of magnitude higher than normal. Nonetheless, normal fertilization and development proceeded at least to gastrulation, giving rise to the conclusion that cyclic AMP was not involved in the regulation of mitosis of these cells

(Rebhun, White, Sander, and Ivy, 1973; Nath and Rebhun, 1973a,b). Likewise, the growth of HTC hepatoma cells (van Wijk et al., 1972; Granner et al., 1968) or normal human diploid lung WI-38 fibroblasts (Ryan and Durick, 1972) is unaffected by cyclic AMP or dibutyryl cyclic AMP. We have also found that prolonged, 48-hr stimulation of SV40 transformed WI-38 fibroblasts with low concentrations of PGE_1 had no apparent effect upon the growth rate of these cells as determined by cell counting. Furthermore, the morphology of these cells remained abnormal, and did not revert to a fibroblastic appearance as judged by light and electron microscopy (Kelly, Butcher, and Chlapowski, *unpublished observations*) (Fig. 6). Under the latter conditions of prolonged PGE_1 stimulation, the intracellular cyclic AMP concentrations, as well as accumulation in the medium, increased to much higher than normal levels, and remained at higher than normal levels throughout the experimental period.

5. Summary

It is well documented that physiological or experimental manipulations which change intracellular cyclic AMP levels also signal differentiated functions in *those cells capable of responding*. This is known as the "second messenger" concept (Robison et al., 1971). In some cells such as thymocytes (Whitfield et al., 1973) the response may involve growth and in others, death (Bourne et al., 1975). In yet others, such as neurons (Seeds et al., 1970; Roisen et al., 1972; Roisen and Murphy, 1973), the differentiated response may be incompatible with growth. Therefore, the observation that *some* neoplastic or transformed cells may respond to cyclic AMP with a reduction of growth and, in some cases, a recovery of "normal microscopic appearance" or a recovery of a differentiated response, may indicate that these cells retain, at least in part, the components of the second messenger system. That some neoplastic cells are not capable of responding to cyclic AMP with a cessation of growth may indicate either that these cells do not normally respond in such a manner or that a loss of some crucial component of the cyclic AMP system has taken place. It is our contention that the experiments in the literature indicate that while experimental transformation or clinical neoplasia are often accompanied by abnormalities in cyclic AMP metabolism, there is no consistent pattern. Rather, as we previously pointed out, there may be no change, a decrease, an absence, or an increase in intracellular cyclic AMP levels.

C. The Enzymes of Cyclic AMP Metabolism in Malignant Cultured Cells and Tissues

Since, as we have seen, some types of proliferating neoplastic cells contain abnormally high or low basal intracellular levels of cyclic AMP and respond inappropriately or not at all to certain hormones, it seems likely that in these cells a defect exists in one or more of the enzymes controlling the level of cyclic AMP.

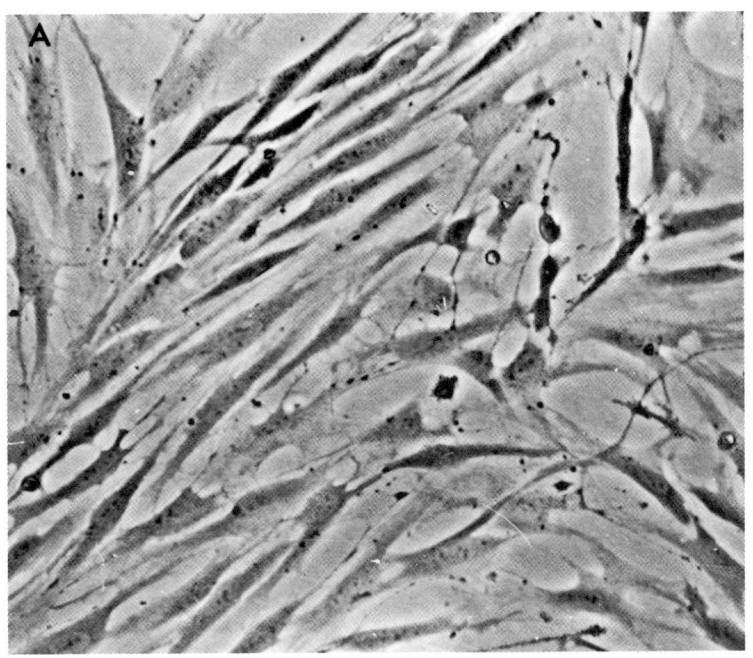

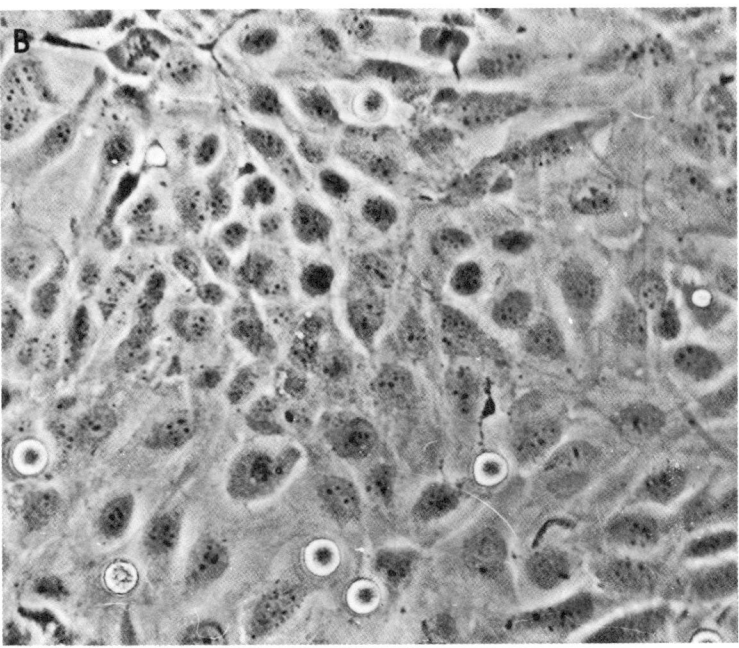

FIG. 6. Phase-contrast micrographs of growing cultures of WI-38 human fibroblasts (A) and SV40 transformed WI-38 fibroblasts (B) following 24-hr exposure to 10 mM PGE_1. Note the rounded, dividing cells. Also note the lack of effect of the prolonged PGE_1 treatment on the morphology of either culture (compare to Fig. 1, A and B). Similar results were obtained after 48-hr of treatment. ×200.

Given the currently accepted model of cyclic AMP metabolism, the two most likely candidates for study are adenylyl cyclase and phosphodiesterase. On the other hand, since some transformed cells show no ability to respond to experimentally increased levels of cyclic AMP, either when the latter enzyme systems are circumvented by use of derivatives or even when they are apparently functional, it is also likely that cyclic AMP-dependent protein phosphokinase activities may be affected.

1. *Adenylyl Cyclase*

A malfunctioning of adenylyl cyclase in neoplastic cells could be due to any one or a combination of defects, including those in the following categories:

a. An aberration in the catalytic site so that more or less cyclic AMP is produced under either basal or stimulated conditions.

b. A loss of, defect in, or an excess of one or more of the specific hormonal receptors. Also, the presence of inappropriate, or "ectopic" hormone receptors would be included in this category.

c. A partial or complete change in the plasma membrane localization of the enzyme.

d. A complete loss of the functional enzyme.

a. *Aberrations in the catalytic site.*

Alterations in basal activity may be apparent in enzymatic measurements. When compared to normal counterparts, the specific activities of unstimulated adenylyl cyclases of SV40 transformed 3T3 cells, MS/Mu LV transformed 3T3 cells, SV/PY transformed 3T3 cells (Peery et al., 1971), rat hepatocarcinoma (Christoffersen, Morland, Osnes, and Elgjo, 1972), rat adrenocortical carcinoma 494 (Ney et al., 1969), Rous sarcoma transformed baby hamster kidney fibroblasts (Burk, 1968), and human glioma cells (Perkins et al., 1971b), all showed sizable increases. On the other hand, the basal specific activities of PY transformed 3T3 cells (Peery et al., 1971), PY transformed BHK cells (Burk, 1968; Peery et al., 1971), adenovirus type 12 transformed BHK cells (Raska, 1973), SV40 transformed hamster astrocytes (Weiss, Skein, and Snyder, 1971), human lymphatic leukemia cells (Polgar, Vera, Kelley, and Rutenberg, 1973) and some Morris rat hepatomas (Allen, Munshower, Morris, and Weber, 1971) were decreased. Basal activities of yet other neoplastic cells or tissues such as some Morris rat hepatomas (Allen et al., 1971) remained about the same as normal.

In those cells where the basal production of cyclic AMP is either higher or lower than normal, a defect in the catalytic site as described in category "a" is possible. However, the reason for continuous, "steady-state" basal production of cyclic AMP in cells is not understood. It could well represent the minimal level of cyclic AMP production that adenylyl cyclase is capable of achieving. In that case, measurements of basal activities might well represent measurements of the integrity or numbers of catalytic sites. In contrast, so-called unstimulated activities may actually be due to stimulation by some component, cofactor, or other factor in membranes which is lower or higher in various transformed cells. In the latter case, "basal" activities would not necessarily represent a measure of the state of the catalytic sites but instead might be indicative of the level of such a factor present.

A number of transformed cells demonstrate increased or decreased responsiveness of adenylyl cyclase to NaF stimulation. The interpretation of this phenomenon is difficult since the mode of action of NaF is not known and since responsiveness to this agent may or may not be correlated with basal or hormone-stimulated changes in the enzyme. For example, both basal and NaF-stimulated adenylyl cyclase activities increase in a rat hepatocarcinoma (Christoffersen et al., 1972) and 3T3 fibroblasts transformed by either MSV/MuL or SV40 viruses (Peery et al., 1971); and both types of activities decrease in rat hepatoma 484A (Emmelot and Bos, 1971) and PY transformed 3T3 fibroblasts (Christoffersen et al., 1972). In contrast, basal activities increase or remain about the same, while NaF activities decrease in mouse hepatoma 147042 (Emmelot and Bos, 1971), rat adrenocortical carcinoma 494 (Ney et al., 1969), and SV40 transformed WI-38 human fibroblasts (Kelly et al., 1974). For these reasons we are not in agreement with the suggestion of Ryan and Heidrick (1974) that "the catalytic site of adenylate cyclase which is stimulated by NaF is apparently unaltered in transformation or malignancy. . . ." The evidence suggests to us that it is not only possible, but probable, that the catalytic site of adenylyl cyclase is altered in certain neoplastic cells.

b. *Loss of or defect in hormonal receptors.*

Observations on the rat adrenocortical carcinoma 494 reveal a situation where the normal adrenal hormonal response to ACTH is not evident in intact cells, and yet ACTH stimulation of adenylyl cyclase occurs in homogenates (Ney et al., 1969). Similar findings have emerged in our laboratory on HEp-2, a cultured cell derived from a human laryngeal carcinoma (Kelly and Butcher, 1975). In intact HEp-2 cells, epinephrine produced only a very small rise in intracellular cyclic AMP, while in homogenates epinephrine stimulated adenylyl cyclase activity several-fold. The phosphodiesterase inhibitor MIX permitted the appearance of modest elevations in intracellular cyclic AMP levels, even though the phosphodiesterase activity of the cells was low. In any event, in both of these malignant tissues, intact cells responded to hormones with little or no increased production of cyclic AMP whereas homogenized cell preparations responded well. These results suggested that transformation to a neoplastic state caused an apparent occlusion of a hormonal receptor site, although alternate interpretations abound. Since preliminary cell fractionation has revealed that in HEp-2, the adenylyl cyclase is still primarily localized in the plasma membranes, this type of defect would appear to fall into category "b." SV40 transformed WI-38 human fibroblasts show a decrease in responsiveness to PGE_1 and an increase in responsiveness to epinephrine, perhaps indicating deletions and additions of the numbers of each type of receptor (Kelly et al., 1974).

Another type of modification which may occur is the emergence of ectopic receptors in neoplastic cells. For example, adenylyl cyclase in rat adrenocortical carcinoma 494 cells appears to acquire the ability to respond to epinephrine, norepinephrine, and thyroid-stimulating hormone (Schorr and Ney, 1971). By contrast, adenylyl cyclase in normal adrenal cortical cells is responsive only to ACTH.

Thus, alterations of receptors may lead to a decrease or even to a loss of the specific hormonal responsiveness of the cells, and in other cases an excessive or nonspecific response. It seems likely that the attendant loss of control of certain differentiated functions would ensue.

c. *Translocation of adenylyl cyclase.*

Changes in the subcellular location of adenylyl cyclase are apparently detectable by cell fractionation. Tomási, Réthy, and Trevisani (1973) have reported that the chemically induced Yoshida ascites rat hepatoma had both plasma membrane-bound adenylyl cyclase and soluble adenylyl cyclase activities. During growth of the tumor, the enzyme activity decreased in the membranes and increased in the cytosol. Both the particulate and the soluble adenylyl cyclases responded to hormonal or NaF stimulation, although responses in all cases were markedly lower than those found in normal hepatocytes. In nontumorous liver, active adenylyl cyclase is thought to be present only in plasma membranes and, possibly, plasma membrane precursors such as the Golgi complex (Cheng and Farquhar, 1974). Réthy, Váczi, Tóth, and Boldogh (1973) have reported similar observations with SV40 transformed hamster embryo fibroblasts, as well as with spleens of Balb/c mice infected with Raucher leukemia virus. These observations suggest that a partial change in the intracellular site of adenylyl cyclase had occurred and are indicative of the type of defect in category "c."

d. *Loss of functional enzyme.*

Bourne, Coffino, and Tomkins (1974) have isolated a strain of S49 mouse lymphosarcoma cells which appears to lack adenylyl cyclase. The cells did not accumulate cyclic AMP and had no measurable adenylyl cyclase activity. Ordinarily, S49 cells have an easily measurable basal adenylyl cyclase activity which can be stimulated by isoproterenol, PGE_1, or NaF. The adenylyl cyclase "minus" strain was capable of responding to dibutyryl cyclic AMP with a cessation of growth, followed eventually by death, had measurable protein kinase activity, and showed an induction of phosphodiesterase in response to dibutyryl cyclic AMP. In contrast, HTC hepatoma cells, which had barely measurable adenylyl cyclase activity (Granner et al., 1968; Makman, 1971a), were not responsive to exogenously added cyclic AMP or dibutyryl cyclic AMP (van Wijk, et al., 1972).

Thus, current data indicate that adenylyl cyclase may be functionally absent or modified in a number of ways in malignant or transformed cells.

2. *Phosphodiesterase*

The distribution of cyclic AMP phosphodiesterase has not been examined comprehensively in many normal cells or tissues, let alone neoplastic cells. In most instances where such examination has been carried out, differential centrifugation techniques have been utilized. Based upon these procedures, a variable amount of phosphodiesterase, ranging from very little to about one-half the cellular total, has been found to be particulate (Appleman et al., 1973). Likewise, few rigorous comparisons of the total levels of soluble and particulate phosphodiesterases in normal and neoplastic counterparts have been reported. Schroder and Plagemann (1972) have reported that 30 to 40% of the phosphodiesterase in Novikoff rat hepatoma, mouse L, and HeLa cells was particulate, being found in the $600 \times g$ and $104,000 \times g$ fractions. It is not known whether this represents "trapped" phosphodiesterase or a truly particulate enzyme. No differences in substrate level kinetics or theophylline inhibition kinetics were observed between the soluble and the particulate enzymes.

In most transformed cells or malignant tissues examined, soluble phosphodiesterase activities decreased when compared to their normal counterparts. This holds true in Novikoff rat hepatoma cells (Schroder and Plagemann, 1972), SV40 transformed 3T3 cells (D'Armiento et al., 1972), SV40 transformed WI-38 cells (Kelly et al., 1974), rat adrenocortical carcinoma 494 (Ney et al., 1969), and a variety of Morris hepatomas (Rhoads, Morris, and West, 1972). While in most instances both the high and low substrate activities decreased, D'Armiento et al. (1972) reported that only the low substrate phosphodiesterase activity remained in SV40 transformed 3T3 cells. Somewhat similarly, high substrate activity decreased and low substrate activity increased in some Morris hepatomas examined (Clark, Morris, and Weber, 1973). An exception to the above reports was that of Weiss et al. (1971), who reported little difference between the phosphodiesterase activities of normal astrocytes and astrocytoma cells.

The observation that phosphodiesterase activities are lower than normal in many transformed cells fits well with the reports that cyclic AMP levels are high in some malignant tissues. In rat adrenocortical carcinoma 494 basal adenylyl cyclase activity increases about 40% while phosphodiesterase activity decreases by approximately 75%. Thus, the almost doubling of basal cyclic AMP levels in this tumor might be accounted for by these changes in enzymatic activity. However, in some Morris hepatomas, both adenylyl cyclase activities and phosphodiesterase activities tend to decrease. For example, in hepatoma 5123-TC, adenylyl cyclase activity decreased by about 30% (Allen et al., 1971), and phosphodiesterase activity by about 50% (Rhoads et al., 1972). In this case, perhaps, the increased basal cyclic levels in the hepatoma (as compared to normal liver) may be due to the differentially larger decrease in phosphodiesterase activity. Nonetheless, without a greater understanding of the characteristics of these enzymes in both normal and neoplastic cells, such interpretations are highly speculative.

Certainly, the correspondingly large decreases in phosphodiesterase activities, when compared to the small decrease or increase in basal adenylyl cyclase activities in SV40 transformed WI-38 (Kelly et al., 1974) or SV40 transformed 3T3 cells (Peery et al., 1971), respectively, are not, at least superficially, compatible with low basal intracellular levels of cyclic AMP reported in these fibroblasts.

Notwithstanding the latter data, accumulation of intracellular cyclic AMP in response to hormones in SV40 transformed WI-38 fibroblasts is much greater than that observed in normal cells (Kelly et al., 1974)—exactly as one might expect with low phosphodiesterase activities. These results seem to suggest that changes in basal cyclic AMP levels may be related to changes in phosphodiesterase activities in some transformed or malignant cells but not in others.

The apparent lack of correlation in some transformed cells between adenylyl cyclase and phosphodiesterase activities as measured in broken-cell systems on the one hand, and basal intracellular levels of cyclic AMP on the other, illustrates the futility of attempting to draw conclusions based upon examining only a few of the components of a complex system. For example, we have little idea of the role of cyclic AMP escape in normal and malignant cells. In WI-38 fibroblasts, for example, it is of considerable magnitude in hormone-stimulated cells (Section II.B). Further, it seems clear that the markedly decreased escape of cyclic AMP from SV40 transformed WI-38 (as compared to WI-38) may in part explain the higher levels of cyclic AMP found in the transformed cell with hormonal stimulation. However, this still tells us little about the mechanism of escape or how transformation lowers the rates of escape in transformed cells (Kelly and Butcher, *unpublished observations*). Furthermore, most studies neglect to examine the relationship between phosphodiesterase and protein kinase levels and activities. Our inability to explain its importance notwithstanding, the experimental data do suggest that phosphodiesterase activities and substrate-level specificities may be altered in the neoplastic state. How this relates to changes in regulation of intracellular cyclic AMP can only be hypothesized at this time.

As in nonmalignant cells, cyclic AMP-dependent induction of increased phosphodiesterase levels has been reported to occur in SV40 transformed 3T3 cells (D'Armiento et al., 1972), neuroblastoma cells (Prasad and Kumar, 1973), astrocytoma cells (Uzunov et al., 1973), and C-6 rat glioma cells (Schwartz and Passonneau, 1974), leading Ryan and Heidrick (1974) to suggest that "this inducibility does not appear to be altered in the malignant state." However, Bourne et al. (1975) reported that while phosphodiesterase levels are inducible in S49 mouse lymphosarcoma cells and in an adenylyl cyclase-minus strain of S49 cells (both of which have a functional and apparently normal protein kinase), phosphodiesterase levels are low and not inducible in a protein kinase-minus strain of S49 cells. Thus, in malignant cells or strains derived from them, there seem to be exceptions to every rule. The results of Daniel et al. (1973b) have been interpreted by those authors as indicating that phosphodiesterase may not be inducible in malignant cells where the protein kinase system is absent or malfunctioning.

3. Protein Phosphokinase

Cyclic AMP-dependent protein kinase activity has been reported in a number of neoplastic cells, including neuroblastoma cells (Kuo and Greengard, 1969), human astrocytoma 1181N1 cells (Perkins et al., 1971a,b), rat glioma C-6 cells (Jard, Premont, and Benda, 1972), HTC hepatoma cells (Granner, 1972), FSV transformed bovine fibroblasts (Troy et al., 1973), S49 mouse lymphoma cells (Daniel et al., 1973b), and SV40 transformed WI-38 cells (Miller, Kelly, and Butcher, 1975, *unpublished*). Comparisons of protein kinase have been made between some of these cells and their parent cells or tissues.

The activities of cyclic AMP-independent and cyclic AMP-dependent protein kinase of FSV transformed bovine fibroblasts were virtually identical to those in untransformed bovine fibroblasts (Troy et al., 1973). As mentioned previously, both cyclic AMP binding protein and protein kinase activities were deficient in a cyclic AMP-resistant strain of S49 mouse lymphoma cells and perhaps correlated to this is an absence of cyclic AMP-dependent phosphodiesterase induction (Daniel et al., 1973b).

In HTC hepatoma cells total protein phosphokinase activity (i.e., cyclic AMP-dependent plus cyclic AMP-independent) is quite similar to normal rat liver (Granner, 1972). However, the hepatoma cells have a deficiency in functional binding protein and, consequently, a somewhat unregulated kinase. The addition of partially purified binding protein from liver restored cyclic AMP regulation to partially purified hepatoma kinase. Studies by Granner (1974) on a variety of cultured hepatomas further elucidated the differences between normal liver protein kinase activity and that of cultured hepatoma cells. While liver-binding protein has two apparent binding sites with different affinities for cyclic AMP ($K_{d_I} \approx 3.4 \times 10^{-7}$ and $K_{d_{II}} \approx 1 \times 10^{-8}$), the binding protein of H4-II-E, RLC, and HTC hepatoma cells had only a low-affinity binding property ($K_d \approx 3.5 \times 10^{-7}$). Furthermore, no heat-stable, specific inhibitors of binding were apparent in the cell sap of the cultured hepatoma cells.

These data indicate that cyclic AMP-dependent protein kinase and/or its regulatory subunit may be altered in certain neoplastic cells.

4. Comments

The data reviewed in the last few sections indicate that an apparent defect in some component or components of cyclic AMP metabolism exists in most neoplastic cells examined. In those cells where no alterations have been revealed, all components of the cyclic AMP system have not been examined, allowing for the possibility that aberrations in cyclic AMP metabolism exist in these cells also. Obviously, the use of homogeneous populations of tissue culture cells may be helpful in proving or disproving this possibility.

The abnormalities uncovered in the cyclic AMP metabolism of various neoplastic cells may include either the adenylyl cyclase system, the cyclic AMP phosphodiesterase, the protein phosphokinase system or, apparently, any combination of these components compatible with viability. The lesions within each enzyme system itself are variable, resulting in effects that range from a deficiency in activity to hyperactivity. Intuitively, one is led to speculate that further investigations comparing normal and transformed cells will yield a complete spectrum of possible lesions in the cyclic AMP-metabolizing system. However, the data now available in the literature suggest that no common underlying anomaly of the cyclic AMP system exists in all neoplastic cells, save that some part of the system is usually different in neoplastic cells. For this reason we find it difficult, at this time, to ascribe the lack of regulation of growth seen in all neoplastic cells to discrete changes in the metabolism of cyclic AMP.

The plethora of defects in the enzymes of cyclic AMP metabolism in various neoplastic cells is indeed perplexing. Yet the evidence surely seems to indicate that this situation must be viewed as a consequence of, rather than a cause of, the neoplastic state until it might be proven otherwise.

D. *Enzyme Changes and Cyclic AMP Levels*

While the study of the enzymes of cyclic AMP metabolism in neoplastic cells may or may not provide a mechanism for comprehending the etiology of neoplasia, it seems to have potential as a powerful tool for understanding cyclic AMP metabolism. By comparing the changes in cyclic AMP levels in cells with specific lesions, a greater appreciation of the fine control of cyclic AMP flux may be possible.

The K_m values for cyclic AMP phosphodiesterase(s) range between 10^{-8} and 10^{-3} M (Appleman et al., 1973). The K_m of the binding protein portion of kinase has been reported to be about 10^{-8} M (Walton and Garren, 1970). These data have been interpreted as indicating that essentially all the cyclic AMP in unstimulated cultured cells is bound to the regulator units of protein kinase (Ryan and Heidrick, 1974). Since attachment of cyclic AMP to binding protein renders it resistant to degradation by phosphodiesterase, it must follow that phosphodiesterase can only play an important role in controlling cyclic AMP levels in hormonally stimulated cells, where the binding protein is saturated with cyclic AMP.

However, this idea seems to run contrary to what little is know about the kinetics of cyclic AMP flux. In all adenylyl cyclase preparations examined to date, an unstimulated or basal production of cyclic AMP occurs. Indeed, if these results are indicative of the situation in living cells, then a certain rate of production of cyclic AMP is ongoing. Since a steady-state, basal level of intracellular cyclic AMP is apparent in cells, the rate of production of cyclic AMP must equal the rate of hydrolysis and/or excretion (if present) of cyclic AMP. Thus, phosphodiesterase, particularly the low-K_m activity, may play a role in maintaining basal cyclic AMP levels under ordinary circumstances. This point is demonstrated by several studies which report that basal cyclic AMP levels are often increased by the incubation of cells in medium containing phosphodiesterase inhibitors. Of course, in some cells such inhibitors may affect cyclic AMP levels by other mechanisms in addition to phosphodiesterase inhibition (see Section II.C.3). The amount of the binding moiety of protein kinase present in cells, unless very large, should not influence the turnover rate of cyclic AMP, but may be important in determining the measurable basal level of cyclic AMP in cells. Thus, the basal level of cyclic AMP may be considered a function of the production and destruction of cyclic AMP resulting in a cellular pool of the nucleotide, which in turn is in equilibrium with binding protein. However, this simplistic scheme is more than likely complicated by other factors, known and unknown, which may ultimately influence cellular cyclic AMP levels, such as escape, ATP availability, and compartmentalization.

In this regard, the work of Beavo, Bechtel, and Krebs (1974) on the protein kinase system of rabbit skeletal muscle is most enlightening. This authors reported that the cyclic AMP concentration necessary for half-maximal stimulation of protein kinase is about 0.2 to 0.3 μM, which is approximately the same concentration as that of the cyclic AMP present in resting muscle. The concentration of protein kinase in rabbit skeletal muscle is also about 0.23 μM. Thus, a more or less equivalent amount of binding sites and ligand exists in these cells. However, the basal activity of the kinase is not high, contrary to what may be anticipated based upon these data. According to Beavo et al. (1974), the basal activity may be due to a number of factors which affect the equilibrium between cyclic AMP and protein-binding sites and between the regulatory and catalytic subunits of the kinase. These factors would include ATP levels, which in physiological amounts increase the dissociation constant of binding, and a heat-stable inhibitor of the kinase. In addition to these factors, these authors also observed that the kinase activation constant is higher if casein or glycogen synthetase is used as a substrate instead of histone.

While to our knowledge no studies on any cell type are yet complete enough to draw conclusions concerning the aberrations in neoplastic cells, an assembling of some of the available data should at least allow us to see what further questions must be asked. In Table 12, a comparison of the relative activities or amounts of adenylyl cyclase, cyclic AMP phosphodiesterase, protein kinase, and intracellular cyclic AMP levels in a number of cells is presented.

In HTC hepatoma cells, both basal and epinephrine-stimulated adenylyl cyclase activities were low when compared to liver (Makman, 1971a). Phosphodiesterase activity was low (Manganiello and Vaughan, 1972b), and cyclic AMP binding was only about 10% of its normal value (Granner, 1972). Correspondingly, both basal and epinephrine-stimulated cyclic AMP levels were extremely low (Makman, 1971a; Granner, 1972). In fact, no rise in cellular cyclic AMP levels could be observed upon hormonal stimulation unless phosphodiesterase was inhibited (Manganiello and Vaughan, 1972b). Therefore, in HTC cells an across-the-board decrease in all three factors, adenylyl cyclase, phosphodiesterase, and binding protein, resulted in an anticipated decrease in cyclic AMP levels. The absence of a rise in cyclic AMP levels upon hormonal stimulation without phosphodiesterase inhibitors seems to be due to the fact that the decrease in phosphodiesterase activity ($\sim$70%) was less than that observed in either adenylyl cyclase or binding protein ($\sim$90%).

TABLE 12. A comparison of relative activities[a] of adenylyl cyclase, phosphodiesterase, binding protein, and protein kinase to relative amounts of intracellular cyclic AMP levels in selected cells

Cell	Adenylyl cyclase		Phosphodiesterase		Cyclic AMP binding protein	Protein kinase		Cyclic AMP levels[c]	
	Basal	Hormone stimulated[b]	High K_m	Low K_m		−cAMP	+cAMP	Basal	Hormone stimulated[b]
Rat liver	1	4	1	—	1	1	10	1	4
Rat HTC hepatoma cells	0.1	0.5[d]	0.3[e]	—	0.1[f]	3.5	10[f]	0.18	0.18[d,e]
S49 lymphoma cells	1	9	1	1	1	1	1.7	1	88
Cyclic AMP resistant S49 lymphoma cells	1	9[g]	0.14	0.42[g,h]	0.2[g,i]	0.8	0.8[g]	1.2	308[h]
WI-38 human fibroblasts	1	3	1	1	—	1	5	1	4
SV40 transformed WI-38 human fibroblasts	0.8	9[j]	0.05	0.11[j]	—	0.17	5[k]	0.5	64[j]

[a] Relative activities: Basal specific activity of adenylyl cyclase, high and low K_m specific activities of phosphodiesterase, binding activity of cytosol protein, specific activity of the cyclic AMP-independent form of partially purified cytosol protein kinase, and basal cyclic AMP levels of parent cell lines or tissue were arbitrarily set equal to 1.
[b] Hormone stimulated refers to maximal stimulation with epinephrine, or in the case of lymphoma cells, isoproterenol.
[c] Cyclic AMP levels were determined in the absence of phosphodiesterase inhibitors.
References: [d] Makman, 1971a; [e] Manganiello and Vaughan, 1972b; [f] Granner, 1972; [g] Bourne et al., 1975; [h] Daniel et al., 1973a; [i] Daniel et al., 1973b; [j] Kelly et al., 1974; [k] Miller (personal communication).

In the cyclic AMP-resistant strain of S49 lymphoma cells, as compared to the cyclic AMP-sensitive parent strain, protein-binding activity and phosphodiesterase activity were lowered to about the same extent (Daniel et al., 1973b). One might suspect that these two effects might offset one another to produce a close-to-normal basal level of cyclic AMP, since the adenylyl cyclase activity appeared to be unchanged (Bourne et al., 1975). In fact, basal cyclic AMP levels were only slightly higher than normal. Correspondingly, hormonally stimulated cyclic AMP levels are very high due to the low phosphodiesterase activity (Table 12).

However, our ability to predict results does not hold up when comparing SV40 transformed WI-38 cells to their normal counterparts. In both cells the protein kinases appear to be normal in that they can be stimulated with prostaglandins to full activity (Miller, *personal communication*). However, the cyclic AMP-independent form of protein kinase is much less than in untransformed cells, perhaps indicating that less cyclic AMP is bound under basal conditions. This assumes that the binding protein, which has not yet been studied in detail, is normal. While these results correspond to the low (50% normal) basal intracellular levels of cyclic AMP, they do not correspond to the enzymatic data. In SV40 transformed cells, basal adenylyl cyclase activity is about 80% of normal values, while phosphodiesterase activity is only 5 to 14% of normal values. Given this situation, one would expect a higher-than-normal basal cyclic AMP level. If indeed the binding protein proves to be normal, then a factor or factors other than the interplay of adenylyl cyclase, phosphodiesterase, and binding protein might be responsible for this phenomenon. Although escape of cyclic AMP seems to be a significant means of disposing of the nucleotide in these cells, it is much slower in the transformed strain. The high levels of cyclic AMP in hormonally stimulated SV40 transformed cells do correspond well to the low phosphodiesterase activity and the somewhat less effective cell escape mechanism reported (Kelly et al., 1974.)

Obviously, our analysis of Table 12 is speculative and may in fact be completely spurious. However, for the sake of continuity let us proceed as if it were logical. Some of the questions raised by the data, and which must be answered if we are to understand the control of cyclic AMP metabolism, include the following:

(1) What factors other than adenylyl cyclase activity, phosphodiesterase activity, and binding protein activity may be responsible for cyclic AMP levels seen in intact cells? This question assumes that the enzyme activities measured in broken-cell preparations adequately represent the enzyme activities as they exist in intact cells, when in fact this may not be true. This leads to another question.

(2) Are the enzyme activities in intact cells similar to those measured in broken-cell preparations? This question is difficult to answer and may only be answerable indirectly by the use of inhibitors, the examination of substrate availability, reconstitution experiments, and the like.

(3) What is the relative importance of each component of cyclic AMP metabolism to basal cyclic AMP levels? For example, it is possible that the level of binding protein, despite its apparent ability to bind cyclic AMP avidly and confer

protection to the nucleotide in a test tube, is not an important determinant of effective cyclic AMP levels in living cells. Alternatively, one might speculate that specific intracellular inhibitors prevent protein kinase from functioning over a range of basal cyclic AMP levels (Beavo et al., 1974).

E. *Cyclic GMP Levels and Cell Growth*

As the transient lowering of cyclic AMP levels has been proposed as a "trigger" responsible for the initiation of growth and cell division (Burger, 1969; Burger et al., 1972b), so also has a transient increase in cyclic GMP levels been suggested to be a mitogenic mediator.

Hadden, Hadden, Haddox, and Goldberg (1972) reported that purified concanavalin A or phytohemagglutinin increased cyclic GMP levels and subsequent proliferation of peripheral human lymphocytes, without increasing cyclic AMP. Cyclic GMP or its derivatives, 8-bromo-cyclic GMP and $N^6,2'$-O-dibutyryl cyclic GMP, also have been reported to increase the incorporation of ^{3}H-uridine and ^{3}H-thymidine into the nucleic acids of cultured Balb/c mouse spleen lymphocytes (Weinstein, Chambers, Bourne, and Melmon, 1974). However, concanavalin A increased ^{3}H-thymidine incorporation to a much greater extent. Further, only the concanavalin A-stimulated DNA synthesis, but not the slight cyclic GMP-stimulated DNA synthesis, was inhibited by simultaneous administration of dibutyryl cyclic AMP. Based upon the latter data, Weinstein et al. (1974) proposed that cyclic GMP affected a subpopulation of splenic lymphocytes other than that stimulated by concanavalin A. Seifert and Rudland (1974) have reported that both a transient decrease in cyclic AMP levels, as well as a transient increase in cyclic GMP levels, occurs upon serum addition to a quiescent culture of Balb/c 3T3 mouse fibroblasts, with DNA synthesis occurring approximately 20 hr later. These authors also reported that the inclusion of cyclic GMP or its butyrated analogues in the medium, along with hydrocortisone and bovine serum albumin, can mimic the action of fresh serum somewhat. The nucleotides induced ^{3}H-thymidine incorporation into about 18% of the cell nuclei, while fresh 20% serum produced a labeling of approximately 85% of the cell nuclei.

As mentioned earlier on in this review, Gospodarowicz (1974) reported the isolation of a fibroblast growth factor (FGF) which is present in bovine brain, pituitary gland, and serum. FGF, which is a small-molecular-weight polypeptide, has been reported to be capable of replacing the serum requirement of Balb/c 3T3 fibroblasts, when used in conjunction with a glucocorticoid and bovine serum albumin (Gospodarowicz, 1974). The growth factor induces growth in Balb/c 3T3 cells in a manner identical to that reported for serum with one important exception. That is, while a transient increase in cyclic GMP levels occurs, little or no change in cyclic AMP levels takes place (Rudland et al., 1974b). If insulin is used instead of FGF, a decrease in cyclic AMP is observed and a small increase in cyclic GMP occurs. Rudland et al. (1974b) further reported that FGF stimulated the membrane-bound guanylyl cyclase, but not the soluble form of the enzyme. Insulin, on the other hand, reduced adenylyl cyclase activity, while

FGF did not. Thus, using FGF as described, it was possible to stimulate and maintain cell division without the so-called trigger effect of lowered cyclic AMP levels (Burger, 1969; Burger et al., 1972b). Intracellular levels of cyclic GMP reportedly rose only in the early G_1 phase of the cell cycle of growing cells as opposed to the various fluctuations of cyclic AMP throughout the cell cycle (Rudland et al., 1974b).

Rudland, Seeley, and Seifert (1974c) compared cyclic AMP and cyclic GMP levels in a variety of fibroblasts during serum activation, growth, and confluency. In Balb/c 3T3 mouse, 3T3-4A Swiss mouse, secondary mouse embryo, BSC-1 monkey, BHK hamster, and 3T6 Swiss mouse fibroblasts, cyclic AMP/cyclic GMP ratios were very low during the period of serum activation, slightly higher during growth, and very high in density-dependent inhibited confluent cultures. This, according to the authors, showed that cyclic GMP and cyclic AMP concentrations were "closely correlated with the growth states of different cultured fibroblasts and that parallel but reciprocal changes are observed in the cyclic nucleotide concentrations in transition between these growth states." However, since cyclic GMP levels can be raised with subsequent cell division and without any concomitant lowering of cyclic AMP, and since raising cyclic AMP levels can block that effect, Rudland et al., (1974c) suggested that cyclic AMP acts as a negative signal while cyclic GMP acts as a positive signal "controlling the growth of cultured cells."

In contrast to untransformed fibroblasts, SV40 transformed 3T3 and PY transformed BHK fibroblasts showed no changes in cyclic AMP/cyclic GMP ratios, independent of whether the cells were serum-activated, sparsely growing, or growing in confluent monolayers (Rudland et al., 1974c). Furthermore, PY transformed Balb/c 3T3 fibroblasts, as well as temperature-sensitive strains of Balb/c 3T3 fibroblasts growing at the permissive temperature, were not stimulated by FGF plus hydrocortisone (Rudland, Eckhard, Gospodarowicz, and Seifert, 1974a). On the other hand, temperature-sensitive strains growing at the nonpermissive temperature (normal) were stimulated by FGF and hydrocortisone. FGF also was reported to be incapable of activating membrane-bound guanylyl cyclase in SV40 transformed Balb/c 3T3 cells. The latter authors indicated that these results suggest "that the temperature-sensitive lesion in mutant cells is either wholly or in part caused by a membrane change."

In summary, cyclic GMP has been postulated to be a positive intracellular signal involved in the regulation of growth of a variety of fibroblasts and peripheral lymphocytes and cyclic AMP a negative intracellular signal in the growth process of these cells (Hadden et al., 1972; Seifert and Rudland, 1974; Rudland et al., 1974c). The rise in cyclic GMP occurs only early in the G_1 phase of the cycle of fibroblasts and is thought to be due to FGF stimulation of membrane-bound guanylyl cyclase in the presence of a glucocorticoid (Rudland et al., 1974a). In contrast, those transformed fibroblasts examined were not stimulated to increased growth by FGF, and maintained a high intracellular cyclic GMP level, independent of the state of growth (Rudland et al., 1974a).

These studies concerned with the relationships of cyclic GMP and the regulation or lack of regulation of growth in normal and transformed fibroblasts are relatively recent and sparse. When enough data become available, it will be necessary to ascertain the validity of the conclusion drawn and whether the data are applicable to cell types other than fibroblasts. For example, studies on a large number of different Morris hepatomas indicate that while basal cyclic AMP levels in the tumors were elevated over normal, cyclic GMP levels were either very greatly or only slightly elevated (Thomas et al., 1973). Despite the rise in levels of nucleotides in many of the tumors examined, the cyclic AMP/cyclic GMP ratios are quite similar to normal liver, with no correlation between growth rates of the tumors and cyclic nucleotide levels being apparent.

IV. ACKNOWLEDGMENTS

The authors are grateful to Drs. H. R. Bourne, R. B. Clark, D. L. Friedman, A. G. Gilman, and T. B. Miller for their personal communications of unpublished works. We also would like to thank Drs. L. Hayflick and A. Girardi, who provided the WI-38 and SV40 transformed WI-38 fibroblasts used in some of the work described. We express our deep appreciation to Helen Smith for her understanding and care in preparing the manuscript.

Part of the work described was supported by research grants AM-13904 and GM-18332 from the National Institutes of Health, U.S. Public Health Service.

V. REFERENCES

Abercrombie, M., and Ambrose, E. J. (1962): The surface properties of cancer cells. *Cancer Research,* 22:525–548.

Allen, D. O., Munshower, J., Morris, H. P., and Weber, G. (1971): Regulation of adenyl cyclase in hepatomas of different growth rates. *Cancer Research,* 31:557–560.

Allen, L. B., Eagle, N. C., Huffman, J. H., Shuman, D. A., Meyer, R. B., Jr., and Sidwell, R. W. (1974): Enhancement of interferon antiviral action in L-cells by cyclic nucleotides. *Proceedings of the Society for Experimental Biology and Medicine,* 146:580–584.

Anderson, W. B., Johnson, G. S., and Pastan, I. (1973a): Transformation of chick-embryo fibroblasts by wild-type and temperature-sensitive Rous sarcoma virus alters adenylate cyclase activity. *Proceedings of the National Academy of Sciences (U.S.),* 70:1055–1059.

Anderson, W. B., Russell, T. R., Carchman, R. A., and Pastan, I. (1973b): Interrelationship between adenylate cyclase activity, cyclic AMP phosphodiesterase activity, cyclic AMP levels, and growth of cells in culture. *Proceedings of the National Academy of Sciences (U.S.),* 70:3802–3805.

Appleman, M. M., Thompson, W. J., and Russell, T. R. (1973): *Advances in Cyclic Nucleotide Research,* 3:65–98.

Armato, U., Andreis, P., Belloni, A., and Draghi, E. (1974): Long term stimulatory effects of adenosine 3′,5′-cyclic monophosphate on nuclear and cyctoplasmic protein synthesis of rat hepatocytes in primary culture. *Acta Anatomica,* 88:456–480.

Bacalao, J., and Rieber, M. (1972): Modified distribution of some cyclic AMP phosphodiesterase activities following growth of L-cell fibroblasts in the presence of prostaglandin E_1. *Experimental Cell Research,* 75:518–521.

Barnett, C. A., and Wicks, W. D. (1971): Regulation of phosphoenolpyruvate carboxykinase and tyrosine transaminase in hepatoma cell cultures I. *Journal of Biological Chemistry,* 246:7201–7206.

Beavo, J. A., Bechtel, P. J., and Krebs, E. G. (1974): Activation of protein kinase by physiological concentrations of cyclic AMP. *Proceedings of the National Academy of Sciences (U.S.),* 71:3580–3583.

Beavo, J. A., Rogers, N. L., Crofford, O. B., Hardman, J. G., Sutherland, E. W., and Newman, E. V. (1970): Effects of xanthine derivatives on lipolysis and on adenosine 3′,5′-monophosphate phosphodiesterase activity. *Molecular Pharmacology*, 6:597–602.

Bloom, W., and Fawcett, D. W. (1968): *A Textbook of Histology*, 9th ed., W. B. Saunders Company, Philadelphia, p. 142.

Blume, A. J., Dalton, C., and Sheppard, H. (1973): Adenosine-mediated elevation of cyclic 3′,5′-adenosine monophosphate concentrations in cultured mouse neuroblastoma cells. *Proceedings of the National Academy of Sciences (U.S.)*, 70:3099–3102.

Bombik, B. M., and Burger, M. M. (1973): cAMP and the cell cycle: Inhibition of growth stimulation. *Experimental Cell Research*, 80:88–94.

Bonnai, S., and Sheppard, J. R. (1974): Cyclic AMP, ATP and cell contact. *Nature*, 250:62–64.

Bourne, H. R., Coffino, P., Melmon, K. L., Tomkins, G. M., and Weinstein, Y. (1975): Genetic analysis of cyclic AMP (cAMP) in a mammalian cell. *Advances in Cyclic Nucleotide Research*, 5:771–786.

Bourne, H. R., Coffino, P., and Tomkins, G. (1974): Selection of a mutant lymphoma cell deficient in adenylate cyclase. *Science*, 187:750–752.

Bourne, H. R., Tomkins, G. M., and Dion, S. (1973b): Regulation of phosphodiesterase synthesis: Requirement for cyclic adenosine monophosphate-dependent protein kinase. *Science*, 181:952–954.

Brooker, G., Thomas, L. J., Jr., and Appleman, M. M. (1968): The assay of adenosine 3′,5′-cyclic monophosphate and guanosine 3′,5′-cyclic monophosphate in biological materials by enzymatic radioisotopic displacement. *Biochemistry*, 12:4177–4181.

Browning, E. T., Groppi, V., and Kon, C. (1974a): Papaverine. A potent inhibitor of respiration in C6 astrocytoma cells. *Molecular Pharmacology*, 10:175–181.

Browning, E. T., Schwartz, J. P., and Breckenridge, B. McL. (1974b): Norepinephrine-sensitive properties of C6 astrocytoma cells. *Molecular Pharmacology*, 10:162–174.

Brunton, L. L., and Guerrant, R. L. (1974): *V. cholerae* and *E. coli* enterotoxins and cyclic AMP (CAMP) metabolism in cultured cells. *Federation Proceedings*, 33:507.

Brush, J. S., Sutliff, L. S., and Sharma, R. K. (1974): Metabolic regulation and adenyl cyclase activity of adrenocortical carcinoma cultured cells. *Cancer Research*, 34:1495–1502.

Buonassisi, V., Sato, G., and Cohen, A. I. (1962): Hormone-producing cultures of adrenal and pituitary origin. *Proceedings of the National Academy of Sciences (U.S.)*, 48:1184–1190.

Burger, M. M. (1969): A difference in the architecture of the surface membrane of normal and virally transformed cells. *Proceedings of the National Academy of Sciences (U.S.)*, 62:994–1001.

Burger, M. M., Bombik, B. M., Breckenridge, B. M., and Sheppard, J. R. (1972a): Growth control and cyclic alterations of cyclic AMP in the cell cycle. *Nature New Biology*, 239:161–163.

Burger, M. M., Bombik, B. M., and Noonan, K. D. (1972b): Cell surface alterations in transformed tissue culture cells and their possible significance in growth control. *Journal of Investigative Dermatology*, 59:24–26.

Burger, M. M., and Goldberg, A. R. (1967): Identification of a tumor-specific determinant on neoplastic cell surfaces. *Proceedings of the National Academy of Sciences (U.S.)*, 57:359–366.

Bürk, R. R. (1968): Reduced adenyl cyclase activity in a polyoma virus transformed cell line. *Nature*, 219:1272–1275.

Burstin, S. J., Renger, H. C., and Basilico, C. (1974): Cyclic AMP levels in temperature sensitive SV40 transformed cell types. *Journal of Cellular Physiology*, 84:69–74.

Butcher, F. R., Becker, J. E., and Potter, V. R. (1971): Induction of tyrosine amino transferase by dibutyryl cyclic AMP employing hepatoma cells in tissue culture. *Experimental Cell Research*, 66:321–328.

Butcher, F. R., Bushnell, D. E., Becker, J. E., and Potter, V. R. (1972): Effect of cordycepin on induction of tyrosine amino transferase employing hepatoma cells in tissue culture. *Experimental Cell Research*, 74:115–123.

Butcher, R. W., and Baird, C. E. (1968): Effects of prostaglandins on adenosine 3′,5′-monophosphate levels in fat and other tissues. *Journal of Biological Chemistry*, 243:1713–1717.

Butcher, R. W., and Sutherland, E. W. (1962): Adenosine 3′,5′-phosphate in biological materials. I. Purification and properties of cyclic 3′,5′-nucleotide phosphodiesterase and use of this enzyme to characterize adenosine 3′,5′-phosphate in human urine. *Journal of Biological Chemistry*, 237:1244–1250.

Carchman, R. A., Johnson, G. S., Pastan, I., and Scolnick, E. M. (1974): Studies on the levels of cyclic AMP in cells transformed by wild-type and temperature-sensitive Kirsten sarcoma virus. *Cell*, 1:59–64.

Casola, L., DiMatteo, J., and Augusti-Tocco, G. (1974): Neuroblastoma cells in culture: ^{32}P-Phosphoprotein labeling and protein kinase activity. *Experimental Neurology*, 44:417–423.
Castaneda, M., and Tyler, A. (1968): Adenyl cyclase in plasma membrane preparations of sea urchin eggs and its increase in activity after fertilization. *Biochemical and Biophysical Research Communications*, 33:782–787.
Chader, G. J. (1971): Hormonal effects on neural retina: Induction of glutamine synthetase by cyclic-3',5'-AMP. *Biochemical and Biophysical Research Communications*, 43:1102–1105.
Channing, C. P. (1974): Temporal effects of LH, LCG, FSH and dibutyryl cyclic 3',5'-AMP upon luteinization of rhesus monkey granulosa cells in culture. *Endocrinology*, 94:1215–1223.
Cheng, H., and Farquhar, M. G. (1974): Presence of adenylate cyclase in Golgi fractions from rat liver. *Journal of Cell Biology*, 63:58a.
Christoffersen, T., Morland, J., Osnes, X. X., Jr., and Elgjo, K. (1972): Hepatic adenyl cyclase: Alterations in hormone response during treatment with a chemical carcinogen. *Biochimica et Biophysica Acta*, 279:363–366.
Clark, J. F., Morris, H. P., and Weber, G. (1973): Cyclic adenosine 3',5'-monophosphate phosphodiesterase activity in normal, differentiating, regenerating, and neoplastic liver. *Cancer Research*, 33:356–361.
Clark, R. B., Gross, R., Su, Y.-F., and Perkins, J. P. (1974): Regulation of adenosine 3':5'-monophosphate content in human astrocytoma cells by adenosine and the adenine nucleotides. *Journal of Biological Chemistry*, 249:5296–5303.
Clark, R. B., and Perkins, J. P. (1971): Regulation of adenosine 3':5'-cyclic monophosphate concentration in cultured human astrocytoma cells by catecholamines and histamine. *Proceedings of the National Academy of Sciences (U.S.)*, 68:2757–2760.
Clark, R. B., and Perkins, J. P. (1972): The effect of adenosine on the formation of cyclic AMP (cAMP) in cultured human astrocytoma cells. *Federation Proceedings*, 31:573a.
Clark, R. B., Su, Y.-F., Ortmann, R., Cubeddu, X., Johnson, G. L., and Perkins, J. P. (1975): Factors influencing the effect of hormones on the accumulation of cyclic AMP in cultured human astrocytoma cells. *Metabolism*, 24:343–358.
Coffino, P., Bourne, H. R., and Tomkins, G. M. (1975): Somatic genetic analysis of cyclic AMP action: Selection of unresponsive mutants. *Journal of Cellular Physiology (in press)*.
Creighton, M. O., and Trevithick, J. R. (1974): Effect of cyclic AMP, caffein and theophylline on differentiation of lens epithelial cells. *Nature*, 249:767–768.
Curtis, G. L., Elliott, J. A., Wilson, R. B., and Ryan, W. L. (1973): Adenosine-3',5'-cyclic monophosphate and cell volume. *Cancer Research*, 33:3273–3277.
Daniel, V., Bourne, H. R., and Tomkins, G. M. (1973a): Altered metabolism of cyclic AMP in cultured cells deficient in cyclic AMP-binding protein. *Nature New Biology*, 244:167–169.
Daniel, V., Litwack, G., and Tomkins, G. M. (1973b): Induction of cytolysis of cultured lymphoma cells by adenosine 3',5'-cyclic monophosphate and the isolation of resistant variants. *Proceedings of the National Academy of Sciences (U.S.)*, 70:76–79.
D'Armiento, M., Johnson, G. S., and Pastan, I. (1972): Regulation of adenosine 3',5'-cyclic monophosphate phosphodiesterase activity in fibroblasts by intracellular concentrations of cyclic adenosine monophosphate. *Proceedings of the National Academy of Sciences (U.S.)*, 69:459–462.
D'Armiento, M., Johnson, G. S., and Pastan, I. (1973): Cyclic AMP and growth of fibroblasts. Effect of environmental pH. *Nature New Biology*, 242:78–80.
Daughaday, W. H., Hall, K., Raben, M. S., Salmon, W. D., Jr., Van den Brande, J. L., and Van Wyk, J. J. (1972): Somatomedin: Proposed designation of sulfation factor. *Nature*, 235:107.
Davoren, P. R., and Sutherland, E. W. (1963): The cellular location of adenyl cyclase in the pigeon erythrocyte. *Journal of Biological Chemistry*, 238:3016–3023.
DeAsua, L. J., Surian, E. S., Flawia, M. M., and Torres, H. N. (1973): Effect of insulin on the growth pattern and adenyl cyclase activity of BHK fibroblasts. *Proceedings of the National Academy of Sciences (U.S.)*, 70:1388–1392.
deVellis, J., and Brooker, G. (1972): Effect of catecholamines on cultured glial cells: Correlation between cyclic AMP levels and lactic dehydrogenase induction. *Federation Proceedings*, 31:513.
deVellis, J., and Brooker, G. (1974): Reversal of catecholamine refractoriness by inhibitors of RNA and protein synthesis. *Science*, 186:1221–1223.
deVellis, J., Inglish, D., and Brooker, G. (1974): Paradoxical effects of actinomycin D, acetoxycycloheximide, and methylisobutylxanthine on cyclic AMP metabolism. *Proceedings of the National Academy of Sciences (U.S.)*, 33:507.

Dianzani, F., and Zucca, M. (1972): Effect of dibutyryl cyclic AMP on interferon production by cells treated with viral or nonviral inducers. *Proceedings of the Society for Experimental Biology and Medicine,* 140:1375–1378.

Eagle, H. (1971): Buffer combinations for mammalian cell culture. *Science,* 174:500–503.

Edström, A., Kanje, M., and Walum, E. (1974): Effects of dibutyryl cyclic AMP and prostaglandin E_1 on cultured human glioma cells. *Experimental Cell Research,* 85:217–223.

Eker, P. (1974): Inhibition of growth and DNA synthesis in cell cultures by cyclic AMP. *Journal of Cell Science,* 16:301–307.

Emmelot, P., and Bos, C. J. (1971): Studies on plasma membranes. XIV. Adenyl cyclase in plasma membranes isolated from rat and mouse livers and hepatomas, and its hormone sensitivity. *Biochimica et Biophysica Acta,* 249:285–292.

Exton, J. H., Lewis, S. B., Ho, R. J., Robison, G. A., and Park, C. R. (1971): The role of cyclic AMP in the interaction of glucagon and insulin in the control of liver metabolism. *Annals of the New York Academy of Science,* 185:85–100.

Fox, T. O., Sheppard, J. R., and Burger, M. M. (1971): Cyclic membrane changes in animal cells: Transformed cells permanently display a surface architecture detected in normal cells only during mitosis. *Proceedings of the National Academy of Sciences (U.S.),* 68:244–247.

Frank, W. (1971): Cyclic 3′,5′-AMP and cell proliferation in culture of embryonic rat cells. *Experimental Cell Research,* 71:238–241.

Franklin, T. J., and Foster, S. J. (1973a): Leakage of cyclic AMP from human diploid fibroblasts in tissue culture. *Nature New Biology,* 246:119–120.

Franklin, T. J., and Foster, S. J. (1973b): Hormone-induced desensitization of hormonal control of cyclic AMP levels in human diploid fibroblasts. *Nature New Biology,* 246:146–148.

Franks, D. J., MacManus, J. P., and Whitfield, J. F. (1971): The effect of prostaglandins on cyclic AMP production and cell proliferation in thymic lymphocytes. *Biochemical and Biophysical Research Communications,* 44:1177–1182.

Froehlich, J. E., and Rachmeler, M. (1972): Effect of adenosine 3′,5′-cyclic monophosphate on cell proliferation. *Journal of Cell Biology,* 55:19–31.

Froehlich, J. E., and Rachmeler, M. (1974): Inhibition of cell growth in the G_1 phase by adenosine 3′,5′-cyclic monophosphate. *Journal of Cell Biology,* 60:249–257.

Furmanski, P., Silverman, D. J., and Lubin, M. (1971): Expression of differentiated functions in mouse neuroblastoma mediated by dibutyryl-cyclic adenosine monophosphate. *Nature,* 233:413–415.

Gilman, A. G. (1970): A protein binding assay for adenosine 3′:5′-cyclic monophosphate. *Proceedings of the National Academy of Sciences (U.S.),* 67:305–312.

Gilman, A. G. (1974): Effect of 2-Cl-adenosine on cyclic AMP concentration of cultured cells. *Federation Proceedings,* 33:507.

Gilman, A. G., and Minna, J. D. (1973): Expression of genes for metabolism of cyclic 3′:5′-monophosphate in somatic cells. I. Responses to catecholamines in parental and hybrid cells. *Journal of Biological Chemistry,* 248:6610–6617.

Gilman, A. G., and Nirenberg, M. (1971a): Effect of catecholamine on the adenosine 3′:5′-cyclic monophosphate concentrations of clonal satellite cells of neurons. *Proceedings of the National Academy of Sciences (U.S.),* 68:2165–2168.

Gilman, A. G., and Nirenberg, M. (1971b): Regulation of adenosine 3′,5′-cyclic monophosphate metabolism in cultured neuroblastoma cells. *Nature,* 234:356–358.

Gilman, A. G., and Schrier, B. K. (1972): Adenosine cyclic 3′,5′-monophosphate in fetal rat brain cell cultures. *Molecular Pharmacology,* 8:410–416.

Goggins, J. F., Johnson, G. S., and Pastan, I. (1972): The effects of dibutyryl cyclic adenosine monophosphate on synthesis of sulfated acid mucopolysaccharides by transformed cells. *Journal of Biological Chemistry,* 247:5759–5764.

Goldberg, N. D., O'Dea, R. F., and Haddox, M. K. (1973): Cyclic GMP. *Advances in Cyclic Nucleotides Research,* 3:155–233.

Goldberg, N. D., Lust, W. D., O'Dea, R. F., Wei, S., and O'Toole, A. G. (1970): A role of cyclic nucleotides in brain metabolism. In: *Advances in Biochemical Psychopharmacology,* Vol. 3, edited by P. Greengard and E. Costa. Raven Press, New York.

Goodman, D. B. P., Rasmussen, H., DiBella, F., and Guthrow, C. E., Jr. (1970): Cyclic adenosine 3′:5′-monophosphate-stimulated phosphorylation of isolated neurotubule subunits. *Proceedings of the National Academy of Sciences (U.S.),* 67:652–659.

Goridis, C., Massarelli, R., Sensenbrenner, M., and Mandel, P. (1974): Guanyl cyclase in chick embryo brain cell cultures: Evidence of neuronal localization. *Journal of Neurochemistry,* 23: 135–138.

Gospodarowicz, D. (1974): Localization of a fibroblast growth factor and its effect alone and with hydrocortisone on 3T3 cell growth. *Nature,* 249:123–127.

Gospodarowicz, D., and Moran, J. S. (1974): Stimulation of division of sparse and confluent 3T3 cell populations by a fibroblast growth factor, dexamethasone, and insulin. *Proceedings of the National Academy of Sciences, (U.S.),* 71:4584–4588.

Granner, D. K. (1972): Altered regulation in a hepatoma cell line deficient in adenosine 3′,5′-cyclic monophosphate-binding protein. *Biochemical and Biophysical Research Communications,* 46:1516–1522.

Granner, D. K. (1974): Absence of high affinity adenosine 3′,5′-monophosphate binding sites from the cytosol of three hepatic-derived cell lines. *Archives of Biochemistry and Biophysics,* 165:359–368.

Granner, D., Chase, L. R., Aurbach, G. D., and Tomkins, G. M. (1968): Tyrosine aminotransferase: Enzyme induction independent of adenosine 3′,5′-monophosphate. *Science,* 162:1018–1020.

Greengard, P., and Kuo, J. F. (1970): On the mechanism of action of cyclic AMP. In: *Advances in Biochemical Psychopharmacology,* Vol. 3, edited by P. Greengard and E. Costa. Raven Press, New York.

Griffin, M. J., Price, G. H., Bazzell, K. L., Cox, R. P., and Ghosh, N. K. (1974): A study of adenosine 3′:5′-cyclic monophosphate, sodium butyrate, and cortisol as inducers of HeLa alkaline phosphate. *Archives of Biochemistry and Biophysics,* 164:619–623.

Grimes, W. J., and Shroeder, J. L. (1973): Dibutyryl cyclic adenosine 3′,5′-monophosphate, sugar transport, and regulatory control of cell division in normal and transformed cells. *Journal of Cellular Biology,* 56:487–491.

Grossman, A., Boctor, A., and Masuda, Y. (1971): Induction of tyrosine aminotransferase with $N^6,O^{2'}$-dibutyryl adenosine 3′-5′-monophosphate in rat. *European Journal of Biochemistry,* 24: 149–155.

Guidotti, A., Weiss, B., and Costa, E. (1972): Adenosine 3′5′-monophosphate concentrations and isoproterenol induced synthesis of deoxyribonucleic acid in mouse parotid gland. *Molecular Pharmacology,* 8:521–530.

Hadden, J. W., Hadden, E. M., Haddox, M. K., and Goldberg, N. D. (1972): Guanosine 3′:5′-cyclic monophosphate: A possible intracellular mediator of mitogenic influences in lymphocytes. *Proceedings of the National Academy of Sciences (U.S.),* 69:3024–3027.

Hamprecht, B., and Schultz, J. (1973a): Stimulation by prostaglandin E_1 of adenosine 3′:5′-cyclic monophosphate formation neuroblastoma cells in the presence of phosphodiesterase inhibitors. *Federation of European Biochemical Societies,* 34:85–89.

Hamprecht, B., and Schultz, J. (1973b): Influence of noradrenalin. Prostaglandin E_1 and inhibitors of phosphodiesterase activity on levels of the cyclic adenosine 3′:5′-monophosphate in somatic cell hybrids. *Hoppe-Seyler's Physiological Chemistry,* 354:1633–1641.

Harary, I., Hoover, F., and Farley, B. (1973): Catecholamine and dibutyryl cyclic AMP effects on myosin adenosine triphosphatase in cultured rat heart cells. *Science,* 181:1061–1063.

Haslam, R. J., and Goldstein, S. (1974): Adenosine 3′,5′-cyclic monophosphate in young and senescent human fibroblasts during growth and stationary phase *in vitro:* Effects of prostaglandin E_1 and of adrenaline. *Biochemical Journal,* 144:253–263.

Hauschka, P. V., Everhart, L. P., and Rubin, R. W. (1972): Alteration of nucleoside transport of Chinese hamster cells by dibutyryl adenosine 3′:5′-cyclic monophosphate. *Proceedings of the National Academy of Sciences (U.S.),* 69:3542–3546.

Hayflick, L. (1965): The limited *in vitro* lifetime of human diploid cell strains. *Experimental Cell Research,* 37:614–636.

Hayflick, L. (1975): Book reviews. *Science,* 187:339–340.

Heidrick, M. L., and Ryan, W. L. (1970): Cyclic nucleotide on cell growth *in vitro. Cancer Research,* 30:376–378.

Heidrick, M., and Ryan, W. L. (1971a): Metabolism of 3′,5′-cyclic AMP by strains of L cells. *Biochimica et Biophysica Acta,* 237:301–309.

Heidrick, M. L., and Ryan, W. L. (1971b): Adenosine 3′,5′-cyclic monophosphate and contact inhibition. *Cancer Research,* 31:1313–1315.

Hilz, H., and Tarnowski, W. (1970): Opposite effects of cyclic AMP and its dibutyryl derivative on glycogen levels in HeLa cells. *Biochemical and Biophysical Research Communications,* 40:973–981.

Hovi, T., and Vaheri, A. (1973): Cyclic AMP and cyclic GMP enhanced growth of chick embryo fibroblasts. *Nature New Biology,* 245:175–177.

Hsie, A. W., Jones, C., and Puck, T. T. (1971): Further changes in differentiation state accompanying the conversion of Chinese hamster cells to fibroblastic form by dibutyryl adenosine cyclic 3′,5′-monophosphate and hormones. *Proceedings of the National Academy of Sciences (U.S.),* 68: 1648–1652.

Hsie, A. W., and Puck, T. T. (1971): Morphological transformation of Chinese hamster cells by dibutyryl adenosine cyclic 3′,5′-monophosphate and testosterone. *Proceedings of the National Academy of Sciences (U.S.),* 68:358–361.

Inbar, M., and Sachs, L. (1969a): Interaction of the carbohydrate-binding protein concanavalin A with normal and transformed cells. *Proceedings of the National Academy of Sciences (U.S.),* 63: 1418–1425.

Inbar, M., and Sachs, L. (1969b): Structural differences in sites on the surface membrane of normal and transformed cells. *Nature,* 223:710–712.

Jard, S., Premont, J., and Benda, P. (1972): Adenylate cyclase phosphodiesterases and protein kinase of rat glial cells in culture. *Federation of European Biochemical Societies,* 26:344–348.

Johnson, G., and Pastan, I. (1972): $N^6,O^{2'}$ = Dibutyryl adenosine = 3′,5′ = monophosphate induces pigment production in melanoma cells. *Nature New Biology,* 237:267–268.

Kagen, L. J., and Freedman, A. (1974): Studies on the effects of acetylcholine, epinephrine, dibutyryl cyclic adenosine monophosphate, theophylline, and calcium on the synthesis of myoglobin in muscle cell cultures estimated by radioimmunoassay. *Experimental Cell Research,* 88:135–142.

Kantor, H., Tao, P., and Wisdom, C. (1974): Action of *Escherichia coli* enterotoxin: Adenylate cyclase behavior of intestinal epithelial cells in culture. *Infection and Immunology,* 9:1003–1010.

Kaukel, E., Fuhrmann, U., and Hilz, H. (1972a): Divergent action of cAMP and dibutyryl cAMP on macromolecular synthesis in HeLa S3 cultures. *Biochemical and Biophysical Research Communications,* 48:1516–1524.

Kaukel, E., and Hilz, H. (1972): Permeation of dibutyryl cAMP into HeLa cells and its conversion to monobutyryl cAMP. *Biochemical and Biophysical Research Communications,* 46:1011–1018.

Kaukel, E., Mundhenk, K., and Hilz, H. (1972b): N^6-Monobutyryladenosine 3′:5′-monophosphate as the biologically active derivative of dibutyryl-adenosine 3′:5′-monophosphate in HeLa S3 cells. *European Journal of Biochemistry,* 27:197–200.

Kelly, L. A., and Butcher, R. W. (1974): The effects of epinephrine and prostaglandin E_1 on cyclic adenosine 3′:5′-monophosphate levels in WI-38 fibroblasts. *Journal of Biological Chemistry,* 249: 3098–3102.

Kelly, L. A., and Butcher, R. W. (1975): Studies on cyclic AMP metabolism in human epidermoid carcinoma (HEp-2) cells. *Metabolism,* 24:359–368.

Kelly, L. A., Hall, M. S., and Butcher, R. W. (1974): Cyclic adenosine 3′:5′-monophosphate metabolism in normal and SV40 transformed WI-38 cells. *Journal of Biological Chemistry,* 249:5182–5187.

Kirkland, W. L., and Burton, P. R. (1972): Cyclic adenosine monophosphate-mediated stabilization of mouse neuroblastoma cell neurite microtubules exposed to low temperature. *Nature New Biology,* 240:205–207.

Klein, M. I., and Makman, M. H. (1971): Adenosine 3′,5′-monophosphate-dependent protein kinase of cultured mammalian cells. *Science,* 172:863–864.

Koblet, H., Kohler, U., and Wyler, R. (1973): Stimulation of ribonucleic acid synthesis in chick-embryo fibroblasts by exogenous adenosine 3′,5′-monophosphate. *European Journal of Biochemistry,* 37:134–142.

Konigsberg, I. R. (1961): Cellular differentiation in colonies derived from single cell platings of freshly isolated chick embryo muscle cells. *Proceedings of the National Academy of Sciences (U.S.),* 47: 1868–1872.

Kono, T. (1970): Insulin effector system of fat cells: Its destruction with trypsin and subsequent restoration. *Hormone Metabolism Research,* 2; Supplement 2:108–111.

Korinek, J., Spelsberg, T. C., and Mitchell, W. M. (1973): mRNA transcription linked to the morphological and plasma membrane changes induced by cyclic AMP in tumor cells. *Nature,* 236:455–458.

Kowal, J. (1973): Adrenal cells in tissue culture X: On the mechanism of the stimulation of steroidogenesis by cyclic cytidine monophosphate. *Endocrinology,* 93:461–468.

Kowal, J., and Fiedler, R. P. (1969): Adrenal cells in tissue culture II: Steroidogenic responses to nucleosides and nucleotides. *Endocrinology,* 84:1113–1117.

Kowal, J., and Harano, Y. (1974): Adrenal cells in tissue culture: The effect of papaverine and amytal on steroidogenesis, respiration and replication. *Archives of Biochemistry and Biophysics,* 163: 466–475.

Kowal, J., Srinivasan, S., and Saito, T. (1974): Calcium modulation of adrenal steroidogenic responses to ACTH and cholera toxin. In: *Endocrinology, Fifty-Sixth Annual Meeting of the Endocrine Society,* Atlanta, Georgia, Abstract 212.

Koyama, H., Kato, R., and Ono, T. (1972): Induction of alkaline phosphatase by cyclic AMP or its dibutyryl derivative in a hybrid line between mouse and Chinese hamster in culture. *Biochemical and Biophysical Research Communications,* 46:305–311.

Kram, R., Mamont, P., and Tomkins, G. M. (1973): Pleiotypic control by adenosine 3′,5′-cyclic monophosphate: A model for growth control in animal cells. *Proceedings of the National Academy of Sciences (U.S.),* 70:1432–1436.

Kram, R., and Tomkins, G. M. (1973): Pleiotypic control by cyclic AMP: Interaction with cyclic GMP and possible role of microtubules. *Proceedings of the National Academy of Sciences (U.S.),* 70:1659–1663.

Kreider, J. W., Rosenthal, M., and Lengle, N. (1973): Cyclic adenosine 3′,5′-monophosphate in the control of melanoma cell replication and differentiation. *Journal of the National Cancer Institute,* 50:555–558.

Kuo, J. F., and Greengard, P. (1969): Cyclic nucleotide-dependent protein kinases. IV. Widespread occurrence of adenosine 3′,5′-monophosphate-dependent protein kinase in various tissues and phyla of the animal kingdom. *Proceedings of the National Academy of Sciences (U.S.),* 64:1349–1355.

Kuroda, Y. (1974): Inhibition by cyclic AMP and dibutyryl cyclic AMP of aggregation of embryonic quail liver cells in culture. *Experimental Cell Research,* 84:303–310.

Kurtz, M. J., Polgar, P., Taylor, L., and Rutenburg, A. M. (1974): The role of adenosine 3′:5′-cyclic monophosphate in the division of WI-38 cells. *Biochemical Journal,* 142:339–344.

Kwan, C., and Wishnow, R. (1974): *Escherichia coli* enterotoxin-induced steroidogenesis in cultured adrenal tumor cells. *Infection and Immunology,* 10:146–151.

Lim, R., and Mitsunobu, K. (1972): Effect of dibutyryl cyclic AMP on nucleic acid and protein synthesis in neuronal and glial tumor cell. *Life Sciences,* 11:1063–1077.

Loomis, W. F., Jr., Wahrmann, J. P., and Luzzati, D. (1973): Temperature-sensitive variants of an established myoblast line. *Proceedings of the National Academy of Sciences (U.S.)* 70:425–429.

Lowry, O. H., Rosebrough, N. J., Farr, A. L., and Randall, R. J. (1951): Protein measurements with the folin phenol reagent. *Journal of Biological Chemistry,* 193:265–275.

Lyle, L. R., and Parker, C. W. (1974): Cyclic adenosine 3′,5′-monophosphate responses to concanavalin A in human lymphocytes. Evidence that the response involves specific carbohydrate receptors on the cell surface. *Biochemistry,* 13:5415–5420.

Macintyre, E. H., Wintersgill, C. J., Perkins, J. P., and Vatter, A. E. (1972): The responses in culture of human tumour astrocytes and neuroblasts to $N^6,O^{2'}$-dibutyryl adenosine 3′,5′-monophosphoric acid. *Journal of Cell Science,* 2:639–667.

MacManus, J. P., Braceland, B. M., Youdale, T., and Whitfield, J. F. (1973): Adrenergic antagonists and a possible link between the increase in cyclic adenosine 3′5′ monophosphate and DNA synthesis during liver regeneration. *Journal of Cellular Physiology,* 82:157–164.

MacManus, J. P., Franks, D. J., Youdale, T., and Braceland, B. M. (1972): Increases in rat liver cyclic AMP concentrations prior to the initiation of DNA synthesis following partial hepatectomy or hormone infusion. *Biochemical and Biophysical Research Communications,* 49:1201–1207.

MacManus, J. P., and Whitfield, J. F. (1974): Cyclic AMP, prostaglandins and the control of cell proliferation. *Prostaglandins,* 6:475–485.

Maguire, M. E., Sturgill, T. W., Anderson, H. J., Minna, J. D., and Gilman, A. G. (1974): Hormonal control of cyclic AMP metabolism in parental and hybrid somatic cells. *Advances in Cyclic Nucleotide Research,* 5:699–718.

Makman, M. H. (1970): Adenyl cyclase of cultured mammalian cells: Activation by catecholamines. *Science,* 170:1421–1423.

Makman, M. H. (1971*a*): Conditions leading to enhanced response to glucagon, epinephrine, or prostaglandins by adenylate cyclase of normal and malignant cultured cells. *Proceedings of the National Academy of Sciences (U.S.),* 68:2127–2130.

Makman, M. H. (1971*b*): Hormone-sensitive adenyl cyclase of cultured human and mouse fibroblasts and of cells of malignant origin. *Federation Proceedings,* 30:458.

Makman, M. H., Dvorkin, B., and Keehn, E. (1974): Hormonal regulation of cyclic AMP in aging and virus transformed human fibroblasts and comparison with other cultured cells. In: *Control of Proliferation in Animals Cells*, edited by B. Clarkson and R. Baserga. Cold Spring Harbor Laboratory Press, New York.

Makman, M. H., and Klein, M. I. (1972): Expression of adenylate cyclase, catecholamine receptors and cyclic adenosine monophosphate-dependent protein kinase in synchronized culture of Chang's liver cells. *Proceedings of the National Academy of Sciences (U.S.)*, 69:456–458.

Makman, M. H., and Sutherland, E. W. (1964): Use of liver adenyl cyclase for assay of glucagon in human gastro-intestinal tract and pancreas. *Endocrinology*, 75:127–134.

Mandel, J. L., and Pearson, M. L. (1974): Insulin stimulative myogenesis in a rat myoblast line. *Nature*, 251:618–620.

Manganiello, V. C., and Breslow, J. (1974): Effects of prostaglandin E_1 and isoproterenol on cyclic AMP content of human fibroblasts modified by time and cell density in subculture. *Biochimica et Biophysica Acta*, 362:509–520.

Manganiello, V., Breslow, J., and Vaughan, M. (1972): An effect of dexamethasone on the cyclic AMP content of human fibroblasts stimulated by catecholamines and prostaglandin E_1. *Journal of Clinical Investigation*, 51:60a.

Manganiello, V. C., Murad, F., and Vaughan, M. (1971): Effects of lipolytic and antilipolytic agents on cyclic 3',5'-adenosine monophosphate release in fat cells. *Journal of Biological Chemistry*, 246:2195–2202.

Manganiello, V., and Vaughan, M. (1972a): Prostaglandin E_1 effects on adenosine 3':5'-cyclic monophosphate concentration and phosphodiesterase activity in fibroblasts. *Proceedings of the National Academy of Sciences (U.S.)*, 69:269–273.

Manganiello, V., and Vaughan, M. (1972b): An effect of dexamethasone on adenosine 3',5'-monophosphate content and adenosine 3',5'-monophosphate phosphodiesterase activity of cultured hepatoma cells. *Journal of Clinical Investigation*, 51:2763–2767.

Marks, F., and Grimm, W. (1972): Diurnal fluctuation and beta-adrenergic elevation of cyclic AMP in mouse epidermis *in vivo*. *Nature New Biology*, 240:178–179.

Mawe, R., Doore, B., McCaman, M., Feucht, B., and Saier, M. (1974): Regulation of cyclic AMP excretion in bacteria and cultured animal cells. *Journal of Cell Biology*, 63:211a.

Medoff, J., and Parker, N. (1972): Stimulation of growth by cyclic AMP in cultured embryonic tissues. *Journal of Cell Biology*, 55:173a.

Millis, A. J. T., Forrest, G., and Pious, D. A. (1972): Cyclic AMP in cultured human lymphoid cells: Relationship to mitosis. *Biochemical and Biophysical Research Communications*, 49:1645–1649.

Millis, A. J. T., Forrest, G. A., and Pious, D. A. (1974): Cyclic AMP dependent regulation of mitosis in human lymphoid cells. *Experimental Cell Research*, 83:335–343.

Minna, J. W., and Gilman, A. G. (1973): Expression of genes for metabolism of cyclic adenosine 3':5'-monophosphate in somatic cells. *Journal of Biological Chemistry*, 248:6618–6625.

Nath, J., and Rebhun, L. I. (1973a): Studies on cyclic AMP levels and phosphodiesterase activity during developing sea urchin eggs. *Experimental Cell Research*, 77:319–322.

Nath, J., and Rebhun, L. I. (1973b): Studies on the uptake and metabolism of adenosine 3':5'-cyclic monophosphate and $N^6,O^{2'}$-dibutyryl 3':5'-cyclic adenosine monophosphate in sea urchin eggs. *Experimental Cell Research*, 82:73–78.

Ney, R. L., Hochella, N. J., Grahame-Smith, D. G., Dexter, R. N., and Butcher, R. W. (1969): Abnormal regulation of adenosine 3',5'-monophosphate and corticosterone formation in an adrenocortical carcinoma. *Journal of Clinical Investigation*, 48:1733–1739.

Niewiarowski, S., and Goldstein, S. (1973): Interaction of cultured human fibroblasts with fibrin: Modification by drugs and aging *in vitro*. *Journal of Laboratory and Clinical Medicine*, 82:605–610.

Oey, J., Vogel, A., and Pollack, R. (1974): Intracellular cyclic AMP concentration responds specifically to growth regulation by serum. *Proceedings of the National Academy of Sciences (U.S.)*, 71:694–698.

O'Keefe, E., and Cuatrecasas, P. (1974): Cholera toxin mimics melanocyte stimulating hormone in inducing differentiation in melanoma cells. *Proceedings of the National Academy of Sciences (U.S.)*, 71:2500–2504.

Opler, L. A., and Makman, M. H. (1972): Mediation by cyclic AMP of hormone-stimulated glycogenolysis in cultured rat astrocytoma cells. *Biochemical and Biophysical Research Communications*, 46:1140.

Otten, J., Bader, J., Johnson, G. S., and Pastan, I. (1972a): A mutation in a Rous sarcoma virus gene that controls adenosine 3',5'-monophosphate levels and transformation. *Journal of Biological Chemistry,* 247:1632–1633.

Otten, J., Johnson, G. S., and Pastan, I. (1971): Cyclic AMP levels in fibroblasts: Relationship to growth rate and contact inhibition of growth. *Biochemical and Biophysical Research Communications,* 44:1192–1198.

Otten, J., Johnson, G. S., and Pastan, I. (1972b): Regulation of cell growth by cyclic adenosine 3',5'-monophosphate. Effect of cell density and agents which alter cell growth on cyclic adenosine 3',5'-monophosphate levels in fibroblasts. *Journal of Biological Chemistry,* 247:7082–7087.

Patterson, D., and Waldren, C. A. (1973): The effect of inhibitors of RNA and protein synthesis on dibutyryl cyclic AMP mediated morphological transformations of Chinese hamster ovary cells *in vitro. Biochemical and Biophysical Research Communications,* 50:566–573.

Paul, D. (1972): Effects of cyclic-AMP on SV 3T3 cells in culture. *Nature New Biology,* 240:179–181.

Pawelek, J., Wong, G., Sansone, M., and Morowitz, J. (1973): Molecular biology of pigment cells: Molecular controls in mammalian pigmentation. *Yale Journal of Biology and Medicine,* 46:530–443.

Peery, C. V., Johnson, G. S., and Pastan, I. (1971): Adenyl cyclase in normal and transformed fibroblasts in tissue culture. *Journal of Biological Chemistry,* 246:5785–5790.

Penit, I., Jard, S., and Benda, P. (1974): Probenecid sensitive 3',5'-cyclic AMP secretion by isoproterenol stimulated glial cells in culture. *Federation of European Biochemical Societies Letters,* 41:156–160.

Perkins, J. P., Macintyre, E. H., and Riley, W. E. (1971a): Studies on adenyl cyclase, phosphodiesterase and protein kinase from cultured glial cells. *Federation Proceedings,* 30:330abs.

Perkins, J. P., Macintyre, E. H., Riley, W. D., and Clark, R. B. (1971b): Adenyl cyclase, phosphodiesterase, and cyclic AMP dependent protein kinase of malignant glial cells in culture. *Life Sciences,* 10:1069–1080.

Peters, H., Karzel, K., Padberg, D., Schonhofer, P., and Dinnendahl, D. (1974): Influence of prostaglandin E_1 on cyclic 3',5'-AMP levels and glycosaminoglycan secretion of fibroblasts cultured *in vitro. Polish Journal of Pharmacology and Pharmacy,* 26:41–49.

Peytremann, A., and Engel, E. (1973): The cyclic AMP content and karyotype of the somatic hybrids of mouse malignant and nonmalignant cells segregating *in vivo* and *in vitro. Hormone Research,* 4:340–348.

Pierson, R. W., Jr., and Temin, H. M. (1972): The partial purification from calf serum of a fraction with multiplication-stimulating activity for chicken fibroblasts in cell culture and non-suppressible insulin-like activity. *Journal of Cellular Physiology,* 79:319–330.

Polgar, P., Vera, J. C., Kelley, P. R., and Rutenburg, A. M. (1973): Adenylate cyclase activity in normal and leukemic human leukocytes as determined by a radioimmunoassay for cyclic AMP. *Biochimica et Biophysica Acta,* 297:378–383.

Posternak, T., Sutherland, E. W., and Henion, W. R. (1962): Derivatives of cyclic 3',5'-adenosine monophosphate. *Biochimica et Biophysica Acta,* 65:558–560.

Prasad, K. N. (1972a): Morphological differentiation induced by prostaglandin in mouse neuroblastoma cells in culture. *Nature New Biology,* 236:49–52.

Prasad, K. N. (1972b): Cyclic AMP-induced differentiated mouse neuroblastoma cells lose tumourgenic characteristics. *Cytobios,* 6:163–166.

Prasad, K. N., Gilmer, K. N., and Sahu, S. K. (1974): Demonstration of acetyl choline-sensitive adenyl cyclase in malignant neuroblastoma cells in culture. *Nature,* 249:765–767.

Prasad, K. N., and Hsie, A. W. (1971a): Further changes in differentiation state accompanying the conversion of Chinese hamster cells to fibroblastic form by dibutyryl adenosine cyclic 3':5'-monophosphate and hormones. *Proceedings of the National Academy of Sciences (U.S.),* 68:1648–1652.

Prasad, K. N., and Hsie, A. W. (1971b): Morphological differentiation of mouse neuroblastoma cells induced *in vitro* by dibutyryl adenosine 3',5'-cyclic monophosphate. *Nature New Biology,* 233:141–142.

Prasad, K. N., and Kumar, S. (1973): Cyclic 3',5'-AMP phosphodiesterase activity during cyclic-AMP-induced differentiation of neuroblastoma cells in culture. *Proceedings of the Society for Experimental Biology and Medicine,* 142:406–409.

Prasad, K. N., and Sheppard, J. R. (1972): Inhibitors of cyclic nucleotide phosphodiesterase induce morphological differentiation of mouse neuroblastoma cell culture. *Experimental Cell Research,* 73:436–440.

Prasad, K. N., and Vernadakis, A. (1972): Morphological and biochemical study in X-ray and

dibutyryl cyclic-AMP-induced differentiated neuroblastoma cells. *Experimental Cell Research,* 70:27–32.
Puck, T. T., Waldren, C. A., and Hsie, A. W. (1972): Membrane dynamics in the action of dibutyryl adenosine 3':5'-cyclic monophosphate and testosterone on mammalian cells. *Proceedings of the National Academy of Sciences (U.S.),* 69:1943–1947.
Rao, G. J. S., Del Monte, M., and Nadler, H. L. (1971): Adenyl cyclase activity in cultivated human skin fibroblasts. *Nature New Biology,* 232:253–255.
Raska, K., Jr. (1973): Cyclic AMP in G_1-arrested BHK21 cells infected with adenovirus type 12. *Biochemical and Biophysical Research Communications,* 50:35–41.
Rebhun, L. I., White, D., Sander, G., and Ivy, N. (1973): Cleavage inhibition in marine eggs by puromycin and 6-dimethylaminpurine. *Experimental Cell Research,* 77:312–318.
Rein, A., Carchman, R., Johnson, G. S., and Pastan, I. (1973): Simian virus rapidly lowers cAMP levels in mouse cells. *Biochemical and Biophysical Research Communications,* 52:899–904.
Reporter, M. (1972): An ATP pool with rapid turnover within the cell membrane. *Biochemical and Biophysical Research Communications,* 48:598–604.
Réthy, A., Váczi, L., Tóth, F. D., and Boldogh, I. (1973): Abnormal distribution of adenylate cyclase in neoplastic cells. In: *Abstracts of the Fifth Annual Miami Winter Symposia,* Vol. 5, pp. 59–62.
Rhoads, A. R., Morris, H. P., and West, W. L. (1972): Cyclic 3',5'-nucleotide monophosphate phosphodiesterase activity in hepatomas of different growth rates. *Cancer Research,* 32:2651–2655.
Richelson, E. (1973): Stimulation of tyrosine hydroxylase activity in an adrenergic clone of mouse neuroblastoma by dibutyryl cyclic AMP. *Nature New Biology,* 242:175–177.
Rixon, R. H., Whitfield, J. F., and MacManus, J. P. (1970): Stimulation of mitotic activity in rat bone marrow and thymus by exogenous adenosine 3',5'-monophosphate. *Experimental Cell Research,* 63:110–117.
Robison, G. A., Butcher, R. W., and Sutherland, E. W. (1971): *Cyclic AMP.* Academic Press, New York.
Roisen, F. J., and Murphy, R. A. (1973): Neurite development *in vitro:* II. The role of microfilaments and microtubules in dibutyryl adenosine 3',5'-cyclic monophosphate and nerve growth factor stimulated maturation. *Journal of Neurobiology,* 4:397–412.
Roisen, F. J., Murphy, R. A., and Braden, W. G. (1972a): Dibutyryl cyclic adenosine monophosphate stimulation of colcemid-inhibited axonal elongation. *Science,* 177:809–811.
Roisen, F. J., Murphy, R. A., Pichichero, M. E., and Braden, W. G. (1972b): Cyclic adenosine monophosphate stimulation of axonal elongation. *Science,* 175:73–74.
Rozengurt, E., and deAsua, L. J. (1973): Role of cyclic 3':5'-adenosine monophosphate in the early transport changes induced by serum and insulin in quiescent fibroblasts. *Proceedings of the National Academy of Sciences (U.S.),* 70:3609–3612.
Rozengurt, E., and Pardee, A. B. (1972): Opposite effects of dibutyryl adenosine 3',5'-cyclic monophosphate and serum on growth of Chinese hamster cells. *Journal of Cellular Physiology,* 80: 273–279.
Rudland, P. S., Eckhart, W., Gospodarowicz, D., and Seifert, W. (1974a): Cell transformation mutants are not susceptible to growth activation by fibroblast growth factor at permissive temperatures. *Nature,* 251:337–339.
Rudland, P. S., Gospodarowicz, D., and Seifert, W. (1974b): Activation of guanyl cyclase and intracellular cyclic GMP by fibroblast growth factor. *Nature,* 250:741–742, 773–774.
Rudland, P. S., Seeley, M., and Seifert, W. (1974c): Cyclic GMP and cyclic AMP levels in normal and transformed fibroblasts. *Nature,* 251:417–419.
Russell, T., and Pastan, I. (1973): Plasma membrane cyclic adenosine 3':5'-monophosphate phosphodiesterase of cultured cells and its modification after trypsin treatment of intact cells. *Journal of Biological Chemistry,* 248:5835–5840.
Ryan, W. L., and Durick, M. A. (1972): Adenosine 3',5'-monophosphate and N^6-2'-O-dibutyryl-adenosine 3',5'-monophosphate transport in cells. *Science,* 177:1002–1003.
Ryan, W. L., and Heidrick, M. L. (1974): Role of cyclic nucleotides in cancer. *Advances in Cyclic Nucleotide Research,* 4:81–116.
Sahib, M. K., Jost, Y.-C., and Jost, J.-P. (1971): Role of cyclic adenosine 3',5'-monophosphate in the induction of hepatic enzymes III. Interaction of hydrocortisone and N^6, $O^{2'}$-dibutyryl cyclic adenosine 3',5'-monophosphate in the induction of tyrosine aminotransferase in cultured H-4-11-E hepatoma cells. *Journal of Biological Chemistry,* 246:4539–4545.

Sato, G. H., and Yasumura, Y. (1966): Retention of differentiated function in dispersed cell culture. *Transcripts of the New York Academy of Sciences,* Ser. II, 28:1063–1079.
Sattin, A., and Rall, T. W. (1970): The effect of adenosine and adenine nucleotides on the cyclic adenosine 3′,5′-phosphate content of guinea pig cerebral cortex slices. *Molecular Pharmacology,* 6:13–23.
Scher, C. D., Stathakos, D., and Antoniades, H. N. (1974): Dissociation of cell division stimulating capacity for Balb/c-3T3 from the insulin-like activity in human serum. *Nature,* 247:279–281.
Schimmer, B. P. (1969): Phenotypically variant adrenal tumor cell cultures with biochemical lesions in the steroidogenic pathway. *Journal of Cell Physiology,* 74:115–122.
Schimmer, B. P. (1971): Effects of catecholamines and monovalent cations on adenylate cyclase activity in cultured glial tumor cells. *Biochimica et Biophysica Acta,* 252:567–573.
Schimmer, B. P. (1972): Adenylate cyclase activity in adrenocorticotropic hormone-sensitive and mutant adrenocortical tumor cell lines. *Journal of Biological Chemistry,* 247:3134–3138.
Schonhofer, P. S., Peters, H. D., Karzel, K., Dinnendahl, V., and Westhofen, P. (1974): Influence of antiphlogistic drugs on prostaglandin E_1 stimulated cyclic 3′,5′-AMP levels and glycosaminoglycan synthesis in fibroblast tissue cultures. *Polish Journal of Pharmacology and Pharmacy,* 26:51–60.
Schorr, I., and Ney, R. L. (1971): Abnormal hormonal responses of an adrenocortical cancer adenyl cyclase. *Journal of Clinical Investigation,* 50:1295–1300.
Schrier, B. K., and Shapiro, D. L. (1973): Effects of N^6-monobutyrylcyclic AMP on glutamate decarboxylase activity in fetal rat brain cells and glial tumor cells in culture. *Experimental Cell Research,* 80:459–462.
Schroder, C. H., and Hsie, A. W. (1973): Morphological transformation of enucleated Chinese hamster ovary cell by dibutyryl adenosine 3′,5′-monophosphate and hormones. *Nature New Biology,* 246:58–60.
Schröder, J., and Plagemann, P. G. W. (1972): Cyclic 3′,5′-nucleotide phosphodiesterases of Novikoff rat hepatoma, mouse L and HeLa cells growing in suspension cultures. *Cancer Research,* 32:1082–1087.
Schubert, D., Humphreys, S., deVitry, F., and Jacob, F. (1971): Induced differentiation of a neuroblastoma. *Developmental Biology,* 25:514–546.
Schultz, J., and Hamprecht, B. (1973): Adenosine 3′,5′-monophosphate in cultured neuroblastoma cells: Effect of adenosine, phosphodiesterase inhibitors, and benzepines. *Naunyn-Schmeidebergs Archiv. Pharmakologie,* 278:215–225.
Schultz, J., Hamprecht, B., and Daly, J. W. (1972): Accumulation of adenosine 3′:5′-cyclic monophosphate in clonal glial cells: Labelling of intracellular adenine nucleotides with radioactive adenine. *Proceedings of the National Academy of Sciences (U.S.),* 69:1266–1270.
Schwartz, J. P., Morris, N. R., and Breckenridge, B. McL. (1973): Adenosine 3′,5′-monophosphate in glial tumor cells: Alterations by 5-bromodeoxyuridine. *Journal of Biological Chemistry,* 248:2699–2704.
Schwartz, J. P., and Passonneau, J. V. (1974): Cyclic AMP-mediated induction of the cyclic AMP phosphodiesterase of C-6 glioma cells. *Proceedings of the National Academy of Sciences (U.S.),* 71:3844–3848.
Seeds, N. W., Gilman, A. G., Amano, T., and Nirenberg, M. W. (1970): Regulation of axon formation by clonal lines of a neural tumor. *Proceedings of the National Academy of Sciences (U.S.),* 66:160–167.
Seifert, W., and Paul, D. (1972): Levels of cyclic AMP in sparse and dense cultures of growing and quiescent 3T3 cells. *Nature New Biology,* 240:281–283.
Seifert, W. E., and Rudland, P. S. (1974): Possible involvement of cyclic GMP in growth control of cultured mouse cells. *Nature,* 248:138–140.
Selawry, H., Gutman, R., Fink, G., and Recant, L. (1973): The effect of starvation on tissue adenosine 3′,5′-monosphate levels. *Biochemical and Biophysical Research Communications,* 51:198–204.
Shainberg, A., Yagil, G., and Yaffe, D. (1971): Alterations of enzymatic activities during muscle differentiation *in vitro.* *Developmental Biology,* 25:1–29.
Shapiro, D. L. (1973): Morphological and biochemical alterations in fetal rat brain cells cultured in the presence of monobutyryl cyclic AMP. *Nature,* 241:203–204.
Sheppard, J. R. (1971): Restoration of contact inhibited growth to transformed cells by dibutyryl adenosine 3′:5′-cyclic monophosphate. *Proceedings of the National Academy of Sciences (U.S.),* 68:1316–1320.
Sheppard, J. R. (1972): Difference in the cyclic adenosine 3′,5′-monophosphate levels in normal and transformed cells. *Nature New Biology,* 236:14–16.

Sheppard, J. R., Hudson, T. H., Larson, J. R., and Cunningham, W. P. (1974): Cyclic AMP analogues promote a circular morphology of cultured Schwannoma cells. *Journal of Cell Biology*, 63:312a.
Sheppard, J. R., and Prescott, D. M. (1972): Cyclic AMP levels in synchronized mammalian cells. *Experimental Cell Research*, 75:293–296.
Shimamura, T., Sasaki, S., and Makanano, M. (1973): Effect of exogenous adenosine cyclic 3',5'-monophosphate on antibody-producing cells *in vitro* and *in vivo*. *Japanese Journal of Microbiology*, 17:530–532.
Shimizu, H., Daly, J. W., and Creveling, C. R. (1969): A radioisotopic method for measuring the formation of adenosine 3',5'-cyclic monophosphate in incubated slices of brain. *Journal of Neurochemistry*, 16:1609–1619.
Stellwagon, R. H. (1972): Induction of tyrosine aminotransferase in HTC cells by $N^6, O^{2'}$-dibutyryl adenosine 3',5'-monophosphate. *Biochemical and Biophysical Research Communications*, 47:1144–1150.
Stoker, M. G. P., and Rubin, H. (1967): Density dependent inhibition of cell growth in culture. *Nature*, 215:171–172.
Su, Y.-F., and Perkins, J. P. (1974): Effect of norepinephrine on the cyclic AMP content of astrocytoma cells: Biphasic changes during prolonged exposure. *Federation Proceedings*, 33:493.
Sutherland, E. W., and Rall, T. (1960): The relation of adenosine-3',5'-phosphate and phosphorylase to the actions of catecholamines and other hormones. *Pharmacological Review*, 12:265–299.
Tan, K. B., and Sokol, F. (1974): Protein kinase stimulated by cyclic GMP in uninfected and simian virus 40-infected monkey kidney cells. *Journal of Virology*, 13:234–236.
Tashjian, A. J., Jr., and Hoyt, R. F., Jr. (1972): Transient controls of organ-specific functions in pituitary cells in culture. In: *Molecular Genetics and Developmental Biology*, edited by M. Sussman, pp. 353–387. Prentice-Hall, Englewood Cliffs, N.J.
Taunton, O. D., Roth, J., and Pastan, I. (1969): Studies on the adrenocorticotropic hormone-activated adenyl cyclase of a functional adrenal tumor. *Journal of Biological Chemistry*, 244:247–253.
Taylor-Papadimitriou, J. (1974): Effects of adenosine 3',5'-cyclic monophosphoric acid on the morphology growth and cell cycle of Earle's L cells. *International Journal of Cancer*, 13:404–411.
Temin, H. M. (1967): Studies on carcinogenesis by avian sarcoma viruses. VI. Differential multiplication of uninfected and of converted cells in response to insulin. *Journal of Cellular Physiology*, 69:377–384.
Temple, R., and Wolff, J. (1973): Stimulation of steroid secretion by antimicrotubular drugs. *Journal of Biological Chemistry*, 248:2691–2698.
Thomas, E. W., Murad, F., Looney, W. B., and Morris, H. P. (1973): Adenosine 3',5'-monophosphate and guanosine 3',5'-monophosphate: Concentrations in Morris hepatomas of different growth rates. *Biochimica et Biophysica Acta*, 297:564–567.
Tihon, C., and Green, M. (1973): Cyclic AMP amplified replication of RNA tumour virus like particles in Chinese hamster ovary cells. *Nature New Biology*, 244:227–231.
Tilney, L. G. (1971): Origin and continuity of microtubules. In: *Origin and Continuity of Cell Organelles*, edited by J. Reinert and H. Ursprung, pp. 222–260. Springer-Verlag, New York.
Tomási, V., Réthy, A., and Trevisani, A. (1973): Soluble and membrane-bound adenylate cyclase activity in Yoshida ascites hepatoma. *Life Sciences*, 12:145–150.
Troy, F., Vijay, I. K., and Kawakami, T. G. (1973): Cyclic 3',5'-AMP-dependent and independent protein kinase levels in normal and feline sarcoma virus transformed cells. *Biochemical and Biophysical Research Communications*, 52:150–157.
Uzunov, P., Shein, H. M., Weiss, B. (1973): Cyclic AMP phosphodiesterase in cloned astrocytoma cells: Norepinephrine induces a specific enzyme form. *Science*, 180:304–306.
Uzunov, P., and Weiss, B. (1972): Separation of multiple molecular forms of cyclic adenosine 3',5'-monophosphate phosphodiesterase in rat cerebellum by polyacrylamide gel electrophoresis. *Biochimica et Biophysica Acta*, 284:220–226.
van Wijk, R., Wicks, W. D., and Clay, K. (1972): Effects of derivatives of cyclic 3',5'-adenosine monophosphate on the growth, morphology, and gene expression of hepatoma cells in culture. *Cancer Research*, 32:1905–1911.
Varga, J. M., Dipasquale, A., Pawelek, J., McGuire, J. S., and Lerner, A. B. (1974): Regulation of melanocyte stimulating hormone action at the receptor level: Discontinuous binding of hormone to synchronized mouse melanoma cells during the cell cycle. *Proceedings of the National Academy of Sciences (U.S.)*, 71:1590–1593.
von Sallmann, L., and Grimes, P. (1974): Effects of isoproterenol and cyclic AMP derivatives on cell division in cultured rat lenses. *Investigative Opthalmology*, 13:210–218.

Wahrmann, J. P., Luzzati, D., and Winand, R. (1973a): Changes in adenyl cyclase specific activity during differentiation of an established myogenic cell line. *Biochemical and Biophysical Research Communications,* 52:576–581.
Wahrmann, J. P., Winand, R., and Luzzati, D. (1973b): Effect of cyclic AMP on growth and morphological differentiation of an established myogenic cell line. *Nature New Biology,* 245: 112–113.
Walton, G., and Garren, L. (1970): An assay for adenosine 3′,5′-cyclic monophosphate based on the association of the nucleotide with a partially purified binding protein. *Biochemistry,* 9:4223–4229.
Watson, J., Epstein, R., and Cohn, M. (1973): Cyclic nucleotides as intracellular mediators of the expression of antigen-sensitive cells. *Nature,* 246:405–409.
Waymire, J. C., Weiner, N., and Prasad, K. N. (1972): Regulation of tyrosine hydroxylase activity in cultured mouse neuroblastoma cells: Elevation induced by analogs of adenosine-3′,5′-cyclic monophosphate. *Proceedings of the National Academy of Sciences (U.S.),* 69:2241–2245.
Webb, D. R., Stites, D. P., and Perlman, J. D. (1973): Lymphocyte activation: The dualistic effect of cAMP. *Biochemical and Biophysical Research Communications,* 53:1002–1009.
Weinstein, Y., Chambers, D. A., Bourne, H. R., and Melmon, K. L. (1974): Cyclic GMP stimulates lymphocyte nucleic acid synthesis. *Nature,* 251:352–353.
Weiss, B., Shein, H. M., and Snyder, R. (1971): Adenylate cyclase phosphodiesterase activity of normal and SV40 virus-transformed hamster astrocytes in cell culture. *Life Sciences,* 10:1253–1260.
Werner, I., Peterson, G. R., and Shuster, L. (1971): Choline acetyl transferase and acetylcholinesterase in cultured brain cells from chick embryos. *Journal of Neurochemistry,* 18:141–151.
Whitfield, J. F., Rixon, R. H., MacManus, J. P., and Balk, S. D. (1973): Calcium, cyclic adenosine 3′,5′-monophosphate, and the control of cell proliferation: A review. *In Vitro,* 8:257–278.
Wicks, W. D., and McKibbin, J. B. (1972): Evidence for translational regulation of specific enzyme synthesis by $N^6,O^{2'}$-dibutyryl cyclic AMP in hepatoma cells cultures. *Biochemical and Biophysical Research Communications,* 48:205–211.
Willingham, M. C., Carchman, R. A., and Pastan, I. (1973): A mutant of 3T3 cells with cyclic AMP metabolism sensitive to temperature change. *Proceedings of the National Academy of Science (U.S.),* 70:2906–2910.
Willingham, M. C., Johnson, G. S., and Pastan, I. (1972): Control of DNA synthesis and mitosis in 3T3 cells by cyclic AMP. *Biochemical and Biophysical Research Communications,* 48:743–748.
Wolff, J., Temple, R., and Cook, G. H. (1973): Stimulation of steroid secretion in adrenal tumor cells by choleragen. *Proceedings of the National Academy of Sciences (U.S.),* 70:2741–2744.
Wollenberger, A., Krause, E. G., and Heier, G. (1969): Stimulation of 3′,5′-cyclic AMP formation in dog myocardium following arrest of blood flow. *Biochemical and Biophysical Research Communications,* 36:664–670.
Wong, G. and Pawelek, J. (1973): Control of phenotypic expression of cultured melanoma cells by melanocyte stimulating hormones. *Nature New Biology,* 241:213–215.
Yaffe, D. (1968): Retention of differentiation potentialities during prolonged cultivation of myogenic cells. *Proceedings of the National Academy of Sciences (U.S.),* 61:477–483.
Zacchello, F., Benson, P. F., Giannelli, F., and McGuire, M. (1972): Induction of adenylate cyclase activity in cultured human fibroblasts during increasing cell population density. *Biochemical Journal,* 126:27P.
Zalin, R. J. (1973): The relationship of the level of cyclic AMP to differentiation in primary cultures of chick muscle cells. *Experimental Cell Research,* 78:152–158.
Zimmerman, J. E., and Raska, K. (1972): Inhibition of adenovirus type 12 induced DNA synthesis in G_1-arrested BHK21 cells by dibutyryl adenosine cyclic 3′,5′-monophosphate. *Nature New Biology,* 239:145–147.

Author Index

Abe, K., 53
Abell, C.W., 65, 69
Abercrombie, M., 304
Abrahamson, E.W., 43
Adelstein, R.S., 11, 56
Adler, W.H., 74
Ahlquist, R.P., 220
Ahren, K., 142, 147, 148, 151, 165, 166, 181, 185
Akerblom, H.K., 221, 226
Albano, J.D.M., 113
Aledort, L.M., 11
Alford, R.H., 74
Allen, D.O., 209, 314, 317
Allen, L.B., 288
Allison, A.C., 12
Allison, S.P., 221
Allwood, G., 74
Alter, S., 110
Amano, T., 309
Ambrose, E.J., 304
Anderson, E., 204
Anderson, H.J., 294
Anderson, O.F., 146
Anderson, P., 58
Anderson, R.N., 143, 144
Anderson, W.B., 70, 277, 279, 291, 296, 301
Andersson, A., 208
Andersson, R.G.G., 48, 49
Andreatta, R., 103
Angers, M., 104
Antoniades, H.N., 297
Appleman, M.M., 10, 278, 282, 317
Archibald, D., 149
Argent, B.E., 29
Argy, W.P., 39
Ariens, E.J., 115
Arimura, A., 22
Armato, U., 287
Armstrong, D.T., 139, 145, 146, 149, 154, 159, 161, 163, 166-171, 176, 177, 180, 181

Arnold, A., 121
Ashby, D.C., 12, 228
Ashcroft, S.J.H., 25, 208, 212, 216-218, 229
Asherson, G.L., 74
Ashmore, J., 203, 219
Askari, H., 140
Asplund, K, 227
Assem, E.S.K., 60
Aten, B., 204
Atkins, T., 209, 212, 214, 218, 220, 221
Aub, J.C., 68
Augusti-Tocco, G., 283
Auletta, F.J., 177
Aurbach, G.D., 273
Austen, K.F., 58-60
Averner, M J., 74
Ayers, C.R., 104

Babiarz, D., 148
Bacalao, J., 279
Bachmann, E., 9
Bader, J., 291
Baggett, B., 143, 144, 150
Bagnara, J.T., 52
Baird, C.E., 53, 106, 111, 148, 161, 254
Baird, K., 10
Baker, P.F., 6, 8, 14, 17, 21, 72
Balant, L., 206
Balk, S.D., 72, 303
Baniukiewicz, S., 103
Banks, P., 20
Bar, H.-P., 42, 49, 106-108
Barbour, B.H., 104
Barnea, A., 174
Barnett, C.A., 286
Barthe, P., 103
Barton, B.R., 175
Bartoov, B., 155
Bartová, A., 43, 109, 123
Bartter, F.C., 104
Baserga, R., 64, 75
Bassett, J.M., 208
Battenberg, E., 17, 160
Batts, A.A., 204

Battu, R.G., 120
Batzri, S., 36, 37, 38
Baudouin, M., 48
Bauer, R.J., 119
Baukal, A.J., 105, 106
Baum, S., 212
Bauminger, S., 145, 174
Bdolah, A., 37
Beall, R.J., 40, 42, 104, 113-115, 125, 141, 158, 160
Beavo, J.A., 272, 321, 324
Bechtel, P.J., 321
Becker, J.A., 55
Bedwani, J.R., 39
Beeler, E.W., 51
Behnke, O., 56
Behrman, H.R., 148, 167, 168, 177
Beitch, B.R., 180
Bell, E.T., 167
Bell, J.J., 120, 124, 125
Belman, S., 54
Bemano, B.V., 221
Benda, P., 268, 275, 319
Bennett, L.L., 6, 24, 204-206, 231
Bensiger, R., 45
Benson, P.F., 264
Benz, L., 27, 28
Bergstom, S., 148
Berl, S., 12
Berlin, R.D., 69
Bernstein, J., 230
Berridge, M.J., 1, 6-8, 13, 30-34, 109
Berson, S.A., 217
Berthet, L., 100, 138, 163
Bertrand, J., 106
Besley, G.T.N., 39
Bettex-Galland, M., 56
Beyer, K.F., 120, 121, 173
Bianchi, L., 77
Bieck, P., 116, 118
Biggins, R., 20
Bikle, D., 53
Birdsall, N.J.M., 8

AUTHOR INDEX

Birmingham, M.K., 43, 102, 108, 109, 123
Birnbaumer, L., 25, 148, 209
Biron, P., 104
Bishop, R., 20
Bitensky, M.W., 43-45, 53, 101
Black, H., 217
Blair-West, J.R., 104, 112
Blancette-Mackie, J., 107
Blaustein, M.P., 8
Blecher, M., 117
Blinks, J.R., 44
Blondel, B., 222
Bloom, F.E., 17, 160
Bloom, G., 22, 58, 112
Bloom, W., 304
Blume, A.J., 252, 271, 282
Bockaert, J., 39
Bohn, H., 103
Boldogh, I., 316
Bombik, G.M., 70, 297, 298, 307, 308
Bongiovani, A.M., 121
Bonnai, S., 296
Bonting, S.L., 28
Borgeat, P., 22
Borland, B., 22
Borle, A.B., 3, 7, 9, 14, 61
Borowitz, J.L., 11, 12, 18, 20
Bos, C.J., 315
Botelho, S.Y., 34
Bourguet, J., 39
Bourne, H.R., 58, 60, 74, 253, 274, 281, 292, 293, 300, 303, 312, 316, 318, 322-324
Bouvier, C.A., 58
Bowen, V., 212, 215, 217, 218
Bower, F., 119, 162
Boyd, G.S., 120, 123-125, 170-174
Boynton, A.L., 72
Braceland, B.M., 77, 311
Braden, W.G., 309
Brambell, F.W.R., 179
Braun, A.C., 68, 72
Braun, T., 153, 155, 156, 160, 182
Bray, D., 11
Breckinridge, B. McL., 17, 70, 264, 297
Breslow, J., 251, 253, 264, 265, 268
Bressler, R., 218
Bretscher, M.S., 68
Brierley, G.P., 9

Brineaux, J.P., 62
Brinley, F.J., 8
Brisson, G., 24, 26, 208, 219
Brock, M.L, 74
Brock, W.A., 148
Brooker, G., 4, 52, 252, 268, 270, 278, 286
Brooks, J.C., 232
Brouillet, J., 104
Brown, B.L., 113
Brown, C.B., 185
Brown, H.D., 77
Brown, J.E., 44
Brown, P.K., 43
Brown, R.D., 105
Brown, R.F., 77
Brown, W.J., 77
Brownie, A.C., 112, 113, 123, 171
Browning, E.T., 252, 282, 286
Brunswick, D.J., 162
Brunton, L.L., 253, 272
Brush, J.S., 274
Buchanan, K.D., 208
Bucher, N.L.R., 76
Buck, C.A., 68
Buhi, W.C., 218
Bulbring, E., 48
Bumpus, F.M., 105
Buonassisi, V., 102, 118, 292
Burdowski, A., 212, 213, 228
Burgen, A.S.V., 36
Burger, M.M., 69-71, 252, 297-300,303, 307, 308, 324, 325
Burgus, R., 22
Burn, J.H., 17
Burk, R.R., 65, 69, 248, 305, 314
Burke, G., 148
Burr, I.M., 206, 222, 227, 230
Burstein, S.R., 53
Burstin, S.J., 296, 300-302
Burton, P.R., 310
Burzawa-Gerard, E., 143, 152
Buse, J., 221
Buse, M.G., 221
Butcher, F.R., 75
Butcher, R.W., 10, 11, 22, 38, 50, 53, 62, 100, 103, 106, 111, 117, 141, 148, 156, 161, 245, 248, 254, 256-259, 267, 272-275, 282, 286, 287, 315, 318, 319

Bygrave, F.L., 9

Cahill, G.F., Jr., 224
Caldwell, B.V., 148, 177
Caldwell, P.C., 4
Callantine, M.R., 185
Cameron, L.E., 68
Cammer, W., 120
Camp, C.E., 143
Camu, F., 24
Canastar, G.D., 180
Candela, J.L.R., 208
Canick, J.I., 121
Cantin, M., 75
Carafoh, E., 9
Carchman, R.A., 42, 109, 110, 113, 114, 116, 120, 277, 291
Caron, M.G., 171, 173
Carpenter, C.C.J., 104
Carsten, M.E., 48
Case, R.M., 26-29
Casey, P., 166, 167
Casida, L.E., 169
Casola, L., 283
Casper, A.G.T., 104
Casteels, R., 48
Castenda, M., 311
Castro, A.E., 183
Caswell, A.H., 6, 7
Catt, K.J., 105, 106, 141, 153, 156, 157, 159, 160, 162
Cerasi, E., 209, 214-216, 221, 222
Chader, G.J., 45, 288
Chahil, A., 57
Chaikoff, I.L., 120
Challoner, D.R., 208
Chamberlain, M.J., 221
Chambers, D.A., 74, 324
Chance, B., 9
Chandler, J.A., 9
Chandler, W.K., 8
Chang, L.L., 148
Chang, M.M., 10
Channing, C.P., 144, 146-148, 176, 179-181, 288
Chapal, J., 219
Charles, M.A., 215-217, 230
Chasalow, F.I., 145
Chase, L.R., 273
Chasin, M., 118, 119
Chattopadhyay, S.K., 77
Chavancy, G., 22
Chavin, W., 162
Chayoth, R., 77
Cheng, H., 316
Chesney, T. McC., 208, 222
Chez, R.A., 227

AUTHOR INDEX

Chlapowski, F.J., 245
Chobsieng, P., 145, 151
Chow, B.F., 146
Chrissiku, M., 221
Christian, B., 20
Christiansen, R.O., 155, 182, 184
Christoffersen, T., 77, 314, 315
Christophe, J., 28
Chung, D., 103
Cingolani, H.E., 221
Cirillo, V.J., 144-146, 180, 181
Cittadini, A., 9
Civan, M.M., 38
Civen, M., 102, 185
Claesson, L., 167
Clark, J.F., 317
Clark, J.L., 206
Clark, J.N., 172
Clark, R.B., 247, 249, 250, 252, 254, 263, 264, 266, 268, 271, 278
Clausen, T., 27, 29
Clay, K., 308
Clayton, R.B., 121
Clothier, G., 66
Cochrane, D.E., 7, 59
Coffino, P., 274, 292, 316
Coghlan, J.P., 104, 112
Cohen, A.J., 102, 292
Cohen, I., 56, 57
Cohen, J.A., 7
Cohen, M., 290
Cole, B., 55, 57
Combarnous, Y., 174
Cone, R.A., 43
Conn, J.W., 216
Connell, G.M., 153, 159
Constantopoulos, G., 120
Conti, M.A., 11
Cook, B., 175, 176
Cook, G.H., 104, 275
Cook, J.R., 25, 208, 215, 218-220, 223, 230
Cooke, B.A., 153-156, 182
Cooper, D.Y., 120
Cooper, R.H., 25, 208, 215, 218, 223, 230
Cooperman, B.S., 162
Coore, H.G., 204, 205, 217, 219
Corbin, J.D., 62, 159, 168
Costa, E., 20, 37, 311
Costrini, N.V., 226
Courot, M., 183
Courte, C., 202

Cox, J.S.G., 59
Creange, J.E., 121, 172
Creed, K.E., 36, 37
Creighton, M.O., 288
Cresto, J., 215
Creveling, C.R., 269
Crisp, T.M., 179
Crofford, O.B., 272
Cropper, M., 176
Cross, M.E., 73, 74
Crumpton, M.J., 7
Cryer, P.E., 221
Cuatrecasas, P., 62, 104, 273, 276, 288
Cubeddu, X., 268
Cunningham, D.D., 204
Cunningham, W.P., 310
Curran, P.F., 40
Currie, N., 20
Curry, D.L., 206, 223
Curtis, G.L., 307, 308
Cushman, P., 110
Cuthbert, A.W., 39

Dalton, C., 271
Dalton, T., 39
Daly, J.W., 269
Dambach, G., 60, 61
Daniel, E.E., 47
Daniel, V., 253, 284, 292, 300, 303, 318, 319, 322, 323
Danzo, B., 151
D'Armiento, M., 267, 278-280, 317, 318
Daughaday, W.H., 22, 297
Davey, M.J., 74
Davies, P.J.A., 146
Davis, B., 25, 209, 213, 214, 219, 228, 232
Davis, J.O., 104
Davis, W.J., 53
Davis, W.W., 108, 120, 123
Davoren, P.R., 267, 268
Dazord, A., 106
Dean, P.M., 24, 25, 28, 205
Deane, H.W., 166
DeAngelo, A.B., 168
deAsua, L.J., 252, 297
Debbas, G., 8, 48
Deery, D.J., 22
de la Llosa, M.P., 22
Delerue-Lebelle, N., 143, 151
DelMonte, M., 275
DeLorenzo, R.J., 40
Dempsay, M.E., 124, 170
Denton, D.A., 104, 112
Denton, R.M., 167
DePetris, S., 69

DePont, J.J.H.H.M., 28
DeRenzo, E.C., 160
DeRobertis, E.R., 74
Desaulles, P.A., 103
Desautel, M., 155, 182, 184
Detwiler, T.C., 7, 55, 56, 57
deVeliss, J., 252, 268, 270, 286
Devine, C.E., 8, 48
Devis, G., 205
deVitry, F., 309
deVries, A., 56, 57
Devrim, S., 207
Dexter, R.N., 120, 121, 274
Diamond, J., 50
Dianzani, F., 287
DiBella, F., 310
DiBona, D.R., 38
Diczfalusy, E., 167
Diecke, F.P.J., 47
DiMatteo, J., 283
Dion, S., 281
Dipasquale, A., 161
Dobbins, C., 117
Dods, R.F., 212, 213, 228
Doering, C.H., 121
Dollinger, H., 221
Donald, R.A., 22
Donaldson, E.M., 152
Donta, S.T., 104
Doore, B., 268
Dorfman, R.I., 120, 139, 149, 169, 170
Dorrington, J.H., 141, 149, 150-156, 166, 167, 182, 183
Douglas, W.W., 2, 6, 7, 11, 16-21, 36, 59
Doyle, D., 122
Drahota, Z., 9, 60
Dransfeld, H., 9
Dreifuss, J.J., 17
Droller, M.J., 56
Drosdowsky, M., 169
Drummond, G.I., 13
Duell, E.A., 66
Dufau, M.L., 141, 153, 154, 156, 157, 159, 160, 162
Dumont, J.E., 202
Duncan, L., 13
Dunham, E., 15
Dunn, M., 148
DuPont, A., 22
Durbin, R.P., 6
Durham, J.P., 75
Durick, M.A., 305, 312
Dvorkin, B., 248

AUTHOR INDEX

Eagle, H., 267
Eakims, K.E., 180
Earp, H.S., 118
Eckhard, W., 325
Eckstein, B., 27
Edelman, I.S., 39
Edstrom, A., 310
Edwards, J.C., 204, 208
Efendic, S., 214, 216
Eichhorn, J., 120, 164
Eik-Nes, K.B., 138, 143, 153-155, 159, 174
Eimerl, S., 7, 36
Eker, P., 306
Ekins, R.P., 113
El-Fouley, M.A., 175, 176
Elgjo, K., 77, 314
Ellerman, J.E., 204
Elliot, F.H., 108
Elliot, H.L., 104
Elliot, J.A., 307
Ellsworth, L.R., 146, 176, 180, 181
Elster, P., 20
Emmelot, P., 315
Endo, M., 8
Endroczi, E., 148
Engel, E., 302
Engel, L.L., 121
Entman, M.L., 13
Enzmann, E.B., 175
Epel, D., 7, 66
Eppenberger, U., 184
Epstein, R., 290
Epstein, S., 13, 77
Erickson, R.R., 148
Erlichman, J., 213
Eshkol, A., 146
Estabrook, R.W., 120, 121, 164
Esterhuizen, A.C., 208
Ewing, L.L., 168
Exton, J.H., 60, 62, 77, 78, 258
Eytan, E., 38

Fager, R.S., 43
Fajans, S.S., 216
Fanska, R., 215
Farese, R.V., 40, 108, 118, 121, 174
Farquhar, M.G., 316
Farr, A.L., 248
Fast, D., 28
Fawcett, C.P., 23
Fawcett, D.W., 304
Field, J.B., 22
Feinman, R.D., 7, 55, 56, 57
Feinstein, H., 37
Feinstein, M., 20, 116
Feldman, J.M., 221

Ferguson, J.J., Jr., 101, 108, 120, 121, 166, 172
Ferguson, R., 221
Fermandjian, S., 48
Ferrari, R.A., 121
Ferrendelli, J.A., 10, 18
Feucht, B., 268
Fiedler, R.M., 118
Fiedler, R.P., 288
Field, J.B., 77, 148, 165
Field, J.P., 112
Field, M., 104
Fink, C.J., 25, 205
Fink, G., 215, 233, 304
Finkelstein, A., 4
Finn, F.M., 103
Fischer, S., 139
Fisher, B., 76
Fisher, E.R., 76
Fishman, L.M., 120, 121
Fitzpatrick, D.F., 8, 48
Flack, J.D., 111-113
Flawia, M.M., 297
Fleckstein, A., 51
Fleischer, N., 22
Fletcher, R.T., 45
Flint, A.P.F., 149, 167-171
Floyd, J.C., Jr., 216
Foa, P.P., 203, 208
Fontaine, Y.A., 143, 151, 152, 155
Fontaine-Bertrand, E., 151, 152
Forbes, K.K., 145
Forchielli, E., 139, 169, 170
Ford, L.E., 8
Foreman, J.C., 7, 59
Forer, A., 71
Forrest, G., 70, 300
Forsham, P.H., 215
Foster, S.I., 252, 267, 268, 270, 282
Fox, T.O., 71, 298
Frank, W., 73, 307
Franklin, T.J., 252, 267, 268, 270, 282
Franks, D.J., 77, 144, 310, 311
Franzini-Armstrong, C., 8, 48
Frawley, T.F., 218
Frazier, H.S., 38
Free, C.A., 119
Freedman, A., 289
Freeman, D.J., 47
Freeman, J., 45
Friedmann, N., 60-62
Fritz, I.B., 153-156, 182, 183
Fritz, M.E., 34

Froelich, J.E., 70, 296, 301, 307
Fudenberg, H.H., 74
Fuhrer, J.P., 68
Fuhrmann, U., 305
Fujimoto, W.Y., 218
Funder, J.W., 112
Furmanski, P., 286, 309
Fussganger, R.D., 206, 221
Fuwa, K., 20

Gabbay, K.H., 25, 205
Gabbiani, G., 58
Gallant, S., 112, 113
Gamble, R.L., 9
Garay, G., 226
Garcia, A.G., 7
Gárdos, G., 12, 62
Garren, L.D., 40, 101, 108, 120, 122-125, 170, 213, 320
Gawadi, N., 71
Gawienowski, A.M., 172
Gaylor, J.L., 124
Gee, J.D., 6, 16
Genest, J., 104
George, W.J., 14
Gerner, E.W., 71
Gershon, E., 104
Geschwind, I.I., 6, 22
Giannelli, F., 264
Gibbons, I.R., 43
Gilbertson, J.R., 77
Gill, G.N., 40, 101, 125, 170, 213
Gill, J.R., 229
Gillett, A.P., 221
Gilman, A.G., 117, 247-250, 252, 253, 256, 267, 269, 271, 282, 293, 294, 306, 309
Gilmer, K.N., 275
Giordano, N.D., 116
Gitelman, H.J., 117
Glass, R.H., 148
Glick, M.C., 71
Glick, S.M., 208
Glinsmann, W.H., 118
Glossman, H., 105, 106
Goberna, R., 206
Goding, J.R., 104
Goffin, J., 48
Goggins, J.F., 287
Goldberg, A.L., 17
Goldberg, A.R., 298, 308
Goldberg, N.D., 10, 13, 15, 66, 68, 74, 75, 78, 104, 119, 157, 285, 304, 324
Goldfine, I.D., 25, 209, 212, 217, 218
Goldman, A.S., 121

AUTHOR INDEX

Goldring, S., 224
Goldstein, S., 142, 168, 169, 173, 253, 263, 265, 268, 287, 296, 301
Golenhofen, K., 48
Gomez-Acebo, J., 208
Gomperts, B.D., 7
Goodford, P.J., 48, 74
Goodfriend, T.L., 105, 106
Goodman, D.B.P., 3, 310
Goodman, D.S., 111
Gordon, R.S., 219
Goren, E.N., 209
Goridis, C., 45, 290
Gorman, R.E., 43, 101
Gorski, J., 174, 175
Gospodarowicz, D., 146, 290, 297, 324, 325
Graber, A.L., 219
Grahame-Smith, D.G., 103, 107, 110, 113, 115, 122, 274
Granner, D.K., 273, 284, 312, 316, 319, 321, 322
Grant, J.K., 123
Grantham, J., 148
Grau, J.D., 17
Greeff, K., 9
Green, H.M., 7
Green, I.C., 224, 226
Green, L., 53
Green, M., 287, 308
Greenberg, S., 47
Greengard, P., 17, 18, 40, 213, 283, 318
Greenough, W.B., 104
Greep, R.O., 139, 146, 152, 154, 158, 167, 168, 170, 177
Greider, M.G., 205
Grette, K., 58
Grey, N.J., 224
Griffin, M.J., 289
Grill, V., 209, 215
Grimes, P., 311
Grimes, W.J., 289
Grimm, W., 300
Grinwich, D.L., 145, 149, 177
Grodsky, G.M., 6, 124, 204-207, 215, 216, 231
Grondin, G., 29
Gronquist, L., 151
Groppi, V., 282
Gross, R., 266
Grossman, A., 286
Grossman, J., 74
Guerra, F., 40, 123
Guerrant, R.L., 253, 272

Guidotti, A., 20, 37, 75, 311
Guillemin, R., 22
Gunnarsson, R., 204, 229
Gupta, B.L., 40
Gut, M., 120
Guthrow, C.E., Jr., 310
Gutman, R., 233, 304
Gwatkin, R.B.L., 146
Gwilliam, J.M., 59

Hadden, E.M., 66, 285, 324
Hadden. J.W., 15, 66, 74, 285, 324, 325
Haddox, M.K., 10, 15, 66, 68, 119, 157, 285, 324
Hadley, M.E., 52, 54
Haegermark, O., 58
Hafs, H.D., 169
Haist, R.E., 207
Haksar, A., 103, 104, 109, 113, 123
Hales, C.N., 9, 24, 25, 64, 203, 219
Halkerston, I.D.K., 99, 116, 120, 164, 167, 169
Hall, D.E., 12, 59
Hall, K., 297
Hall, M.S., 259
Hall, P.F., 121, 123, 138-140, 153, 154, 159, 163, 169, 173, 174, 181, 182
Ham, E.A., 144-146, 181
Hamberger, A., 147
Hamberger, L., 147, 151, 185
Hamprecht, B., 252, 269, 271, 281, 294
Handler, J.S., 38, 39, 148
Hansel, W., 169
Hansen, E.L., 7
Hanson, J.P., 11
Harano, Y., 253, 282
Harary, I., 289
Harding, B.W., 120, 123-125
Hardman, J.G., 10, 14, 49, 50, 101, 105, 117, 160, 272
Harfield, B., 37
Harrell, E.R., 66
Hartman, R.C., 55
Harwood, J.P., 55, 57
Haslam, R.J., 54-57, 253, 263, 265, 268, 296, 301
Hauschka, P., 287
Hausen, P., 73
Hayano, M., 120, 124

Hayflick, L., 305
Haylett, D.G., 60
Haynes, R.C., 149
Haynes, R.C., Jr., 40, 100, 106, 113, 117, 138, 153, 163, 165, 166
Hays, R.M., 38
Heagan, B.M., 58
Heatley, N.G., 28
Hechter, O., 42, 106-108, 116, 120, 164, 170
Hedeskov, C.J., 229
Heidrick, M.L., 65, 69, 248, 278, 279, 296, 297, 301, 302, 304, 306, 307, 315, 318, 320
Heier, G., 304
Heinze, E., 228
Heisler, S., 28, 29
Hellerstrom, C., 204, 208, 227, 229
Hellman, B., 204, 215-217, 229
Henderson, A., 104
Henderson, H.H., 104
Henion, W., 117, 146, 305
Henney, C.S., 58
Henning, L.C., 77
Herbst, A.L., 167
Herlitz, H., 142, 147, 151, 185
Herman, L., 9
Herman, C.M., 221
Hermier, C., 140, 141, 152, 161, 174
Hern, E.P., 118
Herrera, M.G., 224
Hertelendy, F., 219
Hertig, A.T., 175
Hess, D., 9
Hess, S.M., 119
Hiestand, P.C., 184, 185
Hillarp, N.A., 167
Hillensjo, T., 151
Hilliard, J., 148, 149, 185
Hilton, J.G., 110, 112
Hilz, H., 117, 305, 306
Hinkley, R., 71
Hinz, M., 206, 207
Hirsch, A.H., 209, 213
Hirschfield, I.N., 40, 173
Hirschhorn, R., 74
Hitchcock, S.E., 11
Ho, P.J., 117
Ho, R.J., 258
Hochella, N.J., 274
Hochereau-deRiviero, M., 183
Hodgkin, A.L., 4, 6, 8

AUTHOR INDEX

Hoffer, B.J., 17, 160
Hoffmann, B., 139
Hofmann, K., 103
Hogberg, B., 59, 167
Hokin, L.E., 28
Hollinger, M.A., 155, 183, 185
Holmgren, J., 104
Holsinger, K.K., 218
Honn, K.V., 162
Hooker, C.W., 154
Horino, M., 219
Horne, J.R., 143, 144
Horrell, E., 149
Hostetler, K.Y., 229
Houareau, M.H., 219
Hovi, T., 310
Howell, D.C., 166
Howell, S.L., 22, 25, 201, 202, 204, 205-209, 212-214, 216-219, 221, 222, 224-226, 228, 233. 234
Hoyt, R.F., Jr., 289
Hsie, A.W., 287, 308-310
Huang, F., 185
Hubbard, W.R., 143
Huckins, C., 183, 184
Hudson, T., 74, 310
Hulser, D.F., 73
Hume, R., 124
Humes, J.L., 143-145, 181
Humphrey, R.R., 185
Humphreys, S., 309
Hurko, O., 20
Hurme, P., 229
Hurwitz, L., 8, 48, 49
Husakova, A., 77
Hutchinson, D.L., 227
Huttenen, J.K., 168
Huxley, H.E., 11, 12, 71
Hynie, S., 104

Ichii, S., 139, 163, 166
Idahl, L.A., 215, 229
Ide, M., 103, 113
Ide, T., 219
Illiano, G., 62, 104
Illner, P., 23
Imura, H., 110
Inbar, M., 69, 298, 308
Inesi, G., 7
Inglish, D., 268
Iorio, J.-A.M., 13
Isaacs, R.J., 72
Ishizaka, K., 59
Island, D., 122
Ito, S., 66
Ito, Y., 176
Iversen, J., 208, 220-222
Iverson, R.M., 178
Ivy, N., 312

Jaanus, S.D., 20, 21, 41, 42, 109, 110, 111, 120, 121
Jackanicz, T.M., 149
Jackson, R.L., 68
Jacob, F., 309
Jailer, J.W., 167
James, P., 119, 162
Jard, S., 39, 268, 275, 178, 183, 319
Jarlstedt, J., 185
Jaros, P., 206
Jeanrenaud, B., 208, 216
Jefcoate, C.R., 123-125, 171
Jefferson, L.S., 60
Jenkins, C.S.P., 56
Jenkinson, D.H., 6, 17, 49, 60
Jessup, R., 111
Johnson, A.H., 221
Johnson, D.G., 218
Johnson, D.J., 209
Johnson, G., 288
Johnson, G.L., 268
Johnson, G.S., 11, 69-71, 248, 251, 267, 278, 291, 296, 307
Johnson, L., 104
Johnson, M., 26, 45
Johnson, P.L., 165
Johnson, R.E., 146
Jones, C., 308
Jones, E.E., 176
Jonsson, H.T., 144
Jost, J.-P., 74, 101, 285
Jost, Y.-C., 285
Jungas, R.L., 167, 170
Jungmann, R.A., 42, 111, 116, 168, 184, 185
Junod, A., 216
Jutisz, M., 22, 140, 141, 174

Kagan, A., 208
Kagayama, M., 36
Kagen, L.J., 289
Kahane, I., 68
Kahnt, F.W., 121, 124
Kakiuchi, S., 13
Kaliner, M., 58-60
Kalkhoff, R.K., 226
Kamberi, I.A., 22
Kan, K.W., 124, 170
Kanazawa, Y., 206, 227
Kaneko, T., 22, 112, 148
Kanje, M., 310
Kanno, R., 28, 29
Kanno, T., 19, 59
Kantor, H., 272
Kao, L.W.L., 176
Kaplan, J.G., 73

Kaplan, N.M., 104, 112
Karaboyas, G.C., 120, 154
Karg, H., 139
Karim, S.M.M., 180
Karnovsky, M.J., 69
Karsten, C., 206
Katalin, M., 116
Katsumi, W., 22
Katz, A.M., 11, 13
Kaukel, E., 117, 289, 305-307
Kawakami, T.G., 283
Kawao, K., 154
Kazmi, G.A., 124
Kebabian, J.W., 17, 18
Keehn, E., 248
Keirns, J.J., 43, 45
Kelley, P.R., 314
Kelly, L.A., 107, 108, 245, 248, 253, 256-263, 267, 272, 275, 282, 315, 317, 317, 322, 323
Kelly, W.G., 104
Kemmler, W., 206
Kennedy, T.G., 145
Kensler, P.C., 55, 56
Keyes, P.L., 176, 180, 181
Keynes, R.D., 4
Khairallah, P.A., 105
Khoo, J.C., 169
Kiburz, J., 143
Kilpatrick, R., 141, 149, 150, 152, 166, 167
Kimberg, D.V., 104
Kimmel, G.L., 103
Kimura, T., 120, 121
King, C.A., 104
King, M., 104
Kinlough-Rathbone, R.L., 56, 57
Kinscherf, D.A., 10
Kipnis, D.M., 25, 207, 215, 219, 220, 224
Kirchberger, M.A., 13
Kirkland, W.L., 310
Kirpekar, S.M., 7
Kitabchi, A.E., 116-119, 162
Kitchell, L.C., 119, 162
Klein, M.I., 277, 279, 283
Kuzuya, T., 219
Kwan, C., 253, 272

Labrie, F., 22, 202
Lacy, P.E., 11, 25, 204, 205, 207, 217, 230
Lagnado, J.R., 231
Lamb, J.F., 8
Lambert, A.E., 206, 208, 216, 222, 227

AUTHOR INDEX

Lambotte, L., 60
Lamprecht, S.A., 142, 144, 145, 147, 148, 151, 168, 177
Landon, E.J., 8, 48
Lane, R., 102
Langan, T.A., 12, 125, 185
Langer, G.A., 4, 50, 52
Langr, F., 204
Lankester, A., 68
Laragh, J.H., 104
Laraia, P.J., 60
Lardy, H.A., 7
Larson, J.R., 310
Lasser, M., 37
Lastowecka, A., 19
Laube, H., 221
Lavine, R., 215
Lawecki, J., 216
Laycock, S.G., 110
Lazurus, N.R., 25, 207, 209, 212-215, 217-219, 228, 232
Leaf, A., 38
Lebovitz, H.E., 207, 208, 218, 221
Leclerq-Meyer, V., 208, 222
Lee, A.G., 8
Lee, J.B., 218
Lee, K.S., 8, 9
Lee, M., 7, 28
Lee, M.Y.E., 178
Lee, P.C., 168
Lee, S.L., 185
Lee, T.P., 139
Lee, V., 114
Lefebvre, P.J., 218
Lefkowitz, R.J., 42, 102, 103, 107, 108
Lehmeyer, J.E., 6, 22
Lehninger, A.L., 9
Leier, D.J., 42, 111, 116
LeJohn, H.B., 68
Lemaire, S., 202
LeMaire, W.J., 140, 142, 145, 147, 148, 176, 177, 181
Lemay, A., 22
Lerner, A.B., 53, 161
Lernmark, A., 215
Leterrier, J.F., 231
Levey, G.S., 13, 214, 216, 217, 227
Levin, L., 167
Levine, A.J., 71
Levine, L., 55, 58
Levine, M., 76
Levine, R.A., 208
Levy, D.A., 59
Levy, J.V., 7
Levy, P.L., 224

Lewis, S.B., 77, 258
Li, C.H., 103
Lichtenstein, L.M', 58-60
Liddle, G.W., 53, 105, 117, 120-122, 160
Liddle, R.A., 53
Lieberman, I., 77
Lieberman, M.E., 147, 174
Lieberman, S., 104
Like, A.A., 222
Liljekvist, J., 151
Lim, R., 308
Lin, B.J., 207
Lin, M.C., 68, 123
Lin, S.Y., 105, 106
Linarelli, R.G., 118
Lindley, B.D., 6, 8, 30
Lindner, H.R., 142, 145, 147, 148, 151, 174, 177
Lindsay, R., 8
Litwack, G., 284
Loeffler, L.J., 59
Lomsky, R., 204
Long, J.A., 204
Long, J.P., 47
Lonroth, I., 104
Loomis, W.F., Jr., 311
Looney, W.B., 77, 300
Lopez, C., 15
Loraine, J.A., 167
Lostroh, A., 146
Loten, E.G., 62
Loubatières, A., 219, 221
Lovenberg, W., 59
Lowe, I.P., 22
Lowe, J.P., 112
Lowry, O.H., 248
Loy, R.G., 169
Lubin, M., 309
Luft, R., 214, 216
Lundin, L., 66
Lunenfeld, B., 146, 155, 172-174
Loung, D., 74
Lust, W.D., 104, 304
Luxoro, M., 4
Luyckx, A., 207, 218
Luzio, J.P., 9
Luzzati, D., 273, 311
Lyle, L.R., 310

Ma, R.M., 116
MacDougall, E., 159
Macey, R., 7, 55, 57
Machlin, L.J., 219
Macintyre, E.H., 278, 309, 310
Mackie, C., 40, 113, 114, 116, 122, 125
MacLeod, R.M., 6, 22

MacManus, J.P., 77, 303, 310, 311
Maddrell, S.H.P., 6, 16
Maguire, M.E., 294
Mahaffee, D., 117, 118, 123, 124, 170

Maier, V., 207
Maino, V.C., 7, 74
Majino, G., 58
Major, P., 149, 151, 166
Makman, M.H., 248, 253, 263-265, 267-270, 273, 276-279, 283, 286, 316, 321, 322
Malaisse, W.J., 11, 24-26, 203, 205, 207, 208, 216, 219, 221, 223, 224, 226, 229, 231
Malaisse-Lagae, F., 11, 24, 26, 205, 207, 216, 219, 224, 226
Malamed, S., 39
Malamud, D., 37
Malette, L.E., 60
Mallucci, L., 69
Mamont, P., 296
Mancini, R.E., 183
Mandel, J.L., 301
Mandel, L.R., 181
Mandel, P., 290
Manganiello, V., 251, 253, 256, 264, 265, 268, 269, 273, 278, 280, 281, 321, 322
Marchesi, V.T., 68
Marcks, C., 215
Marcus, F.R., 43, 45
Mariani, M.M., 219, 221
Marks, F., 300
Marks, V., 207, 229
Marliss, E.B., 206, 222
Marquis, N.R., 55-58
Marri, G., 207, 229
Marsh, J.M., 137-145, 147, 148, 153, 155, 156, 160, 165-168, 173, 174, 176, 177
Martelo, O.J., 185
Martin, J.M., 221, 226, 227
Martin, S., 22
Martou, J., 116
Marz, E., 139
Mason, J., 3
Mason, J.I., 121, 170
Mason, J.W., 219
Mason, N.R., 139, 142, 149, 151, 152, 163, 166, 167, 180
Mason, W.T., 43, 44
Massarelli, R., 290

Masui, H., 40, 101, 170
Mathews, E.K., 12, 24, 25, 27, 28, 42
Matschinsky, F.M., 17, 207, 224
Matsuba, M., 166
Matsukura, S., 110
Matsuyama, H., 110
Matty, A.J., 209, 212, 214, 218, 220, 221
Maudsley, D.V., 103, 104
Mawe, R., 268
Mayer, S.E., 160, 168
Mayhew, D.A., 203, 207
Mazia, D., 66
McAfee, D.A., 18
McCaman, M., 268
McCann, S.M., 22, 148
McCarthy, J.L., 40, 121, 123
McCauley, R., 110
McCoy, K.E., 124, 170
McCune, R.W., 121, 172, 174
McGraw, E.F., 226
McGuire, J., 53
McGuire, J.S., 161
McGuire, M., 264
McKibbin, J.B., 286
McKune, R.W., 121
McPherson, J.C. III, 148
McWilliams, N.B., 204
Meador, C.K., 122
Means, A.R., 159, 161, 168, 181-184
Medoff, J., 310
Meech, R.W., 12, 62
Meinertz, T., 52
Melmon, K.L., 58, 60, 74, 292, 324
Mendelson, C., 162
Meng, H.C., 117
Menon, K.M.J., 143, 168-170
Menon, T., 107
Metcalfe, J.C., 8
Meves, H., 6
Meyer, P., 48
Miele, E., 109
Mieno, M., 154, 159
Miki, N., 43, 45
Milani, A., 124
Miledi, R., 16, 17
Miller, E.A., 209, 215
Miller, J.B., 176, 180, 181
Miller, J.E., 221
Miller, L.S., 139
Miller, R.E., 219
Miller, T.B., 60
Miller, W.H., 43, 44
Milligan, J.V., 23, 230
Millis, A.J.T., 70, 230

Mills, D.C.B., 58
Mills, T.M., 146-148, 151, 176, 177
Milner, R.D.G., 24, 25, 216, 219
Minna, J.W., 252, 253, 293, 294
Mintz, D.H., 214, 227
Mitchell, W.M., 308
Mitchison, J.M., 64
Mitsunobu, K., 308
Miya, T.S., 11
Miyake, T., 110
Miyamoto, E., 213, 228
Miyamoto, M.D., 17
Moellmann, G., 53
Monahan, T.M., 65, 69
Mongar, J.L., 7, 59
Monn, E., 155, 182, 184
Montague, W., 25, 201, 202, 208, 209, 212-221, 223, 224, 226, 228, 230, 233, 234
Moolten, F.L., 76
Moran, J.S., 297
Morel, F., 39
Morera, M., 106
Morgan, A.P., 224
Morgan, M.E., 50
Morgan, W.D., 71
Morgat, J.-L., 48
Morisaki, M., 166
Morisset, J.A., 28
Morland, J., 77, 314
Moroder, L., 103
Morowitz, J., 272
Morris, H.P., 77, 300, 314, 317
Morris, M.D., 120
Morris, N.R., 264
Morton, H.J., 72
Morton, I.K.M., 49
Moses, H.L., 120
Moskalewski, S., 204
Moskowitz, J., 55, 57
Moudgal, N.R., 154, 158, 168
Moyle, W.R., 113-115, 119, 120, 125, 141, 154, 156-161, 167, 168, 170, 172, 174
Mrotek, J., 122
Muhlstock, B., 102
Muir, T.C., 75
Mukerji, S., 167
Muller, R.U., 4
Mulrow, P.J., 101
Mundhenk, K., 117, 205
Muneyama, K., 119
Munro, J.A., 104
Munshower, J., 314
Munson, P.L., 112, 118

Murad, R., 77, 143, 153, 154, 182, 256, 300
Murer, E., 9
Murphy, R.A., 309, 310, 312
Mustard, J.E., 56, 57

Nadler, H.L., 275
Nagata, N., 3, 13
Naghshineh, S., 169
Naim, E., 37
Nakamura, M., 66
Nakamura, N., 103, 113, 114
Nakano, J., 112
Nakazato, Y., 7, 17
Nalbandov, A.V., 175, 176, 180
Namm, D.H., 50, 160
Narasimhulu, S., 120
Nath, J., 311, 312
Nathans, A.H., 119, 162
Nawrath, H., 52
Nedeljkovic, 110
Neher, R., 121, 124
Nekola, M., 175, 180
Netter, A., 140
Neufeld, A.H., 43-45
Neve, P., 202
Neville, A.M., 121
Newberry, W.M., 74
Newman, E.V., 272
Ney, R.L., 103, 106, 108, 117-121, 123, 125, 162, 170, 274, 300, 302, 314-316
Ng, T.S., 148
Nicholson, G.L., 69
Nicholson, W.E., 53, 105, 117
Nicklas, W.J., 12
Nierle, C., 207
Niewiarowski, S., 287
Niki, E., 219
Nilsson, K., 48, 49
Nilsson, L., 142, 151, 185
Nirenberg, M., 247-250, 252, 309
Nishiyama, A., 36
Nonaka, K, 208
Noonan, K.D., 69-71, 297
Nordmann, J.J., 17
Norland, J.F., 145
Northrop, G., 221
Novales, R.R., 53, 54
Nowaczynski, W., 104
Nunez, J., 231

O'Dea, R.F., 10, 119, 285, 304
O'Dea, R.T., 157

AUTHOR INDEX

Oey, J., 70, 296, 299-301
Ogawa, Y., 8
O'Grady, J.P., 148, 177
Okabayashi, T., 103, 113
O'Keefe, E., 273, 276, 288
Okinaka, S., 219
Oldham, S.B., 120
Oliver, A.P., 17
Ong, S.H., 119, 162
Opler, L.A., 286
Orange, R.P., 59, 60
Orci, L., 25, 205, 222, 227
Orczyk, G.P., 148, 177
Ord, M.G., 73, 74
Orloff, J., 38, 39, 148
Orr, T.S.C., 12, 59
Ortavant, R., 183
Ortmann, R., 268
Oschman, J.L., 30, 34, 40
Osnes, J.B., 77, 314
O'Toole, A.G., 304
Otten, J., 69, 70, 251-253, 291, 296, 297, 301, 302
Oyama, S., 208
Øye, I., 50, 77, 100
Oyer, P.E., 206

Pace, C.S., 25
Packham, M.A., 56
Padnos, D., 174, 175
Page, I.H., 105
Paik, V.S., 119
Paimre, M., 20
Pannbacker, R.G., 43, 45
Pappano, A.J., 17
Pardee, A.B., 287, 307
Park, C.R., 60-62, 77, 159, 258
Parker, C.W., 74, 160, 310
Parker, N., 310
Parlow, A.F., 141
Parrilla, R., 208
Parry, D.A.D., 11
Passonneau, J.V., 252, 280, 318
Pastan, I., 11, 42, 69-71, 102, 107, 108, 248, 251, 267, 277, 278, 288, 291, 286, 307
Patanelli, D.J., 143
Patterson, D.A., 310
Paul, D., 252, 296, 297, 301, 308
Pawelek, J., 161, 253, 272, 284, 288
Peach, M.J., 20
Peake, G.T., 22
Peaker, M., 36

Pearlmutter, 117, 118
Pearson, M.L., 301
Peery, C.V., 248, 267, 275, 314, 315, 317
Penardi, R., 149
Peng, T.C., 112
Penhos, J.C., 233
Penit, I., 268, 278
Pennington, S.N., 77
Perdue, J.F., 8, 11, 71
Perkins, J.P., 10, 247, 249, 250, 252, 254, 263, 264, 267, 268, 270, 271, 278, 283, 309, 314, 319
Perklev, I., 142, 151, 181
Perlman, J.D., 74, 310
Perlman, R., 212
Péron, F.G., 40, 100, 103, 104, 108, 109, 120-123, 138
Peters, H., 287
Petersen, M.J., 39
Petersen, O.H., 28, 29, 34-37
Peterson, R.E., 104
Petersson, D., 204
Peytreman, A., 105, 117, 160, 302
Pfeiffer, E.F., 206, 207, 221
Pharriss, B.B., 145, 148
Pichichero, M.E., 309
Pierce, N.F., 104
Pierson, R.W., Jr., 297
Pincus, G., 175
Pinto, L.H., 44
Pious, D.A., 70, 300
Pittman, R.C., 169
Plagemann, P.G.W., 278, 279, 281, 317
Poch, G., 49
Podolsky, R.J., 8
Poisner, A.M., 6, 16, 20, 36, 230
Polgar, P., 264, 314
Pollack, R., 70, 296
Pollard, T.D., 11, 56
Pollock, D., 75
Pomerantz, S., 147, 148, 151
Pooler, K., 207, 208
Porte, D., Jr., 219-221
Porter, K.R., 53
Portzehl, H., 4
Posternak, T., 117, 140, 305
Poulsen, J.H., 34, 36
Prasad, K.N., 275, 209, 310, 318
Prat, J.C., 7
Premont, J., 275, 319

Prescott, D.M., 298, 300
Prescott, R.R., 70
Pressman, B.C., 6, 7
Preston, A.S., 38
Price, I., 173
Price, S., 25
Pricer, W., 107
Prince, W.T., 6-8, 13, 30, 32-34
Probst, E., 56
Puck, T.T., 308-310
Puig, M., 109, 120
Pulsinelli, W.A., 143, 155
Purvis, J.L., 120, 121
Puszkin, S., 12

Quastel, M.R., 73

Raben, M.S., 297
Rachmeler, M., 70, 296, 301, 307
Raeymaekers, L., 48
Raff, M.C., 69
Rahwan, R.G., 11, 12, 18, 21
Rajerison, R., 39
Rall, T.W., 100, 107, 138, 271, 295
Ramachandran, J., 103, 113, 114, 141, 156-158, 160, 161
Ramseyer, J., 124
Ramwell, P.W., 111, 112, 144
Randall, R.J., 248
Randle, P.J., 25, 203-205, 208, 212, 217, 219, 229
Rao, G.J.S., 275
Rapino, E., 117
Rappaport, L., 231
Raptis, S., 206, 221
Raska, K., 252, 307, 314
Rasmussen, H., 3, 7, 12, 13, 37, 38, 62, 230, 231, 310
Rebhun, L.I., 11, 311, 312
Recant, L., 207, 215, 233, 304
Reddi, A.H., 168
Reddington, M., 231
Reed, P.W., 7
Reichard, G.A., Jr., 224
Reid, W.D., 55
Reiman, E.M., 168, 213
Rein, A., 252
Reitz, R.C., 123, 170
Renold, A.E., 206-208, 216, 222, 227
Repke, D.I., 11, 13

Reporter, M., 301, 311
Réthy, A., 316
Reuter, H., 8, 51, 52
Rhoads, A.R., 317
Rice, B.F., 139, 140
Richardson, M.C., 40, 102, 113, 122, 125, 158, 159
Richelson, E., 286
Richman, R., 117
Richter, A.W., 60
Rickenberg, H.V., 101
Ridderstrap, A.S., 28
Ridgeway, E.B., 6
Rieber, M., 279
Riede, U.N., 77
Riley, W.D., 278
Ritter, M.C., 124, 170
Rivkin, I., 118
Rixon, R.H., 303, 311
Robberecht, P., 28
Roberts, S., 121, 172
Robertson, R.P., 221
Robinson, J., 172
Robinson, J.W., 165, 166
Robison, G.A., 10, 11, 38, 43, 46, 50, 53, 55, 60, 100, 101, 110, 111, 125, 156, 162, 258, 273, 295, 303, 305, 312
Rogers, N.L., 272
Rohlich, P., 58
Rohr, H.P., 77
Roisen, F.J., 309, 310, 312
Romero, P.J., 12, 62
Rommerts, F.F.G., 153, 154, 156, 159
Rondell, P., 176, 177
Rondot, A.M., 219
Rosberg, S., 142, 151
Rosebrough, N.J., 248
Rosen, O.M., 209, 212, 213, 216-218
Rosenblith, J.Z., 69
Rosenstein, M.J., 41, 109
Rosenthal, A.S., 120
Rosenthal, O., 120
Rossi, C.S., 9
Rossini, A.A., 218
Rosson, G.M., 54, 57
Roth, J., 25, 40, 102, 107, 108, 209, 212, 248
Rothenbuchner, G., 221
Roy, A., 29, 39
Rozengurt, E., 252, 287, 307
Rubenstein, A.H., 206
Rubin, R.P., 2, 11, 12, 16, 18, 19-21, 24, 37, 41, 42, 109-111,

Rubin, R.P., (contd.) 119-121
Rubinstein, L., 185
Rudinger, J., 115
Rudland, P.S., 73, 252, 285, 290, 296, 297, 324, 325
Ruegg, J.C., 4
Rull, J., 216
Russell, J.T., 7
Russell, T., 278
Russell, T.R., 10, 17, 277, 282
Rutten, W.J., 28
Rutenberg, A.M., 264, 314
Ryan, G.B., 58
Ryan, W.G., 221
Ryan, W.L., 65, 69, 248, 278, 279, 296, 297, 301, 302, 304-307, 312, 315, 318, 320

Saba, N., 120
Sachs, L., 69, 298, 308
Saez, J.M., 106
Saffran, M., 42, 117
Sahib, M.K., 285, 287
Sahu, S.K., 275
Saier, M., 268
Saito, T., 268
Sakai, A., 76
Salmon, C., 151, 152
Salmon, W.D., Jr., 297
Salomon, Y., 37, 38
Salzman, E.W., 55, 56, 58
Samli, M.H., 6, 22
Samols, E., 207, 229
Sampson, S.R., 19
Sams, D.J., 212, 215-218
Samuels, L.T., 120, 121, 173
Sanders, G., 312
Sandler, R., 153, 154, 159
Sandow, A., 2, 7, 46
Sanner, J.M., 180
Sansome, M., 272
Santos, A.A., 140
Saruta, T., 112
Sato, G., 102, 118, 292
Sato, G.H., 102, 154, 159, 174, 175, 292
Sato, S., 148
Sattin, A., 271
Saunders, R.A., 221
Savard, L., 138-142, 145, 148, 149, 152, 153, 165-167, 173-175, 177
Savi, G., 221
Sawyers, C., 148, 149

Sayers, G., 40, 42, 104, 108, 113-116, 118, 125, 141, 158, 160
Sayers, M.S., 118
Scarpa, A., 7, 9
Schaefer, D.E., 104
Schaffer, R.J., 142
Schally, A.V., 22
Schams, D., 139
Schar, D., 103
Schatz, H., 207
Schatzmann, H.J., 8, 9
Schenk, A., 227
Scher, C.D., 297, 298
Schiller, P., 104
Schimke, R.T., 122
Schimmer, B.P., 102, 253, 274, 288, 292
Schmid, F.G., 215
Schmidt, W.M.T., 214
Schneider, H.P.G., 22, 148
Schneider, M.F., 8
Schneyer, C.A., 36, 75, 76
Schneyer, L.H., 36
Schofield, J.G., 22, 208, 222
Scholz, H., 52
Schonhofer, P.S., 253, 287
Schorn, A., 9
Schorr, I., 106, 316
Schramm, A., 7, 36-38
Schreier, B.K., 252, 271, 286
Schroder, C.H., 310
Schroder, J., 278, 279, 281, 317
Schroeder, T.E., 11, 71
Schubart, U., 212, 213, 215, 234
Schubert, D., 309
Schuetz, A.W., 175, 177
Schulster, D., 40, 102, 113, 116, 122, 125, 158, 159
Schultz, G., 10, 13, 37, 49, 50
Schultz, J., 252, 269, 271, 281, 282, 294
Schultz, K., 10
Schwartz, F.L., 144
Schwartz, I.L., 39
Schwartz, J.P., 252, 264, 270, 280, 318
Schwartz, T.B., 221
Schweppe, J.S., 184, 185
Schwyzer, R., 103, 104
Scian, L.F., 110, 112
Scoggins, B.A., 112
Scolnick, E.M., 291
Scoon, V., 149, 166

AUTHOR INDEX

Scratcherd, T., 26, 28, 29
Seeds, N.W., 309, 310, 312
Seeley, M., 73, 296, 325
Seelig, S., 42, 104, 114, 115
Segrest, J.P., 68
Sehlin, J., 216
Seifart, K.H., 169
Seifert, W., 252, 290, 296
Seifert, W.E., 73, 252, 285, 296, 297, 301, 324, 325
Seiguer, A.C., 183
Seitz, H., 8, 52
Selawry, A., 233
Selawry, H., 215, 216, 219, 304
Selinger, R.C.L., 102
Selinger, Z., 7, 36-38
Selstam, G., 142, 151, 165, 166
Selye, H., 75
Sensenbrenner, M., 290
Sepsenwol, S., 153, 155, 156, 160, 182
Setsuda, T., 110
Seymour, J.F., 146, 179
Shainberg, A., 311
Shapiro, D.L., 286
Sharma, R.K., 116-118, 123, 274
Sharp, G.W.G., 104
Shearer, G.M., 58
Shein, H.M., 279
Shelton, V.L., 144
Sheppard, J.R., 70, 71, 271, 296-301, 303, 308-310
Sherratt, H.S.A., 26
Shield, B., 110
Shimamura, T., 289
Shimizu, H., 269
Shimizu, K., 120
Shimizu, T., 154
Shin, B.C., 8, 9
Shin, S., 154, 159, 174, 175
Short, J., 77
Shroeder, J.L., 289
Shuman, D.A., 119
Sidhu, K.S., 143
Siegel, F.L., 232
Siggins, G.R., 17, 18
Silverman, D.J., 309
Simmer, H.H., 149
Simon, N., 119
Simonis, A.M., 115
Simpson, E.R., 120, 123, 124, 164, 171, 174
Simpson, L.L., 16
Singer, B., 38

Singer, J.J., 17
Sircar, B., 104
Six, K.M., 112
Sjoerdsma, A., 59
Sjogren, A., 185
Skein, H.M., 314
Skrypack, L.M., 168
Slater, J.D.H., 104
Sloper, K., 104
Smeby, R.R., 105
Smith, B.M., 151
Smith, D.F., 204
Smith, J.B., 58
Smith, J.W., 74
Snart, R.S., 39
Sneddon, J.M., 57
Sneyd, J.G.T., 62
Snyder, R., 314
Sode, J., 221
Soderling, T.R., 159
Soeldner, J.S., 224
Somers, G., 205
Somlyo, A.P., 8, 60
Somlyo, A.V., 8, 60
Sorel, G., 221
Spellacy, W.N., 218
Spelsberg, T.C., 308
Spencer, T., 9
Speroff, L., 144, 148, 177
Spigelman, L., 204
Squire, J.M., 11
Srinivasan, C.N., 172, 173, 174
Srinivasan, S., 268
Staehlin, M., 103
Stamenovic, B.A., 17
Stansfield, D.A., 143, 144, 165, 166
Stark, E., 116
Stathakos, D., 297
Statkov, P., 58
Statland, B.E., 58
Staub, M.S., 219
Stauffacher, W., 206, 222
Stawiski, M., 66
Steffen, H., 124
Stein, H., 73
Steinberg, D., 168
Steinberger, E., 181
Steiner, A.L., 17, 22, 74, 119, 160, 162
Steiner, D.F., 204, 206, 207, 216
Steinhardt, R.A., 7, 66
Steinke, J., 224, 228
Stellwagon, R.H., 287
Stephenson, J.H., 68
Sternlicht, E., 28
Stevenson, P.M., 172
Stites, D.P., 74, 310
Stock, K., 116
Stoker, M.G.P., 304

Stollar, V., 118
Stone, D., 120
Stormorken, H., 12
Stowe, N.W., 119, 162
Strauch, B.S., 143
Streeto, J.M., 13
Sturgill, T.W., 294
Su, Y.-F., 266, 268, 270
Sulimovici, S., 155, 172-174
Surian, E.S., 297
Sussman, K.E., 25, 207, 208
Sutherland, E.W., 10, 11, 38, 49, 50, 60, 62, 100-103, 107, 117, 138, 140, 141, 156, 160, 165, 267, 268, 272, 273, 282, 295, 305
Sutherland, R.M., 74
Sutliff, L.S., 274
Svennerholm, L., 104
Symchowicz, S., 219
Szabo, M., 148
Szent-Gyorgyi, A.G., 11
Szuch, P., 76

Tabachnick, I.I.A., 219
Tada, M., 13
Tait, J.F., 113, 122
Tait, S.A.S., 113, 122
Taljebal, I.-B., 25, 212, 216
Tanaka, A., 103, 113
Tanaka, M., 8
Tanese, T., 207
Tao, P., 272
Tarnoff, J., 143, 144
Tarnowski, W., 305, 306
Tashjian, A.J., Jr., 289
Taunton, O.D., 102, 107, 248, 274
Tavormina, P.A., 56, 57
Taylor, A., 55-57
Taylor, J., 219
Taylor, K.W., 204, 205, 207, 208, 226, 227
Taylor, L., 264
Taylor-Papadimitrious, J., 307
Tchen, T.T., 120
Telser, A., 71
Temin, H.M., 297
Temple, R., 104, 253, 275, 289
Tenenhouse, A., 28, 37, 231
Tenner, A.J., 178
Teo, T.S., 13, 52
Teshima, Y., 13
Tesser, G.I., 103

AUTHOR INDEX

Thomas, E.S., 77, 78, 300, 302, 326
Thomas, L.J., Jr., 278
Thompson, W.J., 10, 209, 218, 219, 221, 282
Thorn, A., 7, 9, 16, 17, 36, 39
Tieslau, C., 68
Tihon, C., 287, 308
Tillson, S.A., 148
Tilney, L.G., 53, 309
Timourian, H., 66
Toepel, W., 118
Tomasi, V., 316
Tomita, T., 44, 48
Tomkins, G.M., 273, 274, 281, 284, 290, 292, 296, 316
Toomey, R.E., 142, 180
Toon, P.A., 8
Toren, D., 169
Torres, H.N., 297
Toth, F.D., 316
Traikov, H., 102
Travis, G., 77
Treadwell, C.R., 168
Trevisani, A., 316
Trevithick, J.R., 288
Trifaro, J.M., 19
Tritthart, H., 51
Troll, W., 54
Troy, E., 283, 319
Trzeciak, W.H., 123, 125
Tsafiri, A., 142, 145, 147, 148, 186, 174-178, 181
Tsang, C.P.W., 123
Tsuruhara, T., 153
Turner, C.J., 75
Turtle, J.A., 25, 207, 219, 220
Tyler, A., 311

Udem, L., 212
Ueda, K., 102
Uenishi, K., 13
Ukena, T.E., 69
Ulberg, L.C., 144
Underwood, L., 117
Ungar, F., 124, 170, 171
Unger, R.H., 203
Utiger, R.D., 22
Uvnas, B., 58, 59
Uzunov, P., 279, 318

Vaczi, L., 316
Vaheri, A., 310
Vahouny, G.V., 168
Vale, W., 22
Valere, P.H.L., 108
Van Beurden, W.M.O., 153, 156
Vance, J.E., 208, 218, 219, 222, 224
Van Den Brande, J.L., 297
Van der Kemp, J.W.C.M., 153, 154, 156
Van der Molen, H.J., 153, 154, 156, 159
Van der Veerdonk, F.C.G., 53
Van Dyke, H.B., 146
van Heyningen, W.E., 104
Van Obberghen, E., 205
Van Sande, J., 202
Van Wijk, R., 308, 312, 316
Van Wyk, J., 117, 297
Vapaatalo, H., 118
Varga, B., 116
Varga, J.M., 161, 288
Vargas-Cordon, M., 218
Vatter, A.E., 309
Vaughan, M., 62, 143, 251, 253, 256, 269, 273, 278, 280, 281, 321, 322
Vaughn, G.D., 25, 205, 207, 208
Vcella, C., 204
Vecchio, D., 207
Veilleux, R., 74
Vera, C., 314
Verity, M.A., 77
Vernadakis, A., 309
Vernon, R.G., 182
Vesely, D.L., 54
Vigdahl, R.L., 55-57
Vijay, I.K., 283
Villazon, M., 217
Vincenzi, F.F., 8, 9
Virmaux, N., 45
Vlodausky, I., 69
Vogel, A., 70, 296
Voina, S., 117
Volkmann, R., 51
Von Sallman, L., 311
Voorhees, J.J., 66
Vortel, V., 204
Voyles, N., 233

Wahlquist, Y., 229
Wahrmann, J.R., 273, 289, 301, 311
Wald, G., 43
Waldren, C.A., 310
Walker, M.D., 11, 205
Wall, B.J., 40
Wallace, R.A., 178
Walsh, D.A., 12, 168, 213, 228
Walton, G.M., 40, 101, 125, 170, 320
Walton, K.G., 40
Walun, E., 310
Wang, J.H., 13, 52
Warner, A.E., 72
Warren, B.T., 59
Warren, G.B., 8
Warren, L., 68, 78
Watanabe, K., 153, 156
Watson, B., 118
Watson, B.S., 118
Watson, J., 290
Waymire, J.C., 286
Webb, D.R., 74, 310
Weber, G., 314, 317
Webster, P.D., 28
Wedner, H.J., 160
Wei, S., 304
Weiner, N., 20
Weinstein, Y., 58, 74, 292, 324
Weismann, G., 74
Weiss, B., 37, 279, 311, 314, 317
Weiss, R., 51
Wendell, E.N., 160
Werner, I., 286
West, W.L., 317
Westermann, D.C., 112
Westermann, E., 116, 118
Wherry, F.E., 219
Whitaker, B.D.L., 17
White, D., 312
White, J.G., 56, 58
Whitefield, J.F., 72
Whitfield, J.F., 72, 303, 310-312
Whitfield, M., 204, 209
Whitley, T.J., 119
Whitney, R.B., 74
Whittam, R., 12, 62
Whittey, T.H., 162
Whysner, J.A., 124
Wicks, W.D., 101, 122, 125, 174, 286, 308
Widnell, C.C., 103
Wilber, J.F., 22
Wilkerson, R.D., 14
Wilkinson, G.H., 143, 144
Wilks, J.W., 145
Willems, C., 202
Williams, H., 165, 229
Williams, J.A., 7, 27-29, 158, 230
Williams, K.I., 57
Williams, R.H., 208, 209, 218, 219
Williams, T.F., 62
Williams,-Ashman, H.G., 168
Willingham, M.C., 291, 307

AUTHOR INDEX

Wilson, D.B., 116, 119, 162
Wilson, D.F., 17
Wilson, J.A.F., 36, 37
Wilson, L.D., 120
Wilson, R.B., 307
Winand, R., 273
Wingender, W., 103
Wintersgill, C.J., 309
Wintour, M., 104
Wirz, A., 77
Wisdom, C., 272
Wisnewsky, C., 140
Wisnow, R., 253, 272
Wolfe, S.M., 56
Wolff, D.J., 232
Wolff, J., 104, 230, 253, 275, 289
Wollenberger, A., 304
Wollheim, C.B., 222
Wong, E.H.A., 60, 62
Wong, G., 288
Wong, K.K., 219, 220

Wong, P.Y.D., 39
Woodbury, L.A., 118
Woods, H.N., 68
Wright, M.R., 53
Wright, P.H., 203, 209, 219, 224, 226
Wright, R.D., 112
Wulff, V.J., 43
Wurtman, R.J., 20
Wuytack, F., 48

Yaffe, D., 311
Yagil, G., 311
Yago, N., 166, 169, 173
Yakovac, W.C., 121
Yalow, R.S., 217
Yamashita, K., 154
Yamazaki, R., 13
Yanez, E., 4
Yang, N.S.T., 145, 148
Yang, Y., 11
Yasumasu, I., 66
Yasumura, Y., 292
Yin, H.H., 69

York, D.H., 22
Yoshioka, H., 66
Youdale, T., 72, 77, 311
Young, D.A., 25, 205
Young, D.G., 123, 139, 163
Young, J.A., 36
Young, P.L., 121, 172
Yuen, M., 7, 55, 57
Yun, J., 121, 172

Zacchello, F., 264
Zahand, G.R., 207
Zalin, R.J., 311
Zemel, R., 77
Zenser, T.V., 74
Zimbelman, R.G., 169
Zimmermann, J.E., 307
Zor, H., 22, 23, 142, 148, 151
Zor, U.T., 112, 177
Zucca, M., 287
Zucker, M.B., 54, 57, 58

SUBJECT INDEX

A23187, 6, 17, 28, 32, 33, 36, 37
    blood platelets and, 54-58
    cell division and, 66
    histamine release, 59
    phytohemagglutinin mimick, 74
Acetylcholine, 5, 14, 18, 66, 271, 275
    control in adrenal medulla, 18-21
    enzyme secretion in exocrine pancreas, 26-30
    insulin and, 214, 219, 233
    in mammalian salivary, 34-38
    smooth muscle contraction, 47
Acinar cells
    enzyme release, 26-30
    in mammalian salivary, 34
ACTH, See Adrenocorticotropic hormone
Actin, 47
Actinomycin D, 112, 122, 175, 178, 225, 270
Adenine, 271
Adenohypophysis, 230
Adenosine, 57, 262, 267-271
Adenosine diphosphate, 179
    blood platelets and, 54-58
    steroidogenesis and, 118, 119, 150
3'-Adenosine monophosphate, steroidogenesis and, 140, 150
5'-Adenosine monophosphate, 165, 179, 280
    cultured cells, 284
    ovum maturation and, 178
    steroidogenesis and, 118, 119, 140, 150

Adenosine 3',5'-monophosphate, (cAMP)
    ACTH and, 40-43, 103, 104, 108-111
    adrenal cortex function and, 40-43, 99-126, 170
    adrenergic receptors, 220, 221
    angiotensin and, 105
    blood platelets and, 54-58
    calcitonin and, 2
    calcium and, 2-79, 108-111, 231, 232
    catecholamines and, 17, 18
    cell division and, 64-79
    cell genetics and, 290-294
    cell growth regulation, 291-293, 295-326
    cholera toxin and, 104
    cholesterol transport and, 169-172
    cultured cells and, 245-326
    EGTA in adrenal cortex, 110
    exocrine pancreas control, 26-30
    exogenous
        cell growth and, 304, 305
        steroidogenesis, 117, 118, 148, 149, 174
    glucagon and, 1, 208
    glucose effect on, 215, 216
    glucose metabolism, 229, 230
    gonadal function and, 137-185
    heart function and, 50-52
    histamine release, 58-60
    hormonal stimulation in adrenal cortex, 113-116
    5-hydroxytryptamine, 30-34

SUBJECT INDEX

Adenosine 3',5'-monophosphate (contd.)
   insulin release, 23-26
      Islets of Langerhans, 201-235
   level regulation, 9-11
   liver function and, 60-62, 76-78
   luteinization and, 178-181
   luteinizing hormone
      in corpus luteum, 139-145
      in Graafian follicle, 146-148
   mast cells and, 58-60
   melanophore control, 52-54
   microtubules and, 230
   nucleic acid synthesis, 184, 185
   ovarian tissue and, 137-152
   ovulation and, 177-179
   ovum maturation and, 179
   parathyroid hormone and, 2
   photoreceptors and, 43-45
   prostaglandin and, 106, 111-113, 144, 145
   protein kinase and, 40
   protein synthesis and, 174, 175
   salivary gland function
      insect, 30-34
      mammalian, 34-38
   second messenger, 11, 12
   smooth muscle control, 46-50
   spermatogenesis and, 181-184
   testes and, 153-163
   theophylline and, 116, 117
   toad bladder function, 38-40
Adenosine triphosphate, 173, 209, 212
   angiotensin and, 105
   insulin and, 203, 208
   luteinization and, 179
   steroidogenesis and, 118, 119, 140, 150
Adenylyl cyclase, 4, 5, 50
   ACTH effect on, 274, 275
   adrenal cortex, 100-126
   cAMP regulation, 9-11
   anterior pituitary, 22
   blood platelet, 55-58
   calcium and, 13, 42, 108-111
   catecholamines and, 274, 275
   cholera toxin and, 104
   in corpus luteum, 143, 144
   cultured cells, 245-326
   epinephrine and, 143
   fluoride effect on, 274, 275
   glucose effect on, 214, 215
   insect salivary function, 32
   insulin release, 25, 62
      Islets of Langerhans, 201-235
   interstitial ovarian tissue, 150
   Islets of Langerhans, 201-235
   light inactivation, 45
   malignant cells, 314-316
   melanocyte-stimulating hormone, 276
   pancreas, 26-30
   prostaglandin and, 106, 144, 145
   steroidogenesis, 142-144
   sulfonylurea effect on, 217
   surface lesions in cells, 69

Adenylyl cyclase (contd.)
   testes and, 153-163
   thyroid tissue and, 148
   toad bladder, 38-40, 148
   whole ovarian tissue, 151
Adenylyl imidodiphosphate, 209
Adrenal cortex, 5, 12, 40-43, 166, 173
   adenylyl cyclase and, 106-111
   cAMP and, 99-126
   angiotensin and, 104-106
   cholera toxin and, 104
   cholesterol metabolism in, 169
   prostaglandin receptors, 106
Adrenal medulla, 7, 230, 232
Adrenergic receptors in Islets of Langerhans, 219-222
Adrenocorticotropic hormone, 143, 150, 155, 159, 162, 163, 170
   adenylyl cyclase and, 274, 315
   adrenal cortex receptors, 102
   calcium and, 5, 40-43, 108-111
   cAMP and adrenal cortex, 100-126
   cAMP and, in cultured cells, 262, 268, 274
   cell transformation, 292
   cholesterol transport and, 169-172
   insulin and, 214, 218, 219
   phosphodiesterase inhibitors, 116, 117
   phosphorylase and, 166
   pregnenolone and, 173, 174
   prostaglandin and, 106, 111-113
   protein synthesis and, 174
Adrenodoxin, 120, 173
Adrenodoxin reductase, 173
Agarose, insoluble carrier of ACTH, 102
Aldosterone
   angiotensin and, 104-106
   prostaglandin and, 112
Amino acids, insulin and, 216, 219
p-Aminobenzoyl cellulose, 102
Aminoglutethimide, 121, 168, 169
Aminophylline, 20, 298
   progestin stimulation, 179
   steroidogenesis and, 149
Amylase, calcium and, 7, 28, 29, 35, 37
Androstenedione, 140
Angiotensin, adrenal cortex and, 104-106
Anterior pituitary, 6, 202
   calcium and control in, 21-23
Antidiuretic hormone (ADH) in toad bladder, 38-40
Arginine, 107, 214, 216, 233
Arotic acid, 108
Ascorbic acid, prostaglandin and, 112
ATPase, 143, 155
   calcium and, 8, 9, 11, 52, 58
Atropine, 20, 275
Avian salt gland, 36

Beta cells, 6, 23-26
Bicarbonate ion, 26-28
Blood platelets
   calcium and cyclic nucleotides, 7, 54-58

Bovine serum albumin, 143
8-Bromo cAMP, 118, 307
5-Bromo deoxyuridine, 270
8-Bromo cGMP, 74
Bulbous acini, 34

Caffeine, 10, 20, 21, 165
   insulin and, 207, 208, 217
Calcitonin, 2
Calcium
   Cyclic nucleotide interaction, *See* Outline, 1
   adenylyl cyclase, and, 107
   cAMP and adrenal steroidogenesis, 108
   cAMP and Islets of Langerhans, 231, 232
   prostaglandin and, 106, 112
Cancer, cAMP and, 10
Carbamylcholine, 249
Catecholamines
   blood platelets and, 54-58
   calcium in, 7, 17-21
   heart function, 50-52
   Islets of Langerhans, 220, 221
   liver function, 60-62
Catatonic compound, 48, 59
Cell division and growth
   cAMP and, 291, 295-326
   calcium and cyclic nucleotide control, 64-79
Centroacinar cells, fluid secretion, 26-30
Chloramphenicol, 174
Chloride ion in insect salivary fluid, 30-34
Chlorpromazine, 20
Cholera toxin, 74
   adrenal cortex, cAMP and, 104
   cultured cells, 272, 275, 276
Cholesterol, 40, 163
   biosynthesis, 120-126
   leuteinizing hormone and, 167, 168
   oxygen in side-chain cleavage, 174
   pregnenolone synthesis from, 172, 173
   progesterone synthesis from, 139-145
   prostaglandin and, 112
   testosterone synthesis from, 154
   theophylline and, 116
   transport of, 169-172
Cholesterol arachidonate, 111
Cholesterol esterase, 123, 125, 167-169
Chromaffin cells, 18-21
Chromosomes
   calcium and, 11, 71
   movement and cleavage, 71, 72
Chymotrypsin, 296
Citrate, 230
Cobalt, calcium and, 4, 6
Cofactors, cAMP and steroidogenesis, 163-167
Colchicine, 205, 230, 310
Collagen, blood platelets and, 54-58
Collagenase, 176
Concanavalin A, 68-70, 74, 285, 298, 208

Control systems
   bidirectional, 15, 16, 46-64
   cell division, 64-79
   monodirectional, 15, 16-46, 63
Corpus luteum, 179
   phosphorylase in, 165
   steroidogenesis and cAMP, 139-145, 172
Corticosterone, 102, 103
   ACTH and, 114, 115
   prostaglandin and, 111-113
   theophylline and, 116
Cortisol, angiotensin and, 104
Crown-gall tumor cells, 68
Cultured cells
   calcium and cell division, 68-73
   cyclic nucleotides, *See* Outline, 245
Cyanoketone, 121, 174, 176, 178
Cycloheximide, 122, 169, 170, 174, 178, 225, 270, 272, 280
Cytidine 3',5'-monophosphate (cCMP)
   cultured cells, 284
   luteinization and, 179
   steroidogenesis and, 118, 119
Cytidine triphosphate, 179
   adenylyl cyclase, 209
   steroidogenesis and, 118, 119
Cytochalasin B, 205
Cytochrome P-450, 120, 121, 124, 125, 171, 173
Cytoplasm
   calcium and, 9
   cholesterol transport and, 170

D-600, calcium and 4-6, 17
Deoxycorticosterone, 124
2-Deoxyglucose, 229
Deuterium oxide, 205
Dexamethasone, 62, 251, 281
Diazoxide, 214, 233
Dibutyryl cAMP, 17, 20, 38, 40, 42, 54, 58, 59, 103, 109, 146, 154, 170, 229, 231, 290
   cell division and, 70-72, 74, 298, 305-311
   cultured cells, 280, 284, 305, 306
   diet, insulin and, 224-226
   glucagon and, 208
   hormone release and, 22, 28
   insulin and, 207, 208, 219
   luteinization and, 179-181
   ovulation and ovum maturity, 177-178
   steroidogenesis and, 116-118, 126, 140, 159
   theophylline and, 116
Dibutyryl cGMP, 233
Dichloroisoproterenol, 280
Dihydroergotamine, 219
Dinitrophenol, 268
Dipyridamole, 268
Disodium cromoglycate, 59
DNA, 122, 290, 297, 298, 300, 306, 307, 311

DNA (contd.)
   calcium and, 7, 66
   liver cell division, 77
Dopamine, 18, 271, 275
Ductus deferens, 49

E. coli enterotoxin, 272
EGTA, 42, 72, 109
   adenylyl cyclase and, 107
   cAMP and, 110
   prostaglandin and, 106
Endocrine pancreas, calcium role in, 23-26
Endoplasmic reticulum
   calcium and, 9, 14, 121
   proinsulin biosynthesis, 204-206
   steroidogenesis and, 121
Enzymes
   cell division and, 70
   corticosteroid biosynthesis, 120
   secretion in exocrine pancreas, 26-30
   secretion in mammalian salivary, 34-38
Epidermal cells, 66
Epinephrine, 107, 125, 151, 161, 317
   adenylyl cyclase and, 143
   blood platelets and, 54-58
   cAMP and, in cultured cells, 256-260, 263-277
   glucagon and, 208, 222
   heart function and, 51
   insulin and, 214, 219, 220, 223, 233, 234
   histamine release and, 60
   liver and, 60, 62
Epithelial cells in toad bladder, 38-40
Estradiol, 140, 177
Exocrine pancreas, calcium and cyclic nucleotides in, 5, 8, 26-30

Feedback mechanisms, 12-15
   insect salivary gland, 30-34
Fertilization
   calcium and, 7
   cell division, 66
Fetal pancreas, 227, 228
Fibroblast growth factor, 290, 297
Fibroblasts, 70, 71
   cAMP levels in, and cell growth, 251, 263-267, 277, 279, 296-326
Ficin, 297
Flufenamic acid, 145
Fluid, secretion of, 26-38
Fluoride, 150, 233
   adenylyl cyclase
      adrenal cortex, 107
      cultured cells, 274, 275, 315
      Islets of Langerhans, 209, 212
   cAMP in cultured cells, 260, 262, 267, 277, 293
Follicle-stimulating hormone, 176
   cAMP and, 138
      in Graafian follicle, 147, 148
      in interstitial ovarian tissue, 150
      in testes, 155

Follicle-stimulating hormone (contd.)
      in whole ovarian tissue, 151, 166
   luteinization and, 179, 181
   ovum maturation and, 177
   protein synthesis and, 174
   spermatogenesis and, 181-184
Fungal cells, calcium and, 68

GABA, 249
Glibenclamide, 214, 217, 233
Glucagon, 107, 143
   in cultured cells, 262, 275, 277
   glucose metabolism, 229
   insulin stimulation and, 23-26, 233
      Islets of Langerhans, 201-235
   liver and, 60-62, 76
Glucose
   adrenergic receptor activity and, 221, 222
   calcium role in insulin stimulation, 23-26
   insulin secretion and, 206, 214-216, 223, 233
   liver cell division, 76
   metabolism in Islets of Langerhans, 229
   synthesis, 60
Glucose-6-phosphate, 108, 123, 164, 229
Glycine, 229
Glycolysis
   ACTH and, in adrenal cortex, 109
Glycoprotein, cell growth and, 68-70
Golgi complex, 204-206
Gonad, cAMP in function of, 137-185
Graafian follicle
   cAMP and, 142, 146-148, 175-181
   prostaglandin and, 145
Granules
   calcium and, 1, 7, 11, 12, 18-20, 38
   Islets of Langerhans, 208
   melanophores, 53
Granulosa cells, 176
   luteinization in, 178-181
Growth hormone, 156
   cGMP and, 22
5'-Guanylylimidophosphate, 105
Guanine, 271
Guanosine, 271
5'-Guanosine monophosphate
   angiotensin and, 105
Guanosine 3',5'-monophosphate (cGMP)
   angiotensin and, 105
   cell growth and, 324-326
   cultured cells, 284-290
   Islets of Langerhans, 233, 234
   luteinization and, 179
   steroidogenesis and, 118, 119
   testes and, 157, 162
Guanosine triphosphate, 233
   adenylyl cyclase and, 209, 212
   angiotensin and, 105
   luteinization and, 179
Guanylyl cyclase, 5, 290
   calcium and, 14
   cAMP regulation, 10

SUBJECT INDEX

Guanylyl cyclase (contd.)
    growth hormone and, 22
    Islets of Langerhans, 233, 234

Heart, calcium and cyclic nucleotides, 50-52
Heparin, 77
Hexamethonium, 20
Histamine, 7
    cAMP and in cultured cells, 249-253, 262, 269, 271
    calcium and release from mast cells, 58-60
Horse serum anticholeragenoid, 272
Human chorionic gonadotropin, 140-143, 150, 156, 157, 161
    cAMP and, 185
    cholesterol transport and, 172
    ovum maturation and, 177
Human placenta, 170
8-Hydroxy-cAMP, 119
20-alpha-Hydroxycholesterol, 124
20-alpha-Hydroxypregn-4-en-3-one, 177
20-beta-Hydroxypregn-4-en-3-one
    biosynthesis, 139, 141, 150, 151, 152
    LH and, 149
17-Hydroxyprogesterone, 140
Hydroxysteroid dehydrogenase, 172, 173
5-Hydroxytryptamine
    blood platelets and, 54-58
    cAMP and, in cultured cells, 249-253, 271
    insect salivary gland function, 30-34

ICI 63197, 282
Imidazole, 161
    insulin and, 223, 225
    luteinization and, 179
    progesterone synthesis and, 141
Indomethacin, 112, 113, 145
Inosine 3',5'-monophosphate (cIMP)
    steroidogenesis and, 118, 119
Inosine triphosphate, 105
Insulin, 62, 107
    calcium role in secretion, 23, 26
    cAMP and, in cultured cells, 262, 297
    cAMP and, Islets of Langerhans, 201-235
    dietary changes and, 224-226
    hormonal changes and, 226, 227
    in liver, 76
    potassium and, 24
    pregnancy and, 225, 226
Intestine, cholera toxin, adenylyl cyclase and, 104
Ionophores, calcium, 5-8, 17, 28, 32, 33, 36, 54, 58, 59
Islets of Langerhans, cAMP and physiology, See Outline, 201
3-Isobutyl-1-methylxanthine, 49, 117, 208, 215, 217, 218, 230
    cAMP in cultured cells, 260, 262, 271, 272, 281, 307, 315
Isopropylnorepinephrine, 262

Isoproterenol, 37, 38, 220, 222, 311
    cAMP in cultured cells, 249-253, 264, 267, 270, 275, 292-294, 307
    cell division and, 75

Kirsten sarcoma virus, 291

Lactogen, 226
Lanthanum, calcium and, 4, 5, 17, 42, 48
Leucine, 214, 216, 223, 229, 233
Leukocytes, histamine release, 58
Leydig cells, 154, 161, 168, 170, 172, 175, 182
Linolenic acid, 177
Liver
    calcium and cyclic nucleotide control, 60-62, 76-78
    cell division, 66, 76-78, 295
Luteinization, cAMP and, 179-181
Luteinizing hormone (LH), 167
    cAMP and
        in corpus luteum, 138-145, 166, 172
        in Graafian follicle, 146-148
        in whole ovarian tissue, 150-152, 165
    cholesterol and, 167, 168
    ovulation and ovum maturation, 176-179
    prostaglandin and, 144, 145
    protein synthesis, 174
    spermatogenesis, 181-184
Lymphocyte, 285
    calcium and, 7
    cell division, 66, 73-75

Magnesium, calcium and, 4
Manganese, 4-6, 17, 233
Mannoheptulose, 229
Mannose, insulin and, 223
Mast cells, 7
    calcium and cyclic nucleotides, 58-60
Melanocyte-stimulating hormone, 276
Melanophore, calcium and cyclic nucleotide control, 52-54
Melatonin, 53
Mepp frequency, 17
Methacholine, 37
1-Methyl-3-isobutylxanthine, See under Isobutyl
8-Methylthio cAMP, 119, 298
Methylxanthines, 5, 10, 20, 21
    insulin and, 214, 217, 218
Microfilaments, 12, 53, 56, 205
Microsomes, 172
    calcium and, 8, 37, 49
Microtubules, cAMP and, 230, 231
Mitochondria
    calcium and, 7, 9, 14, 21, 23, 39, 61
    cholesterol and, 169-172, 173
    pregneolone efflux, 173, 174
    steroidogenesis and, 120-126, 156
Monobutyryl cAMP, 117, 118, 306, 309, 311
    hormone release and, 22
    muscle contraction, 6-8

Mutant cells, cAMP and, 291-293
Myosin, 47

Nervous system, calcium and control, 16-18
Neuraminidase, ACTH receptor and, 103, 104
Neurohypophysis, 7, 16, 17
Nicotinamide adenine dinucleotide phosphate (NADP), 108, 120
Norepinephrine, 17, 317
    blood platelets and, 54-58
    cAMP and, in cultured cells, 249-253, 264, 269-272, 274, 275, 280-282
    glucagon and, 222
    insulin and, 214, 219, 220, 223, 233
    in mammalian salivary function, 34-38
    smooth muscle contraction, 47
Nucleic acid synthesis, 184, 185

Oligomycin, calcium and, 9
Ornithine decarboxylase, 117
Ouabain, heart function and, 50-52
Ovarian tissue, steroidogenesis and cAMP in, 139-152, 165
Ovulation and ovum maturation, 176-179
7-Oxa-13-prostynoic acid, 144-145

Pancreas, 7
    Islets of Langerhans, 201-235
Pancreozymin, 26-30, 219, 233
    insulin and, 214
Papain, 297
Papavarine, 49, 262, 271, 282
Parathyroid hormone, 2, 107
Parotid gland, 7, 36, 37
    cell division, 66, 75
Peripheral ganglia, 17
Permeability
    calcium, 4, 6, 12, 17, 19, 20, 24, 32, 72
    chloride, 33, 34
    sodium, 19, 20, 36
    potassium, 34-38, 49, 66
    in photoreception, 43-45
Peroxidase, 48
Phenoxybenzamine, 221
Phentolamine, 55, 57, 220, 221, 222
Phosphodiesterase, 24, 56, 131, 140
    amino acids, effect on, 216
    calcium and, 5, 13, 19, 68
    cAMP regulation, 10, 100
    cell division and, 68
    cultured cells, 260-263, 268-270, 272, 278-283, 316-318, 320-324
    glucose effect on, 215, 216
    in corpus luteum, 143
    histamine release and, 59
    inhibitors and cultured cells, 281
    inhibitors and steroidogenesis, 116, 117
    insulin and, 62, 207, 208
    Islets of Langerhans, 201-235
    luteinization and, 179

Phosphodiesterase *(contd.)*
    luteinizing hormone and, 143
    in photoreception, 45
    steroidogenesis in ovarian tissue, 142-144, 150
    sulfonyl ureas, effect on, 217
    in testes, 155, 156, 182, 183
Phospholamban, 13
Phospholipase, A, 173
Phosphoprotein phosphatase, 213
Phosphorylase, 47, 165
Photoreceptors, calcium and, 43-45
Phytohemagglutinin, 73-75, 284
Pilocarpine, 209
Polyacrylamide, diazotized, 102
Polyphloretin phosphate, 180
Potassium
    calcium and, 5-7, 16-24, 29-38, 49-52, 60-62
    cell division, 66
    heart function and, 50, 52
    insect salivary gland and, 30-34
    insulin stimulation, 24, 62
    liver function and, 60-62
    lymphocytes and, 73
    mammalian salivary gland and, 34-38
    steroidogenesis and, 110
Pregnancy, insulin secretion, 225, 226
Pregnant mare serum gonadotropin, 141, 142
Pregnenolone, 40
    biosynthesis, 120-126, 154, 163, 168, 172, 173
    efflux from mitochondria, 173, 174
Probenecid, 268
Progesterone, 40, 226
    biosynthesis, 121
    in ovarian tissue, 139-141, 149, 150, 165, 167, 174-181
    ovum maturation and, 177
Progestin, 179
Proinsulin biosynthesis, 204-207
Prolactin, 143, 156, 177, 180
Pronase, cell division and, 70
Propranolol, 55, 220, 221, 251, 280
Prostaglandins, 161, 310
    adrenal cortex, 100, 106, 111-113
    anterior pituitary, 21-23
    blood platelets, 54-58
    cAMP and
        in cultured cells, 249-274, 277, 280-290, 293, 294, 298
        in Graafian follicle, 147
        luteinization, 180
        in whole ovarian tissue, 150
    corpus luteum, 144, 145
    insulin and, 214, 218
    ovulation and ovum maturation, 177, 178
    pancreas, 28
Protein kinase, 12, 13, 23, 38, 159
    amino acid effect on, 216

Protein kinase *(contd.)*
    cAMP and, 40, 41, 125, 142
    cGMP dependent, 234
    cultured cells, 283-285, 318, 319, 324
    in corpus luteum, 142, 145, 168, 169, 173
    glucose effect on, 216
    Islets of Langerhans, 212, 213, 228
    ovum maturation, 178
    spermatogenesis, 184
    sulfonyl ureas, effect on, 217
    tubulin substrate, 230, 231
    in whole ovarian tissue, 151
Protein synthesis
    cell restoration, 309
    steroidogenesis and, 174, 175
Purkinje cells, 17
Puromycin, 112, 122, 141, 142, 174, 178
Pyroantimonate, 9
Pyruvate, 207, 230

Regulation of cell calcium conc., 4
Rhodopsin, 43
RNA, 40, 111, 178, 308
    cAMP and, 185
    in lymphocytes, 73
    spermatogenesis, 182
    steroidogenesis and, 122, 175
RO 7-2956, 262
RO 20-1724, 280, 282
Rous sarcoma virus, 70, 72, 273, 291
Ruthenium red, calcium and, 4

Salivary gland
    cell division, 66, 75, 76
    insect, 5-8, 30-34
    mammalian, 34-38, 60, 75
Sarcoplasmic reticulum, 7, 8, 13, 48-52
Sea urchin egg, 7, 311
Second messenger, calcium and, 11, 12
Secretin
    insulin and, 214, 219, 233
    pancreatic fluid and, 26-30
Seminiferous tubules, 182
Serotonin, 5
Sertoli cells, 183
Sialic acid, 103, 104
Skeletal muscle, calcium and control, 48
Smooth muscle, calcium and cyclic nucleotide control, 46-50
Sodium, 8, 9, 17-20, 27, 29, 42, 66
    heart function and, 50-52
    mammalian salivary function, 34-38
    photoreception, 43-45
    steroidogenesis and, 110
    toad bladder function, 38-40
Sodium butyrate, 280
Somatic cell hybridization, 293, 294
Soybean agglutinin, 298
Spermatogenesis, 181-184
Splanchnic nerve, 18, 19
Squid axon, 6, 21

Steroid
    cholesterol transport and synthesis, 169, 172
    protein synthesis and, 174, 175
    stimulation in ovarian tissue, 138-152
    stimulation in testes, 153-175
    synthesis in adrenal cortex, 40-43, 100-126
    synthesis in cultured cells, 272
Stimulus-division coupling, calcium and, 7, 8, 64
Stimulus-secretion, calcium and, 7, 23
Subtilisin, 297
Succinic dehydrogenase, 155
Sulfonyl ureas, 206, 217
Superior cervical ganglion, 18
Surface lesion in cell growth, 68-70

Taenia coli, 48
Testicular tissue, 182
    cAMP and, 153-163
    epinephrine stimulation of adenylyl cyclase, 143
Testosterone, 308, 309
    biosynthesis, cAMP and, 153-163
    epinephrine stimulation, 143
Tetracaine
    calcium and, 4, 5, 20, 24, 62
    glucose in liver, 62
    insulin and, 24
Tetrodotoxin, calcium and, 6, 17, 42
Theophylline, 10, 17, 19, 20, 24, 26, 54, 101, 183, 308
    ADH mimick, 38
    adrenal steroidogenesis and, 116-118
    cell division, 68
    cultured cells, effect in, 251, 256-263, 271, 280, 294
    glucagon and, 208
    glucose metabolism, 229
    histamine release, 59
    hormone release, 22, 26
    insulin secretion and, 206, 207, 217-219, 223-225
    pancreas and, 26-30
    progesterone synthesis and HCG, 141, 143
    salivary gland and, 30-34
    steroidogenesis and, 149
Thrombin, blood platelets and, 54-58
Thrombosthenin, 55
Thymidine, 74, 277
Thyroid gland, 230
Thyroid-stimulating hormone, 148
Thyrotropin, insulin release, 207
Thyroxine, 226
Toad bladder, 148
    calcium and control function, 32, 38-40
Tolbutamide, 214, 217, 219
Triiodothyronine, 77
Trypsin, 57, 297
    ACTH and, 102

Trypsin *(contd.)*
    in cell culture, 264
    in cell division, 65, 70
Tubulin, 230, 231
Tumor
    adrenyl steroidogenesis, 102, 122
    cholera toxin, adenylyl cyclase and, 104
    cultured cells, 248-326
    Leydig cells, 154, 170, 172, 175,
    liver, 77
Tumor-promoting phorbal ester, 54
Tyrosine hydroxylase, 20

Uridine monophosphate, 118, 119
Uridine 3',5'-monophosphate (cUMP)
    in cultured cells, 284
    steroidogenesis and, 118, 119
Uridine triphosphate, angiotensin and, 105
Uterus, angiotensin and, 105

Valinomycin, 268
Vasopressin, 7, 16, 17, 107, 275
    blood platelets and, 54
    prostaglandins and, 112
Veratridine, 269
Vinblastine, 205
Vincristine, 205
Viral transformation and cAMP, 290, 291

Water transport in toad bladder, 38-40
Wheat germ agglutinin, 298

X-537, 6, 17

Yin-Yang hypothesis, 15, 78